Handbook of Plant Pathology

Handbook of Plant Pathology

Edited by **Chris Frost**

R **C**ALLISTO **R**EFERENCE

New York

Published by Callisto Reference,
106 Park Avenue, Suite 200,
New York, NY 10016, USA
www.callistoreference.com

Handbook of Plant Pathology
Edited by Chris Frost

International Standard Book Number: 978-1-63239-408-8 (Hardback)

Printed in the United States of America.

Contents

Preface

The purpose of the book is to provide a glimpse into the dynamics and to present opinions and studies of some of the scientists engaged in the development of new ideas in the field from very different standpoints. This book will prove useful to students and researchers owing to its high content quality.

Dealing with the nature and causes of plant diseases in the fields of agriculture and forestry requires an applied science which is known as plant pathology. Plant pathology plays an important role in achieving food security and food safety for the world.

At the end, I would like to appreciate all the efforts made by the authors in completing their chapters professionally. I express my deepest gratitude to all of them for contributing to this book by sharing their valuable works. A special thanks to my family and friends for their constant support in this journey.

Editor

1

General Description of
Rhizoctonia Species Complex

Genhua Yang and Chengyun Li
*Key Laboratory of Agro-Biodiversity and Pest Management of Education
Ministry of China, Yunnan Agricultural University, Kunming, Yunnan
China*

1. Introduction

The genus concept of *Rhizoctonia* spp. was established by de Candolle (1815) (Sneh *et al.*, 1998). However, the lack of specific characters led to the classification of a mixture of unrelated fungi as *Rhizoctonia* spp. (Parmeter and Whitney, 1970; Moore, 1987). Ogoshi (1975) enhanced the specificity of the genus concept for *Rhizoctonia* by elevating the following characteristics of *R. solani* to the genus level. Based on this revised genus concept, species of *Rhizoctonia* can be differentiated by mycelia color, number of nuclei per young vegetative hyphal cell and the morphology of their teleomorph. The teleomorph of *Rhizoctonia* spp. belongs to the sub-division Basidiomycota, class Hymenomycetes.

The anamorphs of *Rhizoctonia* are heterogeneous. Moore (1987) placed the anamorphs of *Thanatephorus* spp. in *Moniliopsis*. She reserved the genus *Rhizoctonia* for anamorph of ustomycetous fungi which have septa with simple pores. Moniliopsis species have smooth, broad hyphae with brown walls, multinucleate cells, dolipore septa with perforate parenthesomes and teleomorphs in the genera *Thanatephorus* and *Waitea*. Of the binucleate *Rhizoctonia* spp., the anamorphs of the *R. repens* group (teleomorph *Tulasnella*) were assigned to the new genus *Epulorhiza*. Anamorph of *Ceratobasidium* was assigned to the new genus *Ceratorhiza* (Moore, 1987). Moore's system is taxonomically correct and justified. At present, the concept of genus *Rhizoctonia* has become clear from these taxonomical studies at the molecular level (Gonzalez *et al.*, 2001). However, many researchers (Sneh *et al.*, 1998) in the world still retain the name *Rhizoctonia* for Moore's *Moniliopsis* spp., *Ceratorhiza* spp. and *Epulorhiza* spp.. Hence, I used the name of *Rhizoctonia* in this study.

Affinity for hyphal fusion (anastomosis) (Parmeter *et al.*, 1969; Parmeter and Whitney, 1970; Ogoshi *et al.*, 1983a; Burpee *et al.*, 1980a) has been used to characterize isolates among *R. solani*, *R. zeae*, *R. oryzae*, *R. repens* and binucleate *Rhizoctoni*a spp. with *Ceratobasidium* teleomorphs. To date, isolates of *R. solani* have been assigned to 13 anastomosis groups (AG) and those of *R. zeae* and *R. oryzae* have each been assigned to their own one group (Sneh *et al.*, 1998; Carling *et al.*, 1999, 2002c).

Anastomosis reactions between hyphae of paired isolates of *R. solani* consist of several types; such as perfect fusion, imperfect fusion, contact fusion and no reaction (Matsumoto *et al.*, 1932). At present, four categories of anastomosis (C3 to C0) defined by Carling *et al.* (1996)

have been accepted by many researchers. These are useful for a better understanding of the genetic diversity of *R. solani* populations, because of the background genetically supported by vegetative or somatic compatibility (VC or SC) of confronted isolates (MacNish *et al*, 1997). Each of categories is as follows:

C3: walls fuse; membranes fuse, accompanied with protoplasm connection; anastomosis point frequently is not obvious; diameter of anastomosis point is equal or nearly equal hyphal diameter; anastomosing cells and adjacent cells may die, but generally do not. This category occurs for the same anastomosis group, same vegetative compatibility population (VCP) and the same isolate.

C2: wall connection is obvious, but membrane contact is uncertain; anastomosing and adjacent cells always die. This category occurs in same AG, but not between different VCPs.

C1: wall contact between hyphae is apparent, but both wall penetration and membrane-membrane contact do not occur; occasionally one or both anastomosing cells and adjacent cells die. This category occurs between different AGs or in the same AG.

C0: no reaction. This category occurs between different AGs.

In general, hyphal fusion occurs at a high frequency (50%≥) within members of the same AG, with the exception of non-self-anastomosing isolates (Hyakumachi and Ui, 1988). On the other hand, hyphal fusion among members of different AGs occurs at either a low frequency (≤30%) or no fusion occurs. *Rhizoctonia* isolates giving C3 to C1 reactions in anastomosing test have been taken to be the same AG.

To date, isolates of multinucleate *R. solani* have been assigned to 13 anastomosis groups (AG-1 to AG-13), some of which include several subgroups and isolates of *R. zeae* and *R. oryzae* have been assigned to WAG-Z and WAG-O, respectively (Sneh *et al*., 1998; Carling *et al*., 1999, 2002c). Isolates of binucleate *Rhizoctonia* spp. with *Ceratobasidium* teleomorphs have been reported. A system developed in Japan (Ogoshi *et al*.,1979, 1983 a,b; Sneh *et al*., 1998; Hyakumachi *et al*., 2005) includes 21 anastomosis groups designated AG-A to AG-U, in which at present AG-J and AG-M still are in question as members of binucleate *Rhizoctonia*. Another system developed in the USA (Burpee *et al*., 1980a) includes 7 anastomosis groups designed as CAG-1 to CAG -7. CAG-1 corresponds to AG-D, CAG-2 to AG-A, CAG-3 and CAG-6 to AG-E, CAG-4 to AG-F, CAG-5 to AG-R, and CAG-7 to AG-S (Sneh *et al*., 1998; Ogoshi *et al*., 1983a). At present, the anastomosis system based on AG-A through AG-U used in this review paper is widely accepted by many researchers.

Some homogenous groups of isolates of *R. solani* are well known as bridging isolates (AG-BI) that anastomose with members of different AGs (Carling, 1998). In general, there is no contradiction in the conventional anastomosis grouping system by taking anastomosis frequency into consideration. However, two exceptional cases where anastomosis frequency mismatched with morphological, physiological and pathogenic characteristics have been reported from tobacco (Nicoletti *et al*., 1999) and soybean (Naito and Kanematsu, 1994). These demonstrate the limitations of using hyphal anastomosis as the sole criteria for characterization and identification of closely related fungi. In addition, it is not easy to determine the subgroup of isolates within the same AG because no differences occur in their anastomosis reaction. Thus, in order to determine AGs or subgroups in *R. solani*, genetic analysis using molecular approaches that employ multiple genetic loci is needed.

Isolates of *R. solani* that exhibits DNA base sequence homology and affinities for hyphal anastomosis may represent a diverging evolutionary unit (Kuninaga and Yokosawa, 1980). This hypothesis is supported by analysis of restriction fragment length polymorphisms (RFLPs) and the sequences with in ribosomal RNA genes (rDNA) among different anastomosis groups of *R. solani* (Vilgalys and Gonzalez, 1990; Gonzalez, *et al.*, 2001; Carling *et al.*, 2002b).

As mentioned above, many AGs and subgroups of *R. solani* and binucleate *Rhizoctonia* spp. have been reported as causal of agents Rhizoctonia diseases on a wide range of host species. However, little is known about the Rhizoctonia diseases and the anastomosis groups and subgroups of their causal fungi on vegetables, ornamentals and food crops in the Asian tropics especially the southern parts of China.

2. Characteristics of anastomosis groups and subgroups of *Rhizoctonia solani* and binucleate *Rhizoctonia* spp.

Disease symptoms and host range of each AG and its subgroups are summarized as follows. In this review, the book by Sneh et al., 1998 entitled "Identification of *Rhizoctonia* Species" provided a substitute for the reference before 1998.

2.1 Multinucleate *Rhizoctonia* spp.

1.   **AG-1: IA, IB, IC, ID**

AG-1 IA (Li and Yan, 1990; Sneh *et al.*, 1998; Fenille *et al.*, 2002; Naito, 2004).

Symptoms: sheath blight, foliar blight, leaf blight, web-blight, head rot, bottom rot, and brown patch.

Host: rice (*Oryza sativa* L.), corn (*Zea mays* L.), barley (*Hordeum vulgare* L.), sorghum (*Sorghum vulgare* Pes.), potato (*Solanum tuberosum* L.), barnyard millet, common millet, soybean, peanut (*Arachis hypogaea* L.), lima bean, cabbage, leaf lettuce, Stevia, orchard grass, crimson clover, tall fescue (*Festuca arundiacea* Schreb), turfgrass, creeping bentgrass, perennial ryegrass, gentian (*Gentiana scabra*), and camphor.

Note: This group has a tendency to attack aerial parts of the plants. Basidiospore infection of rice has been reported, but sclerotia are more important as an infection source. The optimum growth temperature is higher than those of AG-1 IB.

AG-1 IB (Sneh *et al.*, 1998; Naito, 2004;Yang *et al.*, 2005b).

Symptoms: sheath blight, leaf blight, foliar blight, web-blight, root rot, damping-off, head rot, and bottom rot.

Host: corn, sugar beet, gay feather (*Liatris* spp.), common bean, fig (*Ficus* L.), adzuki bean, soybean, cabbage, leaf lettuce, redtop, bentgrass, orchard grass, leaf lettuce, apple (*Malus pumila* Mill), Japanese pear, European pear, lion'ear (*Leonotis leonurus*), hortensia (*Hydrangea* spp.), *Larix* spp., gazania (*Gazania* spp.) *Cotoneaster* spp., Egyptian atar-cluster (*Pentas lanceolata*), Chinese lantern plant (*Physalis alkekeng* var. franchetii), *Hypericum patulum*, marigold, *Acacia* spp., rosemary, *Eucalyptus* spp., pine (*Pinus* L.), *Larix* spp., cypress (*Cupressus* spp.), and elephant foot (*Amorphophallus Konjac*).

AG-1 IC (Sneh *et al.*, 1998; Naito, 2004).

Symptoms: damping-off, summer blight, foot rot, crown rot canker, and root rot.

Host: sugar beet, carrot (*Daucus carota* L.), buckwheat (*Eriogonum* Michx), flax (*Linum usitatissimum* L.), soybean, bean (*Phaseolus* L.), cabbage, pineapple (*Ananas comosus* (Linn.) Merr.), panicum (*Panicum* spp.), spinach (*Spinacia oleracea* L.), and radish (*Raphanus sativus* Linn).

AG-1 ID (Priyatmojo *et al.*, 2001).

Symptom: leaf spot.

Host: coffee (*Coffea* Linn).

Note: this subgroup was recently reported in the Philippines (Priyatmojo *et al.*, 2001)

Undetermined subgroup: buckwheat, flax, spinach, radish, and durian (*Durio zibethinus* Murr.).

2.   **AG-2: 2-1, 2-2 IIIB, 2-2 IV, 2-2 Lp, 2-3, 2-4, 2-BI.**

AG-2-1 (Satoh *et al.*, 1997; Camporota and Perrin, 1998; Sneh *et al.*, 1998; Rollins *et al.*, 1999; Khan and Kolte, 2000; Naito, 2004)

Symptoms: damping-off, leaf rot, leaf blight, root rot, foot rot, bottom rot, and bud rot.

Host: sugar beet, wheat (*Triticum aestivum* Linn.), potato, cowpea (*Vigna unguiculata* (Linn.) Walp), canola, rape (*Brassica napus* Linn.), cauliflower (*Brassica oleracea* var. *botrytis* Linn.), mustard (*Sinapis* Linn.), turnip (*Brassica rapa* Linn.), pepper (Piper Linn.), *Silene armeria*, spinach, leaf lettuce, strawberry (*Fragaria ananassa* Duchesne), tulip (*Tulipa gesneriana* Linn.), tobacco (*Nicotiana* Linn.), clover (*Medicago* Linn.), and table beet.

Note: This group includes the AG-2-1 tulip strain (former AG-2t) and the AG-2-1 tobacco strain (former homogenous Nt-isolates) (Kuninaga *et al.*, 2000).

AG-2-2 III B (Sneh *et al.*, 1998; Priyatmojo *et al.*, 2001; Naito, 2004).

Symptoms: brown sheath blight, dry root rot, root rot, brown patch, large patch, black scurf, stem rot, stem blight, *Rhizoctonia* rot, damping-off, stem rot, collar rot, and crown brace rot.

Host: rice, soybean, corn, sugar beet, edible burdock (*Arctium lappa*), taro (*Colocasia esculenta*), *Dryopteris* spp., elephant foot, crocus, saffron (*Crocus sativus* Linn.), redtop, bentgrass, St. Augustine grass, turf, balloon flower (*Platycodon grandiflorum*), Christmas-bells (*Sandersonia aurantiaca*), *Hedera rhombea*, mat rash, *gladiolus*, ginger, and *Iris* Linn..

AG-2-2 IV: (Sneh *et al.*, 1998; Naito, 2004).

Symptoms: leaf blight, foliage rot, root rot, and stem rot.

Host: sugar beet, carrot, eggplant (*Solanum* Linn), pepper, spinach, stevenia (*Stevenia* Adams et Fisch), and turfgrass.

AG-2-2 LP: (Aoyagi *et al.*, 1998).

Symptoms: large patch.

Host: Zoysia grass.

AG 2-3: (Naito and Kanematsu, 1994; Sumner *et al.*, 2003).

Symptoms: leaf blight and root rot.

Host: soybean.

Note: basidiospores cause leaf spot of soybean.

AG-2-4: (Sumner, 1985).

Symptoms: crown rot, brace rot, and damping-off.

Host: corn and carrot.

AG-2-BI: (Carling *et al.*, 2002b).

Symptoms;:nonpathogenic.

Host: isolates, obtained only from soils and plants in forests.

Note: former name is AG-BI.

Undetermined subgroup: sesame (*Sesamum* Linn.), white mustard (*Sinapsis alba*), primrose (*Primula* spp.), white lace flower (*Ammi majus*), carnation, baby's-breath (*Gypsophila paniculata*), rusell prairie gentian (*Eustoma grandiflorum*), snap bean, lima bean, and Chinese radish.

3.   **AG 3: PT, TB** (Sneh *et al.*, 1998; Kuninaga *et al.*, 2000).

Symptoms: black scurf, leaf spot, target leaf spot, and damping-off.

PT: potato with black scurf symptoms.

TB: tobacco with target leaf spot symptoms.

Note: Undetermined subgroup: eggplant, sugar beet, tomato, and wheat. Their pathological and ecological information is less.

4.   **AG-4: HG-I, HG-II, HG-III** (Baird, 1996; Holtz *et al.*, 1996; Sneh *et al.*, 1998; Fenille *et al.*, 2002; Ravanlou and Banihashemi, 2002; EI Hussieni, 2003; Kuramae *et al.*, 2002, 2003; Naito, 2004; Yang *et al.*, 2005c).

Symptoms: damping-off, root rot, stem canker, fruit rot, and stem rot.

Host: pea, sugar beet, melon, soybean, adzuki bean, common bean, snap bean, lima bean, carrot, spinach, taro, tomato (*Lycopersicon esculentum* Mill.), potato, alfalfa (*Medicago sativa* Linn.), elephant foot, arrowleaf clover, beans, barley, buckwheat, cabbage, canola, turnip, carnation, cauliflower, Chinese chive, chrysanthemum, corn, cotton (*Gossypium* Linn.), table beet, tobacco, turfgrass, wheat, white lupine, parsley (*Petroselinum* Hill), *Cineraria* Linn., stock, poinsettia, primrose, hybrid bouvardia, *Citrus* Linn., cauliflower, *Euphorbia* spp., geranium (*Pelargonium* spp.), Russel prairie gentian, statice (*Limonium* spp.), baby's-breath, and *Astragalus membranaceus*

5.   **AG-5** (Li, *et al.*, 1998; Demirci, 1998; Sneh *et al.*, 1998; Ravanlou and Banihashemi, 2002; Eken and Demirci, 2004; Naito, 2004).

Symptoms: root rot, damping-off, black scurf, brown patch, and symbiosis (orchids).

Host: soybean, adzuki bean, apple, barley, chickpea, common bean, lima bean, potato, strawberry, sugar beet, table beet, tobacco, turfgrass, wheat, and white lupine.

6. **AG-6: HG-I, GV** (Mazzola, 1997; Meyer *et al.*, 1998; Sneh *et al.*, 1998; Carling *et al.*, 1999; Pope and Carter, 2001; Naito, 2004)

Symptom: root rot, crater rot, and symbiosis (orchids).

Host: apple, wheat, carrot, and carnation.

Note: all isolates from forests are nonpathogenic.

7. **AG-7:** (Naito, *et al.*, 1993; Baird and Carling, 1995; Carling, 1997, 2000; Carling *et al.*, 1998)

Symptoms: damping-off, root rot, and black scurf.

Host: carnation, cotton, soybean, watermelon (*Citrullus lanatus* (Thunb.) Mansfeld), *Raphanus* Linn., and potato.

8. **AG-8:** (Sneh *et al.*, 1998; Naito, 2004).

Symptoms: bare patch.

Host: barley, cereals, green pepper, potato, and wheat.

9. **AG-9:** (Sneh *et al.*, 1998; Naito, 2004).

Symptoms: black scurf.

Host: potato, crucifers, wheat, and barley.

10. **AG-10:** (Sneh *et al.*, 1998.)

Symptoms: weak pathogenic.

Host: barley and wheat.

11. **AG-11:** (Kumar *et al.*, 2002).

Symptoms: damping-off and hypocotyls rot.

Host: barley, lupine, soybean, and wheat.

Note: this group is considered as bridging isolates (anastomose with each members of AG-2-1, AG-2 BI, AG-8) (Carling *et al.*, 1996).

12. **AG-12:** (Kumar *et al.*, 2002).

Symptoms: symbiosis (orchids).

Host: *Dactylorhiza aristata* (Orchidaceae).

13. **AG-13:** (Carling *et al.*, 2002a).

Symptoms: none.

Host: cotton.

2.2 Binucleate *Rhizoctonia* spp.

1.   **AG-A:** (Mazzola, 1997; Sneh *et al.*, 1998).

Symptoms: root rot, damping-off, browning, and tortoise shell.

Host: strawberry, sugar beet, bean, pea, sunflower (*Helianthus annuus* Linn.), tomato, melon, cucumbear (*Cucumis sativas* Linn.), leaf lettuce, spinach, peanut, potato, *Solanum tuberosum*, and apple.

Note: Some isolates in this group form mycorrhizal associations with orchids.

2.   **AG-B: a and b.**

AG-Ba (Sneh *et al.*, 1998).

Symptoms: grey sclerotium disease, sclerotium disease, gray southern blight.

Host: rice, *Echinochloa crugalli* subsp. *submitica* var. typica, and foxtail millet.

AG-Bb (Sneh *et al.*, 1998).

Symptoms: brown sclerotium disease, grey sclerotium disease, and sheath spot.

Host: fox tail, millet, and rice.

3.   **AG-C** (Sneh *et al.*, 1998; Hayakawa *et al.*, 1999).

Symptoms: symbiosis (orchids).

Host: orchids, sugar beet seedlings, subterranean clover, and wheat.

Note: No important pathogens have been reported.

4.   **AG-D: I, II** (Sneh *et al.*, 1998; Toda *et al.*,1999).

Symptoms: sharp eye spot, yellow patch, foot rot, Sclerotium disease, snow mold, root rot, damping-off, lesions on stems, and winter stem rot.

Host: cereals, turf grass, wheat, barley, sugar beet, clove, pea, onions (*Allium cepa* Linn.), potato, cotton, bean, soybean, mat rush, foxtail millet, and subterranean clover.

Note: Recently this group is classified into subgroup AG-D (I) that causes Rhizoctonia patch and winter patch diseases. AG-D (II) causes elephant footprint disease.

5.   **AG-E** (Sneh *et al.*, 1998).

Symptoms: web-blight, damping-off, seedlings, and symbiosis (orchids).

Host: bean, pea, radish, onion, leaf lettuce, tomato lima bean, snap bean, soybean, peanut, cowpea (*Vigna Savi*), flax, sugar beet, *Rhododendron* Linn., long leaf pine (*Pinus palustris* Mill.), slash, lobolly pine (*Pinus taeda* Linn.), and rye (*Secale cereale* Linn.).

6.   **AG-F** (Sneh *et al.*, 1998; Eken and Demirci, 2004).

Symptoms: none.

Host: bean, pea, radish, onion, peanut, leaf lettuce, tomato, subterranean clover radish, tomato, cotton, taro, strawberry (source: DDJB), and *Fragaria x ananassa*.

7.   **AG-G** (Mazzola, 1997; Sneh *et al.*, 1998; Leclerc *et al.*, 1999; Martin, 2000; Botha *et al.*, 2003; Fenille *et al.*, 2005).

Symptoms: damping-off, root rot, and browning.

Host: strawberry, sugar beet, bean, pea, tomato, melon, sunflower, peanut, yacoon, apple, *Rhododendron* Linn., and *Fragaria x ananassa*.

Note: Non-pathogenic binucleate *Rhizoctonia* spp. provide effective protection to young bean seedlings against root rot caused by *R. solani* AG-4 (Leclerc *et al.*, 1999).

8.   **AG-H** (Hayakawa *et al.*, 1999).

Symptoms: symbiosis (orchids).

Host: *Dactylorhiza aristata* (Orchidaceae).

9.   **AG-I** (Mazzola, 1997; Sneh *et al.*, 1998; Ravanlou and Banihashemi, 2002)

Symptoms: root rot and symbiosis (orchids).

Host: strawberry, sugar beet, wheat, apple, orchids, and *Fragaria x ananassa*.

10.   **AG-J**: (Sneh *et al.*, 1998).

Symptoms: none.

Host: apple.

11.   **AG-K** (Demirci, 1998; Li *et al.*, 1998; Sneh *et al.*, 1998; Ravanlou and Banihashemi, 2002).

Symptoms: none.

Host: sugar beet, radish, tomato, carrot, onion, wheat, maize, *Allium cepa* (source: DDJB),

Pyrus communis (pear) (source: DDJB), and Fragaria x ananassa.

12.   **AG-L**: No special diseases have been reported (Sneh *et al.*, 1991).
13.   **AG-N**: No special diseases have been reported (Sneh *et al.*, 1991).
14.   **AG-O**: No special diseases have been reported (Mazzola, 1997; Sneh *et al.*, 1998).

Host: apple.

15.   **AG-P**: (Sneh *et al.*, 1998; Yang *et al.*, 2006).

Symptoms: black rot and wirestem.

Host: tea (*Camellia* Linn.), red birch.

16.   **AG-Q**: (Sneh *et al.*, 1998).

Symptoms: none.

Host: (Bentgrass).

17.   **AG-R**: (Sneh *et al.*, 1998; Yang *et al.*, 2006).

Symptoms: wirestem

Host: bean, pea, radish, onion, leaf lettuce, tomato, lima bean, snap bean, soybean, cowpea, peanuts, red birch, and azalea.

18. **AG-S** (Demirci, 1998; Sneh *et al.*, 1998).

Symptoms: no specific diseases.

Host: azalea, wheat, barley, and azalea.

19. **AG-T:** (Hyakumachi *et al.*, 2005).

Symptoms: stem rot and root rot.

Host: miniature roses.

20. **AG-U:** (Hyakumachi *et al.*, 2005).

Symptoms: stem rot and root rot.

Host: miniature roses (*Rosa rugosa* Thunb.).

3. Summary

In this chapter, we described the classification of *Rhizoctonia* spp. complex. Mutinucleate *Rhizoctonia* spp. included 13 anastomosis, of which AG 1-4 were strong pathogenic on many plants and AG 6-10 were orchid mycorrhizae. Binucleate *Rhizoctonia* spp. included 18 anastomosis groups, but AG-U belonged to AG-P and AG-T belonged to AG-A (Sharon et al., 2008), which were weak or nonpathogenic to plants and some AGs were orchid mycorrhizae.

4. Acknowledgments

This work was supported by the National Basic Research Program (No. 2011CB100400) from The Ministry of Science and Technology of China and the National Natural Science Funds, China (30660006, and 31160352), respectively.

5. References

Aoyagi, T., Kageyama, K. & Hyakumachi, M. (1998) Characterization and survival of *Rhizoctonia solani* AG2-2 LP associated with large patch disease of zoysia grass. Plant Dis. 82, 857-863.

Baird, R. E. (1996) First report of *Rhizoctonia solani* AG-4 on canola in Georgia. Plant Dis. 80(1), 104.

Baird, R. E. & Carling, D. E. (1995) First report of *Rhizoctonia solani* AG-7 in Indiana. Plant Dis. 79(3), 321.

Botha, A., Denman, S., Lamprecht, S. C., Mazzola, M. & Crous, P. W. (2003) Characterization and pathogenicity of *Rhizoctonia* isolates associated with black root rot of strawberries in the Western Cape Province, South Africa. Australasian Plant Pathol. 32(2), 195-201.

Burpee, L. L., Sanders, P. L., Cole, H. Jr. & Sherwood, R. T. (1980a) Anastomosis groups among isolates of *Ceratobasidium cornigerum* and related fungi. Mycologia 72, 689-701.

Camporota, P. & Perrin, R. (1998) Characterization of *Rhizoctonia* species involved in tree seedling damping-off in French forest nurseries. Appl. Soil Ecol. 10(1/2), 65-71.

Carling, D. E. (1997) First report of *Rhizoctonia solani* AG-7 in Georgia. Plant Dis. 82(1), 127.

Carling, D. E. (2000) Anastomosis groups and subsets of anastomosis groups of *Rhizoctonia solani*. Abstract in Proceedings of the Third International Symposium on *Rhizoctonia*. National Chung Hsing University, Taichung, Taiwan .14 pp.

Carling, D. E., Baird, R. E., Gitaitis, R. D., Brainard, K. A. & Kuninaga, S. (2002a) Characterization of AG-13, a newly reported anastomosis group of *Rhizoctonia solani*. Phytopathology 92(8), 893-899.

Carling, D. E., Brainard, K.A., Virgen-Calleros, G. & Olalde-Portugal, V.F. (1998) First report of *Rhizoctonia solani* AG-7 on potato in Mexico. Plant Dis. 82(1), 127.

Carling, D. E., Kuninaga, S. & Brainard, K. A. (2002b) Hyphal anastomosis reactions, rDNA-internal transcribed spacer sequences, and virulence levels among subsets of *Rhizoctonia solani* anastomosis group-2 (AG-2) and AG-BI. Phytopathology 92, 43-50.

Carling, D. E., Pope, E. J., Brainard, K. A. & Carter, D. A. (1999) Characterization of mycorrhizal isolates of *Rhizoctonia solani* from an orchid, including AG-12, a new anastomosis group. Phytopathology 89(10), 942-946.

Carling, D. E. (1996) Grouping in *Rhizoctonia solani* by hyphal anastomosis reaction. In: Sneh, B., Jabaji-Hare, S., Neat, S. and Dijst, G. *et al.*, (eds). *Rhizoctonia* species: Taxonomy, Molecular Biology, Ecology, Pathology and Disease Control. Kluwer Academic Publishers, Dordecht, The Netherlands, pp 37-47.

Carling, D. E., Baird, R. E., Gitaitis, R. D., Brainard, K. A. & Kuninaga, S. (2002c) Characterization of AG-13, a newly reported anastomosis group of *Rhizoctonia solani*. Phytopathology 92,893-899.

de Candolle, A. 1815. Uredo rouille des cereals In Forafran caise, famille des champigons p.83.

Demirci, E. (1998) *Rhizoctonia* species and anastomosis groups isolated from barley and wheat in Erzurum, Turkey. Plant Pathol. 47(1), 10-15.

Eken, C. & Demirci, E. (2004) Anastomosis groups and pathogenicity of *Rhizoctonia solani* and binucleate *Rhizoctonia* isolates from bean in Erzurum, Turkey J. Plant Pathol. 86(1), 49-52.

Fenille, R.C., Ciampi, M.B., Souza, N.L., Nakataniand, A.K. and Kuramae, E.E.(2005) Binucleate Rhizoctonia sp. AG-G causing rot rot in yacon (Smallanyhus sonchifolius) in Brazail. Plant patho. 54,325-330.

Fenille, R. C., Luizde S. N. & Kuramae, E. E. (2002) Characterization of *Rhizoctonia solani* associated with soybean in Brazil. Eur. J. of Plant Pathol. 108(8), 783-792.

Gonzalez, D., Carling, D. E., Kuninaga, S., Vilgalys, R. & Cubeta, M. A. (2001) Ribosomal DNA systematics of *Ceratobasidium* and *Thanatephorus* with *Rhizoctonia* anamorphs. Mycologia 93 (6), 1138-1150.

Hayakawa, S., Uetake, Y. & Ogoshi, A.(1999) Identification of symbiotic rhizoctonia from naturally occurring protocorms and roots of *Dactylorhiza aristata* (Orchidaceae). J. of the Faculty of Agriculture Hokkaido University 69(2), 129-141.

Holtz, B. A., Michailides, T. J., Feguson, L., Hancock, J. G. & Weinhold, A. R. (1996) First report of *Rhizoctonia solani* (AG-4) on pistachio rootstock seedlings. Plant Dis. 80(11), 1303.

Hyakumachi, M., Priyatmojo, A., Kubota, M. & Fukui, H. (2005) New anastomosis groups, AG-Tand AG-U, of binucleate *Rhizoctonia* spp. causing root and stem rot of cut-flower and miniature roses. Phytopathology 95,784-792.

Hyyakumachi, M. and Ui,T. (1988)Development of the teleomorph of non-self-anastomosing isolates of Rhizoctonia solani by a buried-slide method plant pathol. 37(3):438-440

Khan, R. U. & Kolte, S. J. (2000) Some seedling diseases of rapeseed-mustard and their control. Indian Phytopathol. 55(1), 102-103.

Kumar, S., Sivasithamparam, K. & Sweetingham, M. W. (2002) Prolific production of sclerotia in soil by *Rhizoctonia solani* anastomosis group (AG) 11 pathogenic on lupin. Annal. Appl. Biol. 141(1), 11-18.

Kuninaga, S. & Yokosawa, R. (1980) A comparison of DNA compositions among anastomosis groups in *Rhizoctonia solani* Kühn. Ann. Phytopathol. Soc. Japan 46,: 150-158.

Kuninaga, S., Nicoletti, R., Lahoz, E. & Naito, S. (2000). Ascription of Nt-isolates of *Rhizoctonia solani* to anastomosis group 2-1 (AG-2-1) on account of rDNA-ITS sequence similarity. J. Plant Pathol. 82, 61-64.

Leclerc, P. C., Balmas, V., Charest, P. M. & Jabaji,H. S. (1999) Development of reliable molecular markers to detect non-pathogenic binucleate *Rhizoctonia* isolates (AG-G) using PCR. Mycol. Res. 103(9), 1165-1172.

Li, H.R. & Yan, S.Q. (1990) Studies on the strains of pathogens of sheath blight of rice in the east and south of Sichuan Province. Acta Mycologica Sinica 9: 41-9. (Chinese with English abstract).

Li, H.R., Wu, B.C. & Yan, S. Q. (1998) Aetiology of *Rhizoctonia* in sheath blight of maize in Sichuan. Plant Pathol. 47 (1), 16-21.

MacNish, G. C., Carling, D. E. & Brainard, K. A. (1997) Relationship of microscopic vegetative reactions in *Rhizoctonia solani* and the occurrence of vegetatively compatible populations (VCP) in AG-8. Mycol. Res. 100, 61-68.

Martin, F. N. (2000) *Rhizoctonia* spp. recovered from strawberry roots in central coastal California. Phytopathology 90, 345-353.

Mazzola, M. (1997) Identification and pathogenicity of *Rhizoctonia* spp. isolated from apple roots and orchard soils. Phytopathology 87, 582-587.

Meyer, L., Wehner, F. C., Nel, L. H. & Carling, D. E. (1998) Characterization of the crater disease strain of *Rhizoctonia solani*. Phytopathology 88, 366-371.

Moore, R. T. (1987) The genera of *Rhizoctonia*-like fungi: *Asorhizoctonia, Ceratorhiza* gen. nov., *Epulorhiza* gen. nov., *Moniliopsis* and *Rhizoctonia*. Mycotaxon 29, 91-99.

Naito, S. (2004) Rhizoctonia diseases: Taxonomy and population biology. Proceeding of the International Seminar on Biological Control of Soilborne Plant Diseases, Japan-Argentina Joint Study, Buenos Aires, Argentina, p.18-31.

Naito, S. & Kanematsu, S. (1994) Characterization and pathogenicity of a new anastomosis subgroup AG-2-3 of *Rhizoctonia solani* Kühn isolated from leaves of soybean. Ann. Phytopath. Soc. Japan 60(6), 681-690.

Naito, S., Mohamad, D., Nasution, A &. Purwanti, H. (1993) Soilborne diseases and ecology of pathogens on soybean roots in Indonesia. JARQ 26, 247-253.

Nicoletti, R., Lahoz, E., Kanematsu, S., Naito, S. & Contillo, R. (1999) Characterization of *Rhizoctonia solani* isolates from tobacco fields related to anastomosis groups 2–1 and BI (AG 2–1 and AG BI). J. Phytopathology 147 (2), 71-77.

Ogoshi, A. (1975) Grouping of *Rhizoctonia solani* Kühn and their perfect stages. Rev. Plant. Protect. Res. 8, 93-103.

Ogoshi, A. & Ui, T. (1979) Specificity in vitamin requirement among anastomosis groups of *Rhizoctonia solani* Kühn. Ann. Phytopathol.Soc. Jpn. 45, 47-53.

Ogoshi, A., Oniki, M., Araki, T. & Ui, T. (1983a) Anastomosis groups of binucleate *Rhizoctonia* in Japan and North America and their perfect states. Mycol. Soc. Japan 24, 79-87.

Ogoshi, A., Oniki, M., Araki, T. & Ui, T. (1983b) Studies on the anastomosis groups of binucleate *Rhizoctonia* and their perfect states. J. Fac. Agr. Hokkaido Univ. 61, 244-260.

Parmeter, J. R. J., Sherwood, R. T. & Platt, W. D. (1969) Anastomosis grouping among isolates of *Thanatephorus cucumeris*. Phytopathology 59, 1270-1278.

Parmeter, J. R. Jr. & Whitney, H. S. (1970) Taxonomy and nomenclature of the imperfect state. Pages 7-19 in: J. R. Parmeter Jr., (ed.) Biology and Pathology of *Rhizoctonia solani*. University of California Press, Berkeley. 255 pp.

Pope, E. J. & Carter, D. A. (2001) Phylogenetic placement and host specificity of mycorrhizal isolates belonging to AG-6 and AG-12 in the *Rhizoctonia solani* species complex. Mycologia 93(4), 712-719.

Priyatmojo, A., Escopalao, V. E., Tangonan, N. G., Pascual, C. B., Suga, H., Kageyama, K. & Hyakumachi, M. (2001) Characterization of a new subgroup of *Rhizoctonia solani* anastomosis group 1 (AG-1 ID), causal agent of a necrotic leaf spot on coffee. Phytopathology 91,1054-1061.

Ravanlou, A. & Banihashemi, Z. (2002) Isolation of some anastomosis groups of *Rhizoctonia* associated with wheat root and crown in Fars province. Iranian J. Plant Pathol. 38(3-4), 67-69.

Rollins, P. A., Keinath, A. P. & Farnham, M. W. (1999) Effect of inoculum type and anastomosis group of *Rhizoctonia solani* causing wirestem of cabbage seedlings in a controlled environment. Can. J. Plant Pathol. 21(2), 119-124.

Satoh, Y., Kanehira, T. & Shinohara, M. (1997) Occurrence of seedling damping-off of Jew's mallow, *Corchorus olitorius* caused by *Rhizoctonia solani* AG-2-1. Nippon Kingakukai Kaiho 38(2): 87-91 (Japanese with English abstract).

Sharon, M., Kuninaga, S., Hyakumachi, M., Naito, S., & Sneh B. (2008) Classification of *Rhizoctonia* spp. using rDNA-ITS sequence analysis supports the genetic basis of the classical anastomosis grouping. *Mycoscience* 49, 93–114.

Sneh, B., Burpee, L. and Ogoshi, A. (1998) Identification of *Rhizoctonia* species. The APS, St. Paul, Minesota.

Sumner, D. R. (1985) First report of *Rhizoctonia solani* AG-2-4 on carrot in Georgia. Plant Dis. 69, 25-27.

Sumner, D. R., Phatak, S. C. & Carling, D. E. (2003) Characterization and pathogenicity of a new anastomosis subgroup AG-2-3 of *Rhizoctonia solani* Kuün isolated from leaves of soybean. Plant Dis. 87(10), 1264.

Toda, T., Hyakumachi, M., Suga, H., Kageyama, K., Tanaka, A. & Tani, T. (1999) Differentiation of *Rhizoctonia* AG-D isolates from turfgrass into subgroups I and II based on rDNA and RAPD analyses. Eur. J. Plant Pathol. 105, 835-846.

Vilgalys, R. & Gonzalez, D. (1990) Ribosomal DNA restriction fragment length polymorphism in *Rhizoctonia solani*. Phytopathology 80, 151-158.

Yang G. H., Chen H. R., Naito, S., Wu, J. Y., He, X. H. & Duan, C. F. (2005b) Occurrence of foliar rot of pak choy and Chinese mustard caused by *Rhizoctonia solani* AG-1 IB in China. J. Gen. Plant Pathol.71, 377–379.

Yang, G. H., Naito, S., Ogoshi, A. & Dong, W. H. (2006) Identification, isolation frequency and pathogenicity of *Rhizoctonia* spp. causing the wirestem of red birch in China. J. Phytopathology 154(2), 80-83.

2

Taxonomic Review of and Development of a Lucid Key for Philippine Cercosporoids and Related Fungi

M. Mahamuda Begum, Teresita U. Dalisay and
Christian Joseph R. Cumagun
Crop Protection Cluster, College of Agriculture,
University of the Philippines Los Baños,
Philippines

1. Introduction

The genus *Mycosphaerella* Johanson, contains more than 3000 names (Aptroot 2006), and has been linked to more than 30 well-known anamorphic genera (Crous 2006a and 2006b). It has a worldwide distribution from tropical and subtropical to warm and cool regions (Crous 1998; Crous et al. 2000 and 2001). *Mycosphae*rella, however, has been associated with at least 27 different coelomycete or hyphomycete anamorph genera (Kendrick and DiCosmo, 1979), 23 of which were accepted by Crous et al. (2000). More than 3000 names have already been published in *Cercospora* (Pollack, 1987). The genus *Cercospora* Fresen., which is one of the largest genera of hyphomycetes, has been linked to *Mycosphaerella* teleomorphs (Crous et al., 2000). *Cercospora* was first monographed by Chupp (1954), who accepted 1419 species. Subsequent workers such as, F.C. Deighton, B.C. Sutton and U. Braun divided *Cercospora* in to almost 50 different genera which are morphologically similar and distinct with each other (Crous and Braun, 2003).

Cercosporoid fungi are a collective term for a group of fungi belonging, to the genus *Cercospora* and its allied genera, namely *Pseudocercospora, Passalora, Asperisporium, Corynespora, Cladosporium*. Differences among them are based mainly on a combination of characters that include the structure of conidiogenous loci (scars) and hila, presence or absence of pigmentation and ornamentation in conidiophores and conidia, geniculate or non-geniculate conidiophore, and rare presence of additional or unique features such as knotty appearance of conidiophores.

Cercospora Fresen. is one of the largest genera of Hyphomycetes. Saccardo (1880) defined *Cercospora* as having brown conidiophores and vermiform, brown, oliivaceous or rarely subhyaline conidia. Deighton continuously studied the *Cercospora* species (Deighton, 1967a, 1967b, 1971, 1973, 1974, 1976, 1979, 1983, and 1987) and reclassified numerous species into several allied genera based mainly on two distinct taxonomic categories: thickened conidial scars occur in the *Cercospora* and in allied genera such as, *Passalora* and *Stenella*, while unthickened scars are characteristics of the genera *Pseudocercospora* and *Stigmina*.

Cercosporoid fungi are commonly associated with leaf spots (Wellman, 1972) ranging from slight, diffuse discolorations to necrotic spots or leaf blight (Shin and Kim, 2001). Cercosporoid fungi can also cause necrotic lesions on flowers, fruits, bracts, seeds and pedicels of numerous hosts in most climatic regions (Agrios, 2005). They are responsible for great damages to beneficial plants. Furthermore, other than important pathogens of major agricultural crops such as cereals, vegetables, ornamentals, forest trees, grasses and many others species are also known to be hyperparasitic to other plant pathogenic fungi (Goh and Hsieh, 1989).Cercosporoid fungi are known to cause some of the economically important diseases worldwide. One of the most important and common diseases associated with this fungus is the black sigatoka caused by *Mycosphaerella fijiensis* which was first discovered and considered to have caused epidemics in the Valley of Fiji (Stakman and Harrar, 1957).

The Cercosporoid fungi of Philippines are insufficiently known. There have been no comprehensive studies on this group of fungi in Philippines. Welles (1924 and 1925) worked with physiological behavior of Philippine Cer*cosporas* on artificial media and the extent of parasitism. There were 87 species of *Cercospora* reported in the Philippines from 1937 onwards (Quimio and Abilay, 1977). Teodoro (1937) had enumerated 65 species of *Cercospora* in the Philippines. In most cases, however, the causal species have only been cited but not characterized. No attempt was made to determine the host range of each of the species. Naming of the species was based mainly on Chupp's monograph (Chupp, 1953), which together with Vasudeva's book (Vasudeva, 1963) book on Indian *Cercosporae*, as the main reference books used by Quimio and Abilay (1977).

Identification of fungal plant pathogens is commonly done using one of several well-illustrated dichotomous keys by Ellis (1971 and 1976), Sutton (1980), Hanlin (1990), and Barnett & Hunter (1998). Multi-access keys for identifying biological agents are very useful, especially for the non-specialist, as it is not necessary to be able to detect all of the fine distinctions usually found in dichotomous keys. The disadvantage of those printed keys is that they require the user to be able to scan a series of tables of numbers and select those that are common to the specimen being examined (Michaelides *et al.* 1979; Sutton 1980). This task is ideally suited to computers. The Lucid system developed by the Centre for Biological Information Technology (CBIT), University of Queensland (Norton 2000) allows development of multi-access computer-based keys that can also incorporate graphics and text. The result is a very powerful tool. Although some keys have previously been developed for fungi using Lucid, they have generally been for specific groups such as rainforest fungi of Eastern Australia (Young 2001). The main purpose in developing these Lucid identification systems has been to contribute to taxonomic capacity building in two ways - by enabling identification keys to be easily developed and by increasing the availability and usefulness of these keys by making them available on CD or via the Internet. A Lucid was used for identifying genera for identifying genera of plant pathogenic Cercosporoid fungi of Philippine crops. The key was comprised of many characters, which has the potential for being rather cumbersome. For simplicity, the characters were placed in groups and states relating to the structures like the morphology of conidiophores, the stromata, conidia, and fruiting bodies and the names of host family and genus.

The primary objective of this study was to identify Cercosporoid fungi of the Philippines, use recent taxonomic information to amend or rename species, formulate taxonomic keys, and develop Lucid key for identification. An existing computer based software was applied to the development of morphological Lucid key for their identification. For this purpose, field collections were conducted from 2007 to 2009. Microscopic studies on the association of Cercosporoid species to the diseased leaves were carried out at the Postharvest Pathology Lab, Crop Protection Cluster, College of Agriculture, University of the Philippines Los Baños (UPLB). The field collection was confined mostly within UPLB campus, particularly propagation farm and medicinal plant gene bank, vegetable farm of the Crop Science Cluster, UPLB, the production farm at the Jamboree site, production farm at International Rice Research Institute (IRRI), and some residential gardens in Los Baños, Laguna.

Key to Cercosporoid Genera (Crous and Braun, 2003).

1. Conidiogenous loci inconspicuous or subdenticulate, but always unthickened and not darkened or subconspicuous, i.e., unthickened, but somewhat refractive or rarely very slightly darkened, or only outer rim slightly darkened and refractive (visible as minute rings)--*Pseudocercospora*
1. Conidiogenous loci conspicuous, i.e., thickened and darkened throughout, only with a minute central pore--2
     2. with verruculose superficial secondary mycelium; conidia amero- to scolecosporous, mostly verruculose--*Stenella*
     2. If superficial secondary mecelium present, hyphae smooth or almost so------------3
3. Conidia hyaline or subhyaline, scolecosporous, acicular, obclavate-cylindrical, filiform, usually pluriseptate--*Cercospora*
3. Conidia pigmented or, if subhyaline, conidia non-scolecosporus, ellipsoid-ovoid, short cylindrical, fusoid and only few septa--*Passalora*
     4. Conidiogenous loci protuberant, with a central convex part (dome) surrounded by a raised periclinal rim, conidia in long, often branched chains or solitary, smooth to verruculose ---*Cladosporium*
     4. Conidiogenous loci conspicuously thickened, conidia non-scolecosporous, ellipsoid-ovoid, short subcylindrical, aseptate or only with few septa--------------------
     ---*Asperisporium*
5. Conidia contain from 4-20 pseudo-crosswalls (pseudosepta), the base of the conidium (hilum) conspicuously thickened--- *Corynespora*
5. Conidia without septa or with one or a few transverse septa, conidiophores apical portion sometimes branched-- *Periconiella*

Further descriptions of the genera and species belonging to the genus, as they were associated from the collections were presented. The last column of the table indicates whether the collection is considered a first record or has already been reported.

2. Genus *Cercospora*

Cercospora Fresen. (Crous and Braun, 2003).

Stromata lacking to well developed, subhyaline to usually pigmented; *conidiophores* mononematous, macronematous, solitary to fasciculate, arising from internal hyphae or stomata, erect, continuous to pluriseptate, subhyaline to pigmented; *conidiogenous loci* conspicuous, thickened and darkened, planate; *conidia* solitary to catenate, scolecosporous, obclavate, cylindrical, filiform, acicular, hyaline, smooth or almost so, hila thickened and darkened (Crous and Braun, 2003).

Type species: *Cercospora penicillata* (Ces.) Fresen.

In the present study, 48 *Cercospora* diseases were reported. Among them, 20 species were now considered under a compound species *Cercospora apii s. lat.* (Table 1) and 28 under *Cercospora s. str.* which is host specific with a host range confined to species of a single host genus or some allied host genera of a single family (Table 2).The reported genus *Cercospora* was introduced by Fresenius with *Cercospora penicillata* as the type species. Since then over 1000 species were reported and characterized and were compiled in the book "Monograph of the Genus *Cercospora*" by Chupp (1953). He proposed a broad concept for the genus, simply noting whether hila were thickened or not, and if conidia were pigmented or not, single or in chains. Recently Crous & Braun (2003) recognized four true cercosporoid genera, viz. *Cercospora*, *Pseudocercospora* Speg., *Passalora* Fr. and *Stenella* Syd., and several other morphologically similar genera, based on molecular sequence analyses and a reassessment of morphological characters. They represented a compilation of more than 3000 names that have been published or proposed in *Cercospora*, of which 659 are presently recognized in this genus, with a further 281 being referred to *C.apii s.lat.* They amended the species *C. apii* and it is now a compound species, referred to as *C. apii s.lat.* It infects hundreds of plant species. *Cercospora apii s.lat.* is characterized by having solitary to fasciculate, usually long, brown, septate conidiophores with conspicuously thickened and darkened conidiogenous loci and long, acicular, hyaline, pluriseptate conidia formed singly. *Cercospora s.str* is characterized by having stromata, with numerous, densely arranged rather short conidiophores, small conidiogenous loci, and obclavate-cylindrical conidia with truncate base (Table 2).

Species	Host	Stromata	Conidiophores	Conidia	Ref. Coll. Accession No.	Status of collection
Cercospora amaranti	*Amaranthus viridis* (Amaranth)	lacking	2-15 in a divergent fascicle, pale to olivaceous brown, multiseptate, not branched, straight to slightly geniculate, scars conspicuously thickened, 35-200 x 4-5.5 µm	hyaline, acicular, smooth walled, straight, base-truncate, apex-acute, hilum thickened and darkened, 40-200 x 2-4.5 µm	CALP 11707	FR
Cercospora anonae	*Anona squamosa* (Sugar apple)	present	dense fascicle, pale to olivaceous brown, almost uniform in colour and width, not geniculate, slightly branched, septate, large scar present, 20-180 x 2-5 µm	hyaline, solitary, acicular, straight to curved, septate, base obconically truncate, tip acute, hilum thickened and darkened, 40-200 x 1.5-3 µm	CALP 11734	AR

Species	Host	Stromata	Conidiophores	Conidia	Ref. Coll. Accession No.	Status of collection
Cercospora begoniae	*Begonia* sp. (Begonia)	lacking	2-5 in a fascicle or borne singly, pale to very pale brown in colour, paler and more narrow towards the apex, straight or geniculate, septate, truncate at the apex, scars conspicuously thickened 30-180 x 3-5 μm	hyaline, acicular, straight to mildly curved, indistinctly multiseptate, acute at the apex, truncate at the base, hilum conspicuously thickend, 50-260 x 2.5-4μm	CALP 11676	FR
Cercospora capsici	*Capsicum annum* (Chili)	lacking	5-15 in a fascicle, pale to olivaceous brown, straight to slightly curved, not branched, mildly geniculate, acicular to filiform, straight to mildly curved, scars conspicuously thickened, 30-140 x 3-6 μm	hyaline, solitary, acicular to filiform, straight to mildly curved, multiseptate, apex-acute to subacute, base truncate or obconically truncate, hilum thickened, 50-170 x 3-4.5 μm	CALP 11693	AR
Cercospora citrullina	*Cucurbita moschata* (Squash)	lacking	2-5 in a fascicle, pale to olivaceous brown, straight to slightly bent or curved, geniculate, multiseptate, simple, occasionally swollen at some points, subtruncate at the apex, scars conspicuously thickened, 35-250 x 4-6 μm	hyaline, solitary, acicular/ cylindro-obclavate, straight-curved, multiseptate, apex-subacute to obtuse, base-subtruncate or rounded, hilum thickened and darkened, 20-200 x 2-3.5 μm	CALP 11675	FR

Species	Host	Stromata	Conidiophores	Conidia	Ref. Coll. Accession No.	Status of collection
Cercospora citrullina	Momordica charantia (Bitter gourd)	lacking	2-15 in a fascicle, pale to very pale brown, straight to mildly geniculate, multiseptate, scars conspicuously thickened, 20-180 x 4-5 μm	hyaline, solitary, acicular, multiseptate, apex- acute to subacute, base-subtruncate or rounded, hilum thickened, 45-190 x 2-4 μm	CALP 11688	AR
Cercospora citrullina	Luffa cylindrica (Sponge gourd)	lacking	2-5 in a fascicle, pale olivaceous brown, width irregular, straight to slightly curved, mildly geniculate, septate, conidial scars conspicuous and thickened, 40-250 x 4-5.5 μm	hyaline, solitary, acicular to obclavato-cylindric, straight to mildly curved, multiseptate, obtuse apex, truncate base, hilum conspicuously thickened and darkened, 45-200 x 3.5-5 μm	CALP 11728	FR
Cercospora cruenta	Phaseolus lunatus (Lima bean)	present	10-30 in a divergent fascicle, brown to dark brown at the base and apical portion rather paler, straight to mildly geniculate, multiseptate, scars large and conspicuously thickened, 40-150x 5-6.5 μm	hyaline, acicular/ cylindro-bclavate, straight - curved, multiseptate, apex- subacute, base-subtruncate to truncate, hilum thickened and darkened, 50-250 x 2-3.5 μm	CALP 11710	AR

Species	Host	Stromata	Conidiophores	Conidia	Ref. Coll. Accession No.	Status of collection
Cercospora euphorbiae	*Euphorbia heterophylla* (Milk weed)	lacking	2-5 in a fascicle, pale to olivaceous brown, straight to mildly curved, sometimes branched, rarely geniculate, multiseptate, large conidial scars at the subtruncate tip, 25-100 x 4.5-6 μm	hyaline, solitary, cylindrical to acicular, straight to mildly curved, multiseptate, obconically truncate base, obtuse tip, hilum thickened and darkened, 40-120 x 3-4.5 μm	CALP 11724	FR
Cercospora fukushiana	*Impatiens balsamina* (Balsam plant)	present	divergent fascicle, pale olivaceous brown in colour, apex subtruncate, 1-4 septate, rarely branched, straight to flexuous or geniculate, scars medium sized and thickened, 40-150 x 3-4 μm	hyaline, acicular, straight to mildly curved, indistinctly multiseptate, acute to subacute at the apex, truncate at the base, hilum conspicuously thickend, 40-250 x 3-4 μm	CALP 11678	AR
Cercospora grandissima	*Dhalia variabilis* (Dahlia)	lacking	1-10 in a fascicle, pale to medium brown, straight or mildly geniculate with thickened scars, subtruncate at the apex, 60-100 x 4-5 μm	Hyaline, solitary, acicular, straight to slightly curved, multiseptate, apex-acute, base-truncate, hilum thickened and darkened, 50-250 x 2-4 μm	CALP 11677	AR

Species	Host	Stromata	Conidiophores	Conidia	Ref. Coll. Accession No.	Status of collection
Cercospora hydrangeae	*Hydrangea macrophylla* (Milflores)	present	3-15 in a loose fascicle, brown to deep brown throughout, irregular in width, straight to slightlybcurved, geniculate, not branched, 2-5 septate, obtuse to subtruncate at the apex, conidial scars small and conspicuous, 25-210 x 4-5 µm	hyaline, solitary, acicular to obclavate-cylindric, substraight to mildly curved, 2-13 septate, non-constricted,apex-subacute ,base-truncate, hilum conspicuously thickened, darkened, 35-150 x 3-4 µm	CALP 11733	AR
Cercospora ipomoeae	*Ipomoea triloba* (Little bell)	lacking	dense fascicle, pale olivaceous to medium brown, multiseptate, unbranched, straight, mildly geniculate towards the apex, smooth walled, large conidial scars conspicuously thickened, 40-150 x 5-7.5 µm	hyaline, solitary, obclavate, smooth walled, straight, mildly curved, base-truncate, apex-obtuse, 40-120 x2-4.5µm, hilum thickened & darkened.	CALP 11712	FR
Cercospora ipomoeae	*Ipomoea batatas* (Sweet potato)	lacking	2-5 in a fascicle or borne singly, olivaceous to medium brown, paler upward, rarely branched, mildly geniculate, scar conspicuously thickened, 25-180 x 4-6.5 µm	hyaline, acicular to obclavate, straight to mildly curved, indistinctly multiseptate, truncate at the base, acute at the base, 30-150 x 2-3 µm	CALP 11713	AR

Species	Host	Stromata	Conidiophores	Conidia	Ref. Coll. Accession No.	Status of collection
Cercospora ipomoeae	*Ipomoea aquatica* (Kangkong)	lacking	arise singly or 2-7 in a fascicle, pale yellowish olivaceous to medium, brown, slightly paler and more narrow towards the tip, rearly geniculate, subtruncate at the apex, 15-150 x 4-6.5 μm	hyaline, acicular to obclavate straight to curved, indistinctly multiseptate, subacute at the apex, truncate of subtruncate at the base, 25-130 x 3-5 μm	CALP 11711	FR
Cercospora lagenariae	*Lagenaria vulgari* (Bottle gourd)	lacking	2-5 in a divergent fascicle, pale brown to brown, straight to slightly bent or curved, geniculate, occasionally branched, multiseptate, obtuse to subtruncate at the apex, conidial scars conspicuous, 60-250 x 3-6 μm	hyaline, solitary, acicular, substraight to mildly curved, multiseptate, acute to obtuse at the apex, truncate at the base, hilum thickened and darkened, 40-210 x 2-5 μm	CALP 11700	FR
Cercospora laporticola	*Laportea crenulata* (Laportea)	lacking	2-5 in a loose fascicle, pale to olivaceous brown, septate, unbranched, straight, smooth walled, large conidial scars conspicuously thickened, 30-90 x 5-7.5 μm	hyaline, solitary, acicular, smooth walled, straight, base-truncate, apex-obtuse, hilum thickened and darkened, 40-250 x 3-4.5 μm	CALP 11696	FR

Species	Host	Stromata	Conidiophores	Conidia	Ref. Coll. Accession No.	Status of collection
Cercospora moricola	Morus alba (Mulberry)	lacking	2-15 in a fascicle, pale olivaceous brown, straight, rarely septate and geniculate, unbranched, scars conspicuously thickened,10-50 x 4-5.5 µm	hyaline, solitary, acicular, multiseptate base-truncate, tip-acute hilum thickened and darkened, 40-150 x 2-3.5µm	CALP 11690	FR
Cercospora nicotianae	Nicotiana tabacum (Tobacco)	lacking	2-5 in a fascicle, pale olivaceous to medium brown, paler toward the tip, not branched, multiseptate, mildly-geniculate, scars large and conspicuously thickened, 20-100 x 3-6.5 µm	hyaline, solitary, acicular, straight to mildly curved, multiseptate, acute to subacute at the apex, truncate at the base, hilum thickened and darkened, 25-250 x 2-4.5 µm	CALP 11701	AR
Cercospora zinniae	Zinnia elegans (Zinnia)	lacking	2-20 in a fascicle, pale to medium dark olivaceous brown, not branched, straight or geniculate, scars conspicuously thickened, 15-200 x 4-6 µm	hyaline, solitary, acicular/obclavate, straight to mildly curved, apex-acute / subacute at base-truncate to subtruncate, hilum thickened and darkened, 20-160 x 2-4 µm	CALP 11704	AR

Reference: Chupp (1954); Ellis (1971, 1976); Guo & Hsieh (1995); Guo et al. (1998); Hsieh & Goh(1990); Saccardo (1886); Shin & Kim (2001); Vasudeva (1963).
* AR= already reported, FR= first record.

Table 1. List and descriptions of formerly reported *Cercospora* species that were reclassified in this study as *Cercospora apii s.lat.*

Some former species of *Cercospora* that are morphologically different from *C.apii s.lat.*, are now considered to *Cercospora s.str.* As far as known, *Cercospora s.str.* are host specific or with a host range confined to species of a single host genus or some allied host genera of a single family. This phenomenon is constantly being addressed via molecular studies (Crous et al., 2000, 2001). *Cercospora s.str.* is characterized by having stromata, with numerous, densely arranged rather short, solitary to fasciculate, subhyaline to light or olivaceous brown conidiophores with small conspicuously thickened and darkened conidiogenous loci, and obclavate-cylindrical conidia with obconically truncate base (Table 2). Teodoro (1937) had enumerated 65 species of *Cercospora* in the Philippines while 33 species were reported by Quimio and Abilay (1977). In the present study, 48 hosts exhibiting leaf spots were reported as caused by species of *Cercospora*, 32 were from medicinal plants. There were 30 first records of *Cercospora* leaf spots recorded in this study. All species of *Cercospora* associated with those hosts are known except for a species on *Basella albae*. It has not been described on this host; therefore, it warrants description on a new host record, with proposed species name of *Cercospora basellae-albae* (Begum and Cumagun, 2010).

Species	Host	Stromata	Conidiophores	Conidia	Ref. Coll. Accession No.	Status of collection
Cercospora adiantigena	*Adiantum phillipense* (Maiden hair fern)	present	small to moderately long fascicle, subhyaline, straight, subcylindrical to moderately geniculate to sinuous, unbranched, multiseptate, conidial scar thickened and darkened, 25-150 x 4-10 μm	hyaline, solitary, obclavate-cylindrical, short conidia occasionally fusoid, septate, thin walled, smooth, apex obtuse, base short obconically truncate, hilam thickened and darkened, 40-90 x 4-8 μm	CALP 11715	AR
Cercospora basellae-albae	*Basella alba* (Vine spinach-green)	present	2-15 in a divergent fascicle, light brown, straight to rarely curved, unbranched, thick walled, septate, geniculate, with rounded apex, conidial scars distinct, 30-85 x 4-5 μm	hyaline, acicular to sub-cylindrical, straight to rarely curved, unbranched, smooth walled, septate, with truncate to obconically truncate base and obtuse apex, 20-80 x 2-5 μm	CALP 11735	FR

Species	Host	Stromata	Conidiophores	Conidia	Ref. Coll. Accession No.	Status of collection
Cercospora basellae-albae	*Basella albae* (Vine spinach-purple)	present	1-10 in a fascicle, pale olivaceous brown, fairly uniform in color and width, not branched, straight or mildly geniculate with thickened conidial scars, sparingly septate, 30-75 x 4-6 μm	hyaline, acicular, obclavate, straight to slightly curved, in distinctly multiseptate, rounded apex, truncate at the base with a thickened hilum, 15-70 x 1-4 μm.	CALP 11674	NHR
Cercospora brassisicola	*Brassica pekinensis* (Pechay)	present	2-15 in a divergent fascicle, emerging through stomata, pale olivaceous to medium brown, not branched, multiseptate, mildly geniculate, but rarely geniculate in the upper portion, scars large and conspicuously thickened, 20-200 x 3-5.5 μm	hyaline, solitary, acicular to cylindrical, straight to mildly curved, multiseptate, acute to rounded at the apex, truncate at the base, hilum thickened and darkened, 25-250 x 2-4.5 μm	CALP 11705	AR
Cercospora brassicicola	*Brassica campestris* (Mustard)	present	2-15 in a fascicle, pale olivaceous to medium brown, uniform in colour and width but paler the attenuated tips, rarely branched, multiseptate, mildly geniculate, conidial scars at the subtruncate tip, 25-400 x 3.5-6 μm	hyaline, acicular, straight to curved, indistinctly multiseptate, subacute to acute at the apex, truncate at the base, 25-200 x 2-5 μm	CALP 11703	FR

Species	Host	Stromata	Conidiophores	Conidia	Ref. Coll. Accession No.	Status of collection
Cercospora canescens	*Dolichos lablab* (Lab bean)	present	densely fasciculate, pale to medium dark brown, multiseptate, geniculate, rarely branched, apex-truncate, conidial scars conspicuously thickened, 2-4 µm wide, 20-200 x 3-6.5 µm	hyaline, acicular, straight to curved, indistinctly multiseptate, apex-acute, base-truncate, thickened hilum, 25-200 x 2.5-5.5 µm	CALP 11732	FR
Cercospora carotae	*Daucus carota* (Carrote)	present	3-15 in a fascicle or borne singly, pale olivaceous brown, paler tips, upper portion slightly geniculate, straight, scars conspicuous thickened, 20-40 x 2.5-4 µm	hyaline, filiform to cylindric, solitary, straight to slightly curved, 1-5 septa, rounded base, obtuse apex, hilum thickened and darkened, 25-95 x 3.5-5.5 µm	CALP 11730	AR
Cercospora corchori	*Corchorus olitorius* (Jute)	present	2-7 in a fascicle or borne singly, pale to medium olivaceous brown, paler at the apex, septate, not branched, geniculate, subtruncate at tip, with thickened conidial scars, 40-230 x 2-5.5 µm	hyaline, acicular to obclavate, straight to curved, indistinctly multiseptate, obtuse at the apex, base-obconically truncate , thickened hilum, 25-165 x 2.5-5 µm	CALP 11694	FR

Species	Host	Stromata	Conidiophores	Conidia	Ref. Coll. Accession No.	Status of collection
Cercospora corchori	*Corchorus acutangulus* (saluyot)	present	5-15 in a fascicle, pale to medium brown, slightly paler and more narrow towards the tip, springly septate, not branched, mildly geniculate, almost straight, large conidial scar at subtruncate tip, 30-120 x 4-5.5 μm	hyaline, acicular to obclavate, straight to curved, indistinctly multiseptate, base truncate, tip acute, 40-220 x 2.5-5 μm	CALP 11697	FR
Cercospora daturicola	*Datura metal* (Datura)	lacking	2-15 in a fascicle, pale olivaceous brown, uniform in colour, usually straight, septate, not branched, conidial scars conspicuously thickened and darkened, 30-85 x 3.5-5.5 μm	hyaline, acicular to obclavato-cylindrical, multiseptate, straight to mildly curved, tip acute to subacute and base truncate, hilum thickened and darkened, 50-150 x 3-4.5 μm	CALP 11729	FR
Cercospora eluesine	*Eluesine indica* (Dogs tail)	lacking	2-5 in a small fascicle, pale to olivaceous brown, straight to mildly curved, not branched, sometimes mildly geniculate, multiseptate, large conidial scars conspicuous, 25-85 x 4.5-6 μm	hyaline, cylindrical to obclavate, straight to mildly curved, multiseptate, obconically truncate base, rounded tip, hilum thickened and darkened, 40-120 x 3-4.5 μm	CALP 11720	FR

Species	Host	Stromata	Conidiophores	Conidia	Ref. Coll. Accession No.	Status of collection
Cercospora euphorbiae	*Euphorbia* sp. (Euphorbia)	present	5-15 in a fascicle, pale olivaceous brown, uniform in colour and width, paler the tips, smooth wall, straight to mildly curved, not branched, not geniculate, multiseptate, medium conidial scar thickened and darkened, 15-65 x 4-6 μm	hyaline, solitary, cylindro-obclavate, subobtuse tip, obconically truncate base, straight to curved, multiseptate, hilum thickened and darkened, 40-100 x 3.5-5 μm	CALP 11721	FR
Cercospora gendarussae	*Gendarussa vulgaris* (Gendarussa)	present	densely fasciculate, olivaceous brown, uniform in colour and width, paler the tips, smooth wall, straight to mildly curved, not branched, rarely geniculate, multiseptate, large conidial scar thickened and darkened, 20-120 x 4-5.5 μm	hyaline, solitary, acicular to cylidro-obclavate, acute to rounded tip, obconically truncate base, straight to curved, multiseptate, hilum thickened and darkened, 45-180 x 3-4 μm	CALP 11722	FR
Cercospora guatemalensis	*Ocimum sanctum* (Basil)	lacking	2-10 in a fascicle, pale to olivaceous brown, slightly paler and more narrow towards the tip, septate, not branched, straight to slightly curved, conidial scar conspicuous , 25-100 x 4-5.5 μm	hyaline, cylindric or acicular, straight to mildly curved, indistinctly multiseptate, base truncate to obconically truncate, tip rounded to conic, 45-120 x 2.5-4 μm	CALP 11725	FR

Species	Host	Stromata	Conidiophores	Conidia	Ref. Coll. Accession No.	Status of collection
Cercospora helianthicola	*Helianthus annuus* (Sunflower)	present	1-10 in a fascicle, pale to medium brown, paler and more narrow toward the tip, geniculate, rarely branched, large conidial scar at subtruncate tip, 25-100 x 3-5 µm	hyaline, acicular, sometimes curved, multiseptate, base truncate, tip acute, 40-130 x 2-3 µm	CALP 11718	FR
Cercospora kikuchii	*Glycine max* (Soybean)	present	2-10 in a fascicle, medium dark brown, uniform in colour, multiseptate, not branched, mildly geniculate, subtruncate at the apex, scars large and conspicuously thickened, 35-200 x 4-6 µm	hyaline, acicular, straight to curved, indistinctly multiseptate, hilum thickened and darkened, 45-300 x 2.5-5 µm	CALP 11698	FR
Cercospora labiatacearum	*Pogostemon cablin* (Patchouli)	lacking	5-8 in a small fascicle, pale olivacous brown, paler upwards, smooth wall, straight to mildly curved, not branched, geniculate, large conidial scar thickened and darkened, 45-300 x 5-5.5 µm	hyaline, solitary, acicular-obclavate, subacute tip, truncate base, straight to curved, multiseptate, hilum thickened and darkened, 45-180 x 4.5-5.5 µm	CALP 11706	FR

Species	Host	Stromata	Conidiophores	Conidia	Ref. Coll. Accession No.	Status of collection
Cercospora lactucae-sativae	*Lactuca sativa* (Lettuce)	present	2-10 in a fascicle or borne singly, pale olivaceous brown, slightly paler and narrower towards the apex, multiseptate, not branched, springly geniculate, conidial scar at subtruncate tip, 25-100 x 4-5 μm	hyaline, solitary, acicular to obclavate, straight to curved, indistinctly multiseptate, subacute at the apex, subtruncate at the base, 20-200 x 3-5 μm.	CALP 11699	AR
Cercospora menthicola	*Mentha arvensis* (Apple mint plant)	lacking	2-5 in a small fascicle, pale to olivaceous brown, multiseptate, not branched, straight to curved, mildly geniculate towards the apex, smooth walled, conidial scars conspicuously thickened, 35-120 x 4-6.5 μm	hyaline, solitary, acicular, straight to mildly curved, multiseptate, base-subtruncate to obconically truncate. Apex-acute to rounded, hilum thickened and darkened, 40-120 x 2-4 μm	CALP 11691	FR
Cercospora mikaniicola	*Mikania cordifolia* (Climbing hemp weed)	lacking	2-5 in a fascicle, medium olivaceous brown, closely septate, not branched, straight to mildly geniculate, smooth walled, conidial scars conspicuously thickened, 50-150 x 4-6.5 μm	hyaline, solitary, obclavate, truncate at the base, acute the tip, hilum thickened and darkened, 40-80 x 4-9 μm	CALP 11716	FR

Species	Host	Stromata	Conidiophores	Conidia	Ref. Coll. Accession No.	Status of collection
Cercospora pulcherrimae	*Euphorbia pulcherrema* (Poinsettia)	present	3-12 in a fascicle, pale to medium olivaceous brown, multiseptate, rarely branched, straight, mildly geniculate, conidial scars conspicuously thickened, 20-110 x 4-6 μm	hyaline, acicular, acute to subacute at the apex, truncate at the base with a thickened hilum, 30-130 x 3-4 μm	CALP 11702	AR
Cercospora ricinella	*Ricinus communis* (Castor bean)	present	densely fasciculate, pale olivaceous brown, fairly uniform in colour and width, sparingly septate, not branched, geniculate, large conidial scar present at subtruncate tip, 20-250 x 4-8 μm	hyaline, acicular to obclavate, straight to mildly curved, indistinctly multiseptate, subacute to subobtuse at the apex, subtruncate to truncate at the base, hilum thickened and darkened, 20-100 x 2-4 μm	CALP 11719	AR
Cercospora sesame	*Sesamum orientale* (Sesame)	lacking	2-5 in a small fascicle, olivaceous brown, slightly paler towards the apex, multiseptate, rarely branched, straight, mildly geniculate, large conidial scar present, 20-110 x 4-5.5 μm	hyaline, cylindric, straight to curved, indistinctly multiseptate, acute at the apex, truncate at the base, hilum thickened and darkened, 30-150 x 2-4.5 μm	CALP 11723	AR

Species	Host	Stromata	Conidiophores	Conidia	Ref. Coll. Accession No.	Status of collection
Cercospora sesbaniae	*Sesbania sesban* (Sesbania)	present	2-15 in a fascicle, pale to very pale olivaceous brown, uniform in colour, 1-4 septate, width irregular, not branched, mildly geniculate, conidial scar conspicuously thickened, 20-65 x 2.5-5.5 μm	hyaline, acicular, straight to curved, multiseptate, truncate base, tip obtuse, 25-60 x 2.5-3.5 μm	CALP 11695	FR
Cercospora simulate	*Cassia alata* (Ring worm bush)	lacking	2-15 in a fascicle, dark brown in colour, paler the tip, irregular width, not branched, upper portion mildly geniculate, multiseptate, medium conidial scar at subconic tip, 50-280 x 3.5-6 μm	hyaline, cylindro-obclavate, straight to mildly curved, septate, base obconically truncate, tip obtuse, hilum thickened and darkened, 30-100 x 2.5-4 μm	CALP 11726	FR
Cercospora syndrellae	*Syndrella nodiflora* (syndrella)	lacking	5-10 in a fascicle, pale to olivaceous brown, straight to mildly curved, not branched, mildly geniculate, multiseptate, large conidial scar thickened and darkened, 40-90 x 5-6.5 μm	hyaline, cylindrical to obclavate, straight to curved, multiseptate, obconically truncate base, tip rounded, 45-100 x 2.5-4.5 μm	CALP 11727	FR

Species	Host	Stromata	Conidiophores	Conidia	Ref. Coll. Accession No.	Status of collection
Cercospora tagetis-erectae	Tagetes erecta (Marygold)	present	densely fasciculate, very pale olivaceous, cylindric, erect or sinuous, rarely septate or geniculate, truncate or rounded at the apex, conidial scars thickened conspicuous, 10-50 x 2-3 μm	hyaline, narrowly obclavate or filiform, straight, 3-10 septate, acute at the apex, obconic or long obconically truncate at the base; hilum thickened and darkened, 25-90 x 2-3.5 μm	CALP 11714	FR
Cercospora tithoniae	Tithonia diversifolia (African sunflower)	lacking	2-8 in a fascicle, pale olivaceous brown, uniform in colour, straight to slightly curved, not branched, septate, mildly geniculate, conidial scars conspicuously thickened and darkened, 25-65 x 3-4.5 μm	hyaline, cylindric to obclavate, straight to slightly curved, multiseptate, rounded apex and base obconically truncate, hilum conspicuously thickened, 25-50 x 3.5-4 μm	CALP 11692	FR

Reference: Chupp (1954); Ellis (1971, 1976); Guo & Hsieh (1995); Guo et al. (1998); Shin & Kim (2001); Vasudeva (1963).
* AR= already reported, FR= first record, NHR= new host record.

Table 2. List and descriptions of formerly reported Cercospora species, that were reclassified in this study as of Cercospora s. str.

3. Genus *Pseudocercospora*

Pseudocercospora Speg (Crous & Braun, 2003).

Stromata lacking to well developed, usually pigmented; *conidiophores* are pigmented, pale olivaceous to medium dark brown with *conidiogenous loci* inconspicuous, unthickened and not darkened but somewhat refractive or rarely very slightly darkened, or only outer rim slightly darkened and refractive; *conidia* subhyaline to pigmented, solitary or catenulate, scolecosporous, hila unthickened and not darkened (Table 3).

Type species: *Pseudocercospora vitis* (Lev.) Speg.

Pseudocercospora was introduced by Spegazzini (1910). Deighton (1976) re-introduced this name and widened the concept of this genus considerably to include a wide range of cercosporoids with pigmented conidiophores and inconspicuous, unthickened, not darkened conidiogenous loci. It is the second largest Cercosporoid genus, with more than 300 published names (Kirk *et al.* 2001). In Taiwan, 198 species of *Pseudocercospora* have been recognized by Hsieh & Goh (1990). In the present study, 20 *Pseudocercospora* sp. were reported, of which 14 species caused diseases on medicinal plants (Table 3). There were 12 new records of *Pseudocercospora* diseases in the Philippines.

Species	Host	Stromata	Conidiophores	Conidia	Ref. Coll. Accession No.	Status of collection
Pseudocercospora abelmoschi	*Abelmoschus esculentus* (Okra)	lacking or small	densely fasciculate, pale to medium olevaceous brown, uniform in colour, 0-3 septate, sometimes constricted at the septa, irregular in width or slightly clavate, simple or branched, sparingly geniculate, sinuous, rounded or conically truncate at the apex, 10-50 x 3-5 µm	pale olivaceous to brown, obclavate to cylindric, straight to mildly curved, septate, acute to subobtuse at the apex, obconic to obconicaly truncate at the base, hilum unthickened and inconspicuous, 25-80 x 2.5-4 µm	CALP 11746	AR
Pseudocercospora alternantherae-nodiflorae	*Alternanthera nodiflora* (Alternenthera)	present	10-25 in a divergent fascicle, emerging through the stromata, pale brown, straight to curved, not branched, sometimes geniculate, septate, scars inconspicuous, 10-55 x 4-5 µm	subhyaline to very pale olivaceous, obclavate with gradual attenuation, hyaline to pale brown, straight to mildly curved, tip subobtuse, base obconic, 3-12 septate, sometimes constricted at the septa, hilum unthickened and inconspicuous, 30-110 x 2.5-4.5 µm	CALP 11736	FR

Species	Host	Stromata	Conidiophores	Conidia	Ref. Coll. Accession No.	Status of collection
Pseudocerc ospora atro-marginalis	*Solanum nigrum* (Black night shade)	lacking	4-15 in a divergent fascicle, pale olivaceous to olivaceous brown, longer ones curved, sharply bent or undulate, branched, septate, sometimes slightly constricted at the some septa, rarely geniculate, conic at the apex, scars inconspicuous, 10-50 x 3-5 µm	pale olivaceous, cylindric to obclavato-cylindric, straight to mildly curved, 2-8 septate, subobtuse to broadly rounded at the apex, sharply obconic or obconically truncate at the base, hilum unthickened and not darkened, 15-90 x 2.5-5 µm	CALP 11679	AR
Pseudocerc ospora balsa-minicola	*Impatiens balsamina* (Balsam plant)	well develop ed	10-40 in a divergent fascicle, pale olivaceous brown to brown , irregular in width, substraight to mildly curved, not geniculate, not branched, multiseptate, conidial scars inconspicuous, 15-50 x 2-3 µm	subhyaline, solitary, filiform to narrowly obclavate, straight to mildly curved, septate, subacute at the apex, truncate to obconically truncate at the base, hilum unthickened and not darkened; 30-70 x 1.5-3 µm	CALP 11741	AR

Species	Host	Stromata	Conidiophores	Conidia	Ref. Coll. Accession No.	Status of collection
Pseudocerc ospora blumeae	*Blumea balsamifera* (Sambong)	lacking	5-25 in a fascicle, cylindric, pale to medium brown, uniform in colour and width, straight to curved, not branched, septate, mildly geniculate, rounded to truncate at the apex, conidial scars unthickened, 15-70 x 4-5 µm	pale olivaceous brown, mostly cylindrical, rarely obclavate, straight to slightly curved, septate, subobtuse to broadly rounded at the apex, subtruncate to long obconically truncate at the base, hilum unthickened. 30-110 x 3.5-5.5 µm	CALP 11739	AR
Pseudocerc ospora bixicola	*Bixa orellanae* (Bixa)	lacking	densely fasciculate, pale olivaceous, cylindrical, septate, branched, rarely geniculate, conically rounded at the apex, conidial scars unthickened or inconspicuous, 15-40 x 2.5-4 µm	pale olivaceous, obclavate-cylindric, straight to mildly curved, indistinctly 3-6 septate, subobtuse at the apex, obconic or obconically rounded at the base, hilum unthickened, , 30-60 x 2-3 µm	CALP 11738	FR
Pseudocerc ospora borreriae	*Borreria micrantha* (Borreria)	lacking or small	medium to dark brown, arise singly, uniform in colour, irregular in width, multiseptate, branched, slightly geniculate, curved to tortuous, small spore scars at bluntly rounded tip, 35-220 x 3-5.5 µm	subhyaline to pale or medium olevaceous, cylindric to obclavato-cylindric, straight to mildly curved, 3-9 septate, base obconic to obconically truncate, hilum unthickened and inconspicuous, 30-90 x 2.5-5 µm	CALP 11740	FR

Species	Host	Stromata	Conidiophores	Conidia	Ref. Coll. Accession No.	Status of collection
Pseudocerc ospora chrysanthe micola	*Chrysanthe mum morifolium* (Chrysanth emum)	present	2-20 in a fascicle, emerging through stomata, olivaceous brown, cylindrical, septate, rarely geniculate, very rarely branched, scars inconspicuous, 15-40 x 3-5 µm	pale olivaceous brown, obclavate or obclavato-cylindric, straight to mildly curved, multiseptate, rounded at the apex, obconically truncate at the base, hilum unthickened and inconspicuous, 25-100 x 3-5 µm	CALP 11742	FR
Pseudocerc ospora corchorica	*Corchorus capsularis* (Jute)	present	olivaceous brown, in a dense fascicle, paler and narrower the tips, straight to mildly curved, not geniculate, scars inconspicuous, 18-55 x 3.5-5 µm	subhyaline to very pale olivaceous brown, cylindric to obclavate, straight to mildly curved, septate, apex rounded and base subtruncate, hilum unthickened, reflective, 20-80 x 3-5 µm	CALP 11753	FR
Pseudocerc ospora cruenta	*Phaseolus lunatus* (Lima bean)	small	dense fascicle, subhyaline to pale olivaceous brown, straight to sinuous or mildly geniculate, occationally branched, septate, conic at the apex, scars inconspicuous, 15-60 x 3-5 µm	subhyaline to very pale olivaceous brown, cylindric or cylindro-obclavate, straight to mildly curved, multiseptate, subacute to obtuse at the apex, sharply obconic or obconically truncate at the base, hilum unthickened and inconspicuous, 25-130 x 2-5 µm	CALP 11737	AR

Species	Host	Stromata	Conidiophores	Conidia	Ref. Coll. Accession No.	Status of collection
Pseudocercospora formosana	*Lantana camara* (Lantana)	lacking	small fascicle, very pale olivaceous to brown, sparingly septate, not geniculate, straight to undulate, scars inconspicuous, 25-40 x 3-4 μm	narrowly obclavate, very pale olivaceous, straight to curved, indistinctly multiseptate, base long obconically truncate, tip subacute, hilum unthickened and inconspicuous, 30-100 x 2.5-3.5 μm	CALP 11747	FR
Pseudocercospora fuligena	*Lycopersicon esculentum* (Tomato)	well developed	loosely fasciculate usually 5-15 per fascicle, pale olivaceous to very pale olivaceous brown, uniform in color, straight to sinuous, tip rounded or truncate, sometimes geniculate, not branched, septate, conidial scars unthickened, 15-45 x 3-5 μm	subhyaline to pale olivaceous, cylindric to cylindro-obclavate, straight to mildly curved, rounded to obtuse at the apex, obconically truncate, multiseptate, hilum unthickened, not darkened, 25-110 x 2-4 μm	CALP 11748	AR
Pseudocercospora gmelinae	*Gmelina arborea* (Yemen)	lacking	2.8 in a small fascicle, pale olivaceous brown, straight to mildly geniculate, smooth, unbranched, septate, scars inconspicuous, 30-50 x 3-4.5 μm	subhyaline to pale olivaceous brown, cylindro-obclavate, straight to mildly curved, base attenuated, tip subacute, hilum unthickened and inconspicuous, 30-80 x 3-4.5 μm	CALP 11754	AR

Species	Host	Stromata	Conidiophores	Conidia	Ref. Coll. Accession No.	Status of collection
Pseudocercospora jasminicola	Jasminum grandiflorum (Jasmine)	present	numerous, pale olivaceous, straight or slightly sinuous, sometimes slightly geniculate, smooth, simple, septate, subtruncate at the apex, 5-30 x 2-4 μm, conidial scars unthickened	pale olivaceous, subcylindric to slightly obclavate, straight or slightly curved, smooth, thin-walled, obtuse at the apex, shortly tapered at the base, hilum unthickened and inconspicuous, 18-60 x 1.5-2.5 μm	CALP 11745	FR
Pseudocercospora ocimicola	Ocimum basilicum (Sweet basil)	lacking	densely fasciculate, pale to very pale olivaceous brown, conidial scar unthickened and inconspicuous, 10-40 x 3-5 μm	subhyaline, cylindric to narrowly obclavate, straight to mildly curved, septate, subacute to subobtuse at the apex, truncate at the base, hilum unthickened 25-60 x 3-4 μm	CALP 11744	FR
Pseudocercospora pachyrrhizi	Pachyrrhizus erosus (Turnip)	small	densely fasciculate, pale olivaceous to yellowish brown, not branched, septate, mildly geniculate, conidial scar visible but not thickened, 10-35 x 2.5-4 μm	subhyaline, obclavate, straight to slightly curved, septate, rounded at the apex, obconocally truncate base with unthickened hilum, 35-60 x 3-5.5 μm	CALP 11743	FR

Species	Host	Stromata	Conidiophores	Conidia	Ref. Coll. Accession No.	Status of collection
Pseudocercospora sesbanicola	*Sesbania sesban* (Sesbania)	lacking	5-20 in a fascicle, emerging from stromata, septate, rounded at the apex, scars inconspicuous, 15-65 x 3-4 µm	pale olivaceous brown, cylindric, sometimes obclavato-cylindric, straight or slightly curved, septate, rounded at the apex, truncate at the base, hilum unthickened, 18-30 x 3-4.5 µm	CALP 11752	FR
Pseudocercospora solani-melon-genicola	*Solanum melongena* (Eggplant)	present	5-10 in a fascicle, pale olivaceous brown, paler towards the apex, straight or geniculate, sometimes branched, septate, conidial scars visible but not thickened, 30-60 x 3.5-4.5 µm	olivaceous to pale brown, cylindric to cylindro-obclavate, straight to slightly curved, obtuse at the apex, obconically truncate at the base, hilum visible but not thickened, 30-100 x 3.5-5 µm	CALP 11749	FR
Pseudocercospora synedrellae	*Synedrella nodiflora* (Syndrella)	lacking	2-15 in a fascicle , pale olivaceous brown, simple or branched, straight or slightly undulate, septate, sometimes constricted at the septa, not geniculate, rounded or conical at the apex, 5-30 x 2-4 µm	subhyaline to very pale olivaceous, narrowly obclavate or filiform, straight to slightly curved, indistinctly multiseptate, subacute at the apex, obconically truncate at the base, hilum unthickened and inconspicuous, 15-90 x 2-3 µm	CALP 11750	FR

Species	Host	Stromata	Conidiophores	Conidia	Ref. Coll. Accession No.	Status of collection
Pseudocerc ospora viticis	*Vitex nigundo* (Lagundi)	well develop ed	pale olivaceous, borne singly or densely fasciculate, straight, not branched, not geniculate, 5-35 x 2-3.5 µm, conidial scar inconspicuous	subhyaline to pale olivaceous, cylindric to obclavate, straight to mildly curved, septate, acute to subacute at the apex and obconically truncate at the base, hilum unthickened 18-50 x 2.5-3.5 µm	CALP 11751	AR

Reference: Chupp (1954); Ellis (1971); Guo & Hsieh (1995); Guo *et al.* (1998); Hsieh & Goh (1990); Saccardo (1886); Shin & Kim (2001); Vasudeva (1963); Yen & Lim (1980).
* AR= already reported, FR= first record.

Table 3. List and descriptions of *Pseudocercospora* species found in this study.

4. Genus *Passalora*

Passalora Fr. (Crous & Braun, 2003).

Stromata absent to well developed; *conidiophores* are solitary or loosely to densely fasciculate, unbranched or branched, continuous to pluriseptate, subhyaline to pigmented, conidiogenous loci conspicuous, scars thickened and darkened-refractive, *conidia* solitary to catenate, simple or branched, amerosporous to scolecosporous, pale to distinctly pigmented (if subhyaline, conidia non-scolecosporous), smooth to finely verruculose, with few septa, hila thickened and darkened-refractive, more or less truncate (Crous & Braun, 2003).

Type species: *Passalora bacilligera* (Mont. & Fr.)

4.1 Dichotomous Key to the Species *Passalora*

1. Stromata lacking--- *P. bougainvilleae*
1. Stromata present, sometimes well developed--2
    2. Conidiophores strongly geniculate--- *P. personata*
    2. Conidiophores straight to slightly geniculate---3
3. Conidiophores aseptate--- *P. tinosporae*
3. Conidiophores septate--4
    4. Conidiophores up to 50 um long--- *P. henningsii*
    4. Conidiophores up to 100 um long--- *P. amaranthae*

Approximately 550 names of *Passalora* that have been published or amended in the world, compiled by Crous and Braun (2003). In this study, four already known and recorded *Passalora* species were found, namely *Passalora personata* on *Arachis hypogaea* (Quimio and Abilay, 1977), formerly named as *Cercospora arachidis*, *P. bougainvilleae* causing leaf spot of

Bougainvilla, (Ponaya & Cumagun, 2008), *P. henningsii* on *Manihot esculenta*, formerly named as *Cercospora cassavae* Ellis & Everh. / *C. manihots* Henn. / *C. henningsii* Allesch (Crous and Braun, 2003) and *P. tinosporae* on *Tinospora reticulate*. The same host species and the associated fungi were reviewed in the study and were found to confirm with the descriptions of the genus *Passalora* (Table 4).

Species	Host	Stromata	Conidiophores	Conidia	Ref. Coll. Accession No.	Status of collection
Passalora amaranthae	*Amaranthus viridis* (Amaranth)	well developed	densely fasciculate, pale to olivaceous brown, multiseptate, straight or slightly curved, not branched, moderately geniculate, conspicuously thickened small conidial scars,25-100 x 3.5-6.5 μm	pale olivaceous, cylindric or obclavato-cylindric, straight to slightly curved, 3-7 septate, bluntly rounded at the apex, obconic at the base , small thickened hilum, 25-65 x 3.5-6.5 μm	CALP 11757	NS
Passalora bougainvilleae	*Bougainvillea spectabilis* (Bougain-villa)	lacking	5-8 in a small fascicle, pale to olivaceous brown, smooth, straight to geniculate, aseptate, conidial scar conspicuous and darkened, 10-75 x 5-7.5 μm	pale to olivaceous brown, solitary, smooth, slightly curved, cylindrical to obclavato-cylindric, 3-6 septa, truncate at the base and rounded at the apex, hilum thickened and darkened, 30-65 x 5-10 μm	CALP 11758	AR
Passalora henningsii	*Manihot esculenta* (Cassava)	well developed	densely fasciculate, subhyaline to pale olivaceous brown, uniform in colour and width, 1-3 septate, straight or slightly curved, not branched, mildly geniculate, conidial scars conspicuously thickened, 15-50 x 3.5-5 μm	pale olivaceous, cylindric, straight to slightly curved, 3-6 septate, bluntly rounded at the apex, obconic at the base with a small thickened hilum, 25-60 x 4-6.5 μm	CALP 11756	AR

Species	Host	Stromata	Conidiophores	Conidia	Ref. Coll. Accession No.	Status of collection
Passalora personata	*Arachis hypogea* (Peanut)	well developed	densely fasciculate, pale olivacious, smooth, slightly or strongly geniculate, straight or slightly curved, not branched, 0-3 septate, conidial scars conspicuously thickened, 20-90 x 3.5-6.5 µm	solitary, subhyaline to pale olivaceous, filiform or obclavate or obclavato-cylindric, usually very finely rough-walled, obtuse or broadly rounded at the apex, truncate or obconically truncate at the base, hilum conspicuously thickened and darkened, 2-10 septate, 20-80 x 4-7.5 µm	CALP 11755	AR
Passalora tinosporae	*Tinospora reticulate* (Maka-buhay)	present	densely fasciculate, subhyaline to very pale, not branched, geniculate, multiseptate, conidial scars conspicuous and at bluntly rounded tip, 25-110 x 3.5-5.5 µm	subhyaline to very pale, cylindrical to obclavate, aseptate, straight to curved, obconically truncate base, subobtuse tip, 10-45 x 4-6 µm	CALP 11760	AR

Reference: Chupp (1954); Ellis (1971, 1976); Guo & Hsieh (1995); Guo *et al.* (1998); Hsieh & Goh(1990); Katsuki (1965); Saccardo (1886); Shin & Kim (2001); Vasudeva (1963).
* NS= new species, AR= already reported, FR= first record.

Table 4. List and descriptions of different species of *Passalora* that were formerly classified as *Cercospora* species in the Philippines.

A species associated with *Amaranthus viridis* was believed to be new. Its characteristics are close to *P.henningsii* in terms of having amphigenous colonies and with shorter conidiophores (15-50 µm). However, its characteristics are different from others, having darker septation on conidia, and its association with a new host that warrants a new species. The proposed new species was *Passalora amaranthae* on *Amaranthus viridis*.

4.2 Genus *Asperisporium*

Asperisporium Maublanc. (Crous & Braun, 2003)

Sporodochia punctiform, pulvinate, brown, olivaceous brown or black. *Mycelium* immersed. Stromata usually well-developed, erumpent. *Conidiophores* macronematous, mononematous, closely packed together forming sporodochia, usually rather short, unbranched or occasionally branched, straight or flexuous, hyaline to olivaceous brown, smooth, polyblastic, scars prominent. *Conidia* solitary, ellipsoidal, fusiform, obovoid, pyriform, clavate or obclavate, hyaline to brown or olivaceous brown, smooth or verrucose, with 0-3 septa (Ellis, 1971).

Type species: *Asperisporium caricae* (Speg.) Maubl.

The genus *Asperisporium,* introduced by Maublanc (1913) resembles *Passalora*, but differs in having verrucose conidia (Ellis 1971, 1976; von Arx, 1983). In the present study, only one species of *Asperisporium* was identified (Table 5). It was *Asperisporium moringae* (Thirum. & Govindu) Deighton on *Moringa oleifera*. This disease was reported in the Philippines by Quimio & Abilay (1977), with *Cercospora moringae* (Thirum. & Govindu) as the causal pathogen. The black spot of papaya caused by *Asperisporium caricae* was first recorded and described in the Philippines (Cumagun and Padilla, 2007).

Species	Host	Stromata	Conidiophores	Conidia	Ref. Coll. Accession No.	Status of collection
Asperisporium moringae	*Moringa oleifera* (*Moringa*)	well developed	pale olivaceous brown, straight, geniculate, 10-35 x 4-6 μm	pale olivaceous brown, obclavate conico-truncate at the base, verruculose walled, mostly 1-2 septate, 20-45 x 5.5-7.5 μm	CALP 11761	AR

Reference: Chupp (1954); Ellis (1971); Ellis & Holliday (1972).
*AR= already reported. FR= first record.

Table 5. Characteristics of *Asperisporium moringae* associated with leaf spot of *Moringa oleifera*.

4.3 Genus *Periconiella*

Periconiella Saccardo (Ellis, 1971).

Stromata none; *conidiophores* macronematous, mononematous, each composed of an erect, straight or flexious, brown to dark blackish brown, smooth or verruculose; *conidiogenous cells* polyblastic, integrated and terminal, sympodial, cylindrical, scars often numerous; *Conidia* solitary or occasionally in very short chains, simple, ellipsoidal, obclavate or obovoid, hyaline or rather pale olive or olivaceous brown, without septa or with one or a few transverse septa.

Type species: *Periconiella velutina* (Wint.) Sacc.

In the present study, only one *Periconiella lygodii* on *Lygodium japonicum* was reported (Table 6). Four species of *Periconiella* have been reported to occur on ferns (Braun, 2004). He noted that *P.lygodii* is distinguished from all other species of *Periconiella* on ferns by having long, obclavate-cylindrical, pluriseptate, smooth conidia. This is the first record of *P. lygodii* on *Lygodium japonicum* in the Philippines, (Begum et al. 2009).

Species	Host	Stromata	Conidiophores	Conidia	Ref. Coll. Accession No.	Status of collection
Periconiella lygodii	*Lygodium japonicum* (Japanese climbing fern)	lacking	Medium to medium –dark brown, straight, multiseptate, thick-walled, branched apical portion, two to four times dichotomously or occasionally trichotomously branched, 90-350 x 2-5.5 μm	pale olivaceous or olivaceous brown, obclavate-cylindrical, smooth, conico-truncate at the base, apex obtuse or subobtuse, 1-5 septate, 25-75 x 3-5.5 μm	CALP 11680 and BRIP 52369	FR

Reference: Chupp (1954); Ellis (1971); Braun (2004). * FR =First Record.

Table 6. Characteristics of *Periconiella lygodii* associated with leaf spot of *Lygodium japonicum*.

5. Lucid key for Philippine cercosporoid fungi

Before the advent of computers, the traditional way in which scientists identify identify biological specimens was through the use of printed pathway (or dichotomous) keys. However, with the advent of database and multi-media-software, it is now possible to store large amounts of biological data and to access this information through easy-to-use matrix-based (or multi-access) keys. Lucid is one example of a multi-media matrix key.The Lucid Program from The Centre for Biological Information Technology CBIT, University of Queensland, which was licensed to the third author was used to develop the Lucid key. The activities involved in the process of designing, developing, producing and publishing a Lucid key on CD, DVD or the Internet are outlined as follows: (1) Establishing the scope of the key; (2) designing and scoring the key; (3) sourcing, developing and editing facts sheets, images and other multi-media associated with features and entities; (4) packaging up a prototype on CD or on the Internet; (5) beta testing and user testing of the key; (6) graphic design of CD and CD-insert or web; and (7) packing of key for release.

5.1 Lucid Builder

Lucid key consists of two programs: Lucid Builder and Lucid Player. The first program is a key development tool (Fig. 1) that allows taxonomists to input their knowledge base into a form that is readily accessible by other people. A Lucid Builder enables key developers to easily build their own keys. In the present study, information from 74 Cercosporoid fungi data were entered for developed Lucid key.

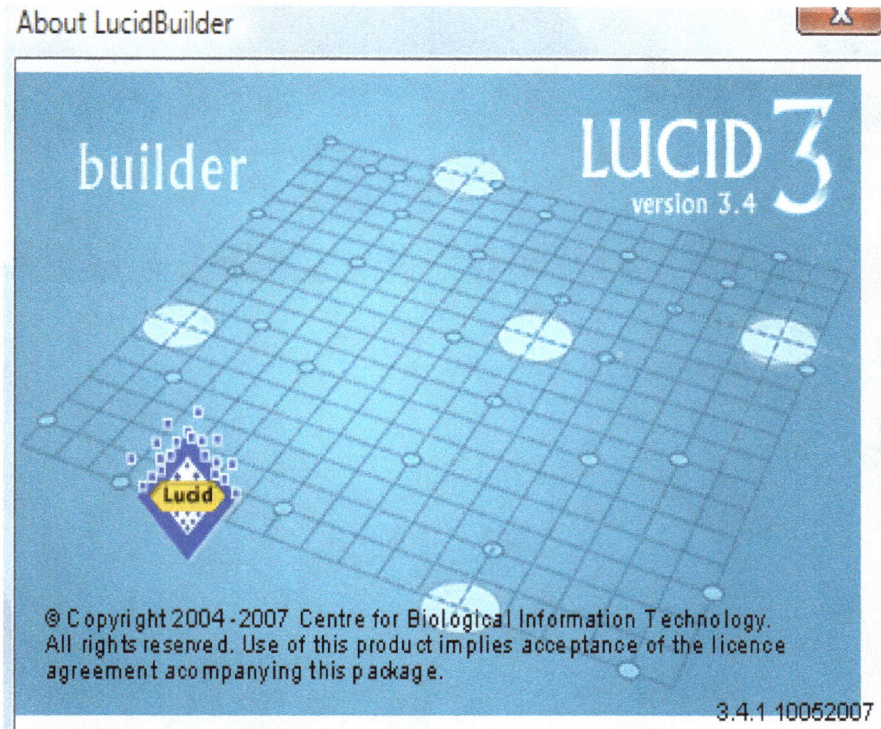

Fig. 1. Screen shot from the Lucid Builder.

In Lucid Builder, data were incorporated for example for *Cercospora adiantigena* on *Adiantum phillipense* (Fig. 2). The right side of the screen shows, all species that were inputted while left side provides the inputted characters for specific species. The information that were entered on the left side of the screen corresponds to the highlighted Cercosporoid species on the right side.

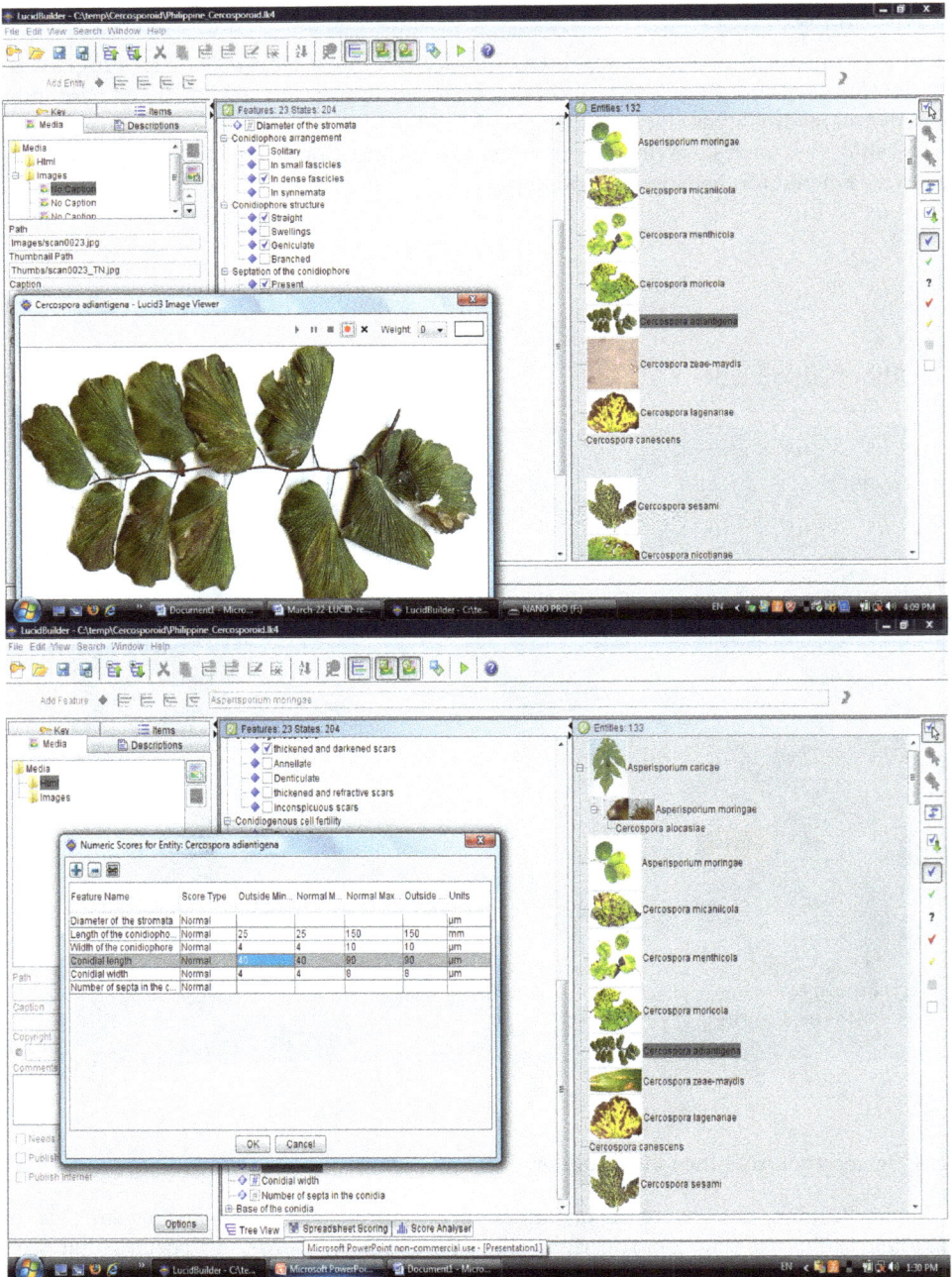

Fig. 2. Screen shot from the Lucid Builder- *Cercospora adiantigena* on *Adiantum phillipense.*
Symptoms of the disease and measurement of morphological characters (Left inset).

During an identification session, Lucid Player allows one to choose any question in its list at any time, but "stepping" through the key in a structured and sensible way will make one task of identification easier. The guidelines for making identification are as follows (1) familiarity with the specimen; (2) note and use of distinctive features; (3) answering easy features first; 4) choosing multiple states; and (5) checking the result.

Familiarity with the characteristics of the specimen to be identified is essential. Briefly reviewing Lucid key and specimen's characteristics before one starts will facilitate the identification. In any key, some taxa may possess particularly distinctive features. Use of these may allow the taxon to be keyed out in a very few steps. At the very least, starting with particularly distinctive or striking features for the first character states selected may quickly reduce the list of entities remaining. One can select any features from any position in the list and start by browsing the list of features available for obvious features that one can quite quickly answer, as opposed to getting stuck on the first one. Lucid is designed to overcome problems associated with difficult and obscure features.Always choose multiple states if one is uncertain which state is the correct one to choose for a particular specimen. One can choose as many states as from any one feature. After a preliminary identification has been made, one can check the other information (notes or image) provided for the taxon.

5.2 Lucid Player

The second program of the Lucid key is key interface or Lucid Player (Fig. 3) through which end-users interact with the Lucid key that has been developed and distributed either as a CD-Rom or via the Internet. The Lucid player enables users to view and interact with the key.

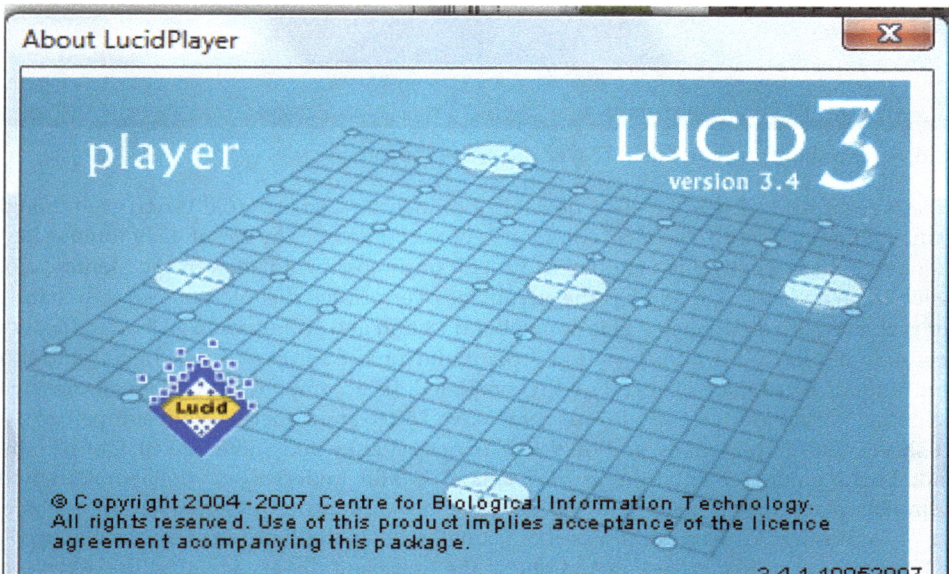

Fig. 3. Screen shot from the Lucid Player.

Lucid Player allows one to input a list of character states that best describe the specimen to be identified. These character states can be selected (or de-selected) in any order, resulting in a shortening of the list of remaining taxa that best match the decribed specimen. The upper left side of the screen shows all characters for a given specimen while it's lower left side indicates the characters that were chosen. The upper right side of the screen provides the possible identity of the specimen while the lower right side shows the discarded taxa from the list (Fig. 4).

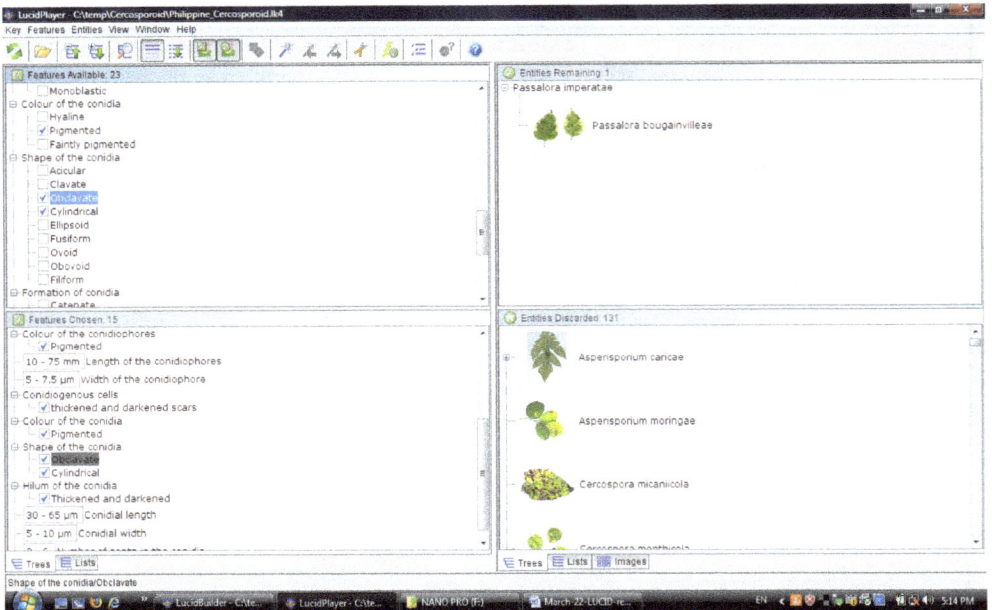

Fig. 4. Screen shot from the Lucid Player- *Passalora bougainvilleae*.

In the present study, Lucid key was developed to identify Cercosporoid fungi, even though the dichotomous keys are the most common keys encountered. The use of dichotomous keys has a major disadvantage: if a couplet is difficult or impossible to answer, the identification session often ends there. Lucid has the advantage over printed dichotomous keys in that the user is able to skip an unanswerable couplet or question and still proceed with identification because Lucid key allows to start at any point and proceed in any order. Lucid guide for smut fungi of Australia has already been completed. Its accompanying CD, incorporating a Lucid Player, provides an easy- to-use, interactive key to smut species, with comprehensive fact-sheets, distribution maps, and over 1000 images (Vanky and Shivas, 2008). On the other hand, Lucid guide for smut fungi of Thailand is still underway in collaboration with Australian plant pathologists (Shivas, personal communication). Gerald (2005) reported that, "Diagnosing Postharvest Diseases of Cantaloupe" is the first Lucid key developed in the U.S. for a set of plant diseases and one of the first plant disease identification keys ever developed in Lucid. A Lucid key was developed for the identification of *Phytophthora species* in USA based on morphological and molecular characters (Ristaino *et al.* 2008). In the present study only the data of the true Cercosporoids like, *Cercospora, Pseudocercospora* and *Passalora* were inputted into the Lucid key.

Identification of Cercosporoid fungi is a difficult task, and the Lucid key was created to help provide individuals with easily accessible tools to distinguish species. Recent experience suggests that computer-based identification keys will become an increasingly important part of the move towards providing taxonomic information on-line.

6. Summary and conclusions

The genus *Cercospora* is one of the largest genera of hyphomycetes. i.e., commonly associated with leaf spots and is responsible for great damages to beneficial plants, such as cereals, vegetables, ornamentals, forest trees, grasses. A total of 105 Cercosporoid diseases were identified, belonging to *Cercospora* (48), *Pseudocercospora* (20), *Passalora* (5), *Asperisporium* (1), *Cladosporium* (30,) and *Periconiella* (1). From the reported *Cercospora* species, 20 were *Cercospora apii s.lat* and 28 were *Cercospora s.str.* The first report of *Cercospora basellae-albae* in the Philippines was observed causing leaf spots on *Basella alba cv. Rubra* (Begum and Cumagun, 2010). Twenty eight first records of Cercospora leaf spots were reported. Among Pseudocercospora leaf spots, 12 first records were reported and all were host specific. A new species of *Passalora amaranthae* found on *Amaranthus viridis* was reported. Only one specimen caused by *Asperisporium moringae* was reported on *Moringa oleifera*.

Lucid key is a powerful and highly flexible knowledge management software application designed to help users with identification and diagnostic tasks. Lucid is one of a multi-media matrix keys, that includes possible the storing of large amounts of information. Lucid key consists of two programs: Lucid Builder and Lucid Player. The first program is a key development tool, which allows developers to easily build their own keys. The second program of the Lucid key is the key interface or Lucid Player, through which end-users interacts with the Lucid key and enables users to view and interact with the key. In the present study, Lucid key was developed to identify Cercosporoid fungi, a total of 74 Cercosporoid fungi and their characters were entered into a program using Lucid Builder. Lucid has the advantage over printed dichotomous keys in that the user is able to skip an unanswerable couplet or question and still proceed with identification. Identification of Cercosporoid fungi is a difficult task, and the Lucid key was created to help provide individuals with easily accessible tools to distinguish species. The end product of the Lucid key of Philippine Cercosporoid fungi, in the future will be useful in teaching, research, and extension work in mycology and plant pathology.

7. References

Agrios, G.N. (2005). Plant Pathology. 5th ed. Academic Press. New York.

Aptroot, A. (2006). *Mycosphaerella* and its anamorphs: 2. Conspectus of *Mycosphaerella. CBS Biodiversity Series* 5:1 –231.

Barnett, H.L. and Hunter, B.B. (1998): Illustrated Genera of Imperfect Fungi 4th ed. APS Press, St Paul, USA. 218.

Begum, M.M., Shivas, R.G. and Cumagun, C.J.R. (2009). First record of *Periconiella lygodii* occurring on *Lygodium japonicum* in the Philippines. *Australasian Plant Disease Notes* 4: 17-18.

Begum, M.M. and Cumagun, C.J.R. (2010). First record of *Cercospora basellae-albae* from the Philippines. *Australasian Plant Disease Notes* 5: 115-116.

Braun, U. (2004). *Periconiella* species occurring on ferns. *Feddes Repertorium* 115: 50-55.

Chupp, C. (1953). A monograph of the fungus genus *Cercospora*. Ithaca New York. 9-20.

Chupp, C. (1954). A monograph of the fungus genus *Cercospora*. Ithaca. New York. 667.

Crous, P.W. (1998). *Mycosphaerella* spp. and their anamorphs associated with leaf spot diseases of *Eucalyptus*. *Mycologia Memoir* 21: 1- 170.

Crous, P.W. and Braun, U. (2003). *Mycosphaerella* and its anamorphs. . Names published in *Cercospora* and *Passalora*. *CBS Biodiversity Series* 1:1 –571.

Crous, P.W. and Groenewald J.Z. (2006a). *Mycosphaerella alistairii*. *Fungal Planet* No. 4. (www.fungalplanet.org).

Crous, P.W. and Groenewald, J.Z. (2006b). *Mycosphaerella maxii*. *Fungal Planet* No. 6. (www.fungalplanet.org).

Crous, P.W. Aptroot, A. Kang, J.C. Braun, U., and Wingfield, M.J.(2000). The genus *Mycosphaerella* and its anamorphs. *Studies in Mycology* 45:107–121.

Crous, P.W., Kang, J.C., and Braun, U. (2001). A phylogenetic redefinition of anamorph genera in *Mycosphaerella* based on ITS rDNA sequence and morphology. *Mycologia* 93: 1081-1101.

Cumagun C.J.R. and Padilla, C.L. (2007). First record of *Asperisporium caricae* causing black spot of papaya in the Philippines. *Australasian Plant Disease Notes* 2:89-90.

Deighton, F.C. (1967a). New names in *Mycosphaerella (M. arachidis and m. pruni-persicae)* and validation of *M. rosicola*. *Trans. Brit. Mycol. Soc.* 50:328-329.

Deighton, F.C. (1967b). Studies on *Cercospora* and allied genera. II. *Passalora, Cercosporidium* and some species of *Fusicladium* on *Euphorbia*. *Mycol. Papers* 112:1-80.

Deighton, F.C. (1971). Studies on *Cercospora* and allied genera. III. *Cercospora*. *Mycol Papers* 124: 1-13.

Deighton, F.C. (1973). Studies on *Cercospora* and allied genera. IV. *Cercosporella* Sacc., *Pseudocercosporella* gen. Nov. and *Pseudocercosporidium* gen.nov. *Mycol Papers*133: 1-62.

Deighton, F.C. (1974). Studies on *Cercospora* and allied genera. V. *Mycovellosiella* Rangel. and a new species of *Ramulariopsis*. *Mycol. Papers* 137: 1-73.

Deighton, F.C. (1976). Studies on *Cercospora* and allied genera. VI. *Pseudocercospora* Speg., *Pantospora* Cif. And *Cercoseptoria* Petr. *Mycol. Papers* 140:1-168.

Deighton, F.C.(1979). Studies on *Cercospora* and allied genera. VII. New Species and redispositions. *Mycol. Papers* 144:1-56.

Deighton, F.C. (1983). Studies on *Cercospora* and allied genera. VIII. Further notes on *Cercoseptoria* and some species and redispositions. *Mycol. Papers* 151: 1-13.

Deighton, F.C. (1987). New species of *Pseudocercospora* and *Mycovellosiella*, and new combinations into *Pseudocercospora* and *Mycovellosiella*. *Trans. Brit. Mycol.soc.* 88: 365-391.

Ellis, M.B. (1971). Dematiaceous Hypomycetes. Kew, UK: Commonwealth Mycological Institute.

Ellis, M.B. (1976). More Dematiaceous Hypomycetes. Kew, UK Commonwealth Mycological Institute.

Ellis, M.B. and Holliday, P. (1972). *Asperisporium caricae*. CMI Descriptions of Pathogenic Fungi and Bacteria No. 347.Kew, UK: CAB International Mycological Institute

Gerald, J.H. (2005). Diagnosing Postharvest Diseases of Cantaloupe. North Carolina State University.

Goh, T. K. & W. H. Hsieh (1989). New species of *Cercospora* and allied genera of Taiwan. *Bot. Bull. Acad. Sinica* 30: 117-132.

Guo, Y.L. and Hsieh, W.H. (1995). The genus *Pseudocercospora* in China. Mycosystema Monographieum Series No. 2. Int. Acad. Pub., Beijing, China. 388.

Guo, Y.L., Liu, Y.J. and Hsieh, W.H. (1998). Flora Fungorum Sinicorum Vol. 9, *Pseudocercospora*. Science Press, Beijing, China.

Hanlin, R.T. (1990). Illustrated Genera of Ascomycetes. 3 Vols. APS Press, St Paul, USA.

Hsieh, W. H. and Goh, T.K. (1990). *Cercospora* and Similar Fungi from Taiwan. Maw Chang Book Co., Taipei. 376.

Katsuki, S. (1965). *Cercospora* of Japan. *Trans. Mycol. Soc. Japan,* Extra Issue No. 1. 100.

Katsuki, S. and Kobayashi. (1975). *Cercospora* of Japan and allied genera (Supplement 3).*Trans. Mycol. Soc.Japan* 16:1-15.

Kendrick, W.B. and Dicosmo, F. (1979). Teleomorph-anamorph connections in ascomycetes. In: The Whole Fungus, Vol.1:283-410. National Museum of Natural Sciences, Ottawa.

Kirk, P.M., Cannon, P.F., David, J.C. and Stalpers, L.A. (2001). Dictionary of the Fungi, 9th edn. CAB International, Oxon.

Maublanc, A. (1913). Su rune maladie de feuilles du papayer "Carica papaya. *Lavoura* 16 : 208-212.

Michaelides, J. Hunter, L. Kendrick, B. and Nag Raj, T.R. (1979). Synoptic Key to 200 Genera of Coelomycetes. University of Waterloo Biology Series 20. University of Waterloo, Waterloo, Canada. 42.

Norton, G.A. (2000). Multi-media keys for identification and diagnostics: the Lucid experience. International Workshop of the Asia-Pacific Advanced Network (APAN) and its Applications: 27-30.

Pollack, F. G. (1987) An annotated compilation of *Cercospora* names. *Mycol. Mem.* 12:1-212.

Ponaya, A.B. and Cumagun, C.J.R. (2008). First record of *Passalora bougainvilleae* causing leaf spot of bougainvillea in the Philippines. *Australasian Plant Disease Notes* 3: 3-4.

Quimio T.H. and Abilay L.E. (1977). Cercospora Species and disease of Philippine Crops. University of the Philippines Los Banos. Philippines.

Ristaino, J.B., Haege, M.J. and Hu, C.H. (2008). Development of a *Phytophthora* Lucid key. *Journal of Plant Pathology*, 90, S2.81-S2.465.

Saccardo, P.A. (1880). Conspectus generum fungorum Italie inferriorum, nempe as Sphaeropsidas, elanconieas et Hyphomycetes pertinentium. Systemate sporologico dispositorum. *Michelia* 2 :1-38.

Saccardo, P. A. (1886). Sylloge fungorum omnium hucusgue cognitorium. Vol. IV. Padova, 810.

Shin, H.D. and Kim, J.D. (2001). *Cercospora* and allied genera from Korea. Plant Pathogens from Korea 7; 1-302.

Spegazzini, C.(1910). Myceters Argentinenses, Ser. V. *An. Mus. Nac. Hist. Nat. Buenos Aires* 20: 309-467.

Stakman E.C. and Harrar J.G. (1957). 362. The Ronald Press Company. New York.

Sutton, B.C. (1980). The coelomycetes: fungi imperfecti with pycnidia, acervuli, and stromata. Commonwealth Mycological Institute, Kew, Surrey, England, 696 p.

Teodoro, N.T. (1937). Enumeration of Philippine fungi. Commonwealth of the Phill. Dept. of Agric. & Commerce. *Tech. Bull.* No. 4. 585.

Vanky, K. and Shivas, R.G. (2008). Fungi of Australia: The Smut Fungi. Australian Biological Resources Study. CSIRO Publishing.

Vasudeva, R.S. (1963). Indian Cercosporae. Indian Council of Agricultural Research, New Delhi. 245.

von Arx, J.A. (1983). *Mycosphaerella* and its anamorphs. Proc. K. Ned. Akad. Wet. Ser. C Biol. Med. Sci. 86, 1: 15-54.

Welles, C.G. (1924). Observation on taxonomic factors used in the genus *Cercospora*. *Science* 59:216-218.

Welles, C.G. (1925). Taxonomic studies in the genus *Cercospora* in the Philippine Islands.*Am.J.Bot.*12:195-220.

Wellman, F.L.(1972). Tropical American Plant Disease. 219 pp. The Scarecrow Press, Inc. New Jersey.

Yen JM, Lim G. (1980) *Cercospora* and allied genera of Singapore and the Malay peninsula. The Gardens' Bulletin Singapore 33, 151–263.

Young, T. (2001) 101 Forest Fungi of Eastern Australia. Knowledge Books and Software,Brighton, Australia. CD ROM.

3

Novel Elicitors Induce Defense Responses in Cut Flowers

Anastasios I. Darras
Department of Greenhouse Cultivation and Floriculture,
Technological Educational Institute of Kalamata
Greece

1. Introduction

Cut flower production and trade in the E.U. and the rest of the world holds the main share within the ornamental horticulture industry. Despite the global economic crisis started in 2008, changes in cut flower trade, such as the merge of the 2 major auctions in Holland (i.e. VBA and FloraHolland), resulted in stabilization or even small increases in stem number sales for the years 2008-2010 (Evans & Van der Ploeg, 2008; Anonymous, 2011). In other words, the importance of cut flower industry in global economy is undisputed, but also reflects the human need for ornamental plant consumption as part of a better life.

Product quality of horticultural crops has been the main area of research the past decades. Growers and sellers have been seeking for best possible product quality and highest possible profits. However, problems in quality after pathogen infections at some point of production, or during storage or transportation eventually result in economic losses (van Meeteren, 2009). *Botrytis cinerea* is the single-most important pathogen infecting ornamental plants and cut flowers postharvest and substantially reduce growers' and sellers' income by increasing product rejections.

B. cinerea Pers. is a common fungal pathogen that infects glasshouse-grown ornamental crops under cool and humid conditions with latent symptoms, which develop during storage or transportation (Elad, 1988). Growers and sellers around the world are deeply concerned by such infection problems. In Europe, for instance, large quantities of *B. cinerea*-infected cut freesias from The Netherlands are rejected in the UK by wholesalers and retailers at certain times of the year (Darras et al., 2004). These rejections result in immediate economic losses and make cooperation between growers and importers problematic. The problem is equally substantial for roses (Elad, 1988; Elad et al., 1993), gerberas (Salinas & Verhoeff, 1995) and Geraldton waxflowers (Joyce, 1993), although species such as chrysanthemum (Dirkse, 1982), narcissus (O'Neill et al., 2004), lisianthus (Wegulo & Vilchez, 2007), dianthus, ranunculus and cyclamen (Seglie et al., 2009) eventually suffer infections by *B. cinerea*, but to a lesser extend.

Infections by *B. cinerea* are usually managed by conventional fungicides applied protectively at certain times of the year and especially during autumn and spring when most infections occur. However, extensive use of fungicides such as dicarboximides, has led to the appearance of resistant *B. cinerea* strains (Pappas, 1997). Alternative methods to control *B. cinerea* disease (i.e. grey mold) within the concept of integrated disease management (IDM) programs are sought by growers and help overcome resistance by the pathogen.

Elicitor-based disease management constitutes an attractive socio-environmentally sound strategy (Joyce & Johnson, 1999). Known activators of plant defence reactions, such as 2,6-dichloroisoniciotinic acid (INA), salicylic acid (SA), 3-aminobutyric acid (BABA), Acibenzolar-S-methyl [ASM; benzo(1,2,3)thiadiazole-7-carbothioic acid S-methyl ester; benzothiadiazole or BTH; CGA 245704] and methyl jasmonate (MeJA), have been shown to enhance natural defence mechanisms or induce systemic defence responses such as SAR or ISR in plants, thereby providing prospects for IDM (Terry & Joyce, 2004a).

1.1 *Botrytis cinerea* infecting cut flowers and ornamental pot plants

Botrytis cinerea Pers. belongs to the Class Deuteromycetes and the Phylum Ascomycota. The disease caused by *B. cinerea* is called grey mold. The fungus is pathogenic to most of the cultivated ornamental pot plants and cut flowers. For example, infection of gerbera (*Gerbera jamesonii*) flowers occurs inside the glasshouse during crop cultivation, but symptoms develop after a latent period at storage or transportation following fluctuations in temperature (Salinas & Verhoeff, 1995). Favourable temperature and relative humidity (RH) for the pathogen after harvest results in rapid disease development (Salinas et al., 1989). Grey mold on gerbera and freesia flowers is observed as small necrotic, dark-brown fleck lesions 'spots'. Similar symptoms developed in the laboratory under controlled conditions following artificial inoculation of gerbera or freesia inflorescences at temperatures ranging from 4 to 25°C (Salinas & Verhoeff, 1995; Darras et al., 2006a). Infection of freesia (*Freesia hybrida*) inflorescences after artificial inoculation occurred in less than 24 h at 12°C and 80-90% RH. Even at the low temperature of 5°C, disease symptoms were evident in a saturated atmosphere (ca. 100% RH) within the first 24 h of incubation.

B. cinerea is also pathogenic to Geraldton waxflower (*Chamelaucium uncinatum*), the Australian native plant which holds a high ornamental and commercial value (Joyce, 1993; Tomas et al., 1995). Geraldton waxflower sprigs artificially inoculated with *B. cinerea* showed increased abscission of flowers from their pedicels.

B. cinerea infects rose (*Rosa hybrida*) flowers and produces necrotic spots or blister-like patches on petal surfaces (Pie & De Leeuw, 1991; Williamson et al., 1995). Infection has been described by Elad (1988) as restricted, brown, volcano-like shaped lesions. *B. cinerea* damages phylloclades of ruscus (*Ruscus aculeatus*) by causing small, dark water soaked necrotic lesions encircled by a faint halo. These lesions later become brown without growing in size (Elad et al., 1993).

Infection of lisianthus (*Eustoma russellianum*) flowers has been recently reported by Wegulo & Vilchez (2007). Significant ($P \leq 0.03$) positive correlations between stem lesion length of naturally infected plants in the glasshouse ($R = 0.74$) and stem lesion length of artificially inoculated ones ($R = 0.62$) with the disease incidence score, and with the percent of necrosis ($R = 0.71$) of detached leaves were reported (Wegulo & Vilchez, 2007). From all the 12 lisianthus cultivars tested, 'Magic Champagne' was suggested as the most resistant and proposed as ideal for commercial cultivation.

In regards to pot plants, *B. cinerea* disease symptoms on geranium (*Pelargonium zonale*) flowers has been described by Strider (1985) as flower blight, leaf blight and stem rot. Martinez et al. (2008) published a detailed report on infection of *Pelargonium x hortorum*, *Euphorbia pulcherrima, Lantana camara, Lonicera japonica, Hydrangea macrophylla*, and *Cyclamen persicum* by *B. cinerea*. They reported that growth of *B. cinerea* isolates in-vitro from the above mentioned ornamentals varied significantly. *B. cinerea* showed a high degree of phenotypical variability among the isolates, not only as regards to visual aspects of the colonies but also to some morphological structures such as conidium length, conidiophores, sclerotia production, and hyphae (Martinez et al., 2008). Increased susceptibility to grey mold from 10% to 80% in stems and from 3% to 14% in leaves was observed after using elevated levels of N supply (i.e. from 7.15 to 57.1 mM) for begonia plant (*Begonia x tuberhybrida* Voss) cultivation (Pitchay et al., 2007).

1.2 Review on host-pathogen interactions and on defence responses

Host-pathogen specificity involves factors that determine the virulence of the pathogen and also factors that confer resistance on the host (Lucas, 1997). Many theories have been proposed concerning mechanisms by which pathogens either achieve or fail to infect host tissue. A model concerning specific gene-for-gene interactions determining the host range of pathogens in wild plant species was first proposed by Flor (1971). In a gene-for-gene system, recognition of the pathogen by the host occurs when a resistance (R) gene of the host interacts with an avirulence (avr) gene of the pathogen (Table 1).

Virulent or avirulent Pathogen genes	Resistant or susceptible genes in the plant	
	R (resistant) dominant	r (susceptible) recessive
A (avirulent) dominant	AR (-)[a]	Ar (+)
a (virulent) recessive	aR (+)	ar (+)

[a] (-) indicate incompatible interaction and, therefore, no infection. (+) indicate compatible interaction and, therefore, infection.

Table 1. Quadratic check of gene combinations and disease reaction types in host-pathogen systems in which the gene-for-gene concept for one gene operates (Lucas, 1997).

According to this model, avr gene products secreted by hyphae or located on the surface of the pathogen bind to a receptor located on the cell membranes of host's epidermal cells. Binding triggers a cascade of defence responses by the host. Every other possible match in the system could lead to infection (Table 1). Thus, a combination of a resistant host gene and a virulent pathogen leads to a compatible host-pathogen interaction. In both cases, when an avr race of the pathogen matches with a susceptible host and a virulent pathogen matches with a susceptible host, the host fails to recognize the pathogen and infection occurs (Flor, 1971).

Culture filtrates or extracts from microbial cells can act as potent inducers of plant defence responses (Chappell & Hahlbrock, 1984; Kombrink & Hahlbrock, 1986; Fritzemeier et al.,

1987; Keller et al., 1999). For instance, extracts from fungal cell walls when applied to plant tissue induced the synthesis and accumulation of phytoalexins (Yoshikawa et al., 1993). Active components in such chemical, biological and physical extracts are referred to as elicitors. This term is now generally used to denote agents, which induce plant defence responses, including accumulation of PR-proteins, cell wall structural (strengthening) changes, and hypersensitive cell death (Kombrink & Hahlbrock, 1986).

1.2.1 Rapid defence responses

The first step in the rapid defence responses by plants is recognition of the infection attempt by the pathogen. Pathogen recognition results in a signalling cascade to neighbouring cells and in initial molecular defence responses (Kombrink & Somssich, 1995). Examples of elicitor-active components produced by pathogenic fungi include the β-glucan elicitor and the 42 kDa glycoprotein derived from the fungus *Phytophthora megasperma*, the oligo-1,4-α-galacturonides from *Cladosporium fulvum* and *Rhynchosporium secalis*, and the harpin protein from *Erwinia amylovora*. These compounds activate defence responses when they bind to host receptors during incompatible host-pathogen interactions (Ebel & Cosio, 1994; Kombrink & Somssich, 1995).

In parsley cells, the existence of a receptor was proposed by Ebel & Cosio (1994). The intracellular changes were subsequent signals activated by the receptor and transported to host plasma membrane. Changes in H^+, K^+, Cl^- and Ca^{2+} fluxes across the plasma membrane and H_2O_2 increase within 2-5 min can occur (Nurnberger et al., 1994; Nurnberger & Scheel, 2001).

The activity of active oxygen species (e.g. O^-, H_2O_2) and the rapidity of their production after invasion characterize the rapid defence response of the host (Dixon et al., 1994; Ebel & Cosio, 1994; Bolwell, 1999). These toxic active oxygen species cause host cell death at the infection site (Kombrink & Somssich, 1995). It has been suggested that reactive oxygen species (ROS) could have a dual function in disease resistance (Kombrink & Schmelzer, 2001). Firstly, ROS participate directly in cell death during HR and, thereby, results in direct pathogen inhibition. Secondly, ROS have a role in signal diffusion for cellular protectant induction and associated defence responses in neighbouring cells (Kombrink & Schmelzer, 2001).

The HR is part of the initial plant defence responses and involves localized cell death at the infection site (Kombrink & Schmelzer, 2001). Thus, the HR is a result of host recognition of infection attempts made by a pathogenic bacterium or fungus. Specific elicitor-molecules comprise signals, which induce these initial defence responses. When pathogenic bacteria are injected inside a non-host plant under artificial conditions they are killed by the HR as a result of being surrounded by dead cells. The HR may occur when either virulent strains of bacteria are injected inside a resistant host or avirulent strains of bacteria are injected inside a susceptible host (Agrios, 1997). HR associated isolation of the pathogen inside necrotic cells causes the pathogen loses its ability to take-up nutrients and grow into adjacent healthy cells (Kombrink & Schmelzer, 2001).

Elicitors which do not cause an HR can also activate defence-related compounds (Schroder et al., 1992; Atkinson, 1993; Kuć 1995; Kombrink & Schmelzer, 2001). Activation of these compounds can be similar for both compatible and incompatible host-pathogen interactions (Schroder et al., 1992; Kombrink and Schmelzer, 2001). However, only with compatible interactions does the pathogen infect and colonize the host. Accumulation of phytoalexins can occur as part of the HR (Dixon et al., 1994). However, it is not clear whether the HR triggers the production of phytoalexins and other antimicrobial compounds or if their accumulation is a direct result of elicitation (Kombrink & Somssich, 1995).

1.2.2 Local acquired resistance

Phytoalexins are low molecular weight antimicrobial compounds produced de-novo by some plants. They accumulate during infection by pathogens or after injury or stress (Ebel, 1986; Isaac, 1992; Kuć, 1995). Accumulation of phytoalexins is mainly observed when fungi, rather than bacteria, viruses or nematodes, try to infect the plant. Accumulation is a result of specific elicitors released either by the fungal cell walls or by the plant cell walls (Ebel, 1986). Elicitors of phytoalexins include a large number of compounds including inorganic salts (Perrin & Cruickshank, 1965), oligoglucans (Sharp et al., 1984), ethylene (Chalutz & Stahmann, 1969), fatty acids (Bostock et al., 1981), and chitosan oligomers (Kendra & Hadwiger, 1984). Over 200 compounds, microorganisms and physiological stresses have been reported to elicit pisatin in pea, phaseollin and kievitone in green bean and glyceollin in soybean (Kuć, 1991).

Most phytoalexins have been isolated from dicot plants, but they are also present in monocots including rice, corn, sorghum, wheat, barley and onions (Kuć, 1995). There is no published work on phytoalexins in cut flower species. Phytoalexins have been found in almost every part of the plant including roots, stems, leaves and fruits (Kuć, 1995). Such plant species produces a characteristic set of phytoalexins derived from secondary metabolism, in most cases via the phenylpropanoid pathway (Ebel, 1986; Kombrink & Somssich, 1995; Kuć, 1995). Phytoalexins belong to a number of key chemical groups including phenolics (e.g. flavonoids and coumarins), polyacetylenes, isoprenes, terpenoids and steroids (Ebel, 1986). They are produced by both resistant and susceptible tissues and resistance appears to be related with the total phytoalexin concentration (Kuć, 1995). Phytoalexins can affect fungal growth by inhibiting germ tube elongation and colony growth (Elad, 1997). The main effect of phytoalexins on fungi is via their cell membranes. Direct contact of phytoalexins with fungal cell walls resulted in fungal plasma membrane disruption and loss of the ultrastructural integrity (Elad, 1997). In compatible interactions, the pathogen apparently tolerates accumulated phytoalexins, detoxifies them, suppresses phytoalexin accumulation and/or avoids eliciting phytoalexin production (Kuć, 1995). Overcoming phytoalexin accumulation is attributed to either suppressor molecules released by the pathogen (i.e. low molecular weight polysaccharides or glycopeptides) or suppression of the intensity and timing of signal genes that could trigger phytoalexin accumulation (Kuć, 1995).

Pathogenesis related proteins (PR-proteins) accumulate either in extracellular space or the vacuole after various types of plant stress, including pathogen infection (Stermer, 1995; Sticher et al., 1997). PR-proteins accumulate at the site of infection as well as in uninfected tissues (Van Loon & Gerritsen, 1989; Ryals et al., 1996). Although healthy plants may contain traces of PR-proteins, the transcription of genes encoding PR-proteins is up-regulated following pathogen attack, elicitor treatment, wounding or stress (Stermer, 1995; Sticher et al., 1997; Van Loon, 1997). Signal compounds responsible for initiating PR-protein production include salicylic acid, ethylene, the enzyme xylanase, the polypeptide systemin and jasmonic acid (Agrios, 1997). The importance of PR-proteins lies in their range of defence activities (Van Loon et al., 1994). A number of PR-proteins release molecules that may act as elicitors (Keen & Yoshikawa, 1983). PR-proteins accumulation has been observed in monocots as well as in dicots (Redolfi, 1983). However, there is no published work on PR-proteins induced in flower species. Eleven families of PR-proteins have been recognized so far (Van Loon et al., 1994). Some inhibit pathogen development during microbial infection by inhibiting fungal spore production and germination. Others are associated with strengthening of the host cell wall via its outgrowths and papillae (Agrios, 1997). Both β-1,3-glucanases and chitinases, PR-2 and PR-3, respectively, are known to have direct antifungal activity (Mauch et al., 1988; Van Loon, 1997). However, many pathogens have evolved mechanisms to reduce the antifungal impact of PR-proteins (Van Loon, 1997). For example, many chitin-containing fungi are not inhibited by host-produced chitinases.

Plant secondary metabolites are divided into the three main categories of terpenes, phenolic compounds and nitrogen containing secondary metabolites (i.e. alkaloids) (Taiz & Zeiger, 1998). All secondary metabolites are produced through one of the major mevalonic, malonic or shikimic pathways (Taiz & Zeiger, 1998). Phenylalanine is a common amino acid produced via the shikimic pathway (Hahlbrock & Scheel, 1989). The most abundant classes of secondary phenolic compounds in plants are derived from phenylalanine via elimination of an ammonia molecule to form cinnamic acid. This reaction is catalyzed by phenylalanine ammonia lyase (PAL), the key enzyme of phenylpropanoid metabolism (Hahlbrock & Scheel, 1989). Derivatives of phenylpropanoid pathway include low-molecular-weight flower pigments, antibiotic phytoalexins, UV-protectants, insect repellents, and signal molecules in plant-microbe interactions (Hahlbrock & Scheel, 1989; Kombrink & Somssich, 1995).

The main phenylpropanoid pathway branches leading to formation of flavonoids, isoflavonoids, coumarins, soluble esters, wall bound phenolics, lignin and suberin. This diverse spectrum of compounds has a wide range of properties (Hahlbrock & Scheel, 1989). For example, the lignin pathway is an important phenylpropanoid pathway branch that produces precursors for lignin deposition (Grisebach, 1981). Various enzymes implicated in the biosynthesis of lignin appeared to be induced in plants in response to infection or elicitor treatment (Matern & Kneusel, 1988). However, not all studies show a role of lignin and cell lignification in disease inhibition (Garrod et al., 1982). Furanocoumarins derived from the furanocoumarin pathway in parsley are considered potent phytoalexins (Beier & Oertli, 1983). Flavonoid and furanocoumarin production as a response to UV light or fungal elicitor treatment respectively was associated with up-regulation of PAL, 4-coumarate: CoA-ligase (4CL) and chalcone synthase (CHS). Up-regulation was based on rapid changes in amounts and activities of the corresponding mRNAs (Chappell & Hahlbrock, 1984).

After pathogen recognition by the host, a cascade of early responses is induced including ion fluxes, phosphorylation events, and generation of active oxygen species (Kombrink & Somssich, 1995). SA acts as a secondary signal molecule and its levels increase during the defence induction process. Thus, SA is required for initiation of synthesis of various defence-related proteins such as the PR-proteins (Van Loon, 1997; Metraux, 2001). SA accumulation endogenously in tobacco and cucumber plants lead to the HR and the SAR responses. However, SA is not necessarily the translocated signal (elicitor) for the onset of SAR. Rather, SA exerts an effect locally (Vernooij et al., 1994; Ryals et al., 1996). Nonetheless,

SA is still required for SAR expression (Van Loon, 1997). The importance of SA in the onset of SAR was determined using transgenic tobacco and Arabidopsis plants engineered to over-express SA-hydroxylase. Transformed plants with the naphthalene hydroxylase G (NahG) gene produced low levels of SA and SAR expression was blocked.

SA is produced from phenylalanine via coumaric and benzoic acid (Mauch-Mani and Slusarenko, 1996; Ryals et al., 1996; Sticher et al., 1997). Biosynthesis of SA starts with the conversion of phenylalanine to trans-cinnamic acid (Sticher et al., 1997). From trans-cinnamic acid, either benzoic acid (BA) or ortho-coumaric acid are produced. Both compounds are SA precursors (Sticher et al., 1997). Pallas et al. (1996) showed that tobacco plants epigenetically suppressed in PAL expression produced a much lower concentration of SA and other phenylpropanoid derivatives when artificially inoculated with tobacco mosaic virus (TMV). This was seen, firstly, due to the lack of resistance to TMV upon secondary infection, and, secondly, to the absence of PR protein induction in systemic leaves (Pallas et al., 1996).

Jasmonic acid (JA) and its methyl ester (MeJA) are derived from linolenic acid. They are cyclopentanine-based compounds that occur naturally in many plant species (Sembdner & Parthier, 1993; Creelman & Mullet, 1997). Linolenic acid levels or its availability could determine JA biosynthetic rate (Farmer & Ryan, 1992; Conconi et al., 1996). The level of JA in plants varies as a function of tissue and cell type, developmental stage, and in response to various environmental stimuli (Creelman & Mullet, 1997). For example, in soybean seedlings, JA levels are higher in the hypocotyls hook (a zone of cell division) and young plumules as compared to the zone of cell elongation and more mature regions of the stem, older leaves and roots (Creelman & Mullet, 1997). High JA levels are also found in flowers and pericarp tissues of developing reproductive structures (Creelman & Mullet, 1997). Jasmonates are wide spread in Angiosperms, Gymnosperms and algae (Parthier, 1991). They can mediate gene expression in response to various environmental and developmental processes (Wasternack & Parthier, 1997). These processes include wounding (Schaller & Ryan, 1995), pathogen attack (Epple et al., 1997), fungal elicitation (Nojiri et al., 1996), touch (Sharkey, 1996), nitrogen storage (Staswick, 1990), and cell wall strengthening (Creelman et al., 1992). Wounding of tomato leaves produced an 18-amino acid polypeptide called systemin, the first polypeptide hormone discovered in plants so far (Pearce et al., 1991). Systemin was released from damaged cells into the apoplast and transported out of the wounded leaf via the phloem (Schaller & Ryan, 1995) (Fig. 1).

Upon herbivore wounding, a systemic signal is delivered from systemin and results in an ABA-dependent rise of linoleic acid. Systemin was believed to bind to the plasma membrane of target cells and thereby initiate JA biosynthesis (Schaller & Ryan, 1995). JA accumulation can also be induced by oligosaccharides derived from plant cell walls and by elicitors, such as chitosans derived from fungal cell walls (Gundlach et al., 1992; Doares et al., 1995; Nojiri et al., 1996). JA also activates gene expression encoding proteinase inhibitors (Creelman & Mullet, 1997). Proteinase inhibitors are known antidigestive proteins that block the action of herbivore proteolytic enzymes (Creelman & Mullet, 1997). Thereby, proteinase inhibitors help the host to avoid consumption by herbivores. Proteinase inhibitors were accumulated in tomato plants after wounding (O'Donnell et al., 1996) and after irradiation with UV-C (Conconi et al. 1996). In response to wounding, ethylene and JA act together to regulate gene expression of proteinase inhibitors (O'Donnell et al., 1996). Exposing tomato leaves to increasing doses of 254 nm UV-C resulted in increased proteinase inhibitors gene expression. Expression of proteinase inhibitors in wounded (Doares et al., 1995; O'Donnell et al., 1996) or UV-C treated (Conconi et al., 1996) tomato leaves was markedly reduced upon treatment with SA. From linoleic acid, jasmonic acid is produced. Ethylene is required in the jasmonic-signalling cascade (O'Donnell et al., 1995).

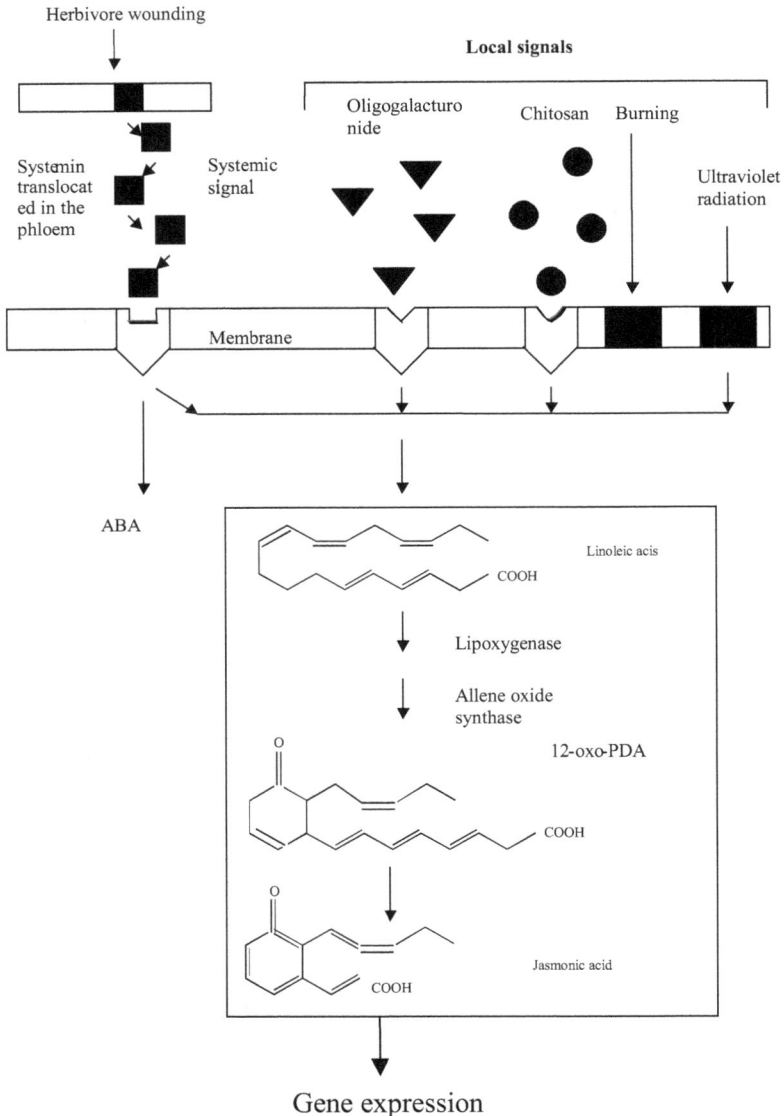

Gene expression

Fig. 1. The octadecanoid-signalling pathway for defence gene expression in tomato (Schaller and Ryan, 1995).

1.2.3 Systemic defence responses (i.e. SAR, ISR) and signalling pathways

SAR is activated following induction of local acquired resistance (LAR). SAR is potentially induced after the HR and after challenge with virulent strains of a pathogen or elicitor treatment. It develops systemically in distant parts of the infected plant (Lawton et al., 1996; Ryals et al., 1996; Metraux, 2001). SAR protects plants from a broad range of potential pathogens (Kessmann et al, 1994).

Specific genes induced in different plant species during SAR have been called SAR-genes (Kessmann et al., 1994; Stermer, 1995; Ryals et al., 1996; Sticher et al., 1997). Most of SAR-genes encode PR-proteins such as those accumulated after inoculation of tobacco with TMV (Ward et al., 1991). These include PR-1 (PR-1a, PR-1b, PR-1c), β-1,3-glucanases (PR-2a, PR-2b, PR-2c), chitinases (PR-3a, PR-3b), hevein-like proteins (PR-4a, PR-4b), thaumatin like proteins (PR-5a, PR-5b), acidic and basic isoforms of class III chitinase, an extracellular β-1,3-glucanase and the basic isoform of PR-1 (Ward et al., 1991). SAR and SAR-gene activation has been observed in various dicots (Kessmann et al., 1994; Ryals et al., 1996). SAR activation involves species specificity (Ryals et al., 1992). For example, acidic PR-1 is only weakly expressed in cucumber. In contrast, acidic PR-1 is the main protein accumulating in tobacco and Arabidopsis. A number of homologous SAR-genes have been identified in monocots. Homologs of the PR-1 family were found in maize and barley and other PR-proteins in maize (Nasser et al., 1988). Gorlach et al. (1996) isolated a group of wheat genes (WCI or wheat chemically induced) induced after chemical treatment with potent SAR inducers. WCI genes seemed to act in a similar manner to SAR-genes in dicots after chemical treatment with plant activators (Gorlach et al., 1996).

Recent research has revealed that JA and ethylene play key roles in signal transduction pathways associated with plant defence responses (Pieterse and van Loon, 1999; Thomma et al., 2000). Inoculation with a necrotizing pathogen resulted predominantly in activation of the SA-dependent SAR response. This response leads to the accumulation of salicylic acid inducible PR-proteins and the expression of SAR (Ryals et al., 1996; Pieterse & van Loon, 1999) (Fig. 2, pathway 2). JA and ethylene inducible defence responses are induced by non-necrotizing rhizobacteria and lead to the ISR phenomenon (Pieterse et al., 1996; Pieterse et al., 1998) (Fig. 2, pathway 1). Both pathways 1 and 2 are regulated in Arabidopsis plants carrying the NPR1 gene.

Depending on the invading pathogen, the composition of defence compounds produced after infection can vary between SA- and JA/ethylene-inducible pathways (Fig. 2, pathways 2 and 3) (Ryals et al., 1996; Epple et al., 1997; Dong, 1998).

Wounding can also result in JA and ethylene inducible defence response activation (Fig. 2, pathway 4) (O'Donnell et al., 1996; Wasternack & Parthier, 1997). However, resultant products of the wounding pathway differ from those induced upon pathogen infection (O'Donnell et al., 1996; Rojo et al., 1999). A second distinct wound-signalling pathway leading to wound responsive (WR) gene expression has been found in Arabidopsis plants (Titarenko et al., 1997; Rojo et al., 1998) (Fig. 2, pathway 6). Upon wounding, Arabidopsis plants carrying the coi1 (JA-insensitive) mutant gene expressed the wound responsive genes choline kinase (CK) and wound responsive (WR3) indicating that the induced pathway was totally independent of JA. UV irradiation of tomato leaves also resulted in induction of the same defensive genes normally activated through the octadecanoid pathway after wounding (Conconi et al., 1996). This response is blocked after SA treatment, confirming the antagonistic regulation of the two distinct pathways (Pena-Cortes et al., 1993; Lawton et al., 1995; Xu et al., 1994; Doares et al., 1995; O'Donnell et al., 1996; Niki et al., 1998; Gupta et al., 2000; Rao et al., 2000).

In the rhizobacteria-mediated induced systemic resistance (ISR) pathway, components from the JA/ethylene response acted in sequence in activating a systemic resistance response that, like pathogen induced SAR, was dependent on the regulatory protein NPR1 (Pieterse & van Loon, 1999). The ISR pathway shares signalling events with pathways initiated upon pathogen infection, but is not associated with the activation of genes encoding plant defensins, thionins or PR-proteins (Pieterse & van Loon, 1999) (Fig. 2, pathway 3). This observation indicates that ISR inducing rhizobacteria, such as P. fluorescens strain WCS417r, trigger a novel signalling pathway leading to the production of so far unidentified defense compounds (Pieterse et al., 1996; Pietrese et al., 1998). Protection of NahG Arabidopsis plants by gaseous MeJA suggested that induction of a SA non-dependent systemic pathway was regulated by JA (Thomma et al., 2000) (Fig. 2, pathway 3). Protection was provided against two necrotrophic fungi, A. brassicicola and B. cinerea.

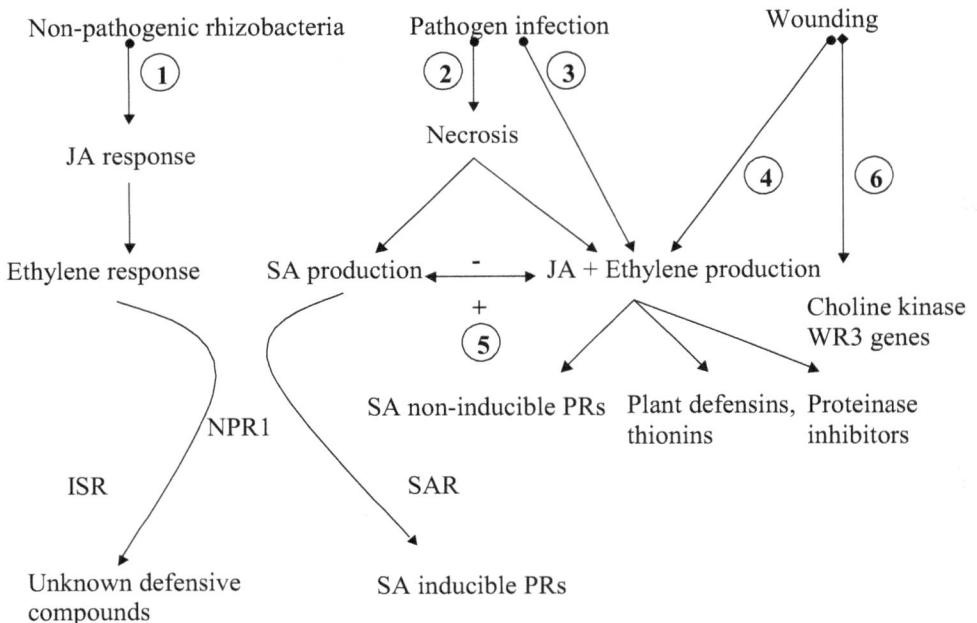

Fig. 2. Model showing systemic signalling pathways that can be induced in plants by non-pathogenic rhizobacteria, pathogen infection and wounding, such as caused by foraging insects. 1: ISR is induced in NPR1 Arabidopsis plants as a result of JA and ethylene responses. 2: SAR is induced in NPR1 Arabidopsis plants after necrosis by pathogenic fungi, bacteria or virus. 3: JA/ethylene pathway is up- regulated after fungal infection. JA/ethylene expression leads to genes encoding plant defensins, thionins, proteinase inhibitors and SA-independent PR-proteins. 4 and 6: A number of genes are regulated after mechanical wounding. JA and ethylene levels rise after mechanical wounding. 5: Cross-talk between SA- and JA- dependent pathways exist. Adopted from Pieterse & van Loon (1999).

1.3 Elicitation of defence responses with chemical activators

Disease management in the past has been achieved by various methods including resistant cultivars, biological control, crop rotation, tillage, and chemical pesticides (Kessmann et al., 1994). Recently, the use of abiotic and/or biotic agents, as well as, synthetic compounds that induce host immune systems have offered a new prospect for disease management.

A chemical is generally characterized as a plant activator when it induces natural and/or systemic defence responses, activate gene expression and provide protection on the same spectrum of diseases exerted by a wild type host (Kessmann et al., 1996; Ruess et al., 1996). Plant activators, normally, do not exert direct antimicrobial activity against pathogens when used for disease control, but rather work through their mutagenic elicitation effect and help eliminate the risk of the development of resistant strains by the pathogen.

1.3.1 Acibenzolar-S-methyl (ASM; benzo[1,2,3]thiadiazole-7-carbothioic acid S-methyl ester; CGA 245704; benzothiadiazole or BTH)

The efficacy of ASM has been tested in field, glasshouse and pot trials (Table 2).

Host	Pathogen	Induced response	ASM concentration	Reference
Apple seedlings cv. Golden Delicious	*Erwinia amylovora*	β-1,3-glucanases, peroxidases	0.1-0.2 g AI L⁻¹	Brisset et al., 2000
Cauliflower (*Brassica oleracea*)	*Peronospora parasitica*	ns	0.0015-0.075 g AI L⁻¹	Godard et al., 1999
Cereals, tobacco	*Erysiphe graminis, Septoria* spp., *Puccinia* spp., *Peronospora hyoscyami* f. sp. *tabacina*	ns	12-30 g AI ha⁻¹	Ruess et al., 1996
Cucumber (*Cucumis sativus* L.)	*Cladosporium cucumerinum*	Acidic peroxidase, class III chitinase and β-1,3-glucanase	32.4 g AI L⁻¹	Narusaka et al., 1999
Cucumber (*Cucumis sativus* L.) and Japanese pear (*Pyrus pyrifolia* Nakai var. culta Nakai)	*Many pathogens*	ns	0.05-0.1 g AI L⁻¹	Ishii et al., 1999
Cauliflower (*Brassica oleracea*)	*P. parasitica*	β-1,3-glucanase. PR-1 and PR-5	0.05 g AI L⁻¹	Ziadi et al., 2001
Grapevine cv Merlot	*B. cinerea*	na	0.3 mM	Iriti et al., 2004
Melons cv. Early Yellow Hami	*Fusarium* spp., *Alternaria* spp., *Rhizopus* spp. *Trichothecium* sp.	ns	0.025 or 0.05 g AI L⁻¹	Huang et al., 2000

Host	Pathogen	Induced response	ASM concentration	Reference
Melon fruit	*Fusarium pallidoroseum*	ascorbate peroxidase, guaiacol peroxidase, PAL, β-1,3-glucanase		Gondim et al., 2008
Parsley cells (*Petroselinum crispum* L.)	With or without elicitor (Pmg)	PAL, coumarins	0.32-6.48 g AI L^{-1}	Katz et al., 1998
Pepper (*Capsicum annuum* L.)	*Xanthomonas campestris* pv. *vesicatoria*	ns	1.25-5 g AI L^{-1}	Buonaurio et al., 2002
Soybean seedlings	*Sclerotinia sclerotiorum*	ns	0.035-0.375 g AI L^{-1}	Dann et al., 1998
Strawberry plants cvs. Elsanta and Andana	*B. cinerea*	ns	0.25-2 g AI L^{-1}	Terry and Joyce, 2000
Strawberry	Microbial populations	chitinase and β-1,3-glucanase	0.05-0.5 g.L^{-1}	Cao et al., 2010
Tobacco plants cv. Kutsaga Mammoth 10	*Pseudomonas syringae* pv *tabaci, Thanatephorus cucumeris, Cercospora nicotianae*	ns	0.05-30 g AI L^{-1}	Cole, 1999
Tomato plants (*Lycopersicon esculentum*)	*Fusarium oxysporum* f.sp. *radicis-lycopersici*	Callose enriched wall appositions phenolic compounds	97.2 g AI L^{-1}	Benhamou & Belanger, 1998
Tomato plants cv. Vollendung	*Cucumber mosaic virus* (*CMV*)	ns	0.1 mM	Anfoka, 2000
Wheat	*Erysiphe graminis* f.sp. *tritici*	WCI genes (1-5)	0.3 mM	Gorlach et al., 1996

Table 2. Effects of ASM on different host-pathogen interactions (ns: not shown, na: not applicable).

Although, most of ASM application were carried out preharvest, there is number of published research on ASM postharvest applications (i.e. Cao et al., 2010). Additionally, a considerable work on postharvest application of ASM on ornamentals has been published the recent years (i.e. Darras et al., 2007). ASM was introduced as a potent inducer of SAR and treated plants were resistant to the same spectrum of diseases as plants activated naturally (Kessmann et al., 1996; Friedrich et al., 1996). Although ASM and its metabolites exhibited no direct antimicrobial activity towards plant pathogens tested, they induced the same biochemical processes in the plant as those observed after natural activation of SAR (Friedrich et al., 1996; Lawton et al., 1996). The compound, which was inactive in plants that do not express the SAR-signaling pathway, required a lag time of approximately 30 days between application and protection (Lawton et al., 1996).

1.3.2 Jasmonates (plant hormones produced through the octadecanoid pathway)

The efficacy of jasmonates has been tested in field, glasshouse and pot trials (Table 3).

Host	Pathogen	Induced response	MeJA concentration	Reference
Arabidopsis (*Arabidopsis thaliana*)	*B. cinerea, A.brassicicola, Plectosphaerella cucumerina*	ns	0.5-50 µM and 0.001-1 µL L⁻¹ air	Thomma et al., 2000
Arabidopsis (*Arabidopsis thaliana*)	*A. brassicicola*	PDF1.2	45 µM	Penninckx et al., 1996
Grapefruit (*Citrus paradisi*) var. 'Marsh Seedless	*Penicillium digitatum*	ns	1-50 µM	Droby et al., 1999
Large number of species	na	PPO	na	Constabel and Ryan, 1998
Loquat fruit	*Colletotrichum acutatum*	chitinase and β-1,3-glucanase	10 µmol L⁻¹	Cao et al., 2008
Potato plants (*Solanum tuberosum*)	*Phytophthora infestans*	phytoalexins	1-10 µM	Il'inskaya et al., 1996
Sweet cherry	*Monilinia fructicola*	PAL, β-1,3-glucanase	0.2 mM	Yao & Tian, 2005
Tobacco cell cultures	na	β-glucuronidase (GUS), osmotin protein	0.045-4550 µM	Xu et al., 1994
Tobacco cv. Xanthi-nc	*Phytophthora parasitica var. nicotianae, Cercospora nicotianae, TMV*	β-glucuronidase (GUS)	45 µM	Mitter et al., 1998
Tomato plants (*Lycopersicon esculentum*)	*Helicoverpa zea, Spodoptera exigua*	PPO, POD, LOX and PIs	0.1-10 mM	Thaler et al., 1996
Tomato plants (*Lycopersicon esculentum*)	*Spodoptera exigua, Pseudomonas syringae* pv. *tomato*	PPO	1 mM	Thaler et al., 1999

Table 3. Effects of MeJA on different host-pathogen interactions. (ns: not shown, na: not applicable).

Although, firstly tested preharvest, JA or MeJA has been extensively used postharvest at different hosts (i.e. fruits, vegetables, cut flowers), application modes (i.e. spray, pulse, gas) and incubation environments (i.e. storage or ambient temperatures). For example, JA and MeJA were tested on grapefruit for suppressing postharvest green mold decay [*Penicillium digitatum* (Pers.:Fr.) Sacc.] (Droby et al., 1999). Studies showed that 50 µM and 1 µM MeJA concentrations were effective against the disease and that the reduction in the decay was the same at incubation temperatures of 2 or 20°C. Moreover, as the in-vitro tests showed no direct antifungal activity of JA and MeJA, it was suggested that the disease suppression was achieved via natural resistance induction (Droby et al., 1999). Treatment of Arabidopsis plants with MeJA reduced *A. brassicicola*, *B. cinerea* and *Plectosphaerella cucumerina* disease development (Thomma et al., 2000). Application of gaseous MeJA to plants resulted in a greater disease reduction compared to that on plants sprayed with MeJA or treated with INA. Gaseous MeJA protected SA-degrading transformant NahG plants, suggesting that gaseous MeJA induced a non-SA dependent systemic response (Thomma et al., 2000). Combination of ASM and JA was tested against bacterial and insect attack on field grown tomato plants (Thaler et al., 1999). Two signaling pathways, one involving SA and another involving JA were proposed to provide resistance against pathogens and insect herbivores, respectively (Thaler et al., 1999).

1.4 Elicitation of defence responses in floriculture

The efficacy of ASM and MeJA on ornamental pot plants and on cut flowers has been tested pre- and postharvest, respectively (Table 4). Most of such tests were carried out in the very recent years and still increasing. For example, pre- and postharvest treatments with MeJA or ASM on cut flowers conferred a variable measure of protection against postharvest infections by *B. cinerea* (Dinh et al., 2007).

JA and MeJA provided systemic protection to various rose cultivars (e.g. Mercedes, Europa, Lambada, Frisco, Sacha and Eskimo) against *B. cinerea* (Meir et al., 1998). MeJA applied as postharvest pulse, significantly reduced *B. cinerea* lesion size on detached rose petals. In the same study, MeJA at concentrations of 100-400 µM showed in-vitro antifungal activity on *B. cinerea* spore germination and germ-tube elongation. Similarly, a postharvest pulse, spray, or vapour treatment with MeJA 200 µM, 600 µM or 1 µL L^{-1}, respectively, significantly reduced petal specking by *B. cinerea* on cut inflorescences of *Freesia hybrida* 'Cote d'Azur' (Darras et al., 2005; 2007). Moreover, 1-100 µL L^{-1} MeJA postharvest vapour treatment reduced *B. cinerea* development on cut Geraldton waxflower 'Purple Pride' and 'Mullering Brook' sprigs (Eyre et al., 2006). Application of gaseous MeJA to fresh cut peonies resulted in the lowest disease severity and in an improvement of vase life compared to the untreated controls (Gast, 2001).

MeJA and ASM, applied preharvest had variable responses against postharvest infection by *B. cinerea*. ASM was not as effective as MeJA in suppressing the development of postharvest *B. cinerea* disease for glasshouse grown freesias (Darras et al., 2006b). Dinh et al. (2007) reported that multiple sprays of ≤1000 µM MeJA to field grown plants significantly reduced *B. cinerea* on Geraldton waxflower 'My Sweet Sixteen' cut sprigs, that were un-inoculated or artificially inoculated with *B. cinerea* (Dinh et al., 2007).

Host	Elicitor	Target pathogen	Application method and timing	Reference
a. Cut flowers				
Rose (*Rosa hybrida*)	ASM	*Diplocarpon rosae*	Spray - preharvest	Suo & Leung, 2002
	MeJA	*B. cinerea*	Pulse, spray - postharvest	Meir et al., 1998; 2005
Gerbera (*Gerbera jamesonii*)	UV-C	*B. cinerea*	Postharvest	Darras et al., 2012
Freesia (*Freesia hybrida*)	MeJA & ASM	*B. cinerea*	Spray - preharvest	Darras et al., 2006b
	MeJA	*B. cinerea*	Pulse, spray, gas - postharvest	Darras et al., 2005; 2007
Sunflower plants (*Helianthus annuus*)	ASM, MeJA, SA, INA	*B. cinerea*	Spray - preharvest	Dimitriev et al., 2003
Sunflower plants (Helianthus annuus)	ASM	*Plasmopara helianthi*	*Spray - preharvest*	*Tosi et al., 1999*
Geraldton waxflower (*Chamelaucium uncinatum*)	SA	*Alternaria* sp., *Epicoccum* sp.	Spray - preharvest	Beasley, 2001
	MeJA	*B. cinerea*	Gas - postharvest	Eyre et al., 2006
	MeJA & ASM	*B. cinerea*	Spray – pre- and postharvest	Dinh et al., 2007; 2008
Peonies (*Paeonia lactiflora*)	MeJA	*B. cinerea*	Gas - postharvest	Gast, 2001
b. Pot plants				
Cyclamen (*Cyclamen persicum*)	ASM	*Fusarium oxysporum* f. sp. *cyclaminis*	Spray - preharvest	Elmer, 2006a
Petunia (*Petunia hybrida*)		*Phytophthora infestans*	Spray - preharvest	Becktell et al., 2005
c. Landscape architecture plants				
Date palm	ASM	*Fusarium oxysporum* f. sp. *albedinis*	Injection in the trunk	Jaiti et al., 2009
d. Propagation material				
Gladiolus corms (*Gladiolus x hortulanus*)	ASM	*Fusarium oxysporum* f. sp. *gladioli*	Dip	Elmer, 2006b

Table 4. Chemical and biological elicitors tested on cut flowers and ornamental pot plants against various pathogens infecting either pre- or postharvest.

Chemical elicitors such as ASM have been applied in pot ornamentals such as petunia (Becktell et al., 2005), cyclamen (Elmer, 2006a) and in gladiolus corms (Elmer, 2006b), but effectiveness varied within the different experimental designs and conditions. In cyclamen, infection by *Fusarium oxysporum* f.sp. *cyclaminis* was reduced with increasing ASM doses (Elmer, 2006a). Additionally, the dry mass of ASM treated cyclamen plants increased with increasing ASM rates. However, as no further assays were carried out to assess possible induction of defence responses, it was not clear whether ASM reduced *F. oxysporum* f.sp. *cyclaminis* via induction of defence mechanisms or via a profound fungitoxic effect. It has been demonstrated in other research that ASM may exert direct toxic activity against *B. cinerea* (Darras et al., 2006b). In addition, ASM did not confer a significant level of protection on gladiolus corms against *F. oxysporum* f. sp. *gladioli*, and compared to conventional fungicides, although, the number of emerging flower spikes increased significantly compared to the ASM-untreated corms (Elmer, 2006b).

2. Elicitation of defence responses in cut *Freesia hybrida* flowers – A typical example

2.1 Background

Infection problems by *Botrytis cinerea* are typical to most geographical areas around the world and concern cut flower industry. Infection of cut flowers by the fungus results in visible lesions on flower petals (petal spotting or petal specking). According to Darras et al. (2004) freesia flower rejections at certain periods of the year (viz. April, May, October) lead in severe economic losses to growers, importers and sellers. Infection by *B. cinerea* of most cut flowers occurs in the glasshouse when a single conidium germinates and penetrates petal epidermal cells. A necrotic lesion appears postharvest after a brief incubation period under favourable environmental conditions (Darras et al., 2006a). Infection is difficult to control as it appears later in handling chain under various conditions during transport or storage.

In most cases, *B. cinerea* disease is controlled by conventional fungicides. However, extensive use of fungicides such as dicarboximides in the glasshouse has led to appearance of fungicide resistance (Pappas 1997). Alternative management methods within the concept of IDM can help overcome such problems.

For this reason, plant defence inducers (i.e. elicitors) such as ASM and MeJA have been tested with applications at various intervals, pre- or postharvest to activate systemic defence responses of the host (Kessmann et al., 1994; Meir et al., 1998; Thomma et al., 2000). For cut freesia flowers postharvest pulse, spray, or gaseous MeJA treatment at 200 µM, 600 µM, or 1 µL L^{-1}, respectively, significantly reduced petal specking by *B. cinerea* on cv. 'Cote d'Azur' inflorescences (Darras et al., 2005; 2007). An apparent induced defence response was recorded by both ASM and MeJA treatment. However, only MeJA conferred constant and significant disease reductions. MeJA vapour at 1 µL L^{-1} significantly reduced lesion numbers and diameters on freesia petals by up to 56% and 50%, respectively (Darras et al., 2005).

2.2 Overview of published research and further discussion

Freesia inflorescences cv 'Cote d'Azur' gassed with 0.1 µL L^{-1} MeJA showed significantly smaller lesions after artificial inoculation with *B. cinerea* (Fig. 3). Gaseous MeJA might have induced a range of defence mechanisms to halt infection development. MeJA applied post-harvest as vapour at 1-100 µL L^{-1} significantly reduced the development of *B. cinerea* on cut Geraldton waxflower 'Purple Pride' and 'Mullering Brook' sprigs (Eyre et al., 2006). In a very recent study, Darras et al. (2011) demonstrated that gaseous MeJA at 0.1 µL L^{-1} significantly increased polythenol oxidase (PPO) activities 24 and 36 h post-treatment. This observation suggests that MeJA-induced defence mechanisms might be associated with the production of quinones (Constabel and Ryan, 1998), which probably helped in *B. cinerea* disease reduction. The effects of PPO in *B. cinerea* disease control have been confirmed for gerbera flowers (Darras et al., 2012). A low dose of UV-C irradiation increased PPO activity and was positively correlated with the reduction of *B. cinerea* disease symptoms on the florets (Darras et al., 2012). This indicates that PPO might play an important role in *B. cinerea* disease control on cut flowers.

Fig. 3. *B. cinerea* necrotic lesions on artificially inoculated freesia cv. 'Cote d'Azur' flowers treated with 0.1 µL L^{-1} gaseous MeJA (left) or left un-treated (control) (right) and incubated for 48 h at 20°C (Darras, 2003).

Lesion diameters on the detached freesia petals were significantly reduced with increasing MeJA spray, pulse or gaseous concentrations (Darras et al., 2007). The first published evidence of postharvest MeJA spray treatments enhancing protection of cut flowers against *B. cinerea* was the work by Meir et al. (2005) on cut roses. According to Meir et al. (2005), simultaneous MeJA pulsing and spraying under handling conditions resulted in suppression of gray mold in seven rose cultivars ('Eskimo', 'Profita', 'Tamara', 'Sun Beam', 'Pink Tango', 'Carmen', 'Golden Gate'). In an earlier study MeJA applied as a pulse variably reduced *B. cinerea* lesion numbers and diameters (Meir et al., 1998). Our findings are in agreement with those by Meir et al. (1998) that disease severity in both artificially inoculated and naturally infected rose flowers was reduced by a MeJA pulse at 0.2 mM at 20°C. On cut Geraldton waxflower 'Purple Pride' and 'Mullering Brook' sprigs, 1-100 μL L^{-1} MeJA postharvest vapour treatment significantly reduced the development of *B. cinerea* (Eyre et al., 2006). However, it also induced flower fall incidence, which was correlated with a systemic resistance-associated up-regulation of ethylene biosynthesis.

Irrespective to the concentration tested, ASM provided no protection to artificially inoculated freesia flowers (Darras et al., 2007). However, natural infection was significantly ($P < 0.05$) reduced after ASM treatment during storage at 5 and at 12°C. On the contrary, postharvest treatments of strawberry cv. Camarosa fruit with ASM failed to reduce natural infection by *B. cinerea* at 5°C (Terry & Joyce, 2004b). Generally, ASM tended to provide protection on freesia flowers at lower incubation temperatures (Darras et al., 2007). However, it was not clear whether such disease reductions were the result of the induction of host's defence responses or a direct fungitoxic activity measured in the same study. Likewise, Terry & Joyce (2000) showed that ASM reduced in-vitro *B. cinerea* mycelial growth on ASM-amended agar. It is possible that the limited disease control on freesia flowers at 5°C was due to direct toxic effect of ASM rather than via SAR induction.

Elicitation of defence responses in cut flowers is an interesting prospect for *B. cinerea* disease control especially as it may offer alternatives to fungicide application. In series of postharvest experiments with freesia inflorescences the potential to induce natural defence mechanisms or directly controlling *B. cinerea* disease by application of biological and chemical elicitors was investigated. Postharvest treatments with ASM, MeJA or UV-C irradiation markedly suppressed *B. cinerea* specking on freesia petals by reducing disease severity, lesion numbers and lesion diameters. However, attempts to further minimise disease damage caused by *B. cinerea* using combined treatments with different plant activators (i.e. both ASM and MeJA), were not successful (Darras et al., 2011).

In summary, ASM was the least effective in reducing *B. cinerea* specking on cut freesia flowers (Fig. 4). In addition, it remained unclear as to whether or not SAR was induced. In contrast, gaseous MeJA reduced disease severity most probably by inducing JA-dependant biochemical responses. These contrasting results tend to concur with observations by Pieterse & van Loon (1999) and Thomma et al. (2001) that SA- versus JA-dependent pathways are effective against different pathogens. The results of a most recent paper (i.e. Darras et al., 2011)) suggested that, SA-dependant pathway and consequently the SAR response was not effective in freesia flowers against *B. cinerea* infection. In contrast, the JA-dependant pathway was apparently induced and suppressive of *B. cinerea* infection (Darras et al., 2011).

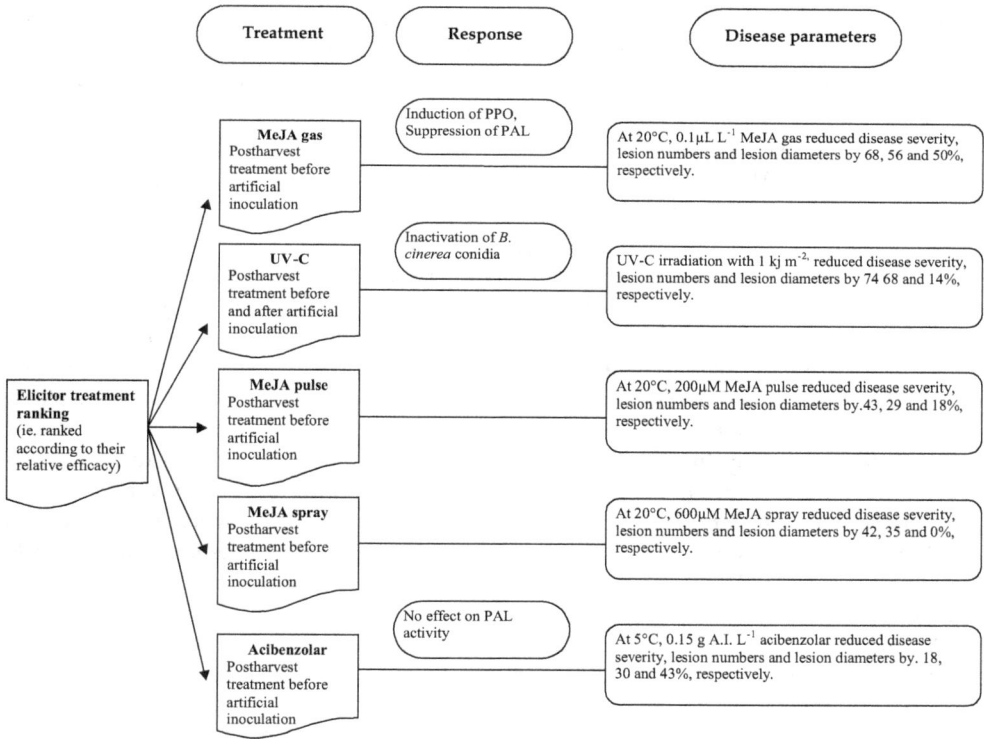

Fig. 4. Ranking, in terms of relative efficacy, of postharvest biological (i.e. UV-C) and chemical (i.e. ASM, MeJA) elicitors tested on cut freesia inflorescences to control *B. cinerea* infection starting with the most effective (Darras, 2003).

3. Conclusions and recommendations for future research

Management of postharvest infection of cut freesia flowers by *B. cinerea*, was, in most cases, successful. ASM was somewhat effective compared to untreated controls mostly when applied preharvest at 1.43 μM. In-vitro studies showed direct antifungal activity of ASM against *B. cinerea* colony growth and conidial germination. Inconsistency of ASM applied pre- or postharvest may be explained by: 1) variability of environmental conditions in the glasshouse, which may affected defence enhancement (Herms & Mattson, 1992; Terry, 2002); and, 2) infection by *B. cinerea* might not necessarily be sensitive to induced SAR responses, and thus ASM treatments may not correspond to *B. cinerea* disease suppression (Thomma et al., 1998; Govrin & Levine, 2002). Friedrich et al. (1996) reported that ASM failed to control *B. cinerea* in tobacco, but was effective against other pathogens. The apparent inability of ASM to control *B. cinerea* was seemingly supported by the observation that PAL activity in ASM treated freesia inflorescences was not higher compared to the untreated controls. Therefore, ASM did not induce biochemical defence processes, such as the production of antifungal secondary metabolites like phytoalexins through the phenylpropanoid pathway (Kombrink & Somssich, 1995; Kuć, 1995).

In contrast to inconsistent effects of ASM, MeJA was markedly effective in suppressing *B. cinerea* specking on cut freesia flowers when applied either pre- or postharvest. MeJA effectiveness was application method and concentration dependent. MeJA applied as gas was more effective compared to pulsing or spraying. It is possible that MeJA may function as an airborne signal which activated disease resistance and the expression of defence related genes in plant tissue (Shulaev et al., 1997). This finding agrees with earlier findings in Arabidopsis presented by Thomma et al. (2000). In Arabidopsis, this effect was mediated via the JA-dependent defence responses (Thomma et al., 2000). MeJA did not exert any direct antifungal activity in-vitro except at the concentration of 600 μM and therefore it is possible that MeJA reduced *B. cinerea* disease on freesia flowers by inducing responses correlated with the JA-dapendent pathway (Darras et al., 2005). PPO levels in freesia flowers after MeJA gaseous treatment increased by 47 and 57% compared to the untreated controls (Darras et al., 2011). However, PAL activity decreased markedly compared 12 h post MeJA application and maintained at minimum level (i.e. ≈ 0). These findings suggest that MeJA might suppress the action of PAL in the phenylpropanoid pathway and consequently reduce or block SA production. Antagonistic regulation of JA- and SA-dependent pathways has been documented in the past by Pena-Cortes et al. (1993), Conconi et al. (1996), Niki et al. (1998), Gupta et al. (2000), and Rao et al. (2000). The apparent suppression of PAL in freesia flowers by MeJA constitute additional evidence of a JA- and SA- antagonistic response.

MeJA applied to freesia plants 28 days before harvest suppressed postharvest flower specking caused by *B. cinerea* in both a temperature and variety dependent fashion. MeJA was highly effective when flowers were incubated at 20°C compared to incubation at 5 or 12°C. It is likely that low incubation temperatures slow down plant's metabolism and also the production of defence related compounds (Jarvis, 1980). Overall, MeJA provided a considerable level of protection against *B. cinerea* when applied preharvest and, thus, could be considered a promising tool in an IDM context. Further study at the molecular level is warranted to help interpret the MeJA mode of action in cut flowers. Also, additional in-planta trials on extra freesia varieties and a wider range of MeJA concentrations may help in better understanding MeJA efficacy.

In view to the promising results using MeJA, it is likely that elicitor based strategies within IDM could be used for the control of Botrytis or other pathogens on freesias and ornamental pot plants, as well as on various cut flowers. In turn, IDM would minimise the risk of pathogens developing resistance to fungicides and also reduce public concerns over extensive fungicide use (Jacobsen & Backman, 1993).

More research could be undertaken into potential synergistic effects of combined pre- and postharvest treatments with plant activators and/or abiotic biological agents (i.e. UV-C irradiation). In due course, pre and/or postharvest use of plant activators could have commercial potential for postharvest disease suppression (Kessmann et al., 1994; Kessmann et al., 1996; Thaler et al., 1996; Meir et al., 1998; Huang et al., 2000; (Darras et al., 2011).

4. References

Agrios, G. N. 1997. *Plant Pathology* (4th edition), Academic Press, London UK

Anfoka, G.H. (2000). Benzo-(1,2,3)-thiadiazole-7-carbothioic acid S-methyl ester induces systemic resistance in tomato (*Lycopersicon esculentum*. Mill cv. Vollendung) to Cucumber mosaic virus. *Crop Protection*, Vol.19, pp. 401-405

Anonymous, (2011). FloraHolland reposts 7% increase in turnover. In *Floraculture International*, 30/08/2011, Available from:
<http://www.floracultureinternational.com/index.php?option=com_content&view=article&id=2196:floraholland-reports-7-increase-inturnover&catid=52:bussiness&Itemid=376>

Atkinson, M.M. (1993). Molecular mechanisms of pathogen recognition by plants. *Advances in Plant Pathology*, Vol.10, pp. 35-64

Beasley, D.R. (2001). Strategies for control of *Botrytis cinerea* on Geraldton waxflower flowers. PhD thesis. The University of Queensland, Australia

Becktell, M.C.; Daughtrey, M.L. & Fry, W.E. (2005). Epidemiology and management of petunia and tomato late blight in the greenhouse. *Plant Disease*, Vol.89(9), pp. 1000-1008.

Beier, R.C. & Oertli, E.H. (1983). Psoralen and other linear furocoumarins as phytoalexins in celery. *Phytochemistry*, Vol.22, pp. 2595-2597

Benhamou, N. & Bélanger, R.R. (1998). Benzothiadiazole-mediated induced resistance to Fusarium oxysporum f.sp. radicis-lycopersici in tomato. *Plant Physiology*, Vol.118, pp. 1203-1212

Bostock, R.; Kuć, J., & Laine, R. (1981). Eicosapentaenoic and arachidonic acids from *Phytophthora infestans* elicit fungitoxic sesquiterpenes in potato. *Science*, Vol.212, pp. 67-69

Bolwell, G.P. (1999). Role of active oxygen species and NO in plant defence responses. *Current Opinion in Plant Biology*, Vol.2, pp. 287-294

Brisset, M-N.; Cesbron, S.; Thomson S.V. & Paulin, J-P. (2000). Acibenzolar-S-methyl induces the accumulation of defense-related enzymes in apple and protects from fire blight. *European Journal of Plant Pathology*, Vol.106, pp. 529-536

Buonaurio, R.; Scarponi, L.; Ferrara, M.; Sidoti, P. & Bertona, A. (2002). Induction of systemic acquired resistance in pepper plants by acibenzolar-S-methyl against bacterial spot disease. *European Journal of Plant Pathology*, Vol.108, pp. 41-49

Cao, S.; Zheng, Y.; Yang Z.; Tang, S.; Jin, P.; Wang, K. & Wang, X. (2008). Effect of methyl jasmonate on the inhibition of *Colletotrichum acutatum* infection in loquat fruit and the possible mechanisms. *Postharvest Biology and Technology,* Vol.49, pp. 301-307.

Cao, S.; Hu, Z.C.; Zheng, Y.H.; Li, X.W.; Wang, H.O. & Pang, B. (2010). Effect of post-harvest treatment with BTH on fruit decay, microbial populations, and the maintenance of quality in strawberry. *Journal of Horticultural Science & Biotechnology*, Vol.85(3), pp. 185-190

Chalutz, E. & Stahmann, M. (1969). Induction of pisatin by ethylene. *Phytopathology*, Vol.59, pp. 1972-1973

Chappell, J. & Hahlbrock, K. (1984). Transcription of plant defence genes in response to UV light or fungal elicitor. *Nature*, Vol.311, pp. 76-78

Cole, D.L. (1999). The efficacy of acibenzolar-S-methyl, an inducer of systemic acquired resistance, against bacterial and fungal diseases of tobacco. *Crop Protection*, Vol.18, pp. 267-273

Conconi, A.; Smerdon, M.J.; Howe, G.A. & Ryan, C.A. (1996). The octadecanoid signaling pathway in plants mediates a response to ultraviolet radiation. *Nature*, Vol.383, pp. 226-229

Constabel, C.P. & Ryan C.A. (1998). A survey of wound and methyl jasmonate induced leaf polyphenol oxidase in crop plants. *Phytochemistry*, Vol.47, pp. 507-511

Creelman, R.A.; Tierney, M.L. & Mullet, J.E. (1992). Jasmonic acid/methyl jasmonate accumulate in wounded soybean hypocotyls and modulated wound gene expression. *Proceedings of the National Academy for Science U.S.A.*, Vol.89, pp. 4938-4941

Creelman R.A. & Mullet J.E. (1997). Biosynthesis and action of jasmonates in plants. *Annual Review of Plant Physiology and Plant Molecular Biology*, Vol.48, pp. 355-381

Dann, E.; Diers, B.; Byrum, J. & Hammerschmidt, R. (1998). Effect of treating soybean with 2,6-dichloroisonicotinic acid (INA) and benzothiadiazole (BTH) on seed yields and the level of disease caused by *Sclerotinia sclerotiorum* in field and greenhouse studies. *European Journal of Plant Pathology*, Vol.104, pp. 271-278

Darras, A.I. (2003). Biology and management of freesia flower specking caused by *Botrytis cinerea*. PhD Thesis, Cranfield University, UK

Darras, A.I.; Joyce, D.C. & Terry, L.A. (2004). A survey of possible associations between pre-harvest environment conditions and postharvest rejections of cut freesia flowers. *Australian Journal of Experimental Agriculture*, Vol.44, pp. 103-108

Darras, A.I.; Terry, L.A. & Joyce, D.C. (2005). Methyl jasmonate vapour treatment suppresses specking caused by *Botrytis cinerea* on cut *Freesia hybrida* L. flowers. *Postharvest Biology and Technology*, Vol.38, pp. 175-182

Darras, A.I.; Joyce, D.C.; Terry, L.A. & Vloutoglou I. (2006a). Postharvest infections of *Freesia hybrida* L. flowers by *Botrytis cinerea*. *Australasian Plant Pathology*, Vol.35, pp. 55-63

Darras, A.I.; Joyce, D.C. & Terry, L.A. (2006b). Acibenzolar-S-methyl and methyl jasmonate of glasshouse-grown freesias suppress postharvest petal specking caused by *Botrytis cinerea*. *Journal of Horticultural Science & Biotechnology*, Vol.81(6), pp. 1043-1051

Darras, A.I.; Joyce, D.C.; Terry, L.A.; Pompodakis, N.E. & Dimitriadis, C.I. (2007). Efficacy of postharvest treatments with acibenzolar-S-methyl and methyl jasmonate against *Botrytis cinerea* infecting cut *Freesia hybrida* L. flowers. *Australasian Plant Pathology*, Vol.36, pp. 332-340

Darras, A.I.; Joyce, D.C. & Terry, L.A. (2011). MeJA and ASM protect cut *Freesia hybrida* inflorescences against *Botrytis cinerea*, but do not act synergistically. *Journal of Horticultural Science & Biotechnology*, Vol.86(1), pp. 74-78

Darras, A.I.; Demopoulos, V. & Tiniakou, C.A. (2012). UV-C irradiation induces defence responses and improves vase life of cut gerbera flowers. *Postharvest Biology and Technology* Vol.64, pp. 168-174

Dinh, S.-Q.; Joyce, D.C.; Irving, D.E. & Wearing, A.H. (2007). Field applications of three different classes of known host plant defence elicitors did not suppress infection of Geraldton waxflower by *Botrytis cinerea*. *Australasian Plant Pathology*, Vol.36, pp. 142-148

Dinh, S.-Q.; Joyce, D.C.; Irving, D.E. & Wearing, A.H. (2008). Effects of multiple applications of chemical elicitors on *Botrytis cinerea* infecting Geraldton waxflower. *Australasian Plant Pathology*, Vol.37, pp. 87-94

Delaney, T.P.; Uknes, S.; Vernooij, B.; Friedrich, L.; Weymann, K.; Negretto, D.; Gaffney, T.; Gut-Rella, M.; Kessmann, H. & Ward, E. (1994). A central role of salicylic acid in plant disease resistance. *Science*, Vol.266, pp. 1247-1249

Dimitriev, A.; Tena, M. & Jorrin, J. (2003). Systemic acquired resistance in sunflower (*Helianthus annus* L.). *TSitologiia i Genetika*, Vol.37, pp. 9-15

Dirkse, FB. (1982). Preharvest treatment of chrysanthemum against *Botrytis cinerea*. *Acta Horticulturae*, Vol.125, pp. 221–226.

Dixon, R.A.; Lamb, C.J. & Harrison, M.J. (1994). Early events in the activation of plant defense responses. *Annual Review of Phytopathology*, Vol.32, pp. 479-501

Doares, S.H.; Syrovers, T.; Weiler, E. & Ryan C.A. (1995). Oligogalacturonides and chitosan activate plant defensive genes through the octadecanoid pathway. *Proceedings of the National Academy of Science U.S.A.*, Vol.92, pp. 4095-4098

Dong, X. (1998). SA, JA, ethylene, and disease resistance in plants. *Current Opinions in Plant Biology*, Vol.1, pp. 316-323

Droby, S.; Porat, R.; Cohen, L.; Weiss, B.; Shapiro, S.; Philisoph-Hadas, S. & Meir, S. (1999). Suppressing green mold decay in grapefruit with postharvest jasmonate application. *Journal of the American Society for Horticultural Science*, Vol.124(2), pp. 184-188

Ebel, J. (1986). Phytoalexin synthesis: the biochemical analysis of the induction process. *Annual Review of Phytopathology*, Vol.24, pp. 235-264

Ebel, J. & Cosio, E.G. (1994). Elicitors of plant defense responses. *International Review of Cytology*, Vol.148, pp. 1-36

Elad, Y. (1988). Latent infection of *Botrytis cinerea* in rose flowers and combined chemical and physiological control of the disease. *Crop Protection*, Vol.7, pp. 361-366

Elad, Y.; Kirshner, B. & Gotlib, Y. (1993). Attempts to control *Botrytis cinerea* on roses by pre- and postharvest treatments with biological and chemical agents. *Crop Protection* Vol.12, pp. 69-73

Elad, Y. (1997). Responses of plants to infection by *Botrytis cinerea* and novel means involved in reducing their susceptibility to infection. *Biological Reviews*, Vol.72, pp. 381-342

Elmer, W.H. (2006a). Efficacy of preplant treatments of gladiolus corms with combinations of acibenzolar-S-methyl and biological or chemical fungicides for suppression of fusarium corm rot [*Fusarium oxysporum* f. sp. *gladioli*]. *Canadian Journal of Plant Pathology*, Vol.28(4), pp. 609-624

Elmer, W.H. (2006b). Effects of acibenzolar-S-methyl on the suppression of Fusarium wilt on cyclamen. *Crop Protection*, Vol.25, pp. 671-676

Epple P.; Apel, K. & Bohlmann, H. (1997). Overexpression of an endogenous thionin enhances resistance of Arabidopsis against *Fusarium oxysporum*. *The Plant Cell*, Vol.9, pp. 509-520

Evans, A. & van der Ploeg, R. (2008). Auctions around the world. *FloraCulture International*, Vol.5, pp. 8-9

Eyre, J.X.; Faragher, J.; Joyce, D.C. & Franz, P.R. (2006). Effects of postharvest jasmonate treatments against *Botrytis cinerea* on Geraldton waxflower (*Chamelaucium uncinatum*). *Australasian Plant Pathology*, Vol.46, pp. 717-723

Farmer, E.E. & Ryan, C.A. (1992). Octadecanoid precursors of jasmonic acid activate the synthesis of wound-inducible proteinase inhibitors. *Plant Cell*, Vol.4, pp. 129-134

Flor, H.H. (1971). Current status of the gene-for-gene concept. *Annual Review of Phytopathology*, Vol.9, pp. 275-296

Friedrich, L.; Lawton, K.; Ruess, W.; Masner, P.; Specker, N.; Gut Rella, M.; Meiers, B.; Dincher, S.; Staub, T.; Uknes, S.; Metraux, J-P.; Kessmann, H. & Ryals, J. (1996). A Benzothiadiazole derivative induces systemic acquired resistance in tobacco. *Plant Journal*, Vol.10, pp. 61-72

Fritzemeier, K-H.; Cretin, C.; Kombrink, E.; Rohwer, F.; Taylor, J.; Scheel, D. & Hahlbrock, K. (1987). Transient induction of phenylalanine ammonia-lyase and 4-coumarate:CoA legase mRNAs in potato leaves infected with virulent or avirulent races of *Phytophthora infestans*. *Plant Physiology*, Vol.85, pp. 34-41

Gast, K. (2001). Methyl jasmonate and long term storage of fresh cut peony flowers. *Acta Horticulturae*, Vol.543, pp. 327-330.

Garrod, B.; Lewis, B.G.; Brittain, M.J. & Davies, W.P. (1982). Studies on the contribution of lignin and suberin to the impedance of wounded carrot toot tissue to fungal invasion. *New Phytologist*, Vol.90, pp. 99-108

Godard, J.-F.; Ziadi, S.; Monot, C.; Le Corre, D. & Silue, D. (1999). Benzothiadiazole (BTH) induces resistance in cauliflower (*Brassica oleracea* var *botrytis*) to downy mildew of crucifers caused by *Peronospora parasitica. Crop Protection*, Vol.18, pp. 397-405

Gondim, D.M.F.; Terao, D.; Martins-Miranda, A.S.; Vansconcelos, I.M. & Oliveira, J.T.A. (2008). Benzo-thiadiazole-7-carbothioic acid S-methyl ester does not protect melon fruits against *Fusarium pallidoroseum* infection but induces defence responses in melon seedlings. *Journal of Phytopathology*, Vol.156(10), pp. 607-614

Gorlach, J.; Volrath, S.; Knauf-Beiter, G.; Hengy, G.; Beckhove, U.; Kogel, K.-H.; Oostendorp, M.; Staub, T.; Ward, E.; Kessmann, H. & Ryals, J. (1996). Benzothiadiazole, a novel class of inducers of systemic acquired resistance, activates gene expression and disease resistance in wheat. *Plant Cell*, Vol.8, pp. 629-643

Govrin, E.M. & Levine, A. (2002). Infection of Arabidopsis with a necrotrophic pathogen, *Botrytis cinerea*, elicits various defense responses but does not induce systemic acquired resistance (SAR). *Plant Molecular Biology*, Vol. 48, pp. 267-276

Grisebach, H. (1981). *Lignins*. In 'The Biochemistry of Plants' (Ed. E.E. Conn). Chapter 7, pp. 457-478, Academic Press, New York, USA

Gundlach, H.; Muller, M.J.; Kutchan, T.M. & Zenk, M.H. (1992). Jasmonic acid is a signal transducer in elicitor-induced plant cell cultures. *Proceedings of the National Academy of Science USA*, Vol.89, pp. 2389-2393

Gupta, V.; Willits, M.G. & Glazebrook, J. (2000). *Arabidopsis thaliana* EDS4 contributes to salicylic acid (SA)-dependent expression of defense responses: evidence for inhibition of jasmonic acid signaling by SA. *Molecular Plant Microbe Interactions*, Vol.13, pp. 503-5011

Hahlbrock, K. & Scheel, D. (1989). Physiology and molecular biology of phenylpropanoid pathway. *Annual Review of Plant Physiology and Plant Molecular Biology*, Vol.40, pp. 347- 369

Herms, D.A. & Mattson, W.J. (1992). The dilemma of plants: to grow or defend. *The Quarterly Review of Biology*, Vol.67, pp. 283-335

Huang, Y.; Deverall, B.J.; Tang, W.H.; Wang, W. & Wu, F.W. (2000). Foliar application of acibenzolar-S-methyl and protection of postharvest rock melons and Hami melons from disease. *European Journal of Plant Pathology*, Vol.106, pp. 651-656

Il'inskaya, L.I.; Goenburg, E.V.; Chalenko, G.I. & Ozeretskovskaya, O.L. (1996). Involvement of jasmonic acid in the induction of potato resistance to Phytophthora infection. *Russian Journal of Plant Physiology*, Vol.43(5), pp. 622-628

Iriti, M.; Rossoni, M.; Borgo, M. & Faoro, F. (2004). Benzothiadiazole enhanves resveratrol and anthocyanin biosynthesis in grapevine, meanwhile improving resistanve to *Botrytis cinerea. Journal of Agricultural and Food Chemistry*, Vol.52, pp.4406-4413

Isaac, S. (1992). *Fungal-Plant Interactions*. Chapman & Hall, London, UK

Ishii, H.; Tomita, Y.; Horio, T.; Narusaka, Y.; Nakazawa, Y.; Nishimura, K. & Iwamoto, S. (1999). Induced resistance of acibenzolar-S-methyl (CGA 245704) to cucumber and Japanese pear diseases. *European Journal of Plant Pathology*, Vol.105, pp. 77-85

Jacobsen, B.J. & Backman, P.A. (1993). Biological and cultural plant disease controls: alternative and supplements to chemicals in IPM systems. *Plant Disease*, Vol.77, pp. 311-315

Jaiti, F.; Verdeil, J.L. & El Hadrami, I. (2009). Effect of jasmonic acid on the induction of polyphenoloxidase and peroxidase activities in relation to date palm resistance against *Fusarium oxysporum* f. sp. *albedinis*. *Physiological and Molecular Plant Pathology*, Vol.74(1), pp. 84-90

Jarvis, W.R. (1977). *Botryotinia and Botrytis species: taxonomy, physiology, and pathogenicity. A guide to the literature*. Research Branch Canada Department of Agriculture, Monograph No 15

Jarvis, W.R. (1980). *Epidemiology*. In 'The Biology of Botrytis' Eds. Coley-Smith, J.R., Verhoeff, K., and Jarvis, W.R. pp. 219-245, Academic Press, York, London

Joyce, D.C. (1993). Postharvest floral organ fall in Geraldton waxflowers (*Chamelaucium uncinatum* Schauer). *Australian Journal of Experimental Agriculture*, Vol.33, pp. 481-487

Joyce, D.C. & Johnson, G.I. (1999). Prospects for exploitation of natural disease resistance in harvested horticultural crops. *Postharvest News and Information*, Vol.10, pp. 45-48

Katz, V.A.; Thulke, O.U. & Conrath, U. (1998). A Benzothiadiazole primes parsley cells for augmented elicitation of defense responses. *Plant Physiology*, Vol.117, pp. 1333-1339

Keen, N.T. & Yoshikawa, M. (1983). β-1,3-endoglucanase from soybean releases elicitor-active carbohydrates from fungus cell walls. *Plant Physiology*, Vol.71, pp. 460-465

Keller, H.; Pamboukdjian, N.; Ponchet, M.; Poupet, A.; Delon, R.; Verrier, J-L.; Roby, D. & Ricci, P. (1999). Pathogen-induced elicitin production in transgenic tobacco generates a hypersensitive response and nonspecific disease resistance. *The Plant Cell*, Vol.11, pp. 223-235

Kendra, D.F. & Hadwiger, L.A. (1984). Characterization of the smallest chitosan oligomer that is maximally antifungal to *Fusarium solani* and elicits pisatin formation in *Pisum sativum*. *Experimental Mycology*, Vol.8, pp. 276-282

Kessmann, H.; Staub, T.; Hofmann, C.; Maetzke, T. & Herzog, J. (1994). Induction of systemic acquired disease resistance in plants by chemicals. *Annual Review of Phytopathology*, Vol.32, pp. 439-459

Kessmann, H.; Oostendorp, M.; Ruess, W.; Staub, T.; Kunz, W. & Ryals, J. (1996). Systemic activated resistance – a new technology for plant disease control. *Pesticide Outlook* pp. 10-13

Kombrink, E. & Hahlbrock, K. (1986). Responses of cultured parsley cells to elicitors fro phytopathogenic fungi. Timing and dose dependency of elicitor-induced reactions. *Plant Physiology*, Vol.81, pp. 216-221

Kombrink, E. & Somssich, I. (1995). Defense responses of plants to pathogens. *Advances in Botanical Research*, Vol.21, pp. 2-34

Kombrink, E. & Schmelzer, E. (2001). The hypersensitive response and its role in local and systemic disease resistance. *European Journal of Plant Pathology*, Vol.107, pp. 69-78

Kuć, J. (1991). *Phytoalexins: perspectives and prospects*. In 'Mycotoxins and phytoalexins' (Eds. Sharma, R., and Salunkhe, D.), . pp. 595-603, Boca Raton: CRC, USA

Kuć, J. (1995). Phytoalexins, stress metabolism and disease resistance in plants. *Annual Review of Phytopathology*, Vol.33, pp. 275-297

Lawton, K.A.; Weymann, K.; Friedrich, L.; Vernooij, B.; Uknes, S. & Ryals, J. (1995). Systemic acquired resistance in Arabidopsis requires salicylic acid but not ethylene. *Molecular Plant-Microbe Interactions*, Vol.8, pp. 863-870

Lawton, K.; Friedrich, L.; Hunt, M.; Weymann, K.; Delaney, T.; Kessmann, H.; Staub, T. & Ryals, J. (1996). Benzothiadiazole induces disease resistance in Arabidopsis by activation of the systemic acquired resistance signal transduction pathway. *Plant Journal*, Vol.10, pp. 71-82

Lucas, J.A. (1997). *Plant Pathology and Plant Pathogens*. 3rd edition, Blackwell Science, UK

Martinez, J.A.; Valdes, R.; Vicente, M.J.; & Banon, S. (2008). Phenotypical differences among *Botrytis cinerea* isolates from ornamental plants. *Communications in Agricultural and Applied Biological Sciences*, Vol.73, pp. 121-129

Matern, U. & Kneusel, R. (1988). Phenolic compounds in plant disease resistance. *Phytoparasitica*, Vol.16, pp. 153-170

Mauch, F.; Hadwiger, L.A. & Boller, T. (1988). Antifungal hydrolases in pea tissue. 2. Inhibition of fungal growth by combinations of chitinase and β-1,3-glucanase in pea pods by pathogens and elicitors. *Plant Physiology*, Vol.76, pp. 607-611

Mauch-Mani, B. & Slusarenko, A.J. (1996). Production of salicylic acid precursors is a major function of phenylalanine ammonia lyase in the resistance of Arabidopsis to *Peronospora parasitica*. *The Plant Cell*, Vol.8, pp. 203-212

Meir, S.; Droby, S.; Davidson, H.; Alsvia, S.; Cohen, L.; Horev, B. & Philosoph-Hadas, S. (1998). Suppression of Botrytis rot in cut rose flowers by postharvest application of methyl jasmonate. *Postharvest Biology and Technology*, Vol.13, pp. 235-243

Meir, S.; Droby, S.; Kochanek, S.; Salim, S. & Philosoph-Hadas, S. (2005) Use of methyl jasmonate for suppression of Botriytis rot in various cultivars of cut rose flowers. *Acta Horticulturae*, Vol.669, pp. 91-98

Métraux, J-P. (2001). Systemic acquired resistance and salicylic acid: current state of knowledge. *European Journal of Plant Pathology*, Vol.107, pp. 13-18

Mitter, N.; Kazan, K.; Way, H.M.; Broekaert, W.F. & Manners, J.M. (1998). Systemic induction of an Arabidopsis plant defensin gene promoter by tobacco mosaic virus and jasmonic acid in transgenic tobacco. *Plant Science*, Vol.136, pp. 169-180

Narusaka, Y.; Narusaka, M.; Horio, T. & Ishii, H. (1999). Induction of disease resistance in cucumber by acibenzolar-S-methyl and expression of resistance-related genes. *Annals of the Phytopathological Society of Japan*, Vol.65, pp. 116-122

Nasser, W.; De Tapia, M.; Kauffmann, S.; Montasser-Kouhsari, S. & Burkard, G. (1988). Identification and characterization of maize pathogenesis related proteins. Four maize PR proteins are chitinases. *Plant Molecular Biology*, Vol.11, pp. 529-538

Niki, T.; Mitsuhara, I.; Seo, S.; Ohtsubo, N. & Ohashi, Y. (1998). Antagonistic effect of salicylic acid and jasmonic acid on the expression o pathogenesis-related (PR) protein genes in wounded mature tobacco leaves. *Plant Cell Physiology*, Vol.39(5), pp. 500-507

Nojiri, H.; Sugimori, M.; Yamane, H.; Nishimura, Y.; Yamada, A.; Shibuya, N.; Kodama, O.; Murofushi, N. & Omori, T. (1996). Involvement of jasmonic acid in elicitor-induced phytoalexin production in suspension-cultured rice cells. *Plant Physiology*, Vol.110, pp. 387-392

Nurnberger, T.; Nennstiel, D.; Jabs, T.; Sacks, W.R.; Hahlbrock, K. & Scheel, D. (1994). High affinity binding of a fungal oligopeptide elicitor to parsley plasma membranes triggers multiple defenses responses. *The Plant Cell*, Vol.78, pp. 449-460

Nurnberger, T. & Scheel, D. (2001). Signal transmission in the plant immune response. *Trends in Plant Science*, Vol.6(8), pp. 372-379

O'Donnell, P.J.; Calvert, C.; Atzorn, R.; Wasternack, C.; Leyser, H.M.O. & Bowles, D.J. (1996). Ethylene as a signal mediating the wound response of tomato plants. *Science*, Vol.274, pp. 1914-1917

O'Neil, T.M.; Hanks, G.R. & Wilson, T.W. (2004). Control of smoulder (*Botrytis narcissicola*) in narcissus with fungicides. *Annals of Applied Biology*, Vol.145, pp. 129-137

Pallas, J.A.; Paiva, N.L.; Lamb, C. & Dixon, R.A. (1996). Tobacco plants epigenetically suppressed in phenylalanine ammonia lyase expression do not develop systemic acquired resistance in response to infection by tobacco mosaic virus. *The Plant Journal*, Vol.10, pp. 281-293

Parthier, B. (1991). Jasmonates, new regulators of plant growth and development: many facts and few hypotheses on their actions. *Botanica Acta* pp. 446-454

Pappas, A.C. (1997). Evolution of fungicide resistance in *Botrytis cinerea* in protected crops in Greece. *Crop Protection*, Vol.16, pp. 257-263

Pearce, G.; Strydom, D.; Johnson, S. & Ryan, C.A. (1991). A polypeptide from tomato leaves induces wound-inducible proteinase inhibitor proteins. *Science*, Vol.253, pp. 895-898

Pena-Cortes, H.; Albrecht, T.; Prat, S.; Weiler, E.W. & Willmitzer, L. (1993). Aspirin prevents wound-induced gene expression in tomato leaves by blocking jasmonic acid biosynthesis. *Planta*, Vol.191, pp. 123-128

Penninckx, I.A.M.A.; Eggermont, K.; Terras, F.R.G.; Thomma, B.P.H.J.; De Samblanx, G.W.; Buchala, A.; Metraux, J-P.; Manners J.M. & Broekaert, W.F. (1996). Pathogen-induced systemic activation of a plant defensin gene in Arabidopsis follows a salicylic acid-independed pathway. *The Plant Cell*, Vol.8, pp. 2309-2323

Perrin, D.R. & Cruickshank, I.A.M. (1965). Studies on phytoalexins VII. The chemical stimulation of pisatin formation in *Pisum sativum* L. *Austrian Journal of Biological Science*, Vol.18, pp. 803-816

Pichay, D.S.; Frantz, J.M.; Locke, J.C.; Krause, C.R. & Fernandez, G.C.J. (2007). Impact of applied nitrogen concentration on growth of elatior Begonia and new guinea impatiens and susceptibility of Begonia to *Botrytis cinerea*. *Journal of the American Society for Horticultural Science*, Vol.132, pp. 193-201

Pie, K. & de Leeuw, G.T.N. (1991). Histopathology of the initial stages of interaction between rose flowers and *Botrytis cinerea*. *Netherlands Journal of Plant Pathology*, Vol.97, pp. 335-344

Pieterse, C.M.J.; van Wees, S.; Hoffland, E.; van Pelt, J.A. & van Loon, L.C. (1996). Systemic resistance in Arabidopsis induced by biocontrol bacteria is independent of salicylic acid accumulation and pathogenesis related gene expression. *The Plant Cell*, Vol.8, pp. 1225-1237

Pieterse, C.M.J.; van Wees, S.; van Pelt, J.A.; Knoester, M.; Laan, R.; Gerrits, H.; Weisbeek, P.J. & van Loon L.C. (1998). A novel signalling pathway controlling induced systemic resistance in Arabidopsis. *The Plant Cell*, Vol.10, pp. 1571-1580

Pieterse, C.M.J. & van Loon, L.C. (1999). Salicylic acid-independent plant defence pathways. *Trends in Plant Science*, Vol.4(2), pp. 52-58

Rao, M.V.; Lee, H-I.; Creelman, R.A.; Mullet, J.E. & Davis, K.R. (2000). Jasmonic acid signaling modulates ozone-induced hypersensitive cell death. *The Plant Cell*, Vol.12, pp. 1633-1646

Redolfi, P. (1983). Occurrence of pathogenesis related and similar proteins in different plant species. *Netherlands Journal of Plant Pathology*, Vol.89, pp. 245-254

Rojo, E.; Titarenko, E.; Leon, J.; Berger, S.; Vancanneyt, G. & Sanchez-Serrano, J.J. (1998). Reversible protein phosphorylation regulates jasmonic acid-dependent and independent wound signal transduction pathways in *Arabidopsis thaliana*. *The Plant Journal*, Vol.13, pp. 153-165

Rojo, E.; Leon, J. & Sanchez-Serrano, J.J. (1999). Cross-talk between wound signaling pathways determines local versus systemic gene expression in *Arabidopsis thaliana*. *The Plant Journal*, Vol.20, pp. 136-142

Ross, A.F. (1961). Systemic acquired resistance induced by localized virus infections in plants. *Virology*, Vol.14, pp. 341-358

Ruess, W.; Mueller, K.; Knauf-Beiter, G.; Kenz, W. & Staub, T. (1996). Plant activator CGA 245704: an innovative approach for disease control in cereals and tobacco. *Brighton Crop Protection Conference-Pests and Diseases*, Vol.2A-6, pp. 53-60, 18-21 November 1996

Ryals, J.A.; Ward, E. & Metraux J.P. (1992). *Systemic acquired resistance: An inducible defense mechanism in plants*. In 'Inducible Plant Proteins: Their Biochemistry and Molecular Biology', (Ed. J.L. Wray), pp. 205-229, Cambridge University Press, Cambridge UK

Ryals, J.A.; Neuenschwander, U.H.; Willits, M.G.; Molina, A.; Steiner, H-Y. & Hunt, M.D. 1996. Systemic Acquired Resistance. *The Plant Cell*, Vol.8, pp. 1809-1819

Salinas, J.; Glandorf, D.C.M.; Picavet, F.D. & Verhoeff, K. (1989). Effects of temperature, relative humidity and age of conidia on the incidence of specking on gerbera flowers caused by *Botrytis cinerea*. *Netherlands Journal of Plant Pathology*, Vol.95, pp. 51-64

Salinas, J. & Verhoeff, K. (1995). Microscopic studies of the infection of gerbera flowers by *Botrytis cinerea*. *European Journal of Plant Pathology*, Vol.101, pp. 377-386

Seglie, L.; Spadano, D.; Devecchi, M.; Larcher, F.; Gullino, M.L. (2009). Use of 1-methylcyclopropene for the control of *Botrytis cinerea* on cut flowers. *Phytopathologia Mediterranea*, Vol.48, pp. 253-261

Schaller, A. & Ryan, C.A. (1995). Systemin – a polypeptide defense signal in plants. *BioEssays*, Vol.18(1), pp. 27-33

Schroder, M.; Hahlbrock, K. & Kombrink, E. (1992). Temporal and spatial patterns of 1,3-β-glucanase and chitinase induction in potato leaves infected by *Phytophthora infestans*. *Plant Journal*, Vol.2, pp. 161-172

Sembdner, G. & Parthier, B. (1993). The biochemistry and the physiological and molecular actions of jasmonates. *Annual Review of Plant Physiology and Plant Molecular Biology*, Vol.44, pp. 569-589

Sharkey, T.D. (1996). Emission of low molecular mass hydrocarbons from plants. *Trends in Plant Science*, Vol.3, pp. 78-82

Sharp, J.; Albersheim, P.; Ossowski, O.; Pilotti, A.; Garegg, P. & Lindberg, B. (1984). Comparison of the structure of elicitor activities of a synthetic and mycelial-wall-derived hexa-(β-D-glucopyranosyl)-D-glusitol. *Journal of Biological Chemistry*, Vol.259, pp. 11341-11345

Shulaev, V.; Silverman, P. & Raskin, I. (1997). Airborne signaling by methyl salicylate in plant pathogen resistance. *Nature*, Vol.385, pp. 718-721

Staswick, P.E. (1990). Novel regulation of vegetative storage protein genes. *The Plant Cell*, Vol.2, pp. 1-6

Stermer, B.A. (1995). *Molecular regulation of systemic induced resistance*. In "Induced Resistance to Disease in Plants". Eds. R. Hammerscmidt and Kuć, J. pp. 111-140, Kluwer Academic Publishers, USA

Sticher, L.; Mauch-Mani, B. & Metraux, J.P. (1997). Systemic Acquired Resistance. *Annual Review of Phytopathology*, Vol.35, pp. 235-270

Strider, D.L. (1985). *Diseases of Floral Crops*. Vol. 2, Praeger Publishers, New York, USA

Suo, Y.; & Leung, D.W.M. (2002). BTH-induced accumulation of extracellular proteins and blackspot disease in rose. *Biologia Plantarum*, Vol.45, pp. 273-279

Taiz, L. & Zeiger, E. (1998). *Plant Physiology*. 2nd ed. pp. 61-80, Sinauer Associates, Inc., Publishers, USA

Terry, L.A. & Joyce, D.C. (2000). *Suppression of grey mould on strawberry fruit with the chemical plant activator acibenzolar*. Pest Management Science, *Vol.56, pp. 989-992*

Terry, L.A. (2002). Natural disease resistance in strawberry fruit and Geraldton waxflower flowers. PhD Thesis, Cranfield University, UK

Terry, L.A. & Joyce, D.C. (2004a) Elicitors of induced disease resistance in postharvest horticultural crops: a brief review. *Postharvest Biology and Technology*, Vol.32, pp. 1-13

Terry, L.A. & Joyce, D.C. (2004b) Influence of growing conditions on efficacy of acibenzolar and botryticides in suppression of *Botrytis cinerea* on strawberry fruit. *Advances in Strawberry Research*, Vol.23, pp. 11-19

Thaler, J.S.; Stout, M.J.; Karban, R. & Duffey, S.S. (1996). Exogenous jasmonates simulate insect wounding in tomato plants (*Lycopersicon esculentum*) in the laboratory and field. *Journal of Chemical Ecology*, Vol.22, pp. 1767-1781

Thaler, J.S.; Fidantsef, A.L.; Duffey, S.S. & Bostock R.M. (1999). Trade-offs in plant defense against pathogens and herbivores: a field demonstration of chemical elicitors of induced resistance. *Journal of Chemical Ecology*, Vol.25, pp. 1597-1609

Thomma, B. P.H.J.; Eggermont, K.; Penninckx, I.A.M.A.; Mauch-Mani, B.; Vogelsang, R.; Cammue, B.P.A. & Broekaert, W. (1998). Separate jasmonate-dependent and salicylate-dependent defense response pathways in Arabidopsis are essential for resistance to distinct microbial pathogens. *Proceedings of the National Academy of Science USA*, Vol.95, pp. 15107-15111

Thomma, B. P.H.J.; Eggermont, K.; Broekaert, W. & Cammue, B.P.A. (2000). Disease development of several fungi on Arabidopsis can be reduced by treatment with methyl jasmonate. *Plant Physiology and Biochemistry*, Vol.38, pp. 421-427.

Thomma, B. P.H.J.; Penninckx, I.A.M.A.; Broekaert, W.F. & Cammue, B.P.A. (2001). The complexity of disease signaling in Arabidopsis. *Current Opinion in Immunology*, Vol.13, pp. 63-68

Titarenko, E.; Rojo, E.; Leon, J. & Sanchez-Serrano, J.J. (1997). Jasmonic acid-dependent and independent signaling pathways control wound-induced gene activation in *Arabidopsis thaliana*. *Plant Physiology*, Vol.115, pp. 817-826

Tomas, A.; Wearing, A.H. & Joyce, D.C. (1995). *Botrytis cinerea*: a causal agent of premature flower drop in packaged Geraldton waxflower. *Australasian Plant Pathology*, Vol.24, pp. 26-28

Tosi, L. & Zazzerini, A. (1999). Benzothiadiazole induces resistance to *Plasmopara helianthi* in sunflower plants. *Journal of Phytopathology*, Vol.147, pp. 365-370

Van Loon, L.C. & Gerritsen, Y.A.M. (1989). Localization of pathogenesis-related proteins in infected and non-infected leaves of Samsun NN tobacco during the hypersensitive reaction to tobacco mosaic virus. *Plant Science*, Vol.63, pp. 131-140

Van Loon, L.C.; Pierpoint, W.S.; Boller, T. & Conejero, V. (1994). Recommendations for naming plant pathogenesis-related proteins. *Plant Molecular Biology Reporter*, Vol.12, pp. 245-264

Van Loon, L.C. (1997). Induced resistance in plants and the role of pathogenesis-related proteins. *European Journal of Plant Pathology*, Vol.103, pp. 753-765

Van Meeteren, U. (2009). Causes of quality loss of cut flowers - A critical analysis of postharvest treatments. *Acta Horticulturae*, Vol.847, pp. 27-36

Vernooij, B.; Friedrich, L.; Morse, A.; Reist, R., Kolditz-Jawhar, R.; Ward, E.; Uknes, S.; Kessmann, H. & Ryals, J. (1994). Salicylic acid is not the translocated signal responsible for inducing systemic acquired resistance but is required in signal transduction. *The Plant Cell*, Vol.6, pp. 959-965

Ward, E.R.; Uknes, S.J.; Williams, S.C.; Dincher, S.S.; Wiederhold, D.L.; Alexander, D.C.; Ahi-Goy, P.; Metraux, J.P. & Ryals, J.A. (1991). Coordinate gene activity in response to agents that induce systemic acquired resistance. *The Plant Cell*, Vol.3, pp. 1085-1094

Wegulo, S.N. & Vilchez, M. (2007). Evaluation of lisianthus cultivars for resistance to *Botrytis cinerea*. *Plant Disease*, Vol.91, pp. 997-1001

Wasternack, C. & Parthier, B. (1997). Jasmonate-signalled plant gene expression. *Trends in Plant Science*, Vol.2, pp. 302-307

Williamson, B.; Duncan, G.H.; Harrison, J.G.; Harding, L.A.; Elad, Y. & Zimand, G. (1995). Effect of humidity on infection of rose petals by dry-inoculated conidia of *Botrytis cinerea*. *Mycological Research*, Vol.99, pp. 1303-1310

Xu, Y.; Linda Chang, P-F.; Liu, D.; Narasimhan, M.L.; Raghothama, K.G.; Hasegawa, P.M. & Bressan, R.A. (1994). Plant defense genes are synergistically induced by ethylene and methyl jasmonate. *The Plant Cell*, Vol.6, pp. 1077-1085

Yao, H. & Tian, S. (2005). Effects of pre- and post-harvest application of salicylic acid or methyl jasmonate on inducing disease resistance of sweet cherry fruit in storage. *Postharvest Biology and Technology*, Vol.35, pp. 253-262

Yoshikawa, M.; Yamaoka, N. & Takeuchi, Y. (1993). Elicitors: their significance and primary modes of action in the induction of plant defense reactions. *Plant Cell Physiology*, Vol.34, pp. 1163-1173

Ziadi, S.; Barbedette, S.; Godard, J.F.; Monot, C.; Le Corre, D. & Silue, D. (2001). Production of pathogenesis-related proteins in the cauliflower (*Brassica oleracea* var. *botrytis*)-downy mildew (*Peronospora parasitica*) pathosystem treated with acibenzolar-S-methyl. *Plant Pathology*, Vol.50, pp. 579-586

4

Phytobacterial Type VI Secretion System – Gene Distribution, Phylogeny, Structure and Biological Functions

Panagiotis F. Sarris[1,2], Emmanouil A. Trantas[2], Nicholas Skandalis[3],
Anastasia P. Tampakaki[4], Maria Kapanidou[1],
Michael Kokkinidis[1,5] and Nickolas J. Panopoulos[1,6]
[1]*Department of Biology, University of Crete, Heraklion*
[2]*Department of Plant Sciences, School of Agricultural Technology,*
Technological Educational Institute of Crete, Heraklion
[3]*Benaki Phytopathological Institute, Kifisia, Athens*
[4]*Department of Agricultural Biotechnology,*
Agricultural University of Athens, Athens
[5]*Institute of Molecular Biology and Biotechnology,*
Foundation for Research and Technology-Hellas, Heraklion
[6]*University of California, Berkeley, CA,*
[1,2,3,4,5]*Greece*
[6]*USA*

1. Introduction

Microbes, and their distant relatives, plants, are thought to have co-evolved during the last 2 billion years. Most of the plant-associated prokaryotes are commensals, found primarily on leaf surfaces or roots, and have no discernible or known effects on plant growth or physiology; others evolved more or less intimate relationships with plants such as N-fixing symbioses, endophytic existence or plant growth-promoting (rhizobacterial) associations; yet others, the minority, wage outright hostility with plants, inciting various diseases.

Although some phytopathogenic bacteria internalize themselves in the plant vascular system, most of them colonize plant tissues extracellularly and target plant cell wall and membrane or internal cellular structures, signaling systems and metabolic machinery from the outside. For targeting they deploy phytotoxic metabolites, hormones, polysaccharides, enzymes for the hydrolysis of cell walls and other catalytic macromolecular effectors (and, exceptionally, DNA) as "ballistic missiles". To accomplish efficient transport of macromolecules across the bacterial and/or the plant cell envelop (plant cell wall and membrane), Gram-negative bacteria possess a suite of specialized transport systems, dedicated to the transport of selected sets of proteins from the bacterial cytoplasm to the external environment or into other living cells. Type I to type VI secretion systems (abbreviated T1SS to T6SS) form channels by assembling oligomeric macromolecular complexes of varying composition and sophistication. These assemblies function as molecular machines, are broadly conserved across Gram-negative bacteria and play important roles in the virulence of pathogens. In general, most of these systems require a component providing energy to the secretion process (usually an ATPase), an outer-membrane protein, and various components involved either in scaffolding the macromolecular complex into the cell envelope or in the specific recognition of secreted substrates. It is noteworthy that certain components of these secretion machines are thought to be derived from other membrane-bound multiprotein structures serving a different purpose (see below).

2. Historical highlights

The T6SS is a relatively recent discovery, first identified as a protein secretion apparatus involved in virulence of *Vibrio cholerae* in the *Dictyostelium* (Mougous et al., 2006) and *Pseudomonas aeruginosa* in mouse models (Pukatzki et al., 2006). In several cases it has been shown to be important for bacterial virulence (host-pathogen interaction) and has attracted strong interest because it has been found via *in silico* analysis in the genomes of a large number of Gram-negative bacteria associated with human and animal diseases. While initially considered an atypical type IV secretion system (T4SS), various lines of evidence have established its identity as a distinct protein transport system (Bladergroen et al., 2003; Roest et al., 1997; Pukatski et al. 2006, Boyer et al., 2009).

An interesting twist to the T6SS story is the discovery of multiple copies of gene clusters coding for T6SS homologs in a large number of sequenced eubacterial genomes, including those of several plant-associated species (KEGG gene database; http://www.genome.jp/kegg-bin/get_htext?ko02044.keg). These species are mostly within the class of *Proteobacteria*, but also within the *Planctomycetes* and *Acidobacteria* (Tseng et al., 2009; Boyer et al., 2009). Several T6SS gene clusters are within "pathogenicity islands", for example, *P. aeruginosa*-HSI (Hcp-secretion island), enteroaggregative *Escherichia coli* (EAEC-*pheU*), *Salmonella typhimurium*-SCI (*Salmonella* **C**entrisome **I**sland), *Francisella tularensis*-FPI (*Francisella* **P**athogenicity **I**sland), *Agrobacterium tumefaciens* (Wu et al., 2008), *Pectobacterium atrosepticum* (Liu et al., 2008) and *Xanthomonas oryzae* (Tseng et al., 2009), which indicates relationship to virulence or survival in the host. Bioinformatic analysis revealed that most of the "avirulent" bacterial species studied (i.e. bacteria that have no known host) lack T6SS orthologs but active protein secretion or the ability to invade hosts await experimental testing (Shrivastava & Mande 2008; Bingle et al., 2008). Many interesting highlights are best expressed by the imaginative titles of many publications cited in this chapter.

3. T6SS genes, proteins, injectisomes

3.1 Gene content and proteins of T6SS clusters

T6SSs are typically encoded by clusters of 12 to over 20 genes, with 13 genes thought to constitute the minimal number needed to produce a functional apparatus (Boyer et al., 2009). They are found mostly in α-, β-, and γ-proteobacteria (about 25% of the sequenced genomes; Bingle et al., 2008). Recently, Chow and Mazmanian (2010) characterized a T6SS in *Helicobacter hepaticus*, which belongs to the ε-(epsilon) subgroup of proteobacteria. These clusters (frequently referred to as T6SS loci in the literature) often occur in multiple, non-orthologous copies/genome (i.e. are not the result of simple duplication), indicating that they have probably been acquired by horizontal gene transfer (Sarris at al., 2011). Detailed studies in a few bacteria further suggest that each T6SSs assumes a different role in the interactions of the harbouring organism with others. However, it is not known if there are T6SSs that can target both prokaryotes and eukaryotes. Unlike the Type III secretion systems (T3SSs), only few T6SS substrates have been identified and experimentally verified to date, but others may merely await identification.

It is now becoming increasingly clear that T6SS probably represents an evolutionary adaptation of a transmembrane protein translocation mechanism and at least some of its core components may share a common ancestor with bacteriophages. Common evolutionary ancestry and similar design are features ostensibly shared between other macromolecular transport systems and other bacterial devices that have evolved to serve entirely different biological functions (e.g. between T2SS and type IV pili, T3SS and flagella, or T4SS and conjugative pili). Indeed, the study of the macromolecular assembly process in these systems cross-feeds our understanding about structure, function and molecular mechanisms of these bacterial nanomachines.

The T6SS appears to be an injectisome, with some of its core component proteins structurally related to the cell-puncturing devices of tailed bacteriophages, and at least in some well-studied cases, have been shown capable of translocating effector proteins into the host cell cytoplasm (Bingle et al., 2008; Cascales, 2008; Filloux et al., 2008, 2011a; 2011b; Shrivastava & Mande 2008; Russell et al., 2011; Zheng et al., 2011), as is the case with the T3SS and T4SS. In the human pathogenic species *V. cholerae* and *P. aeruginosa* T6SS exports haemolysin-coregulated proteins (Hcp) and Valine-glycine repeat (Vgr) proteins; for these proteins the role of effectors associated with cytotoxicity in some *in vitro* models has been proposed (Pukatzki et al., 2006; Mougous et al., 2006). However, VgrG and Hcp display mutual dependence for secretion in *V. cholerae*, *Edwardsiella tarda* and enteroaggregative *E. coli* (Pukatzki et al., 2007; Zheng & Leung, 2007; Dudley et al., 2006), suggesting that these proteins might be not only passengers but also components of the secretion machine, a fact also supported by recent structural studies (see section 3.3). Such "dual function" could be related to distinct protein domains. In particular, the N-terminal domains of Vgr proteins show strong homology with the T4 bacteriophage base plate components gp27 and gp5 (Pukatzki et al., 2007) and a conserved core followed by a highly polymorphic C-terminal domain. The *V. cholerae* VgrG1 protein has a C-terminal domain homologous to the actin cross-linking domain (ACD) of the RtxA toxin, while VgrG1 from *Aeromonas hydrophila* possesses actin ADP-ribosylating activity (Pukatzki et al., 2007; Sheahan et al., 2004; Suarez et al., 2010). Some VgrGs ("evolved" VgrGs) from various bacterial species possess various effector-like C-terminal domains: a) a tropomyosin-like domain, which is thought to manipulate actin filaments during *Yersinia* infections, b) a pertactin-like, YadA-like, mannose-binding-like, or fibronectin-like domains or share similarities with peptidoglycan- or fibronectin-binding sequences and c) homologs of the eukaryotic lysosomal cathepsin D protein (Cascales, 2008). On the other hand, it was suggested that some VgrG orthologs may not be injected but may remain attached to the bacterial cell surface (Pukatzki et al., 2007).

Similarly to T3S systems, Hcp-secreted proteins lack N-terminal hydrophobic signal sequences, indicating secretion in a Sec- or Tat-independent manner, and a probable crossing of the bacterial cell envelope in a single step (Bingle et al., 2008; Pallen et al., 2003). Furthermore, Hcp-secreted proteins seem to have intracellular targets in eukaryotic hosts.

Thus, the Hcp protein of *A. hydrophila* was found in culture supernatants, as well as in the cytosol and the membrane of human epithelial cells after infection. Hcp secretion was independent of the T3SS and the flagellar system and the secreted protein was capable of binding to the murine macrophages from the outside, in addition to being translocated into mammalian model host cells; heterologous expression of this protein in HeLa cells increased the rate of apoptosis mediated by caspase 3 activation (Suarez et al., 2008). These findings are consistent with Hcp being secreted/translocated by T6SS, along with other yet unidentified effectors. The Hcp1 protein from pathogenic *P. aeruginosa* was also shown to be actively secreted in cystic fibrosis patients resulting in Hcp specific antibody production. Likewise, a novel T6SS protein, VasX, which is required for pathogenicity against the amoeboid host model *Dictyostelium discoideum* has recently been described. VasX is unique because it contains a putative pleckstrin homology domain which is typically only found in eukaryotic and not in bacterial proteins. VasX can bind to mammalian membrane lipids, an interaction mediated by the putative pleckstrin homology domain. It has been proposed that this domain may direct VasX to specific targets within the host cell resulting in disruption of host cell signaling (Miyata et al., 2011).

Another hallmark of T6SSs clusters is a gene coding for an AAA+ Clp-like ATPase , named 17 ClpV, belonging to a sub class of ClpB ATPases which comprise hexameric enzymes involved in protein quality control. A possible role of T6SS Clp-ATPase members might be the unfolding of substrates to be secreted, as demonstrated for the T4SS and the T3SS ATPases (Cascales, 2008). However, the *Salmonella enterica* T6SS Clp protein forms oligomeric complexes with ATP hydrolytic activity but fails to unfold aggregated proteins (Schlieker et al., 2005). A study by Bonemann et al. (2009) revealed the involvement of the ClpV protein of *V. cholerae* in remodelling supramolecular assemblies formed by two core components VipA/VipB (synonyms: ImpB/ImpC) which are crucial for T6SS secretion and virulence (see section 3.3). However, some bacterial species (e.g. *Rhizobium leguminosarum* and *Francisella tularensis*) may not contain functional Clp homologs within their T6SS clusters (Filloux et al., 2008).

Another T6SS-linked gene, *icmF* (intracellular multiplication in macrophages), has been previously studied in the context of T4SS secretion and shown to be necessary for efficient secretion. This protein carries three transmembrane domains (Sexton et al., 2004) and is partially required for *Legionella pneumophila* replication in macrophages (Purcell & Shuman 1998). Furthermore, the lack of IcmF resulted in a reduced level of another core protein, DotU, suggesting that the two proteins interact or are co-regulated. It was also shown that the lack of DotU and/or IcmF affected the stability of other core components, which suggests that DotU and IcmF assist in assembly and stability of a functional T6SS (Filloux et al., 2008). Furthermore, IcmF has been proposed to function as a further energizing component (Bonemann et al., 2009). Amino acid sequence analysis predicts that IcmF is located in the inner membrane and consists of a cytosolic and a periplasmic domain. The cytosolic domain has a conserved Walker-A motif, indicating a function as an ATPase during secretion, consistent with the finding that IcmF mutations prevent Hcp secretion (Pukatzki et al., 2006; Zheng & Leung 2007). An *icmF* mutant of avian pathogenic *E. coli* had decreased adherence to and invasion of epithelial cells, as well as decreased intra-macrophage survival and was also defective for biofilm formation on abiotic surfaces (Pace et al., 2011).

3.2 Regulation

Although T6SS gene expression inside macrophages has been demonstrated for several animal and human pathogens (e. g. *Burkholderia pseudomallei, S. enterica, V. cholerae* and *Francisella* (Shalom et al., 2007; Parsons & Heffron 2005; de Bruin et al., 2007), and to a small extent in plant pathogens, the signals triggering T6SS expression are largely unknown. For example, in *V. cholerae*, upon phagocytosis, expression of the T6SS induces cytoskeleton rearrangements through the secretion of the actin cross-linking domain of VgrG (Ma et al., 2009). Likewise, the Hcp1 of *P. aeruginosa* is induced during the infection of cystic fibrosis patients (Mougous et al., 2006), while Hcp3 is expressed upon addition of epithelial cell extracts (Chugani & Greenberg, 2007). No information concerning the expression of Hcp2 is available.

The T6SS regulation involves various transcriptional activators such as AraC, TetR-, and MarR-like proteins, σ^{54}-like factors and heat-stable nucleoid-structural (H-NS) proteins (Bernard et al., 2010). Likewise, two-component systems, like the ferric uptake regulator Fur and the Quorum sensing (QS) related regulators like LuxI, LuxR and acyl homoserine lactones (AHL) are reported to be involved in regulation of T6SS expression (Bernard et al., 2010). It has been also reported (Mougous et al., 2007) that the regulation of the HSI-1 in *P. aeruginosa* PAO1 is influenced by the sensor kinases RedS and the LadS, resulting in opposite patterns of regulation for the type III and type VI secretion systems in this bacterium, as in *S. enterica* (Parsons & Heffron 2005). T6SS gene expression in *P. syringae pv. syringae* B278a is also regulated by the two sensor kinases RetS (negatively) and LadS (positively). Two more proteins, PpkA and PppA, seem to play an important role in T6SS gene regulation (Mougous et al., 2007). PpkA is a kinase, which becomes activated by auto-phosphorylation under certain environmental conditions whereas PppA is a phosphatase, which counteracts with the action of PpkA. Both proteins act on a common protein substrate, Fha1 (*f*ork *h*ead-*a*ssociated domain protein; Mougous et al., 2007). Gene expression in HSI-2 and HSI-3 is proposed to be regulated by two σ^{54} factors which are encoded in the respective T6SS clusters of *P. aeruginosa*, as well as in their *V. cholerae* and *A. hydrophila* homologs.

Recent work reported the identification of Fur as the main regulator of the enteroaggregative *E. coli sci1* T6SS gene cluster. A detailed analysis of the promoter region showed the presence of conserved motifs, which are target of the DNA adenine methylase Dam (Brunet et al., 2011). The authors showed that the *sci1* gene cluster expression is under the control of an epigenetic methylation-dependent switch: Fur binding prevents methylation of a conserved motif, whereas methylation at this specific site decreases the affinity of Fur for its binding box. In other work (Bernard et al., 2011), several clusters were identified (including those of *V. cholerae, A. hydrophila, P. atrosepticum, P. aeruginosa, Pseudomonas syringae* pv. *tomato,* and a *Marinomonas* sp.) as having typical –24/–12 sequences, enhancer binding motifs recognized by the alternate sigma factor σ^{54} which directs the RNA polymerase to cognate promoters and requires the action of a bacterial enhancer binding protein (bEBP), which binds to *cis*-acting upstream activating sequences. The authors further showed that putative bEBPs are encoded within the T6SS gene clusters possessing σ^{54} boxes and, through *in vitro* binding and *in vivo* reporter fusion assays, they demonstrated that the expression of these clusters is dependent on both σ^{54} and bEBPs (Bernard et al., 2011).

A study by Zheng et al. (2011) provides new insights into the functional requirements of secretion as well as killing of bacterial and eukaryotic phagocytic cells by *V. cholerae* by analyzing non-polar mutations (in-frame deletions) in each gene predicted to code for *V. cholera* T6SS components. They grouped 17 proteins into four categories: twelve proteins (VipA, VipB, VCA0109–VCA0115, ClpV, VCA0119, and VasK) are essential for Hcp secretion and bacterial virulence, and thus likely function as structural components of the apparatus; two proteins (VasH and VCA0122) were thought to be regulators that are required for T6SS gene expression and virulence; another two (VCA0121 and VgrG-3) were not essential for Hcp expression, secretion or bacterial virulence, and their functions are unknown; one protein (VCA0118) was not required for Hcp expression or secretion but still played a role in both amoebae and bacterial killing and may therefore be an effector protein. ClpV was required for *Dictyostelium* virulence but was less important for killing *E. coli*. In addition, VgrG-2 which is encoded outside of the T6SS cluster was required for bacterial killing but VgrG-1 was not and several genes in the same putative operon as *vgrG*-1 and *vgrG*-2 also contributed to *Dictyostelium* virulence but had a smaller effect on *E. coli* killing.

3.3 Structure and functions of T6SS proteins and injectisomes

In contrast to other bacterial secretion systems (e.g. T3SS) there is only a small number of experimentally determined structures of T6SS core proteins or of their macromolecular assemblies. These structural studies have allowed within a relatively short time to understand important aspects of T6SS function to a considerable detail. Additional insights into structure-function relationships of T6SS have been deduced from structural/sequence similarities observed between T6SS proteins and a) components of cell puncturing devices utilized by tailed bacteriophages for DNA delivery or b) proteins from other types of bacterial secretion systems; experimental verification of the relationships derived by analogies to other systems remains largely to be delivered.

Medium to high resolution crystal structures exist for the *P. aeruginosa* Hcp1 protein, the N-terminal fragment of VgrG from the uropathogenic *E. coli* CFT073, and EvpC from *E. tarda*. Electron microscopy has been used in the study of the assemblies of various T6SS proteins including *P. aeruginosa* Hcp1, Hcp2 and Hcp3, the *E. coli* CFT073 Hcp, and *V. cholerae* VipA/VipB complex.

The structures of Hcp and VgrG provide the strongest evidence about the evolutionary relatedness between proteins of T6SS and phage tails with extended homologies existing at the levels of structure, assembly and function. The structure of Hcp1 (Mougous et al., 2006) was determined at a resolution of 1.95 Å (PDB ID 1Y12) and was found to be strikingly similar to that of the gpVN tail tube protein of phage lambda, with only minor deviations between the two structures (Pell et al., 2009). In the Hcp1 crystal, symmetry-related molecules assemble into hexameric ring which can be superimposed onto the trimeric pseudohexamer formed by the tube domains of the T4 bacteriophage gp27 trimer. Sequence analyses suggest that Hcp is evolutionary related to a further viral protein, the T4 tail tube protein gp19. Strikingly, the packing of Hcp1 hexamers in the crystals studied produces tube-like structures which are geometrically nearly identical to the T4 tail tube which is composed of stacked gp19 hexamer. The Hcp hexameric rings can also be induced to polymerize *in vitro* into stable nanotubes through the introduction of cysteine mutations capable of engaging in disulfide bridges formation across the hexamers (Ballister et al., 2008).

The crystal structure of the *E. tarda* EvpC protein, an Hcp1 homolog from the virulent protein gene cluster EVP which contains a conserved T6SS, has been recently determined at 2.8 Å resolution (PDB accession code 3EAA) and revealed a high structural similarity with Hcp1 (Jobichen et al., 2010). In solution, EvpC exists as a dimer at low concentrations and as a hexamer at high concentrations. In the crystals symmetry-related EvpC molecules form hexameric rings which stack to form tubes similar to Hcp1. Structure-based mutagenesis has revealed a critical role for EvpC secretion for three negatively charged N-terminal residues, and three positively charged C-terminal ones (Jobichen et al., 2010). This secretion impairment of EvpC decreases the virulence of the T6SS-containing pathogenic bacteria.

The structure of the N-terminal fragment (residues 1-483 out of 824) of the VgrG protein encoded by the *E. coli* CFT073 gene c3393 was determined at a resolution of 2.6 Å (PDB ID 2P5Z). The protein shows striking structural similarities (Leiman et al., 2009) to the structure of the complex (gp5)$_3$-(gp27)$_3$ of the T4 bacteriophage cell-puncturing device (PDB ID 1K28). VgrG can be described as a fusion of T4 gp27 and gp5; at the level of equivalent domains VgrG shares the highest structural homology with gp27 and exhibits only minor modifications relative to gp5. The VgrG structure comprises a domain (residues 380-470) of unknown function which is conserved in all VgrGs. This domain (DUF586) is the equivalent of the oligosaccharide/oligonucleotide-binding (OB)-fold domain of T4 gp5. The secondary structure prediction of the C-terminal part of VgrG (residues 490-820) which follows the OB-fold domain, shows repetitive β-strands (5-10 residues each) flanked by glycines. It is likely, that these strands form a β-helix that is equivalent to the triple stranded β-helix in trimeric gp5 which is involved in membrane penetration. The trimeric structure of the N-terminal VgrG fragment in the crystal, probably indicates that the complete VgrG protein may also adopt a trimeric structure that is equivalent to the (gp5)$_3$-(gp27)$_3$ complex. Sequence analyses suggest that effector domains are fused to the C-termini of many VgrG proteins ('evolved' VgrGs). The gp5 C-terminal β-helix has a 23 KDa extension of unknown function corresponding to a VgrG effector domain. Since many VgrG proteins do not contain an additional C-terminal domain, it may be concluded that different T6SS injectisomes service separate sets of effector proteins.

The structural similarities of Hcp and VgrG to components of the injection apparatus of tailed bacteriophages are highly suggestive that the two proteins might be structural components of T6SS, rather than effector proteins. The absence of 'evolved' VgrGs in many T6SSs suggests also that Hcp/VgrG might act as a conduit for T6SS effectors. The inner diameter of the Hcp tubule (40 Å) would allow the passage of proteins in an unfolded form; delivery of effectors to target cells might involve the cell-puncturing activity of VgrG.

The energy for the translocation of secreted proteins to the extracellular environment is provided by two components of the T6SS that have been introduced in section 3.1: IcmF and ClpV (Bonemann et al., 2009). IcmF is a membrane-embedded component that forms a structure spanning the inner and outer membranes. ClpV does not interact with the exoproteins Hcp and VgrG, but binds specifically with its N-domain to two cytosolic proteins, VipA (COG3516) and VipB (COG3517), that are conserved and essential components of T6SS (Bonemann et al., 2009, 2010). VipA and VipB interact with each other forming a tubular, cogweel-like structure larger than 200 kDa, with a diameter of approximately 300Å and a central channel of 100Å in diameter, and a length ranging from 25 to 500nm. Electron microscopy studies of VipA/VipB suggest that there is an overall resemblance between the VirA/VirB tubules and the T4 tail sheath structure which accommodates the viral tail tube proteins (Aksyuk et al., 2009); the diameter of the inner channel of VirA/VirB tubules is sufficient to encase Hcp tubes. In-frame deletion mutants of *vipA* and *vipB* genes could no longer secrete Hcp and VgrG proteins, although the total levels of the proteins were not affected (Bonemann et al., 2009), thus suggesting a crucial role of VirA and VirB for T6SS function. Importantly, there is no evidence for an interaction of VipA and VipB with the cytosolic proteins Hcp and VgrG. VipA/VipB cogwheel-like tubules are disassembled by ClpV; this ClpV-mediated remodelling of VipA/VipB tubules into smaller complexes (100kDa), has been suggested as an essential step in T6SS secretion, revealing an unexpected role for this ATPase component in a bacterial protein secretion system. The recent characterization of the *P. syringae* pv. *syringae* T6SS proteins ImpB (177aa) and ImpC (500aa) (homologs of VipA and VipB respectively) suggest that the two proteins form supramolecular structures of comparable size to the assemblies of VipA/VipB (M. Kokkinidis, unpublished results). These ongoing studies represent the first structural analysis of the T6SS of a plant pathogen.

4. T6SS role in host colonization and interbacterial interactions

The T6SS was initially thought to play a role primarily in bacterial pathogenicity and host colonization. Roest et al. (1997) and Bladergroen et al. (2003) characterized *Rhizobium* loci (*imp*) that hinder effective nodulation on certain plants. Imp mutants (impaired in nodulation) were deficient in the secretion of an effector protein (RbsB-like) and were mapped in a cluster of 16 genes (from pRL120462 [impN-like] to pRL120480 [vgrG-like], Fig. 1). The later authors predicted that these genes encoded a new protein secretion system, which was later named T6SS by Pukatzki et al. (2006). However, it is also found in environmental isolates and recent findings suggest that they may also function in a broader biological context: to mediate cooperative or competitive interactions between bacteria, including bacterial biofilm formation, or to promote the establishment of commensal or mutualistic relationships between bacteria and eukaryotes (Aschtgen et al., 2008; Hood et al., 2010; Jani & Cotter, 2010; Schwarz et al., 2010a; Russell et al., 2011; Zheng et al., 2011). For example, *P. aeruginosa* is capable of secreting the antibacterial factors Tse (Type VI secretion exported) (Hood et al., 2010). Tse2 is a toxic protein from *P. aeruginosa* (PA2702) and arrests growth of both prokaryotic and eukaryotic cells when expressed intracelularly. It is proposed to be an export substrate of the *P. aeruginosa* HSI-I (Hood et al., 2010). Tse2 expressing bacterial cells produce also an immunity protein, Tsi2 (PA2703), preventing cell death when co-expressed with Tse2. It is noteworthy that *tse2/tsi2* are not found in other, phylogenetically close *Pseudomonas* species (*P. entomophila* and *P. mendocina*), suggesting that it is a species-restricted regulon that responds to specific needs (Sarris & Scoulica, 2011). That the T6SS can target bacteria has also been demonstrated experimentally for *Burkholderia thailandensis* (Hood et al., 2010; Schwarz et al., 2010b) and *V. cholerae* (MacIntyre et al., 2010; Zheng et al., 2011). In *B. thailandensis* two of the five T6SSs assume specialized functions: either in the survival of the organism in a murine host (T6SS-5), or against other bacteria (T6SS-1), since strains lacking the bacterial-targeting T6SS-1 could not persist in a mixed biofilm with competing bacteria (Schwarz et al., 2010b). Miyata et al. (2010) speculate that *V. cholerae* uses its T6SS to outcompete bacterial neighbours as well as eukaryotic predators

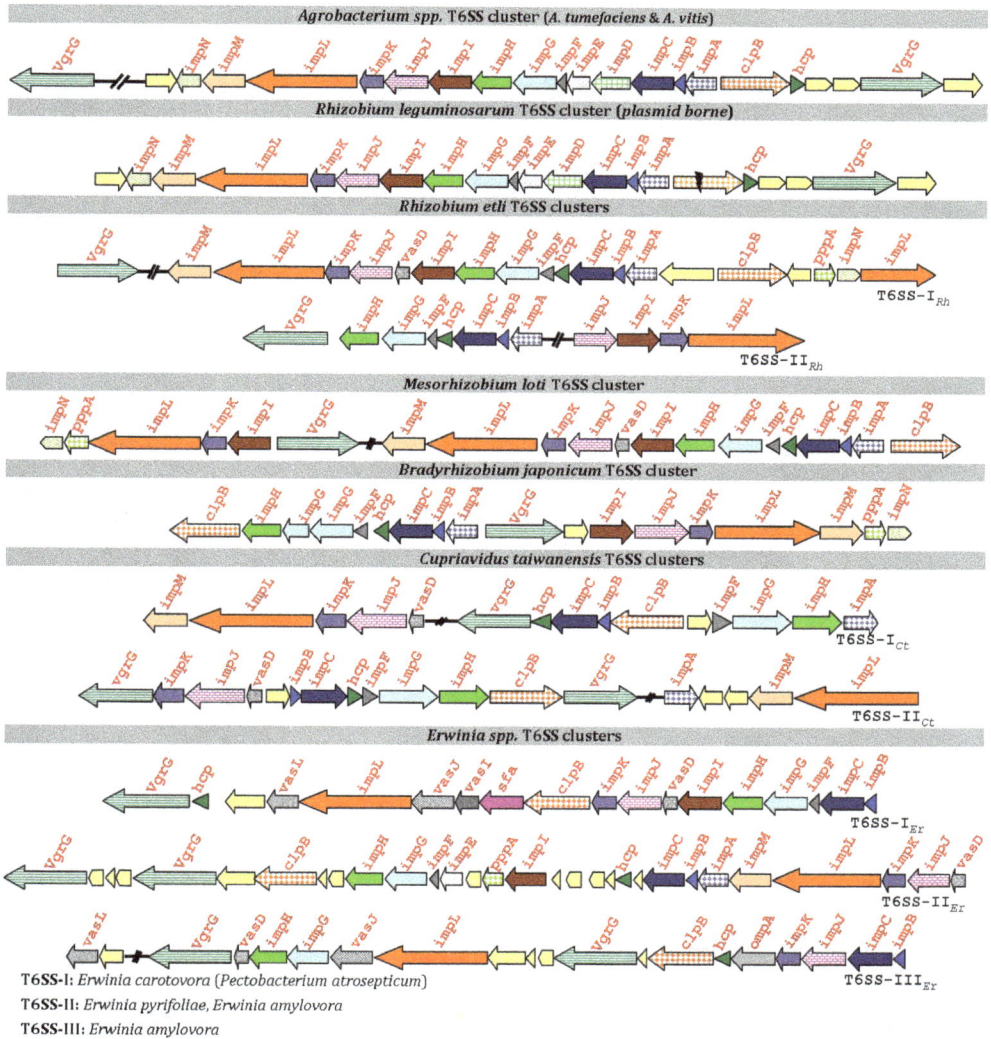

Fig. 1. *Part 1*

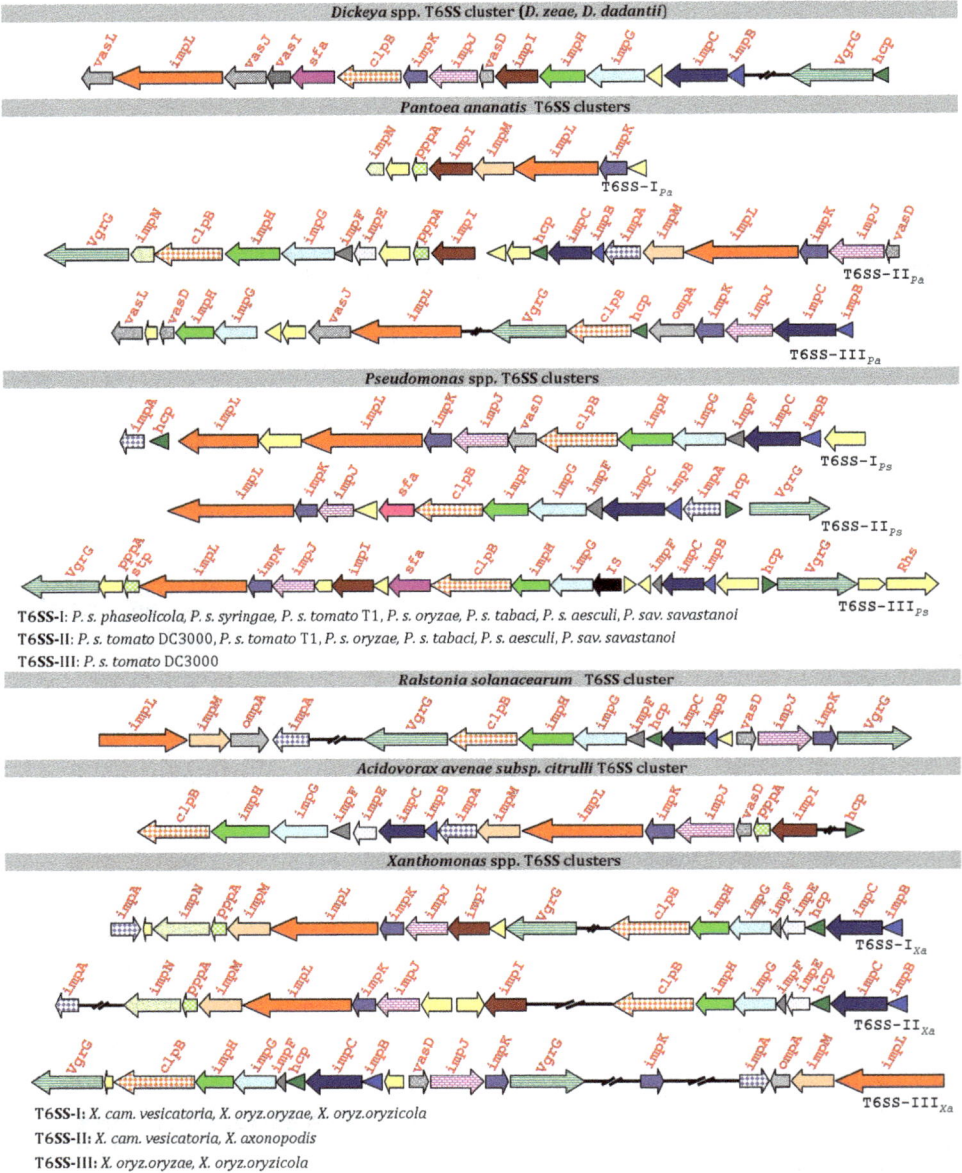

Fig. 1. *Part 2*

Fig. 1. Maps of T6SS clusters of plant-associated bacteria. Orthologs are indicated by the same color. The genes adjacent to or encoded by the T6SS gene clusters but not recognized as orthologs are indicated by light beige arrows. Arrows indicate the transcriptional direction. The gene locus numbers are referred in the text and the published or annotated gene designations are indicated above the genes of each cluster.

like amoebae and mammalian immune cells. Chow & Mazmanian (2010) further propose that pathobionts of the human gastrointestinal tract, such as H. hepaticus, may have evolved a T6SS as a mechanism to actively maintain a non-pathogenic, symbiotic relationship in the GI tract by regulating bacterial colonization and host inflammation; they hypothesize that alteration in the composition of the microbiota, known as dysbiosis, may be a critical factor in various human inflammatory disorders such as inflammatory bowel disease and colon cancer.

Knowledge on the functionalities and biological roles of T6SS in plant-associated bacteria is poor and limited to relatively few systems (reviewed in Records, 2011). With regard to phytopathogens, functionality has been demonstrated for A. tumefaciens, P. atrosepticum, and two pathovars of Pseudomonas syringae. A study with P. s. pv. tomato DC3000 (Wang et al., 2008) has shed some light on the possible role of the T6SSs in pathogenicity. Deletions of the entire T6SS clusters (T6SS-II or T6SS-III; grouping is based on our phylogenetic analysis, see sections 5, and 6) or both copies of icmF (icmF1 and icmF2) caused reduction of bacterial population in Nicotiana benthamiana and milder symptoms on tomato leaves. When either the T6SS-II or T6SS-III cluster was deleted, both symptoms severity and bacterial populations were reduced. However, vgrG1 or vgrG2 deletions had no effect on disease development on tomato or on N. benthamiana. However, an insertional mutant of the clpV/B gene of the T6SS maintained the ability for in planta multiplication and produced disease symptoms similar to those caused by wild-type strain (Records & Gross 2010). RNA transcripts of the icmF homologs of the P. syringae pv. tomato DC3001 and P. syringae pv. phaseolicola 1448a were detected by RT-PCR in both rich and minimal media, indicating that the gene is probably expressed in both strains (Sarris et al., 2010). Microarray analysis showed that the A. tumefaciens T6SS is induced by mildly acidic conditions, such as encountered in plant tissues and in the rhizosphere (Yuan et al., 2008) and deletion of hcp resulted in reduced tumorigenesis on potato tuber slices (Wu et al., 2008). The P. atrosepticum T6SS is induced by potato tuber extracts (Mattinen et al., 2007). Transcriptome profiling (Liu et al., 2008) also indicated regulation of the T6SS of P. atrosepticum by quorum sensing, as deletion of expI, a gene responsible for N-(3-oxohexanoyl)-L-homoserine lactone synthesis. Furthermore, deletions in either ECA3438 (impJ) or ECA3444 (vipB) resulted in slightly reduced virulence in potato stems and tubers. However, mutation of the ECA3432 (icmF) gene resulted in increased potato tuber maceration, indicating that the T6SS may be involved in antipathogenesis activity (Yuan et al., 2008). Whether T6SS mechanisms engage other aspects of P. syringae-host/vector biology, antagonism and predation in the plant or other micro-environments remain open questions. Among the symbiotic N-fixing bacteria, extended symbiosis phenotypes of certain rhizobia have been linked to a T6SS. The presence of T6SS homologs in the sequenced genomes of many rhizobia presents opportunities to further investigate it's role in bacteria-plant symbiosis (Fauvart & Michiels, 2008).

5. Mining phytobacterial for T6SS homologs

Further to our recently published study (Sarris et al., 2010), a genome-wide in silico analysis was carried out for 13 phytobacterial species and more than 30 strains from different genera, to identify conserved gene clusters encoding for T6SSs by BLASTP and reverse BLAST analysis of sequences deposited in various genome databases (e.g. KEGG, NCBI, RizoBase), both complete annotated and draft sequenced phytobacterial genomes. The baits consisted of protein sequences encoded by the P. syringae Hcp secretion islands I, II and III (HSI-I, II,

III, here referred to interchangeably as T6SS-I, T6SS-II and T6SS-III, respectively) (Sarris et al., 2010), as well as their homologs from other known T6SS clusters (Boyer et al., 2009). Clusters containing at least five genes encoding proteins with similarity to known T6SS core proteins were considered as part of a putative T6SS locus. The genomic regions thus identified were then extended by examining four kilo-bases up- and down-stream for putative conserved genes associated with T6SS by "orthologue" and "paralogue finder" analysis against all the KEGG deposited genomes. Maps of the genomic islands were constructed manually in PowerPoint Microsoft office software. For sequence alignment and phylogenetic tree construction, the conserved proteins ImpL, ImpG, IpmC and ImpH from all phytobacterial species deposited at the NCBI and KEGG databases were edited with the DNAman computer package (Lynnon Co) and were included for sequence alignment and tree construction. Phylogenetic relations were inferred using the neighbour-joining method (Saitou & Nei 1987) offered in MEGA4 software (Tamura et al., 2007). In Table 1 and Fig. 1 the clusters are identified by the organsim's initials, and in the phylogenetic trees the organism's name/strain number and the T6SS groupings used in the text and figure legends are given.

6. Phylogenetic analysis of phytobacterial T6SS

6.1 Phylogenies based on overall gene content

Table 1 shows the presence/absence of a T6SS protein homolog (indicated with the plus sign [+] and minus [-] sign, respectively). Homologs of each T6SS-related bait protein exist in members of all phytobacterial species/strains, but with substantial differences among strains in gene content and copy number. Nine core genes, *impK, impB, impC, impG impH, impJ, hcp, impL* and *clpB*, are found in 100% of the species examined, with *clpB* seemingly present as a pseudo-gene in *Rhizobium leguminosarum)*. On the other hand, homologs of genes such as *ompA* and *impD* are present only in 14% of the species/strains examined, a finding of unclear significance. Several instances of multiple T6SS clusters were identified in the same strain, mostly in distant locations with respect to each other. These clusters are depicted in Fig. 1, based on the phylogenetic analysis described in the sections below.

To analyse the evolutionary history of phytobacterial T6SS, a distance tree was initially constructed, depicting the phylogenetic relationships and gene composition among the T6SSs of all species examined based on the data in Table 1. Initially, the evolutionary history was inferred using the Neighbor-Joining method (Saitou & Nei 1987) by scoring each locus as present (+) or absent (-), including all the data (Fig. 2). The constructed tree essentially gives a graphic representation of the data in Table 1 and reveals two distinct T6SS clusters, in terms of the presence or absence of the genetic elements. The tree shows extensive intermixing among the T6SSs of various bacterial species/groups, with several deep and shallow branches, many with low bootstrap values. Noteworthy in this tree is the close relationship between: *a.* the *Xanthomonas* T6SS-I and II and rhizobial T6SSs (except *Rhizobium leguminosarum*), *b.* the two phytopathogenic enterobacteria T6SS-II (*Erwinia amylovora,* and *Erwinia pyrifoliae*) and the *Acidovorax* T6SS, *c.* the sole T6SS of the *Agrobacteria* spp., *Rh. leguminosarum* and *Xanthomonas* spp. T6SS-II *d.* the *Ralstonia solanacearum* and *Xanthomonas* spp. T6SS-III and *e.* between other phypathogenic enterobacterial T6SSs (*E. amylovora* and *P. ananatis* T6SS-III, the *P. atrosepticum* T6SS-I and the T6SS of the two *Dickeya* species).

Gene	COG	%	AAsC	AT	AV	BJ	PA_I***	EA_II	EA_III	EP_I	ML	PS_I	PS_II	PS_III	RS	RE	RL	Xsp_I	Xsp_II	Xsp_III	CT_I	CT_II	DZ	DD	PA_II
Imp N	COG 0515	37	-	+	+	+	-	-	-	+	-	-	-	-	+	+	+	+	-	-	-	-	-	+	-
Imp D	COG 3604	14	-	+	+	-	-	-	-	-	-	-	-	-	-	+	-	-	-	-	-	-	-	-	-
Omp A	—	14	-	-	-	-	-	-	+	-	-	-	-	-	+	-	-	-	+	-	-	-	-	-	+
Sfa	COG 3604	23	-	-	-	-	+	-	-	-	-	-	+	+	-	-	-	-	-	-	-	-	+	+	-
Vas J	COG 3515	18	-	-	-	-	+	-	+	-	-	-	-	-	-	-	-	-	-	-	-	+	+	-	+
Vas L	COG 3515	18	-	-	-	-	+	-	+	-	-	-	-	-	-	-	-	-	-	-	-	+	+	-	+
Imp E	COG 4455	36	+	+	+	-	-	+	-	+	-	-	-	-	-	+	-	+	-	+	-	-	-	+	-
Ppp A	COG 0631	41	+	-	-	+	-	+	-	+	+	-	-	+	-	+	-	+	+	-	-	-	-	+	-
Imp I	COG 3456	50	+	+	+	+	+	+	-	+	++	-	-	-	-	+	+	+	+	-	-	-	+	+	+
Vas D	COG 3521	59	+	-	-	-	-	+	+	+	+	+	-	-	+	+	-	-	-	+	+	+	+	+	+
Imp M	COG 3913	64	+	+	+	+	-	+	-	+	-	-	-	-	+	+	+	+	+	+	+	-	-	+	-
Imp F	COG 3518	86	+	+	+	+	+	+	-	+	+	+	+	+	+	+	+	+	+	+	+	-	-	+	-
Imp A	COG 3515	77	+	+	+	+	-	+	-	+	+	+	+	-	+	+	+	+	+	+	+	-	-	+	-
Vgr G	COG 3501	86	-	++	+	+	+	++	+	++	+	-	+	++	++	+	+	+	-	++	+	++	+	+	+
Clp B	COG 0542	96	+	+	+	+	+	+	+	+	+	+	+	+	-	+	+	+	+	+	+	+	+	+	+
Hcp	COG 3157	100	+	+	+	+	+	+	+	+	+	+	+	+	+	+	+	+	+	+	+	+	+	+	+
Imp J	COG 3522	100	+	+	+	+	+	+	+	+	+	+	+	+	+	+	+	+	+	+	+	+	+	+	+
Imp H	COG 3520	100	+	+	+	+	+	+	+	+	+	+	+	+	+	+	+	+	+	+	+	+	+	+	+
Imp G	COG 3519	100	+	+	+	++	+	+	+	+	+	+	+	+	+	+	+	+	+	+	+	+	+	+	+
Imp C	COG 3517	100	+	+	+	+	+	+	+	+	+	+	+	+	+	+	+	+	+	+	+	+	+	+	+
Imp B	COG 3516	100	+	+	+	+	+	+	+	+	+	+	+	+	+	+	+	+	+	+	+	+	+	+	+
Imp L	COG3 523	100	+	+	+	+	+	+	+	+	+	+	+	+	++	+	+	+	+	+	+	+	+	+	+
Imp K	COG 3455	100**	+	+	+	+	+	+	+	++	+	+	+	+	+	+	++	+	+	+	+	+	+	+	+

Table 1. Gene content of the putative T6SS genes of phytobacterial species, indicating locus names and COG numbers (not available for OmpA). (+): present, (++): present in a second copy, (-): missing. *Species abbreviations: AAsC: *Acidovorax avenae* subsp. *citrulli*; AT: *Agrobacterium tumefaciens*; AV: *Agrobacterium vitis*; BJ: *Bradyrhizobium japonicum*; PA: *Pectobacterium atrocepticum*; EA: *Erwinia amylovora*; EP: *Erwinia pyrifoliae*; ML: *Mesorhizobium loti*; PS: *Pseudomonas syringae*; RS: *Ralstonia solanacearum*; RE: *Rhizobium etli*; RL: *Rhizobium leguminosarum*; Xsp: *Xanthomonas* spp.; CT: *Cupriavidus taiwanensis*; DZ: *Dickeya zeae*; DD: *Dickeya dadantii*; PA: *Pantoea ananatis*. **The numbers in the second row of the table denote the percentage of cases where the gene/protein is present among the species/strains examined. ***Latin numerals I, II, III denote T6SS-I, T6SS-II and T6SS-III, respectively. A table of the locus numbers from the KEGG database is available upon request from P.F. Sarris.

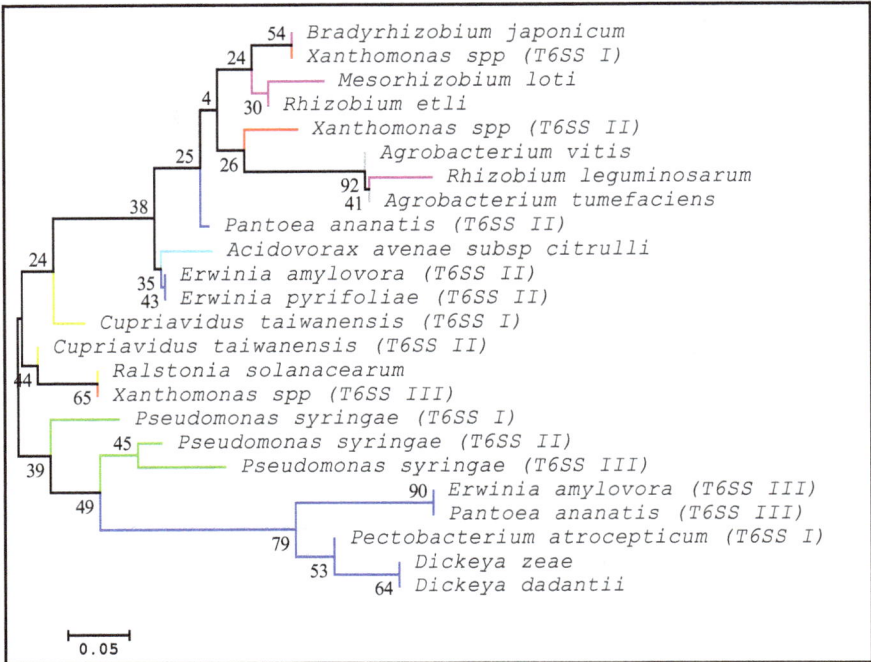

Fig. 2. Distance tree of T6SS of various plant pathogenic bacteria; constructed with data of Table 1, through a matrix where each gene locus was scored as (+) when present or as (-) when not present. The evolutionary history was inferred using the Neighbor-Joining method (Saitou & Nei, 1987). The bootstrap consensus tree inferred from 5000 replicates (Felsenstein, 1985) is taken to represent the evolutionary history of the species analysed (Felsenstein 1985). Branches corresponding to partitions reproduced in less than 50% bootstrap replicates are collapsed. The percentage of replicate trees in which the associated taxa clustered together in the bootstrap test (5000 replicates) is shown next to the branches (Felsenstein, 1985). The tree is drawn to scale, with branch lengths in the same units as those of the evolutionary distances used to infer the phylogenetic tree. The evolutionary distances were computed using the Poisson correction method (Zuckerkandl & Pauling 1965) and are in the units of the number of amino acid substitutions per site. All positions containing gaps and missing data were eliminated from the dataset (Complete deletion option). There were a total of 29 positions in the final dataset. Phylogenetic analyses were conducted with MEGA4 (Tamura et al., 2007).

6.2 Phylogenetic analysis of four core proteins

Subsequently, we carried out a phylogenetic cluster analysis of four highly conserved T6SS core proteins (ImpC, ImpG, IpmH and ImpL-like) by combining the protein sequences and forcing the software to reckon phylogenetic analysis of all four proteins. Data clustering was accomplished by the Neighbor-Joining method (Saitou and Nei, 1987) and are presented as a tree for the evaluation of similarity/distances among each T6SS in each species (Fig, 3) and is used as a basis to infer the T6SS phylogenies discussed below.

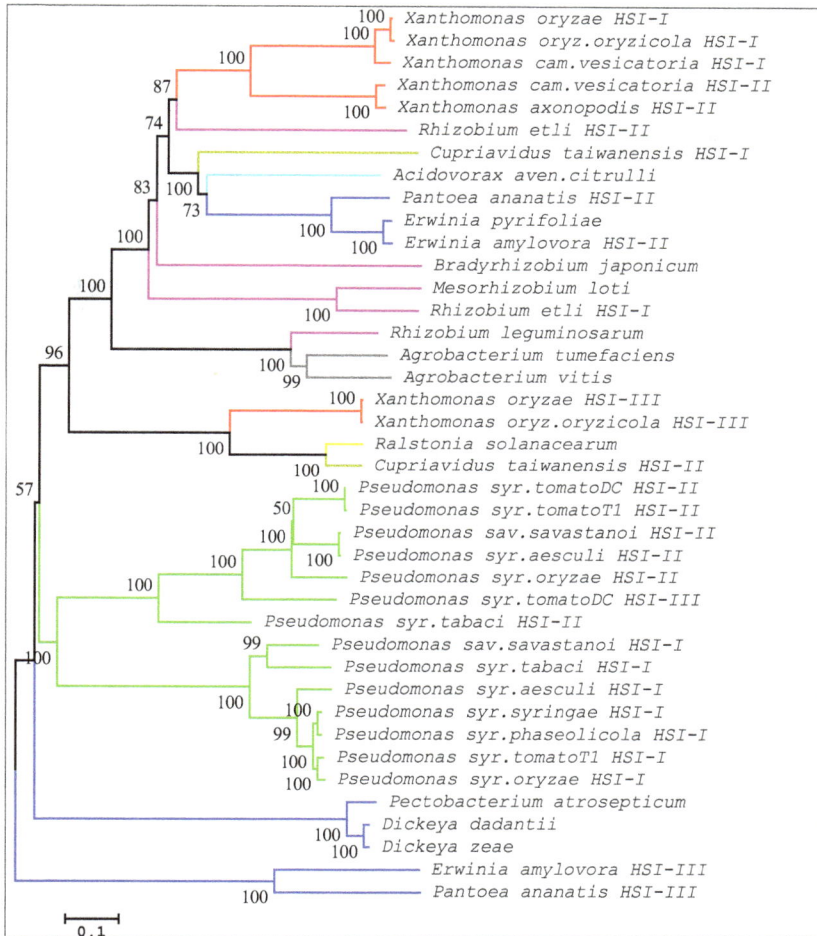

Fig. 3. Phylogenetic clustering of the plant-associated bacterial species based on the sequence of four T6SS core proteins (ImpC ImpG, ImpH and ImpL). The evolutionary history was inferred using the Neighbor-Joining method (Saitou & Nei 1987). The bootstrap consensus tree inferred from 5000 replicates (Felsenstein, 1985) is taken to represent the evolutionary history of the proteins analysed replicates (Felsenstein 1985). For other details see Fig. 2 legend. There were a total of 1729 positions in the final dataset. Phylogenetic analyses were conducted with MEGA4 (Tamura et al., 2007). Difference in tree branch colors indicates the different bacterial species.

The consensus phylogenetic tree obtained shows four deep branches. One branch hosts the majority of the phytobacterial T6SSs, except for those of *P. syringae* pathovars (T6SSs-I, II and III), the *Dickeya* spp. T6SS and *P. carotovorum* T6SS. In this branch several T6SS groups are evident. One group includes the *Xanthomonas* spp. T6SS-I and T6SS-II together with the T6SS of *Rhizobium etli* T6SS-II. A second group includes the *C. taiwanensis* T6SS-I, *A. avenae* subsp. *citrulli*, *P. ananatis* T6SS-II and the *Erwinia* spp. T6SS-II. The *B. japonicum* T6SS appears distinct (with bootstrap value 83) but very close to a subgroup formed by the *M. loti* and *R. etli* T6SS-II. Another group in this phylogenetic branch includes the *R. leguminosarum* and *Agrobacterium* spp. T6SSs. Finally, in the same branch are the *Xanthomonas* spp. T6SS-III, grouped together with the *R. solanacearum* and *C. taiwanensis* T6SS-II. In the second phylogenetic branch are grouped only the *P. syringae* pathovars. As previously reported (Sarris et al., 2010), in this group there are two distinct sub-groups. The first one includes the *P. syringae* T6SS-II and T6SS-III, while the second carries only the *P. syringae* T6SS-I. Finally, two more branches were formed but without bootstrap value. The first one includes the T6SSs of *Dickeya* spp. and *P. atrosepticum*, while the second consists of the *P. ananatis* T6SS-II and *Erwinia* spp. T6SS-III.The phylogenetic relationships of the *P. syringae* T6SSs were further examined by constructing an additional phylogenetic tree of the four core proteins (ImpC, ImpG, IpmH and ImpL-like) by including representatives of fully sequenced non-phytopathogenic fluorescent *Pseudomonas* and the *R. solanacearum* T6SS, *C. taiwanensis* T6SS-II, *P. ananatis* T6SS-II and *Erwinia* spp. T6SS-III, as these appear as reference species based on their distant relationships in the tree of Fig. 3. The *Pseudomonas* tree shows three deep branches (not including out-group species), each including species with high boostrap values (Fig. 4). The first branch includes a group which is formed by the *P. syringae* T6SS-II

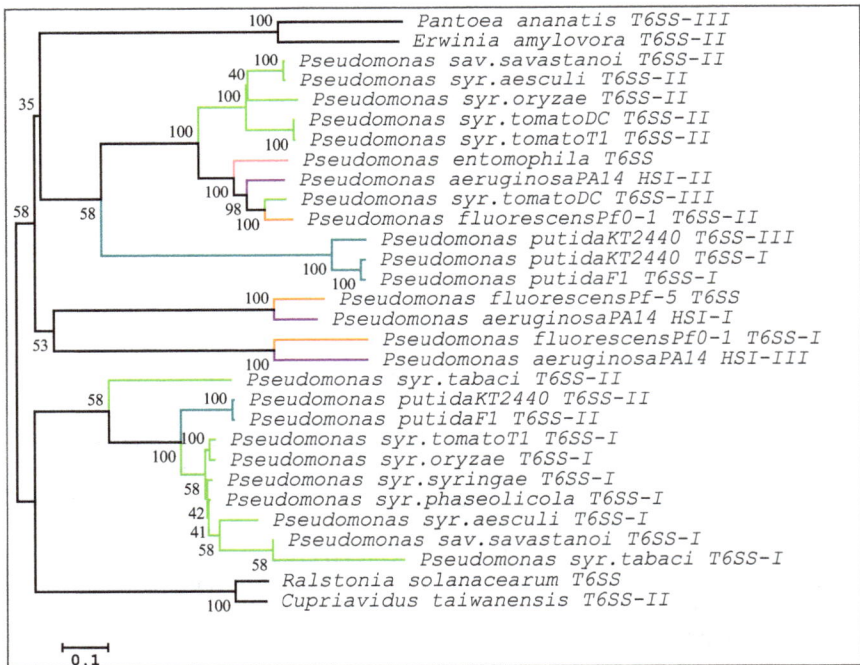

Fig. 4. T6SS evolutionary relationships of 30 fluorescent *Pseudomonas* T6SSs. The evolutionary history was inferred using four (4) T6SS core proteins (ImpC, ImpG, ImpH, ImpL), by the Neighbor-Joining method (Saitou & Nei, 1987). The optimal tree with the sum of branch length = 6.49662029 is shown. For other details see Fig. 2 and Fig. 3 legends). Difference in tree branch colors indicates the different *Pseudomonas* species, while the out-group species are presented with black.

and T6SS-III, the *P. aeruginosa* HSI-II, *P. fluorescens* T6SS-II and the *P. entomophila* four T6SS core proteins. In this group, the *P. syringae* T6SS-III is phylogenetically very close to the *P. aeruginosa* HSI-II and *P. fluorescens* T6SS-II. The *P. syringe* T6SS-III is present only in *P. syringae* pv. *tomato* strain DC3000, and has the same gene order and high protein homology with the T6SS-II of the *P. aeruginosa* PA14 HS-II (data not shown), which reinforces the view of Yan and colleagues (Yan et al., 2008) that the model plant pathogen *P. s.* pv. *tomato* DC3000 is a very atypical tomato strain. Thus, it appears that the *P. syringae* T6SS-III may have been horizontally acquired and maintained through vertical transfer and is remarkably well conserved in *P. s.* pv. *tomato* DC3000, *P. aeruginosa* and *P. fluorescens* species with distant genetic relatedness and very distinct, opportunistic relationships with plants. A second distinct group in the same branch is formed by the T6SSs –I and –III of *P. putida*. The second branch in the four-protein tree is formed by the *P. fluorescens* T6SS-I and the *P. aeruginosa* HSI-I and –III. Finally, the third branch of the four protein tree includes the *P. syringae* T6SS-I grouped with the *P. putida* T6SS-II. Remarkably, in the two *P. fluorescens* strains analysed the sole T6SS in Pf-5 and the T6SS-I and T6SS-II in Pf0-1 are quite distant. Two of the three T6SSs found in the *P. putida* strains studied (T6SS-I in Pf-5 and PKT2440, and T6SS-III in the later strain) are very close relatives while the third one (T6SS-II) is very distant from the others (Fig. 4).

7. T6SS in phytopathogenic bacterial species

7.1 *Pseudomonas* spp. (*P. syringae* and *P. savastanoi*)

The species *P. syringae* (*P.s.*; *γ-proteobacteria*) comprises phytopathogenic members that are placed in infra-sub-specific groupings (pathovars); collectively, they infect a wide range of plant species and can also live as epiphytes in the plant phyllosphere, until conditions are favourable for disease development (Hirano & Upper, 2000). Some non-pathogenic strains of *P. syringae* have been used as biocontrol agents against post-harvest rots (Janisiewicz & Marchi, 1992).

Bioinformatics analysis (Sarris et al., 2010) revealed the presence of multiple T6SS clusters and putative effectors in six fully or partially sequenced genomes of *P.s.* pathovars. In a subsequent study, two more draft genomes of the bleeding cankers pathogen of the horse chestnut trees (*Aesculus hippocastanum*), *P. syringae* pv. *aesculi* (*Pae*), became available in the NCBI gene database (Green et al., 2010). The two strains show genomic differences implicated in host association and fitness: one strain was isolated recently from bleeding stem cankers on European horse chestnut in Britain (E-*Pae*), and the second (I-*Pae*) from leaf spots on Indian horse chestnut in India in 1969 (Green et al., 2010).

Genome comparisons between the sequence assemblies of E-*Pae* and I-*Pae* revealed differences in a number of genomic regions including two T6SS clusters and a large number of putative effectors that are present in I-*Pae* but absent from E-*Pae*. In the phylogenetic trees of Figs 3 and 4, the two *Pae* T6SS clusters group together with the *P. syringae* T6SS-I & II respectively. Analysis of the draft genome sequence of a close relative, *P. savastanoi* pv. *savastanoi* strain NCPPB 3335 (Rodriguez-Palenzuela et al., 2010) indicated also the presence of two putative T6SS gene clusters:, one (AER-0002618 to AER-0002633) highly similar to the *P. syringae* T6SS-I and another (AER-0002971 to AER-0002983) that is more similar in gene order and protein sequence similarity to the T6SS-II (Figs. 3 and 4).

7.2 *Ralstonia solanacearum*

R. solanacearum (*R.s.*, previously known as *Pseudomonas solanacearum*; *β-proteobacteria*), the causative agent of bacterial wilt of solanaceous plants, is a soil-resident bacterium with wide geographical distribution in all warm and tropical areas and the causal agent of vascular wilt disease in over 200 annual and perennial plant species, representing over 50 botanical families, both monocots and dicots. It also causes disease on the model plant *Arabidopsis thaliana*, and along with *P. syringe* pv. *tomato* DC3000, is one of the leading models in the study of plant-microbe interactions.

The genome of *R.s.* strain GMI1000, as was annotated in KEGG gene data base has one T6SS cluster comprising 16 genes, lying between and *impL*-like (RS01945; also designated as RSp0763) and a *vgrG*-like RS01970, also designated as RSp0738). This cluster seems to be interrupted by some apparently T6SS-unrelated putative open reading frames (11 ORFs) indicating possible gene rearrangements. The gene clusters RS01945-46-47 (also designated as RSp0763-62-61) and RS01967-68-69-70 (also designated as RSp0741-40-39-38) have been annotated as putative *impL-impM-ompA* and *vasD-impJ-impK-vgr* respectively, and are found in reverse orientation compared to the middle section of the cluster, which include genes annotated as *vgr-clpB-impH-impG-impF-hcp-impC-impB* [from RS01949 to RS01966 (also designated as RSp0743 to RSp0749)]. Thus far, there is no experimental evidence for a role of the T6SS in the *R.s.* pathogenicity.

The consensus phylogenetic analysis tree of four *R.s.* T6SS core proteins revealed the closest phylogenetic proximity of the *R.s.* T6SS core components to those of the T6SS-II of *Cupriavidus taiwanensis* which belongs to the phylogenetically distant β-Rhizobium group (Fig. 3), with the same gene order, and the next closest relatives being those of the *Xanthomonas* T6SS-III and the *Ralstonia* T6SS studied (also β-proteobacteria), indicating very recent common ancestry.

7.3 *Xanthomonas* spp.

The genus *Xanthomonas* (γ-*Proteobacteria*) consists of 27 plant-associated species, each with many pathovars, which collectively, cause disease on at least 124 monocot species and 268 dicot species, including fruit and nut trees, solanaceous and brassicaceous plants, and cereals (Hayward, 1993; Leyns et al., 1984). *X. campestris* pv. *vesicatoria* (Doidge; syn. *Xanthomonas vesicatoria*) has tomato and pepper as principal hosts. However, various other *Solanaceae*, mainly weeds, have been recorded as incidental hosts. *X. axonopodis* pv. *citri* (*X.a.c.*) is the causal agent of "citrus canker", which affects most commercial citrus cultivars, resulting in significant losses worldwide. (Stall & Seymour, 1983). *X. oryzae* includes the two non-European rice pathogens pvs. *oryzae* and *oryzicola* (*X.o.* & *X.o.o.*). The principal host of both pathovars is rice and high-yielding cultivars are often highly susceptible. *Oryza sativa* subsp. *japonica* is usually more resistant than subsp. *indica* to pv. *oryzicola*. *X.o.* invades the vascular tissue, while *X.o.o.* proliferates in the parenchyma (Nino-Liu et al., 2006).

Experimental evidence for a role of T6SS in *Xanthomonas* pathogenicity or other aspects of its life cycle is limited. Our data base search revealed a variable situation vis-à-vis T6SS homologs among several sequenced strains of *Xanthomonas*; All three *X. campestris* pv. *campestris* (*X.c.c.*) strains (ATCC33913, Xcc8004 and B100) and the avirulent *X. populi* and *X. codiaii* seem to lack T6SS-related genes, while two distinct T6SS loci are found in *X.*

campesrtis pv. *vesicatoria* XCV2120 (*X.c.v.*), one in *X. axonopodis* 4122 (*X.a.c.*) and three in each, *X.o* and *X.o.o.* One T6SS cluster in XCV2120 extends from XCV2120 (*impB*-like) to XCV2143 (*impA*-like) and is referred to here as T6SS-I, while a second cluster is located from XCV4202 (*impA*-like) to XCV4243 (*impB*-like) and referred as T6SS-II (Fig. 1). This cluster contains two interruptions by some apparently T6SS-unrelated putative ORFs, one, between XCV4214 (*impI*-like) and XCV4236 (*clpB*-like), and another between XCV4202 (*impA*-like) and XCV4208 (*impM*-like). Two T6SS clusters are found in *X.c.v.*: one of them is phylogenetically closer to the T6SS-I found in *X.o.* and *X.o.o* while the second seems to more closely related to the sole T6SS locus of *X.a.c* (referred to as T6SS-II in Fig. 3) which spans from XAC4112 (*impA*-like) to XAC4147 (*impB*-like) (Fig. 1). The *X.a.c.* T6SS locus is almost collinear for the T6SS related genes and for the two T6SS-unrelated putative ORF insertions with the T6SS of *X.c.v.* Whole genome comparisons (Potnis et al., 2011) recently enabled an extensive analysis of the presence and distribution of T6SS among *Xanthomonas* strains representing 15 pathovars. *X.o.* and *X.o.o.* carry the T6SS-I, located from XOO3034 (*impA*-like) to XOO3052 (*impB*-like) and Xoryp_12330 (*impA*-like) to Xoryp_12445 (*impB*-like), and a second T6SS which is referred to as T6SS-III in Fig. 1 and is located between XOO3517 (*impL*-like) and XOO3474 (*vgrG*-like) in *X.o.* and Xoryp_06365 (*impL*-like) and Xoryp_06645 (*vgrG*-like) in *X.o.o.* This locus contains a large number of T6SS unrelated ORF insertions, and is phylogenetically distant from the *Xanthomonas* T6SS-I and T6SS-II, while is phylogenetically close to the *R. solanacearum* and the *C. taiwanensis* T6SS-II.

7.4 *Erwinia* spp. (*E. pyrifoliae* and *E. amylovora*)

Bacteria of the genus *Erwinia* (γ-proteobacteria) are plant-associated as pathogens, saprophytes and epiphytes and exhibit considerable heterogeneity, forming four phylogenetic clades that are intermixed with members of other genera, such as *E. coli*, *Klebsiella pneumoniae*, and *Serratia marcescens* (Kwon et al., 1997). *E. amylovora* (*E.a*) originated in North America and has spread to other continents, threatening the native apple germplasm in Central Asia. *E. pyrifoliae* (*E.p.*) is a newly described necrotrophic pathogen initially isolated in the middle 1990's from Japanese pear (*Pyrus pyrifolia*), but could not be found in later years in the previously affected orchards (Kim et al., 2001). *E.p.* causes fire blight symptoms essentially indistinguishable from those of *E.a.* infection (Kim et al., 2001) but has more limited host range. Both pathogens share many common virulence factors, including two distinct type III secretion systems (T3SS) and genes for desferrioxamine biosynthesis. However, *E.p* lacks a third T3SS cluster found in *E. amylovora*.

The genome sequence of the highly virulent strain Ea273 (ATCC 49946) isolated from diseased apple (*Malus pumila* cv. Rhode Island Greening) in New York State, is over 99.99% identical to that of the European isolate CFBP 1430 (ERWAC), indicating minimal divergence since the global dispersion of *E.a.* However, a large-scale rearrangement of the genome resulting in repositioning of two large portions of the chromosome has taken place. Additionally, ERWAC has only the smaller plasmid (pEA29) found in *E.a.* and its two T3SSs on genomic islands PAI-2 and PAI-3 have high homology to the insect endosymbiont *Sodalis glossidinius* str. *morsitans* (SODGM) and to the mammalian pathogens *Salmonella* and *Yersinia* spp., indirectly suggesting a closer insect association than mere passive dispersal (Smits et al., 2010).

E.p. strain Ep1/96 harbors one T6SS cluster which spans the region from EpC_06150 (*vasD*-like) to EpC_06440 (*vgrG2*-like) and shares sequence and gene order homology with the one of the *E.a* T6SS clusters [designated T6SS-II in Fig. 1, 2 and 3), starting from EAMY_3027 (*vasD*-like) to EAMY_3000 (*vgrG*-like)]. These clusters are very close phylogenetic relatives and have as next closest relative the *Pantoea ananatis* T6SS-II (Fig. 3). Furthermore, a small T6SS cluster of four ORFs (not presented in Fig. 1) is present in both *E.p.* and *E.a.* (*E.p.*:EpC_19520-EpC_19550 and *E.a.*: EAMY_1620-EAMY_1623). *E.a.* also harbors a second T6SS cluster (designated T6SS-III in Figs. 1, 2 and 3) which spans from EAMY_3228 (*impB*-like) to EAMY_3201 (*vasL*-like). This cluster exhibits gene order and sequence relatedness to *P. ananatis* T6SS-III (Fig. 1). Similar results were reported for *E.p.* DSM 12163T and *E.a* CFBP 1430 based on whole genome sequence analysis (Smits et al., 2010). There are no experimental data concerning the biological role of the T6SS in *Erwinia* spp. and most genes within the T6SS clusters are uncharacterized.

7.5 *Pantoea ananatis*

The genus *Pantoea* (γ-proteobacteria) consists of both important plant pathogens and clinically relevant species. Clinical isolates have been reported to cause bacteraemia in humans. *P. ananatis* (*P.a.*) is considered an unconventional plant associated species, being associated with plants as an epiphyte, endophyte, pathogen, or symbiont, but can also occupy unusual ecological niches (e.g. contaminating aviation jet fuel tanks). It's ice nucleation activity has been exploited in the food industry and in the biological control of insects (Coutinho & Venter, 2009).

The exact role of T6SS in the adaptive capacity of *P.a.* is not known, but such a role might be considered likely as there has been no evidence of T2SS or T3SS in any of the strains sequenced so far (De Maayer et al., 2010). Our database search in the sequenced *P.a.* strain LMG 20103 revealed one truncated and two entire T6SS clusters named T6SS-I, II and III respectively in Figs 1 and 3. The truncated T6SS-I is located from PANA_1650 (a possible T6SS related protein kinase A; PknA=ImpN) to PANA_1656 (*impK*-like). The obvious absence of some basic T6SS core genes raises questions about the functionality of this T6SS locus. The T6SS-II locus is embedded between PANA_2352 (*vgrG*-like) to PANA_2372 (*vasD*-like) while the T6SS-III locus is divided in two sub-loci the first one is located from PANA_4130 (*vasL*-like) to PANA_4138 (*impL*-like), the second one, being located a few apparently T6SS-unrelated putative ORFs away, from PANA_4144 (*vgrG*-like) to PANA_4151 (*impB*-like). Phylogenetically, the *P.a.* T6SS-I seems to be related to some random distributed T6SS related ORFs in the genome of *E. amylovora* (data not shown). The *P.a.* T6SS-II is phylogenetically closer in protein sequence and the gene order to the *Erwinia* spp. T6SS-II (Figs 1 and 3). Interestingly, the *P.a.* T6SS-III forms a distant phylogenetic branch with the *E.a.* T6SS-III and appears as an out-group in our analysis (Fig. 3).

7.6 *Pectobacterium atrosepticum*

P. atrosepticum (*P.a.*; formerly *Erwinia carotovora* subsp. *atroseptica;* γ-proteobacteria) is a member of the pectolytic *Erwinia* responsible for the soft rot and blackleg of potato stems and tubers. A T6SS locus is found in *P.a.* strain SCRI1043 extending from ECA3427 (*impB*-like) to ECA3445 (*vgrG*-like) and is here referred as *Erwinia* spp. T6SS-I (Fig. 1). One additional solitary locus, designated as *vgrG*-like gene (ECA2104), as well as four other loci, designated as *hcp*-like genes (ECA4275, ECA2866, ECA0456 and ECA3672), are also present. Interestingly, virulence assays, performed with mutants in ECA3438 and ECA3444, in potato stems and tubers, showed significantly reduced virulence compared with the wild type strain in both cases (Liu et al., 2008). In our phylogenetic analysis the *P.a.* T6SS is presented as a member of a distinct phylogenetic branch comprising the *P.a.* and *Dickeya* spp. T6SSs (Fig. 3).

7.7 *Dickeya* spp. (*D. dadantii* and *D. zeae*)

Dickeya dadantii (*D.d.*; formerly *Erwinia chrysanthemi*; γ-proteobacteria) is an opportunistic plant pathogen causing soft-rot, wilt, and blights on a wide range of plant species, such as maize, pineapple, banana, rice, tobacco, tomato, *Brachiaria ruziziensis* and *Chrysanthemum morifolium*. It possesses two O-serogroups, O: 1 and O: 6. *D.d.* is also highly virulent on the pea aphid *Acyrthosiphon pisum*, and possesses four genes encoding homologs of the Cyt family of insecticidal toxins from *Bacillus thuringiensis* (Grenier et al., 2006). *Dickeya zeae* (*D.z.*; formerly *Erwinia chrysanthemi*) was isolated from soft rot and wilt of a various range of plants, such as *Zea mays*, *Ananas comosus*, *Brachiaria ruziziensis*, *Chrysanthemum morifolium*, *Musa* spp., *Nicotiana tabacum*, *Oryza sativa* and *Solanum tuberosum*, as well as from water samples. *D.z.* in contrast to *D.d.* possesses more than nine O-serogroups.

Two strains, *D.d.* Ech586 and *D.z.* Ech1591 (Lucas et al., 2009) that have been examined contain identical T6SS loci consisting of 17 genes lying from Dd586_1304 (*vasL*-like) to Dd586_1272 (*hcp*-like) for *D.d.*, with a disruption of several apparently T6SS-unrelated putative ORF insertions between Dd586_1290 (*impB*-like) and Dd586_1273 (*vgrG*-like) genes (Fig. 1). The *D.z.* T6SS locus spans from Dd1591_2793 (*vasL*-like) to Dd1591_2826 (*hcp*-like) with a disruption of several apparently T6SS-unrelated ORF insertions between Dd1591_2807 (*impB*-like) and Dd1591_2825 (*vgrG*-like) genes (Fig. 1). The two clusters are almost identical and the four core T6SS proteins examined form a distinct phylogenetic branch which includes the *P. atrosepticum* T6SS (Fig. 3).

7.8 *Acidovorax avenae* subsp. *citrulli*

A. avenae subsp. *citrulli* (*A.c.*; β-proteobacteria) is formerly known as *Pseudomonas pseudoalcaligenes* subsp. *citrulli* and is the causal agent of bacterial fruit blotch. It spreads by infested seeds, infected transplants, and occurs naturally in wild hosts. It can be asymptomatic on older plants, which can lead to high numbers of infected young plants early in the planting season. A T6SS cluster is found in *A.c.* strain AAC00-1 consisting of 16 genes between Aave_1482 (*clp*-like ATPase) and Aave_1465 (*hcp*-like) as annotated in Fig. 1. The T6SS locus is contiguous, except of two putative ORF insertions between the Aave_1468 (*fha*-like) and Aave_1465 genes that are apparently T6SS-unrelated. The *A.c.* T6SS cluster lacks a *vgrG* homolog, which potentially raises questions about the system's functionality. To date, there is no experimental evidence concerning a role of this system in *A.c.*-host interactions. Phylogenetically the *A.c.* T6SS cluster forms a sub-group with the *C. taiwanensis* T6SS-I, *P. ananatis* T6SS-II, *Erwinia* spp. T6SS-II (Fig. 3).

7.9 *Agrobacterium* spp. (*A. tumefaciens* and *A. vitis*)

Agrobacterium strains (α-proteobacteria) invade the crown, roots and stems of a great variety of plants via wounds causing overgrowths (crown gall, hairy root, and cane gall). *A.*

tumefaciens (A.t.) have a wide host range among dicotyledonous plants, whereas other possess a very limited host range [*A. rubi (A.r.)* and *A. vitis (A.v.)* form galls on raspberries and grapes, respectively]. *A. rhizogenes* causes hairy roots on many plants. The ability to cause disease is associated with transmissible plasmids which may move from one strain to another. *A.t.* strain C58 is a representative of biovar I, which has also been extensively modified for biotechnological uses. *A.v.* strain S4 is a virulent biovar III isolated from *Vitis vinifera* (grape) cv. Izsaki Sarfeher crown galls in Kecskemet, Hungary in 1981 (Szegedi et al., 1988; 1996). *A.v.* strains not only cause galls on grapevines but also necrosis on grapevine roots and a hypersensitive response on non-host plants.

The two sequenced species, *A.t.* strain C58 and *A.v.* strain S4 seem to harbor almost identical T6SS gene cluster. According to our database search, those loci are lying between Atu4330 (*impN*-like) and Atu4348 (*vgrG*-like) in *A.t.* C58 and between Avi_6039 (*impN*-like) and Avi_6054 (*hcp*-like) in *A.v.* S4 (Fig. 1). *A.v.* seems to lack the *vgrG*-like gene upstream of the *clpV*-like gene (Fig. 1), while *A.t.* carries a *vgrG*-like gene in the solitary locus Atu_3642 outside the T6SS cluster. However, the *A.v.* has five additional *vgrG*-like genes at the solitary loci Avi_1646, Avi_2758, Avi_3254, Avi_7056, Avi_7557, which also are located outside the T6SS cluster. Phylogenetically the *Agrobacterium* spp. T6SSs branch more closely to the *Rhizobium leguminosarum* T6SS phylogenetic sub-group (Fig. 3).

8. T6SS in plant symbiotic bacteria

8.1 *Rhizobium leguminosarum*

R. leguminosarum bv. *viciae* 3841 (*R.l.*; α-proteobacteria) has a genomic portfolio consisting of seven circular DNA modules (totalling about 7,8 Mb): one circular chromosome of about 5 Mb and six plasmids varying in size from 147 kb to 870 kb. A T6SS is one of the many protein secretion systems identified in *R.l.*, (Krehenbrink & Downie, 2008). The T6SS gene cluster was initially reported to contain 14 genes (*impA-impN*); later an *hcp* and a *vgrG* gene homolog (pRL120477 and pRL120480), coding for secreted proteins, were added. Our data mining in the genome of *R.l.*, 3841 leave open the presence of a functional Clp-like ATPase because of multiple sequence insertions and point mutations in the *clpV/B* gene (locus No: pRL120476) (Fig. 1), which is annotated as pseudogene in the KEGG, RhizoBase and NCBI databases. Phylogenetically, the *R.l.* T6SS clusters with the *Agrobacterium* spp. T6SSs and form a distinct phylogenetic branch (Fig. 3).

8.2 *Rhizobium etli*

R. etli (R.e. a-proteobacteria) contributes to a significant proportion of nitrogen coming to the earth through microorganisms and is the prime species found associated with cultivated beans in the Americas. Although there is no experimental evidence for a functional T6SS in *R.e.*, bioinformatic analysis revealed two putative T6SSs, one of which (T6SS-I) is similar in organization to that of *R.l.* (Bladergroen et al., 2003) (Fig. 1), and in our phylogenetic distance tree (Fig. 3) it branches with the *Mesorhizobium loti* T6SS.

Our data mining revealed two interesting features. First, of the two strains of *R.e.* sequenced to date, *R.e.* CFN 42 and *R.e.* CIAT 652, only the latter seems to have T6SS core genes. Second, in this strain there are two T6SS gene clusters (T6SS-I and –II) located in distinct genomic islands (Fig. 1). The T6SS-I spans from RHECIAT_PC0000958 (*vgrG*-like) to RHECIAT_PC0000933 (*impL*-like). While the T6SS-II seems to be divided in two segments with opposite gene orientations, located from RHECIAT_PB0000227 (*impL*-like) to RHECIAT_PB0000224 (*impJ*-like) and from RHECIAT_PB0000217 (*impA*-like) to RHECIAT_PB0000210 (*vgrG*-like). Furthermore, the *Clp*-like ATPase is absent in T6SS-II, which raises questions about the functionality of the cluster (Fig. 1). Phylogenetically, the four core proteins of the *R.e.* CIAT 652 T6SS-I are more closely related to those of the *Mesorhizobium loti* T6SS, forming a distinct sub-group, while those of the T6SS-II branch with the *Xanthomonas* spp. T6SS-I and T6SS-II, forming a separate sub-group (Fig. 3).

8.3 *Bradyrhizobium japonicum*

B. japonicum (*B.j.* α-proteobacteria; *Rhizobium japonicum* in earlier references) utilizes similar mechanisms, as the other symbiotic bacteria, to establish a symbiotic relationship. The *B.j.* strain USDA110 genome consists of a single circular chromosome of about 9 Mbp and has no plasmids. Our *in silico* genome mining results were in agreement to Boyer et al. (2009) and Records et al. (2011) revealing the presence of one T6SS gene cluster consisting of 17 ORFs located between blr3604 (*ImpN*-like) and bll3587 (*clpB*-like) (Fig. 1). Phylogenetically, the *B.j.* T6SS forms a distinct branch from other *Rhizobium* T6SS and clusters closest to the *Xanthomonas* T6SS-I and III (Fig. 3).

8.4 *Mesorhizobium loti* (*Rhizobium loti*)

M. loti (*M.l.* α-proteobacteria), a symbiont of *Lotus japonicus* contains a chromosomal symbiosis island, similar to what is observed with other rhizobacteria. *M.l.* strain MAFF303099 contains one T6SS which is located between mlr2363 (*impN*-like) and mll2335 (*clpB*-like) gene loci (Fig. 1). In contrast to *R.l.*, *M.l.* possesses *vasD* while it apparently lacks *impE* (Figs. 1 and 3).

8.5 *Cupriavidus taiwanensis*

C. taiwanensis (*C.t.* β-proteobacteria; originally called *Ralstonia taiwanensis* or *Wautersia taiwanensis*), belongs to the β-rhizobia group. It was first isolated from a nodule from the legume *Mimosa pudica* in Taiwan. The type strain LMG19424 has a three-replicon genome made up of two chromosomes of 3.5 Mb, and 2.4 Mb, and a 0.5 kb symbiotic plasmid which carries the genes essential for nodulation and nitrogen fixation (Amadou et al., 2008). In our *in silico* genome mining, two T6SS gene clusters were found in the LMG19424, one (T6SS-I) consisting of 15 genes located between RALTA_A0602 (*impM*-like) and RALTA_A0622 (*impA*-like), with six apparently T6SS-unrelated putative ORF insertions between RALTA_A0611 (*vgrG*-like) and RALTA_A0618 (*vasD*-like) genes (Fig. 1). Phylogenetically, the *C.t.* T6SS-I groups together with the T6SSs of *A. avenae* pv. *citrulli*, *P. ananatis* T6SS-II, *E. pyrifoliae* and *E. amylovora* T6SS-II (Fig. 3). The second cluster (T6SS-II) consists of 18 genes located between RALTA_B1008 (*vgrG*-like) and RALTA_B1029 (impL-like) and containing four putative ORF insertions apparently unrelated to T6SS, between RALTA_B1019 (*clpB*-like ATPase) and RALTA_B1025 (*impA*-like) genes (Fig. 1). Based on gene order and sequence similarities, the C. t. T6SS-II and *R. solanacearum* T6SS appear phylogenetically close (Fig. 3). There are no studies concerning the functionality and/or the contribution of these T6SS clusters in host colonization.

9. Prospects

Protein secretion is fundamental to bacterial virulence and several systems mediate pathogenesis and other types of bacteria-host interaction. Beyond the other recognized secretion systems of Gram-negative bacteria with established role in host-pathogen interactions, the T6SS is of particular interest in this respect and has been shown to be important for bacterial virulence and for interaction with the host in various ways, often leading to "anti-pathogenesis" (Jani & Cotter, 2010). Nevertheless, its role and function in most bacteria is not clearly established and formal evidence for protein secretion/translocation into plant cells is scant. At present, we do not fully understand how the T6SS works and are only beginning to understand the biological role/s of the T6SS in plant-associated bacterial life. Multiple copies of T6SS in a single bacterial strain appear to be a frequent phenomenon, and this holds true for many plant associated species. Recent studies (Boyer et al., 2009; Filloux et al., 2008) have established the presence of multiple copies of apparently complete and/or degenerate T6SS loci in about one quarter of the proteobacterial genomes examined, that they generally display different phylogenetic origins and are not a result of recent duplication events, suggesting sustained and constrained mechanisms that favour this trend. Based on our own analysis (Sarris & Scoulica, 2011; Sarris et al., 2011), most strains of *Pseudomonas syringae*, the insect pathogenic *Pseudomonas entomophila* strain L48, the human opportunistic pathogen *Pseudomonas mendocina* strain ymp, and most of the *Pseudomonas* strains studied by Barret et al. (2011) typically carry T6SS from more than one phylogenetic clade and/or additional *vgrG* and *hcp* genes.

Although the majority of the recent studies concern the contribution of T6SS in bacterial pathogenicity (positively), many bacteria with genomes encoding putative T6SS are not known to be pathogens or symbionts, and T6SS may also function in non-pathogenic bacteria-host interactions and/or in interactions not involving eukaryotic partners. In *R. leguminosarum* the T6SS limits host-range and in *S. typhimurium* and *H. hepaticus* the evidence suggests a possible role of T6SS in limiting of bacterial virulence and, therefore, contribution to host colonization (Bladergroen et al., 2003; Parsons & Heffron, 2005; Jani & Cotter, 2010; Chow & Mazmanian, 2010). A relatively new twist in the system's repertoire of biological functions in a broader context is the finding that bacteria engage each other in a T6SS-dependent manner and can provide protection for a bacterium against cell contact-induced growth inhibition caused by other species of bacteria (Hood et al., 2010; Schwarz et al., 2010a). This leads to speculation that this pathway is of general significance to interbacterial interactions in polymicrobial diseases and the environment.

T6SS clusters occur with high frequency, have divergent phylogenies and individual strains or species often possess non-orthologous clusters with distinct or overlapping functions in bacterial interactions with multiple hosts, antagonists or predators unsuspected at present. In a recent study of the ocean metagenome (Persson et al., 2009) the T6SS was more abundant among γ-proteobacteria than other protein transport systems. The weight of present evidence suggests, at least indirectly, an apparently rampant lateral transfer of T6SS clusters/genes in the microbial world, which could be a significant driver for newly emerging pathogens, as proposed for the gastroenteritis agent *E. tarda* (Leung et al., 2011).

Future studies are needed and expected to further advance the T6SS field. It is important to remember that formal evidence of the translocation of effector proteins into plant cells through T6SS is presently lacking, as is also the case for the molecular targets of these effectors. Paraphrasing Schwarz et al. (2010a), outstanding questions for future research on T6SS include the following: What are the physiological role(s) and adaptive significance of T6S-mediated plant cell targeting in disease, symbiosis, and interbacterial interactions in the environment? What is the significance of additional *vgrG* and *hcp* genes Are host- and bacterial cell-targeting T6SSs discernible by sequence or gene content? Are there T6SSs that can target both eukaryotic and prokaryotic cells? Are there other T6S substrates that await identification? Given the resemblance T6SS components to bacteriophage proteins it is also tempting to ask if T6SS transports only protein substrates and/or other macromolecules as well.

It is instructive to point out that the genes coding for several secretion systems, including T6SS, were first identified in phenotypic screens of mutants altered in their interaction with higher eukaryotes. It is conceivable that new secretion systems may be identified in other appropriately designed screens in multi-organism settings. Bioinformatic sourcing of genome, transcriptome and proteome data may point to new potential candidates, as occurred historically with the T6SS. Finally, the striking similarities between secretion systems and type IV pili, flagella, bacteriophage tail, or efflux pumps invite speculation that new systems may even be predicted, "the way Mendeleïev had anticipated characteristics of yet unknown elements" (Filloux, 2011b).

10. Acknowledgments

We apologize to any researchers whose work was not included in this review due to space limitations. This work was supported by PYTHAGORAS, and PENED 03ED375 grants of the Greek General Secretariat for Research and Development implemented within the framework of the "Reinforcement Programme of Human Research Manpower" (PENED) and co-financed by National and Community Funds (25% from the Greek Ministry of Development General Secretariat of Research and Technology and 75% from EU-European Social Fund) and by the EPEAEK graduate programs in Plant Molecular Biology and Biotechnology and Protein Biotechnology of the Greek Ministry of Education, Lifelong Learning and Religious Affairs.

11. References

Aksyuk, A. A., Leiman, P. G., Kurochkina, L. P., Shneider, M. M., Kostyuchenko, V. A., Mesyanzhinov, V. V. & Rossmann, M. G. (2009). The tail sheath structure of bacteriophage T4: a molecular machine for infecting bacteria. *EMBO Journal*, Vol. 28, No 7, pp 821-829.

Amadou, C., Pascal, G., Mangenot, S., Glew, M., Bontemps, C., Capela, D., Carrere, S., Cruveiller, S., Dossat, C., Lajus, A., Marchetti, M., Poinsot, V., Rouy, Z., Servin, B., Saad, M., Schenowitz, C., Barbe, V., Batut, J., Medigue, C. & Masson-Boivin, C.

(2008). Genome sequence of the beta-rhizobium *Cupriavidus taiwanensis* and comparative genomics of rhizobia. *Genome Research*, Vol. 18, No 9, pp 1472-1483.

Aschtgen, M.-S., Bernard, C. S., De Bentzmann, S., Lloubes, R. & Cascales, E. (2008). SciN is an outer membrane lipoprotein required for Type VI secretion in enteroaggregative *Escherichia coli*. *Journal of Bacteriology*, Vol. 190, No 22, pp 7523-7531.

Ballister, E. R., Lai, A. H., Zuckermann, R. N., Cheng, Y. & Mougous, J. D. (2008). In vitro self-assembly of tailorable nanotubes from a simple protein building block. *Proceedings of the National Academy of Sciences*, Vol. 105, No 10, pp 3733-3738.

Barret, M., Egan, F., Fargier E., Morrissey, J. P. & O'Gara, F. (2011). Genomic analysis of the type VI secretion systems in Pseudomonas spp: novel 3 clusters and putative effectors uncovered. Microbiology, vpl.157, No 6, p.1726-1740.

Bernard, C. S., Brunet, Y. R., Gueguen, E. & Cascales, E. (2010). Nooks and Crannies in type VI secretion regulation. *Journal of Bacteriology*, Vol. 192, No 15, pp 3850-3860.

Bernard, C. S., Brunet, Y. R., Gavioli, M., Lloubes, R. & Cascales, E. (2011). Regulation of type VI secretion gene clusters by σ^{54} and cognate enhancer binding proteins. *Journal of Bacteriology*, Vol. 193, No 9, pp 2158-2167.

Bingle, L. E., Bailey, C. M. & Pallen, M. J. (2008). Type VI secretion: a beginner's guide. *Current Opinions in Microbiology*, Vol. 11, No 1, pp 3-8.

Bladergroen, M. R., Badelt, K. & Spaink, H. P. (2003). Infection-blocking genes of a symbiotic *Rhizobium leguminosarum* strain that are involved in temperature-dependent protein secretion. *Molecular Plant-Microbe Interactions*, Vol. 16, No 1, pp 53-64.

Bonemann, G., Pietrosiuk, A., Diemand, A., Zentgraf, H. & Mogk, A. (2009). Remodelling of VipA/VipB tubules by ClpV-mediated threading is crucial for type VI protein secretion. *EMBO Journal*, Vol. 28, No 4, pp 315-325.

Bonemann, G., Pietrosiuk, A. & Mogk, A. (2010). Tubules and donuts: a type VI secretion story. *Molecular Microbiology*, Vol. 76, No 4, pp 815-821.

Boyer, F., Fichant, G., Berthod, J., Vandenbrouck, Y. & Attree, I. (2009). Dissecting the bacterial type VI secretion system by a genome wide *in silico* analysis: what can be learned from available microbial genomic resources? *BMC Genomics*, Vol. 10, No, pp 104.

Brunet, Y. R., Bernard, C. S., Gavioli, M., Lloubes, R. & Cascales, E. (2011). An epigenetic switch involving overlapping fur and DNA methylation optimizes expression of a Type VI secretion gene custer. *PLoS Genetics*, Vol. 7, No 7, pp e1002205.

Cascales, E. (2008). The type VI secretion toolkit. *EMBO Reports*, Vol. 9, No 8, pp 735-741.

Chow, J. & Mazmanian, S. K. (2010). A pathobiont of the microbiota balances host colonization and intestinal inflammation. *Cell Host & Microbe*, Vol. 7, No 4, pp 265-276.

Chugani, S. & Greenberg, E. P. (2007). The influence of human respiratory epithelia on *Pseudomonas aeruginosa* gene expression. *Microbial Pathogenesis*, Vol. 42, No 1, pp 29-35.

Coutinho, T. A. & Venter, S. N. (2009). *Pantoea ananatis*: an unconventional plant pathogen. *Molecular Plant Pathoogyl*, Vol. 10, No 3, pp 325-335.

De Bruin, O., Ludu, J. & Nano, F. (2007). The *Francisella* pathogenicity island protein IglA localizes to the bacterial cytoplasm and is needed for intracellular growth. *BMC Microbiology*, Vol. 7, No 1, pp 1.

De Maayer, P., Chan, W. Y., Venter, S. N., Toth, I. K., Birch, P. R. J., Joubert, F. & Coutinho, T. A. (2010). Genome sequence of *Pantoea ananatis* LMG20103, the causative agent of Eucalyptus Blight and Dieback. *Journal of Bacteriology*, Vol. 192, No 11, pp 2936-2937.

Dudley, E. G., Thomson, N. R., Parkhill, J., Morin, N. P. & Nataro, J. P. (2006). Proteomic and microarray characterization of the AggR regulon identifies a *pheU* pathogenicity island in enteroaggregative *Escherichia coli*. *Molecular Microbiology*, Vol. 61, No 5, pp 1267-1282.

Fauvart, M. & Michiels, J. (2008). Rhizobial secreted proteins as determinants of host specificity in the rhizobium-legume symbiosis. *FEMS Microbiology Letters*, Vol. 285, No 1, pp 1-9.

Felsenstein, J. (1985). Confidence-Limits on Phylogenies - an Approach Using the Bootstrap. *Evolution*, Vol. 39, No 4, pp 783-791.

Filloux, A. (2011a). The bacterial type VI secretion system: on the bacteeriophage trail. *Microbiology Today*, May 2011,, pp 96-101.

Filloux, A. (2011b). Protein secretion systems in *Pseudomonas aeruginosa*: An Essay on diversity, evolution, and function. *Frontiers in Microbiology*, Vol. 2, No, 155 pp1-21.

Filloux, A., Hachani, A. & Bleves, S. (2008). The bacterial type VI secretion machine: yet another player for protein transport across membranes. *Microbiology*, Vol. 154, No 6, pp 1570-1583.

Green, S., Studholme, D. J., Laue, B. E., Dorati, F., Lovell, H., Arnold, D., Cottrell, J. E., Bridgett, S., Blaxter, M., Huitema, E., Thwaites, R., Sharp, P. M., Jackson, R. W. & Kamoun, S. (2010). Comparative genome analysis provides insights into the evolution and adaptation of *Pseudomonas syringae* pv. *aesculi* on *Aesculus hippocastanum*. *PLoS ONE*, Vol. 5, No 4, pp e10224.

Grenier, A. M., Duport, G., Pages, S., Condemine, G. & Rahbe, Y. (2006). The phytopathogen *Dickeya dadantii* (*Erwinia chrysanthemi* 3937) is a pathogen of the pea aphid. *Applied and Environmental Microbiology*, Vol. 72, No 3, pp 1956-1965.

Hayward, A. C. (1993). The host of *Xanthomonas*, In: *Xanthomonas*, Swings, J. G., Civerolo, E. L., eds, pp. 51-54, Chapman & Hall, London, United Kingdom

Hirano, S. S. & Upper, C. D. (2000). Bacteria in the leaf ecosystem with emphasis on *Pseudomonas syringae*-a pathogen, ice nucleus, and epiphyte. *Microbiology and Molecular Biology Reviews*, Vol. 64, No 3, pp 624-653.

Hood, R. D., Singh, P., Hsu, F., Guvener, T., Carl, M. A., Trinidad, R. R., Silverman, J. M., Ohlson, B. B., Hicks, K. G., Plemel, R. L., Li, M., Schwarz, S., Wang, W. Y., Merz, A. J., Goodlett, D. R. & Mougous, J. D. (2010). A type VI secretion system of *Pseudomonas aeruginosa* targets a toxin to bacteria. *Cell Host & Microbe*, Vol. 7, No 1, pp 25-37.

Jani, A. J. & Cotter, P. A. (2010). Type VI Secretion: Not Just for pathogenesis anymore. *Cell Host & Microbe*, Vol. 8, No 1, pp 2-6.

Janisiewicz, W. J. & Marchi, A. (1992). Control of storage rots on various pear cultivars with a saprophytic strain of *Pseudomonas syringae*. *Plant Disease*, Vol. 76, No, pp 555-560.

Jobichen, C., Chakraborty, S., Li, M., Zheng, J., Joseph, L., Mok, Y. K., Leung, K. Y. & Sivaraman, J. (2010). Structural basis for the secretion of EvpC: a key type VI secretion system protein from *Edwardsiella tarda*. *PLoS ONE*, Vol. 5, No 9, pp e12910.

Kim, W. S., Jock, S., Paulin, J.-P., Rhim, S.-L. & Geider, K. (2001). Molecular detection and differentiation of *Erwinia pyrifoliae* and host range analysis of the asian pear pathogen. *Plant Disease*, Vol. 85, No 11, pp 1183-1188.

Krehenbrink, M. & Downie, J. A. (2008). Identification of protein secretion systems and novel secreted proteins in *Rhizobium leguminosarum* bv. *viciae*. *BMC Genomics*, Vol. 9, No 9, pp 55.

Kwon, S. W., Go, S. J., Kang, H. W., Ryu, J. C. & Jo, J. K. (1997). Phylogenetic analysis of *Erwinia* species based on 16S rRNA gene sequences. *International Journal of Systematic Bacteriology*, Vol. 47, No 4, pp 1061-1067.

Leiman, P. G., Basler, M., Ramagopal, U. A., Bonanno, J. B., Sauder, J. M., Pukatzki, S., Burley, S. K., Almo, S. C. & Mekalanos, J. J. (2009). Type VI secretion apparatus and phage tail-associated protein complexes share a common evolutionary origin. *Proceedings of the National Academy of Sciences USA*, Vol. 106, No 11, pp 4154-4159.

Leung, K. Y., Siame, B. A., Tenkink, B. J., Noort, R. J. & Mok, Y. K. (2011). *Edwardsiella tarda* - Virulence mechanisms of an emerging gastroenteritis pathogen. *Microbes and Infection*, Vol. in press, doi:10.1016/j.micinf.2011.08.005.

Leyns, F., De Cleene, M., Swings, J. & De Ley, J. (1984). The host range of the genus *Xanthomonas*. *The Botanical Review*, Vol. 50, No 3, pp 308-356.

Liu, H., Coulthurst, S. J., Pritchard, L., Hedley, P. E., Ravensdale, M., Humphris, S., Burr, T., Takle, G., Brurberg, M. B., Birch, P. R., Salmond, G. P. & Toth, I. K. (2008). Quorum sensing coordinates brute force and stealth modes of infection in the plant pathogen *Pectobacterium atrosepticum*. *PLoS Pathogens*, Vol. 4, No 6, pp e1000093.

Lucas, S., Copeland, A., Lapidus, A., Glavina Del Rio, T., Tice, H., Bruce, D., Goodwin, L., Pitluck, S., Chertkov, O., Brettin, T., Detter, J. C., Han, C., Larimer, F., Land, M., Hauser, L., Kyrpides, N., Ovchinnicova, G., Balakrishnan, V., Glasner, J. & Perna, N. T. (2009). Complete sequence of *Dickeya zeae* Ech1591, In: *EMBL/GenBank/DDBJ databases*.

Ma, A. T., Mcauley, S., Pukatzki, S. & Mekalanos, J. J. (2009). Translocation of a *Vibrio cholerae* type VI secretion effector requires bacterial endocytosis by host cells. *Cell Host & Microbe*, Vol. 5, No 3, pp 234-243.

Macintyre, D. L., Miyata, S. T., Kitaoka, M. & Pukatzki, S. (2010). The *Vibrio cholerae* type VI secretion system displays antimicrobial properties. *Proceedings of the National Academy of Sciences USA*, Vol. 107, No 45, pp 19520-19524.

Mattinen, L., Nissinen, R., Riipi, T., Kalkkinen, N. & Pirhonen, M. (2007). Host-extract induced changes in the secretome of the plant pathogenic bacterium *Pectobacterium atrosepticum*. *Proteomics*, Vol. 7, No 19, pp 3527-3537.

Miyata, S. T., Kitaoka, M., Wieteska, L., Frech, C., Chen, N. & Pukatzki, S. (2010). The *Vibrio cholerae* Type VI Secretion System: Evaluating its role in the human disease cholera. *Frontiers in Microbiology*, Vol. 1, No 117, pp 1-7.

Miyata, S. T., Kitaoka, M., Brooks, T. M., Mcauley, S. B. & Pukatzki, S. (2011). *Vibrio cholerae* requires the type VI secretion system virulence factor VasX to kill *Dictyostelium discoideum*. *Infection and Immunity*, Vol. 79, No 7, pp 2941-2949.

Mougous, J. D., Cuff, M. E., Raunser, S., Shen, A., Zhou, M., Gifford, C. A., Goodman, A. L., Joachimiak, G., Ordonez, C. L., Lory, S., Walz, T., Joachimiak, A. & Mekalanos, J. J. (2006). A virulence locus of *Pseudomonas aeruginosa* encodes a protein secretion apparatus. *Science*, Vol. 312, No 5779, pp 1526-1530.

Mougous, J. D., Gifford, C. A., Ramsdell, T. L. & Mekalanos, J. J. (2007). Threonine phosphorylation post-translationally regulates protein secretion in *Pseudomonas aeruginosa*. *Nature Cell Biology*, Vol. 9, No 7, pp 797-803.

Nino-Liu, D. O., Ronald, P. C. & Bogdanove, A. J. (2006). *Xanthomonas oryzae* pathovars: model pathogens of a model crop. *Molecular Plant Pathology*, Vol. 7, No 5, pp 303-324.

Pace, F., Boldrin, J., Nakazato, G., Lancellotti, M., Sircili, M., Guedes, E., Silveira, W. & Sperandio, V. (2011). Characterization of IcmF of the type VI secretion system in an avian pathogenic *Escherichia coli* (APEC) strain. *Microbiology*, vol 157, Pt 10, pp 2954-2962.

Pallen, M. J., Chaudhuri, R. R. & Henderson, I. R. (2003). Genomic analysis of secretion systems. *Current Opinions in Microbiology*, Vol. 6, No 5, pp 519-527.

Parsons, D. A. & Heffron, F. (2005). sciS, an icmF homolog in *Salmonella enterica* serovar *typhimurium*, limits intracellular replication and decreases virulence. *Infection and Immunity*, Vol. 73, No 7, pp 4338-4345.

Pell, L. G., Kanelis, V., Donaldson, L. W., Howell, P. L. & Davidson, A. R. (2009). The phage lambda major tail protein structure reveals a common evolution for long-tailed phages and the type VI bacterial secretion system. *Proceedings of the National Academy of Sciences USA*, Vol. 106, No 11, pp 4160-4165.

Persson, O. P., Pinhassi, J., Riemann, L., Marklund, B. I., Rhen, M., Normark, S., Gonzalez, J. M. & Hagstrom, A. (2009). High abundance of virulence gene homologues in marine bacteria. *Environmental Microbiology*, Vol. 11, No 6, pp 1348-1357.

Potnis, N., Krasileva, Chow, K. V., Almeida, N. F., Patil, P., Ryan, R., Sharlach, M., Behlau, F., Dow, J. M., White, F., Preston, J., Vinatzer, B., Koebnik, R., Setubal, J. C., Norman, D. J., Staskawicz B. & Jones J. B. (2011). Comparative genomics reveals diversity among xanthomonads infecting tomato and pepper. *BMC Genomics*, Vol. 12, No 146, pp 2-23.

Pukatzki, S., Ma, A. T., Sturtevant, D., Krastins, B., Sarracino, D., Nelson, W. C., Heidelberg, J. F. & Mekalanos, J. J. (2006). Identification of a conserved bacterial protein secretion system in *Vibrio cholerae* using the *Dictyostelium* host model system. *Proceedings of the National Academy of Sciences USA*, Vol. 103, No 5, pp 1528-1533.

Pukatzki, S., Ma, A. T., Revel, A. T., Sturtevant, D. & Mekalanos, J. J. (2007). Type VI secretion system translocates a phage tail spike-like protein into target cells where it cross-links actin. *Proceedings of the National Academy of Sciences USA*, Vol. 104, No 39, pp 15508-15513.

Purcell, M. & Shuman, H. A. (1998). The *Legionella pneumophila* icmGCDJBF genes are required for killing of human macrophages. *Infection and Immunity*, Vol. 66, No 5, pp 2245-2255.

Records, A. R. & Gross, D. C. (2010). Sensor kinases RetS and LadS regulate *Pseudomonas syringae* type VI secretion and virulence factors. *Journal of Bacteriology*, Vol. 192, No 14, pp 3584-3596.

Records, A. R. (2011). The type VI secretion system: a multipurpose delivery system with a phage-like machinery. *Molecular Plant-Microbe Interactions*, Vol. 24, No 7, pp 751-757.

Rodriguez-Palenzuela, P., Matas, I. M., Murillo, J., Lopez-Solanilla, E., Bardaji, L., Perez-Martinez, I., Rodriguez-Moskera, M. E., Penyalver, R., Lopez, M. M., Quesada, J. M., Biehl, B. S., Perna, N. T., Glasner, J. D., Cabot, E. L., Neeno-Eckwall, E. & Ramos, C. (2010). Annotation and overview of the *Pseudomonas savastanoi* pv. *savastanoi* NCPPB 3335 draft genome reveals the virulence gene complement of a tumour-inducing pathogen of woody hosts. *Environmental Microbiology*, Vol. 12, No 6, pp 1604-1620.

Roest, H. P., Mulders, I. H. M., Spaink, H. P., Wijffelman, C. A. & Lugtenberg, B. J. J. (1997). A *Rhizobium leguminosarum* Biovar *trifolii* locus not localized on the Sym plasmid hinders effective nodulation on plants of the pea cross-inoculation group. *Molecular Plant-Microbe Interactions*, Vol. 10, No 7, pp 938-941.

Russell, A. B., Hood, R. D., Bui, N. K., Leroux, M., Vollmer, W. & Mougous, J. D. (2011). Type VI secretion delivers bacteriolytic effectors to target cells. *Nature*, Vol. 475, No 7356, pp 343-347.

Saitou, N. & Nei, M. (1987). The neighbor-joining method: a new method for reconstructing phylogenetic trees. *Molecular Biology and Evolution*, Vol. 4, No 4, pp 406-425.

Sarris, P. F., Skandalis, N., Kokkinidis, M. & Panopoulos, N. J. (2010). *In silico* analysis reveals multiple putative type VI secretion systems and effector proteins in *Pseudomonas syringae* pathovars. *Molecular Plant Pathology*, Vol. 11, No 6, pp 795-804.

Sarris, P. F. & Scoulica, E. V. (2011). *Pseudomonas entomophila* and *Pseudomonas mendocina*: Potential models for studying the bacterial type VI secretion system. *Infection, Genetics and Evolution*, Vol. 11, No 6, pp 1352-1360.

Sarris, P. F., Zoumadakis, C., Panopoulos, N. J. & Scoulica, E. (2011). Distribution of the putative type VI secretion system core genes in *Klebsiella* spp. *Infection Genetics and Evolution*, Vol. 11, No 1, pp 157–166.

Schlieker, C., Zentgraf, H., Dersch, P. & Mogk, A. (2005). ClpV, a unique Hsp100/Clp member of pathogenic proteobacteria. *Biological Chemistry*, Vol. 386, No 11, pp 1115-1127.

Schwarz, S., Hood, R. D. & Mougous, J. D. (2010a). What is type VI secretion doing in all those bugs? *Trends in Microbiology*, Vol. 18, No 12, pp 531-537.

Schwarz, S., West, T. E., Boyer, F., Chiang, W.-C., Carl, M. A., Hood, R. D., Rohmer, L., Tolker-Nielsen, T., Skerrett, S. J. & Mougous, J. D. (2010b). *Burkholderia* Type VI Secretion Systems have distinct roles in eukaryotic and bacterial cell interactions. *PLoS Pathogen*, Vol. 6, No 8, e1001068. doi:10.1371/journal.ppat.1001068.

Sexton, J. A., Miller, J. L., Yoneda, A., Kehl-Fie, T. E. & Vogel, J. P. (2004). *Legionella pneumophila* DotU and IcmF are required for stability of the Dot/Icm complex. *Infect Immun*, Vol. 72, No 10, pp 5983-5992.

Shalom, G., Shaw, J. G. & Thomas, M. S. (2007). In vivo expression technology identifies a type VI secretion system locus in *Burkholderia pseudomallei* that is induced upon invasion of macrophages. *Microbiology*, Vol. 153, No Pt 8, pp 2689-2699.

Sheahan, K.-L., Cordero, C. L. & Fullner Satchell, K. J. (2004). Identification of a domain within the multifunctional *Vibrio cholerae* RTX toxin that covalently cross-links actin. *Proceedings of the National Academy of Sciences USA*, Vol. 101, No 26, pp 9798-9803.

Shrivastava, S. & Mande, S. S. (2008). Identification and functional characterization of gene components of Type VI Secretion system in bacterial genomes. *PLoS ONE*, Vol. 3, No 8, pp e2955.

Smits, T. H., Rezzonico, F., Kamber, T., Blom, J., Goesmann, A., Frey, J. E. & Duffy, B. (2010). Complete genome sequence of the fire blight pathogen *Erwinia amylovora* CFBP 1430 and comparison to other *Erwinia* spp. *Molecular Plant-Microbe Interactions*, Vol. 23, No 4, pp 384-393.

Stall, R. E. & Seymour, C. P. (1983). Canker, a threat to citrus in the gulf-coast states. *Plant Disease*, Vol. 67, No 5, pp 581-585.

Suarez, G., Sierra, J. C., Sha, J., Wang, S., Erova, T. E., Fadl, A. A., Foltz, S. M., Horneman, A. J. & Chopra, A. K. (2008). Molecular characterization of a functional type VI secretion system from a clinical isolate of *Aeromonas hydrophila*. *Microbial Pathogenesis*, Vol. 44, No 4, pp 344-361.

Suarez, G., Sierra, J. C., Kirtley, M. L. & Chopra, A. K. (2010). Role of Hcp, a type 6 secretion system effector, of *Aeromonas hydrophila* in modulating activation of host immune cells. *Microbiology*, Vol. 156, No 12, pp 3678-3688.

Szegedi, E., Czako, M., Otten, L. & Koncz, C. S. (1988). Opines in crown gall Tumors induced by biotype 3 isolates of *Agrobacterium tumefaciens*. *Physiological and Molecular Plant Pathology*, Vol. 32, No 2, pp 237-247.

Szegedi, E., Czako, M. & Otten, L. (1996). Further evidence that the vitopine-type pTi's of *Agrobacterium vitis* represent a novel group of Ti plasmids. *Molecular Plant-Microbe Interactions*, Vol. 9, No 2, pp 139-143.

Tamura, K., Dudley, J., Nei, M. & Kumar, S. (2007). MEGA4: Molecular Evolutionary Genetics Analysis (MEGA) software version 4.0. *Molecular Biology and Evolution*, Vol. 24, No 8, pp 1596-1599.

Tseng, T. T., Tyler, B. M. & Setubal, J. C. (2009). Protein secretion systems in bacterial-host associations, and their description in the Gene Ontology. *BMC Microbiology*, Vol. 9 Suppl 1, No, pp S2.

Wang, Y.-Y. (2008). Characterization of type six secretion systems in *Pseudomonas syringae* pv. *tomato* DC3000: National University of Taiwan, MSc thesis.

Wu, H. Y., Chung, P. C., Shih, H. W., Wen, S. R. & Lai, E. M. (2008). Secretome analysis uncovers an Hcp-family protein secreted via a type VI secretion system in *Agrobacterium tumefaciens*. *Journal of Bacteriology*, Vol. 190, No 8, pp 2841-2850.

Yan, S., Liu, H., Mohr, T. J., Jenrette, J., Chiodini, R., Zaccardelli, M., Setubal, J. C. & Vinatzer, B. A. (2008). Role of recombination in the evolution of the model plant pathogen *Pseudomonas syringae* pv. *tomato* DC3000, a very atypical tomato strain. *Applied and Environmental Microbiology*, Vol. 74, No 10, pp 3171-3181.

Yuan, Z. C., Liu, P., Saenkham, P., Kerr, K. & Nester, E. W. (2008). Transcriptome profiling and functional analysis of *Agrobacterium tumefaciens* reveals a general conserved response to acidic conditions (pH 5.5) and a complex acid-mediated signaling involved in *Agrobacterium*-plant interactions. *Journal of Bacteriology*, Vol. 190, No 2, pp 494-507.

Zheng, J. & Leung, K. Y. (2007). Dissection of a type VI secretion system in *Edwardsiella tarda*. *Molecular Microbiology*, Vol. 66, No 5, pp 1192-1206.

Zheng, J., Ho, B. & Mekalanos, J. J. (2011). Genetic analysis of anti-amoebae and anti-bacterial activities of the Type VI secretion system in *Vibrio cholerae*. *PLoS ONE*, Vol. 6, No 8, pp e23876.

Zuckerkandl, E. & Pauling, L. (1965). Evolutionary divergence and convergence in proteins, In: *Evolving Genes and Proteins*, Bryson, V., Vogel, H. J., eds, pp. 97-166, Academic Press, New York.

5

Functional Identification of Genes Encoding Effector Proteins in *Magnaporthe oryzae*

Jing Yang and Chengyun Li
*Key Laboratory of Agro-Biodiversity and Pest Management of Education
Ministry of China, Yunnan Agricultural University, Kunming, Yunnan
China*

1. Introduction

In the course of coevolution of plants and pathogens for many millions of years, plants possessed many kinds of recognition and resistance mechanisms to prevent or limit pathogen infection. At the same time, pathogen also initiated many pathogenicity mechanisms, such as development of specialized infection structures, secretion of hydrolytic enzymes, production of host selective toxins, and detoxification of plant antimicrobial compounds (Idnurm and Howlett, 2001; Talbot, 2003; Randall et al., 2005), to avoid or overcome plant resistance mechanism. However, filamentous pathogen including fungi and oomycete could secrete a diverse array of effector proteins into the plant cell to manipulate the plant innate immunity, which facilitates the pathogen to successfully colonize and reproduce (Birch et al., 2006; Chisholm et al., 2006; Kamoun, 2006; O'Connell and Panstruga, 2006; Catanzariti et al., 2007; Kamoun, 2007). Several studies have shown that effector proteins could play dual role as both toxins and inducers of host resistance. Effector proteins were regarded as functioning primarily in virulence, but they also could elicit innate immunity in plant varieties carrying corresponding resistance protein.

Rice blast caused by *Magnaporthe oryzae* (Couch and Kohn, 2002) is the most devastating fungal disease of rice (*Oryza sativa*; Zeigler et al., 1994; Talbot, 2003). Functional identification of *M. oryzae* effectors can elucidate some pathogenicity mechanisms of the blast fungus, providing a clue to better manage blast disease. Several *Avr* genes have been cloned and characterized from *M.oryzae*, such as *Avr-Pita* (Orbach et al., 2000; Valent et al., 1991), *Avr1-CO39* (Farman and Leong, 1998), *Ace1* (Bohnert et al., 2004; Collemare et al., 2008) and the *Pwl* effectors (Kang et al.,1995; Sweigard et al., 1995). The *Avr-Pita* effector appears homologous to fungal zinc-dependent metalloproteases and is dispensable for virulence on rice (Jia et al., 2001; Orbach et al., 2000). *Avr-Pita* interacts with the cognate resistance protein *Pi-ta* (Jia et al., 2000). *Avr-Pita1* (*Avr-Pita*) including *Avr-Pita2* and *Avr-Pita3* are *Avr-Pita* family (Khang et al., 2008). *Avr-Pita2* acts as an elicitor of defense responses mediated by *Pi-ta*, while *Avr-Pita3* does not. Members of *Avr*-Pita family are detected among blast isolates isolated from different kinds of hosts by PCR-based method.

Ace1 effector is a putative cytoplasmic fusion polypeptide containing a polyketide synthase (PKS) and a nonribosomal peptide synthetase (NRPS), two distinct classes of enzymes that are involved in the production of microbial secondary metabolites (Bohnert et al., 2004;

Collemare et al., 2008). *Ace1* is thought to function as avirulence indirectly by producing a secondary metabolite that activates Pi33. *Ace1* is only expressed in appressoria, suggesting that the secondary metabolite produced might have a role in virulence (Bohnert et al., 2004; Fudal et al., 2005).The *Pwl* effectors, encoded by the *Pwl* (pathogenicity on weeping lovegrass, *Eragrostis curvula*) gene family, are rapidly evolving, and they are small glycine-rich secreted proteins that commonly distributed in *M.oryzae*. Presently, four genes such as *Pwl1-Pwl4* have been determined in *M. oryzae*, and the four genes confer species-specific avirulence on weeping lovegrass and finger millet, but not on rice (Kang et al., 1995; Sweigard et al., 1995). Yoshida *et al.* (2009) examined DNA polymorphisms of 1032 putative secreted proteins from the genome sequence of isolate 70-15 among 46 isolates, and found no association with Avr function on a set of differential rice cultivars carrying different *R* genes, indicating that the isolate 70-15 might have lost several functional *Avr* genes through sexual recombination.

After fungal effector proteins are secreted into plant cells, the question arises: do they mediate virulence or avirulence on host? To discover pathogenicity mechanism of the pathogen, it is indispensable to identify the function of effector proteins. Here we will introduce our studies on functional identification of effector proteins from *M. oryzae*.

2. Screening candidate effector-encoding genes from *M. oryzae*

Whole-genome sequence of fungal pathogens has provided an enormous amount of data that can be analyzed for mining putative secreted effector proteins. *M. oryzae* genome sequence has been available online, which contribute many novel effector-encoding genes. Some online softwares could be aided to predict some features such as secretion, domain and homology of effector proteins. Presently, secreted proteins are categorized into two classes based on their secreted pathway, one is classically secreted proteins, with N-terminal signal peptide, and the other is non-classically secreted proteins, whose secreted pathway is known as leaderless secretion (Nickel, 2003). Combination of SignalP v3.0, TargetP v1.01, big-PI predictor and TMHMM v2.0 (http://www.cbs.dtu.dk/ services/) are used to predict classically secreted effector proteins. Non-classical secreted proteins were further predicted using SecretomeP 2.0 Server (http://www.cbs.dtu.dk/services/).

Total of 12,595 putative proteins including 1,486 small proteins from *M. oryzae* genome database were predicted. Of which, 1,134 putative proteins were predicted for classically secreted proteins with N-terminal signal peptide. Here, we will focus on small secreted proteins (amino acid length <100), there were 119 classically secreted proteins among 1,486 small proteins. Among 119 effector proteins, 116 effectors had a Sec-type signal peptides, and had common A-X-A motif, X stand for any amino acid residue, C-domain of the signal peptide could be cleaved by one of the various type I SPase of *B.Subtilis* (Tjalsma et al., 1997; 1998; 1999). In C-domain, uncharged residues were present at the -1 and -3 positions, high frequency of leucine (29%) was at -2 position. Frequency of alanine at -1 position was 71%, and the other 19 amino acid residues occurred at +1 position except cysteine (C).Most of secretory proteins with this signal peptide are secreted into the extracellular environment. Length of signal peptides of 116 secretory proteins centralized in 16~22 amino acid residues, signal peptides with 18 amino acid residues reached the highest amounts, the second was signal peptide with 19 amino acid residues. Signal peptide with the most length was composed of 36 amino acid residues, and the shortest signal peptide was composed of 15 amino acid residues (Figure 1).

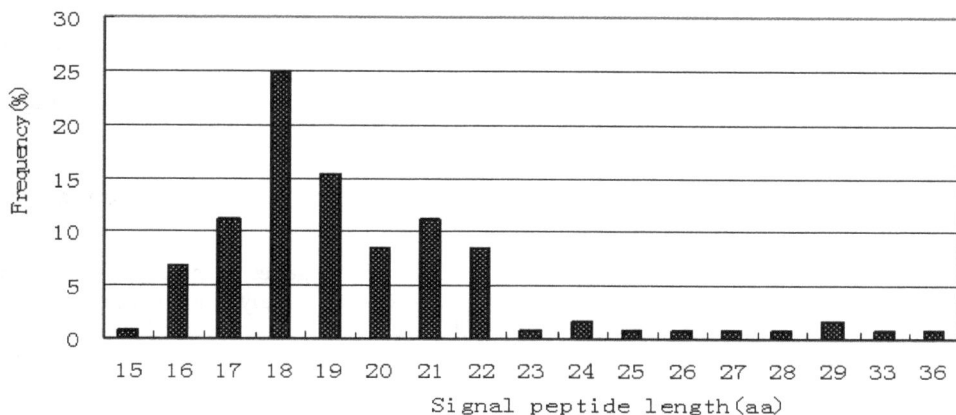

Fig. 1. Length distribution of predicted secretory (Sec-type) signal peptide of short protein in *M. oryzae*.

Because signal peptide has common application value in exogenous gene expression, it was necessary to analyze their composition and structure. We analyzed amino acid composition of 116 Sec-type signal peptides sequence. Frequency of 20 kinds of amino acids in 116 signal peptides sequences of secreted proteins were analyzed (Figure 2). Result showed that nonpolar amino acids such as alanine, leucine, proline and valine have the highest frequency(45.79%), followed by negatively charged acidic amino acids including aspartic acid, glutamic acid, phenylalanine, histidine, isoleucine, threonine, methionine, tryptophan and tyrosine (33.49%). The frequency of polar amino acids (glycine, asparagines, glutarnine, serine) positively charged basic amino acids such as arginine and lysine had the lowest frequency (15.83%).

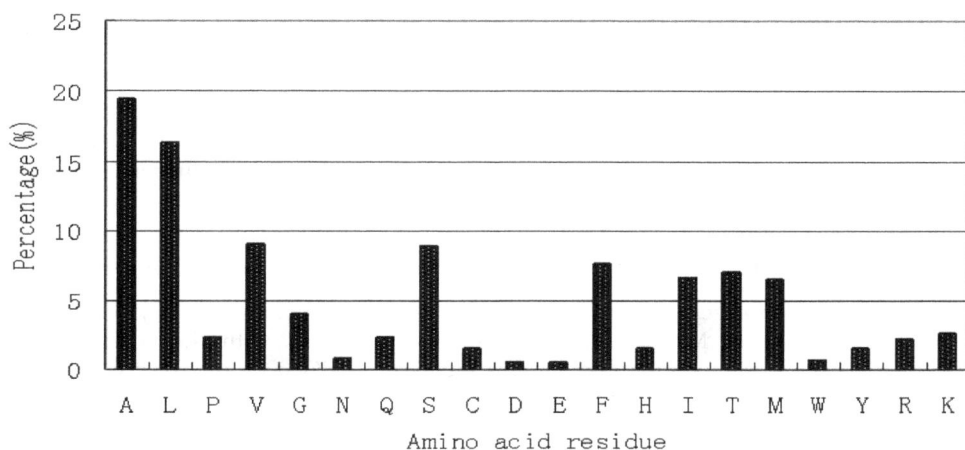

Fig. 2. Frequency of amino acid in predicted signal-peptide-containing short proteins.

Subcellular location of secreted effector proteins can provide important information to explain plant pathogen interactions. SubLoc v1.0 was used to predict subcellular location of 116 secreted effector proteins. Result showed that 50 proteins were secreted extracellularly, 30 proteins were transported into nucleus, 25 effectors were transferred into mitochondria and 11 were translocated into cytoplasm.

3. Polymorphism of effector-encoding genes in blast isolates from Yunnan, China

To analyze polymorphism distribution of effector-encoding genes in blast isolates from Yunnan, 45 ones from 116 genes were selected as candidates for analyzing polymorphism in 21 isolates from Yunnan. The result showed that each gene appeared in different distribution among 21 isolates, For example, MGS0001.1 was presented in 16 isolates, but not in five isolates. MGS0011.1 was distributed in 21 isolates. Although MGS0351.1 was distributed in 21 isolates with the PCR product size ranging from 350 to 400 base pairs, the reference sequence size of the gene was 412 bp. To explain sequence difference between PCR product and reference sequence of gene, PCR products of MGS0001.1 and MGS0351.1 from three different isolates were cloned and sequenced, respectively, the sequence analysis showed that PCR products sequences of MGS0001.1 from three different isolates appeared high identical with the reference sequence. While PCR product sequences of MGS0351.1 from three different isolates showed fragment- deletion of GTTGTTTTGTTGTT and GTTGTT, comparing with reference sequence, but the deletion occurred in intron region of the gene.

There was three-type polymorphism distribution of 45 genes in 21 blast isolates. The type I included 18 ones among 45 genes, which distributed in 21 isolates, the type II consisted only of MGS0351.1 which was present in 21 isolates, but PCR products showed fragment deletion comparing to reference sequence. The type III consisted of 26 genes that were randomly present in 21 isolates, while not all genes were distributed in 21 isolates. Among 45 effector-encoding genes, MGS0123.1 had the lowest frequency of 52.4%. Many genes could be examined in each isolate, except in isolate 21. Nineteen genes were not determined in the isolate 21. More than 40 genes could be determined in other 44 isolates, and all the 45 genes distributed in isolate 7, 14 and 15 (Table 1).

The results indicated that 45 effector-encoding genes not only had the polymorphism distribution but also appeared conserved in 21 blast isolates. Some genes were not determined in isolates, the reason might be the result of gene evolution during plant-pathogen interaction. Thus, it is essential to analyze their function for conserved or varied genes.

4. Effector-encoding gene cloning and *in vitro* functional identification

We selected 10 effector genes as candidates to clone and functionally identify. The cloned nucleotide sequence of MgNIP04 from Y99-63c was aligned with the short protein, MGS0004previously sequenced.MgNIP04 was identical to MGS0004.MgNIP04 encoding a 96 amino acid protein with unknown domains or motifs. MgNIP04 contained a signal peptide of 20 amino acids at the N-terminus. Subcellular localization prediction suggested that it was a cytoplasmic protein. However, it is not homologous to Nep1-like proteins, a novel class of necrosis-inducing proteins found in a variety of taxonomically unrelated microorganisms (Clare et al., 2004). The plasmids of pMALMgNIP04 and pMAL were induced to express of fusion proteins of MBP-MgNIP04 andMBP.

Gene code	Type	1	2	3	4	5	6	7	8	9	10	11	12	13	14	15	16	17	18	19	20	21	Frequency (%)
MGS0011.1		+	+	+	+	+	+	+	+	+	+	+	+	+	+	+	+	+	+	+	+	+	100
MGS0253.1		+	+	+	+	+	+	+	+	+	+	+	+	+	+	+	+	+	+	+	+	+	100
MGS0255.1		+	+	+	+	+	+	+	+	+	+	+	+	+	+	+	+	+	+	+	+	+	100
MGS0274.1		+	+	+	+	+	+	+	+	+	+	+	+	+	+	+	+	+	+	+	+	+	100
MGS0338.1		+	+	+	+	+	+	+	+	+	+	+	+	+	+	+	+	+	+	+	+	+	100
MGS0662.1		+	+	+	+	+	+	+	+	+	+	+	+	+	+	+	+	+	+	+	+	+	100
MGS0703.1		+	+	+	+	+	+	+	+	+	+	+	+	+	+	+	+	+	+	+	+	+	100
MGS0718.1		+	+	+	+	+	+	+	+	+	+	+	+	+	+	+	+	+	+	+	+	+	100
MGS0992.1	I	+	+	+	+	+	+	+	+	+	+	+	+	+	+	+	+	+	+	+	+	+	100
MGS0997.1		+	+	+	+	+	+	+	+	+	+	+	+	+	+	+	+	+	+	+	+	+	100
MGS1033.1		+	+	+	+	+	+	+	+	+	+	+	+	+	+	+	+	+	+	+	+	+	100
MGS1035.1		+	+	+	+	+	+	+	+	+	+	+	+	+	+	+	+	+	+	+	+	+	100
MGS1195.1		+	+	+	+	+	+	+	+	+	+	+	+	+	+	+	+	+	+	+	+	+	100
MGS1242.1		+	+	+	+	+	+	+	+	+	+	+	+	+	+	+	+	+	+	+	+	+	100
MGS1298.1		+	+	+	+	+	+	+	+	+	+	+	+	+	+	+	+	+	+	+	+	+	100
MGS1344.1		+	+	+	+	+	+	+	+	+	+	+	+	+	+	+	+	+	+	+	+	+	100
MGS1382.1		+	+	+	+	+	+	+	+	+	+	+	+	+	+	+	+	+	+	+	+	+	100
MGS1473.1		+	+	+	+	+	+	+	+	+	+	+	+	+	+	+	+	+	+	+	+	+	100
MGS0351.1	II	+	+	+	+	+	+	+	+	+	+	+	+	+	+	+	+	+	+	+	+	+	100
MGS0001.1		+	+	-	+	+	+	+	+	-	-	+	+	+	+	+	+	-	+	+	+	-	76.2
MGS0004.1		+	+	+	+	+	+	+	+	+	+	+	+	+	+	+	+	+	+	+	+	-	95.2
MGS0074.1		+	+	+	-	+	+	+	+	+	+	+	+	+	+	+	+	+	+	+	+	-	90.5
MGS0123.1		+	-	+	-	+	+	+	-	+	-	-	-	-	+	+	-	-	+	+	-	+	52.4
MGS0140.1		+	+	+	+	-	+	+	-	+	+	+	+	+	+	+	+	+	+	+	+	+	90.5
MGS0149.1		+	+	+	+	-	+	+	-	+	+	+	+	+	+	+	+	+	+	+	+	+	90.5
MGS0398.1		+	+	+	-	+	+	+	+	+	+	+	+	+	+	+	+	+	+	+	+	+	95.2
MGS0415.1		+	+	-	+	+	+	+	+	+	+	+	+	+	+	+	+	+	+	+	+	+	95.2
MGS0431.1		+	+	-	+	+	-	+	+	+	+	+	-	+	+	+	+	+	+	+	+	+	85.7
MGS0621.1		+	+	+	+	+	+	+	+	+	+	+	+	+	+	+	+	+	+	+	+	-	95.2
MGS0698.1		-	+	+	+	+	-	+	+	-	+	+	+	+	+	+	+	+	+	+	+	+	85.7
MGS0879.1		+	+	+	+	+	+	+	+	+	+	+	+	+	+	+	+	+	+	+	+	-	95.2
MGS1011.1	III	+	+	+	+	+	+	+	+	+	+	+	+	+	+	+	+	+	+	-	-	-	85.7
MGS1041.1		+	+	+	+	+	+	+	+	+	+	+	+	+	+	+	+	+	+	+	+	-	95.2
MGS1070.1		+	+	+	+	+	+	+	+	+	+	+	+	+	+	+	+	+	+	+	-	-	90.5
MGS1078.1		+	+	+	+	+	+	+	+	+	+	+	+	+	+	+	+	+	+	+	+	-	95.2
MGS1117.1		+	+	+	+	+	+	+	+	+	+	+	+	+	+	+	+	-	-	-	-	-	76.2
MGS1172.1		+	+	+	+	+	+	+	+	+	+	+	+	+	+	+	+	+	+	+	-	-	90.5
MGS1276.1		+	+	+	+	+	+	+	+	+	+	+	+	+	+	+	+	+	+	+	-	-	90.5
MGS1322.1		+	+	+	+	+	+	+	+	+	+	+	+	+	+	+	+	+	+	+	+	-	95.2
MGS1361.1		+	+	+	+	+	+	+	+	+	+	+	+	+	+	+	+	+	+	+	+	-	95.2
MGS1392.1		+	+	+	+	+	+	+	+	+	+	+	+	+	+	+	+	+	+	+	-	-	90.5
MGS1439.1		+	+	+	+	+	+	+	+	-	+	+	+	+	+	+	+	+	+	+	-	-	90.5
MGS1460.1		+	+	+	+	+	+	+	+	+	+	+	+	+	+	+	+	+	+	+	+	-	95.2
MGS1470.1		+	+	+	+	+	+	+	+	+	+	+	-	+	+	+	+	+	+	+	+	-	90.5
MGS1477.1		+	+	+	+	+	+	+	+	+	+	+	+	+	+	+	+	+	+	+	+	-	95.2
Frequency (%)		97.7	97.7	93.3	93.3	95.6	95.6	100	93.3	95.6	93.3	97.7	95.6	95.6	100	100	97.7	95.6	95.6	97.7	82.2	57.8	-

"+" mean gene was examined in isolates, "-"mean gene was not examined in isolates.

Table 1. Frequency of 45 genes in 21 isolates of M. oryzae from Yunnan, China.

To determine if MgNIP04 could interact with rice proteins, *E. coli*-expressing MBP-MgNIP04 was inoculated on wounded rice leaves. MBP expressed from pMAL-c2X was used as a control. The concentration of expression products of MBP-MgNIP04 and MBP were determined following the procedure of Bradford (1976) using bovine serum albumin as a standard. The expressed products of pMAL-MgNIP04 and pMAL-c2X at a concentration of 2.0 µg/µl were inoculated on wounded rice seedling leaf tissues. Necrotic specks formed around the inoculation site of leaf tissues that were inoculated with fused MBP-Mg04, while no necrotic specks appeared when inoculated with MBP. The results demonstrated that the protein encoded by MgNIP04 could directly interact with rice proteins. We also performed other experiments such as MgNIP04 *in vitro* induced H_2O_2 production, induced callose deposition in rice leaves and roots which could be automatically transported in rice root cells, when rice suspension-cultured cells were treated by MBP-MgNIP04.

The vector of pCAMbia-MgNIP04 was transformed into blast isolate of Y98-16. The differences of conidiation, germination, appressorium and pathogenicity between wild type strain of Y98-16 and transformant harboring MgNIP04::GFP were identified. The result revealed =no obvious difference in conidiation, germination and appressorium between Y98-16 and the transformant.The pathogenicity was further assayed for Y98-16 and transformant, rice cultivar Lijiangxintuanheigu that was almost susceptible to all races of blast fungus was challenged with blast strains. The result showed the transformant caused less symptom on Lijiangxintuanheigu than Y98-16. These data indicated *MgNIP04* did not influence blast fungus pathogenicity qualitatively but quantitatively, and virulence level of blast fungus decreased along with increasing of copy number of *MgNIP04* (there was one more copy at least in transformant than in Y98-16 although the gene copy number was not analyzed).

In order to test infecting ability difference of Y98-16 and transformant to rice roots, we used Y98-16 and transformant to inoculate the roots of Lijiangxintuanheigu. Brown symptom appeared on rice roots when blast fungus infected rice roots for 7 days. Brown symptom on the roots caused by Y98-16 regardless of areas and amounts of brown lateral roots regardless of areas and amounts of brown lateral roots was more apparent than by transformant. To determine whether *MgNIP04* was expressed in mycelia colonizing the roots, the roots infected by transformant and Y98-16 were observed using laser scanning confocal microscopy, respectively. The result showed that MgNIP04::GFP was observed in mycelia colonized roots. Along with infecting time elongation, the colonizing mycelia gradually increased. The frozen slices from brown- and no brown-root tissue were observed using laser scanning confocal microscopy. The results showed that mycelia not only had infected the epidermal cell but also colonized along root cell intervals.

5. Expression pattern of effector protein-encoding genes from *M. oryzae*

Many studies have used quantitative polymerase chain reaction (PCR) to evaluate fungal growth during the infection process (Hu et al., 1993; Mahuku et al., 1995; Groppe and Boller, 1997; Judelson and Tooley, 2000).Therefore, we detected the expression pattern of candidate novel genes *MGNIP10*, *MGNIP18*, *MGNIP24*, *MGNIP34*, *MGNIP38*, *MGNIP53*, *MGNIP74*, *MGNIP97* and *MgNIP04* in different isolates from Yunnan, China, the same isolate grown under nitrogen-starvation medium and complete medium and different time points when Lijiangxintuanheigu was challenged with blast fungus using real-time fluorescence quantitative PCR.

All expression levels of candidate genes were normalized by *actin* housekeeping gene and quantified by both the comparative threshold method and standard curve method. The results showed that expression levels of all candidate genes were significantly different in isolates of 94-64-1b Y99-63, 95-23-4a, Y98-16 and 94-64-1b. When two isolates of Y98-16 and Y99-63 grown under complete medium and nitrogen-starvation medium, relative expression quantity of genes was different. Expression of more genes was detected when two isolates grew under nitrogen starvation for 24 h, than when the two isolates grew under complete medium (Figure 3).

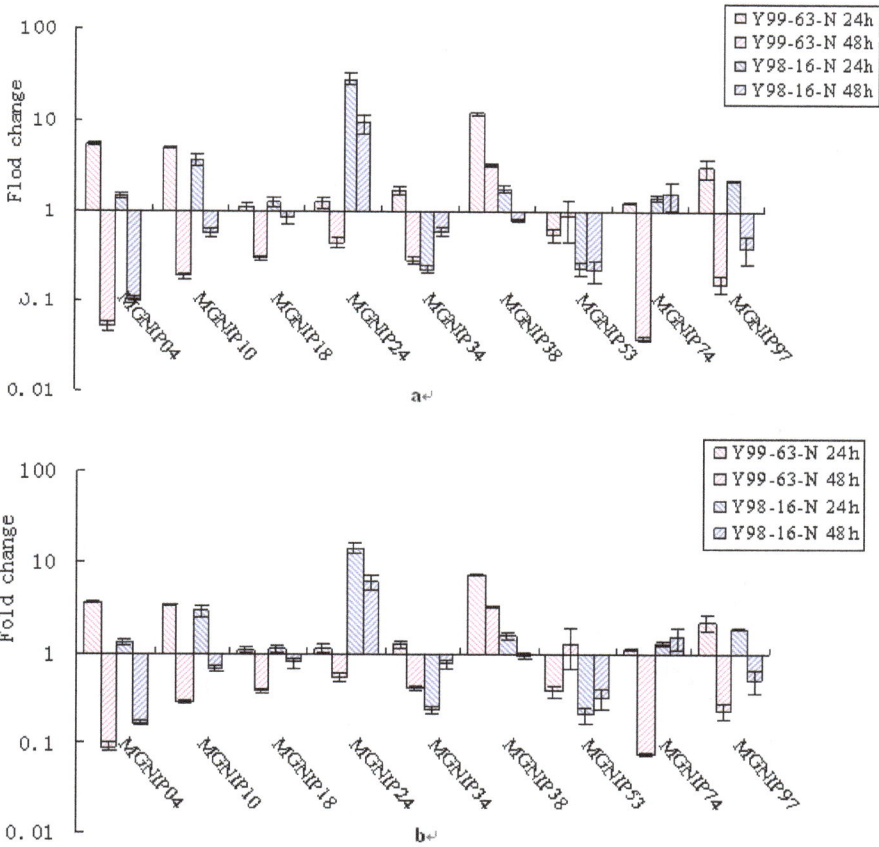

Fig. 3. Expression pattern of some predicted effector protein-encoding genes from *M.oryzae*.
a: Relative expression quantity of target gene of Y99-63 and Y98-16 cultured in different mediums using $2^{-\Delta\Delta Ct}$ method
b: Relative expression quantity of target gene of Y99-63 and Y98-16 cultured in different mediums using the standard curve method.

We detected expression level of all candidate genes at 24 hpi, 48 hpi, 72 hpi, 96 hpi and 168 hpi. Results revealed that all gene expression levels were apparently up-regulated, achieved the maximum at 48hpi, but decreased after 72 hpi (Table 2 and Figure 4).

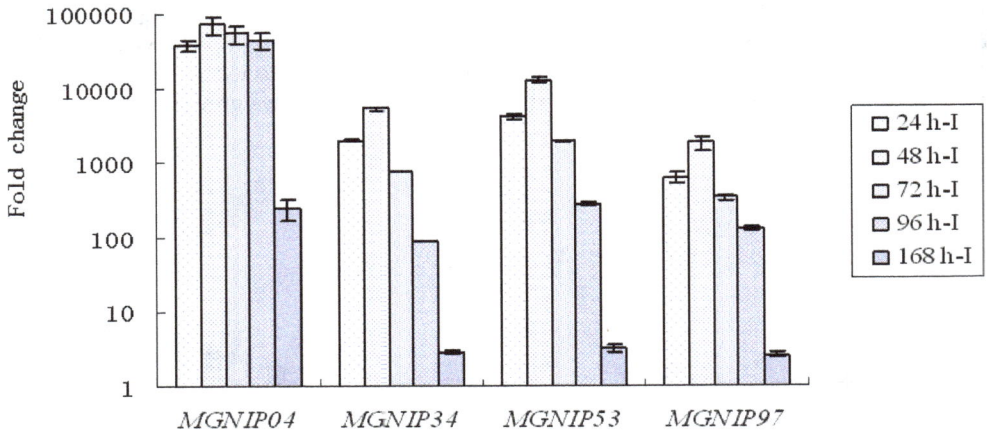

a

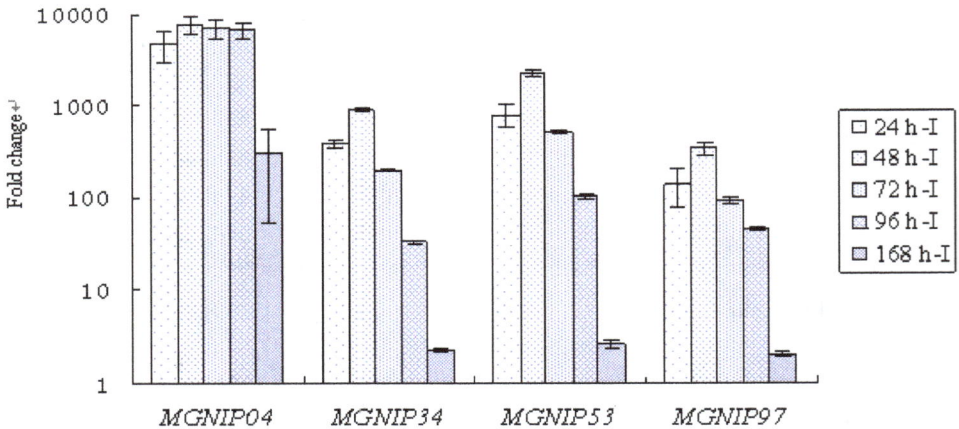

b

Fig. 4. Expression pattern of some predicted effector protein-encoding genes in infected rice leaves.
a: Relative expression amount of target gene in infected rice leaves at different stages post inoculation using $2^{-\Delta\Delta Ct}$ method
b: Relative expression amount of target gene in infected rice leaves at different stages post inoculation using stand curve method.

Isolates Y99-63		24 h-I		48 h-I		72 h-I		96 h-I		168 h-I	
Targeted gene	Control	$2^{-\Delta\Delta Ct}$ Fold change	Standard curve Fold change	$2^{-\Delta\Delta Ct}$ Fold change	Standard curve Fold change	$2^{-\Delta\Delta Ct}$ Fold change	Standard curve Fold change	$2^{-\Delta\Delta Ct}$ Fold change	Standard curve Fold change	$2^{-\Delta\Delta Ct}$ Fold change	Standard curve Fold change
MGNIP04	1	37602± 6231	4800± 854	71139± 19038	7660± 1802	54388± 14248	7101± 1626	44878± 10524	6812± 1360	243± 76.8	291± 237
MGNIP10	1	4.88± 0.22	2.24± 0.08	8.92± 1.04	3.63± 0.37	0.92± 0.04	0.6± 0.02	0.29± 0.01	0.25± 0.00	0.1± 0.01	0.13± 0.02
MGNIP18	1	1799± 201	334± 28.1	2386± 743	414± 95.7	359± 36.3	95± 7.37	76± 4.36	29± 1.26	5± 0.48	3± 0.24
MGNIP24	1	236± 37.5	77± 15.49	287± 14.8	77± 3.26	68± 9.84	25± 2.95	19± 0.89	10± 0.36	1± 0.06	0.6± 0.04
MGNIP34	1	2012± 68.3	376± 11.7	5401± 297	879± 44.2	773± 8.74	192± 1.99	88± 1.51	32± 0.51	3± 0.11	2± 0.08
MGNIP38	1	1323± 121	263± 21.4	1959± 512	354± 79.7	282± 20.7	78± 4.98	100± 17.0	36± 5.35	8± 0.57	5± 0.35
MGNIP53	1	4117± 306	776± 56	12888± 1242	2198± 209	1931± 87.3	501± 22.5	277± 14.2	101± 5.10	3± 0.27	3± 0.22
MGNIP74	1	9± 2.00	5± 0.82	17± 5.22	8± 1.87	5± 0.39	3± 0.19	3± 0.17	2± 0.09	0.1± 0.01	0.1± 0.01
MGNIP97	1	634± 98.7	135± 18.5	1825± 385	325± 59.7	337± 34.3	89± 7.86	130± 9.00	45± 2.65	3± 0.19	2± 0.13

Table 2. Relative expression quantity of target gene in infected rice leaves at different stages post inoculation.

6. Rice defense-related gene expression pattern

Based on the results, we knew that *MgNIP04 in vitro* expression products could induce callose deposition of rice suspension cells, up-taken by rice root without presence of pathogen and quantitatively influence virulence level of blast fungus Y98-16. Are there differences of defense-related gene expression pattern between Y98-16 and transformant infecting Lijiangxintuanheigu, respectively? We used RT-PCR to analysis gene expression of selected defense-related genes when rice cultivar of Lijiangxintuanheigu was inoculated by Y98-16 and transformant, respectively. The genes of *APXa* (AY254495.1), *Chia4a* (AB096140.1), *CHS* (AB058397.1), *OsPAL* (AX16099.1), *OsPHGPX* (AJ270955.1), *OsPR1a* (AJ278436.1) and *PR-10a* (AF274850.1) were expressed from 24hpi to 168hpi when Lijiangxintuanheigu was challenged with Y98-16 and transformant, respectively, and rice β-actin gene (CT831215.1) was control. The expression of *OsGST2* (AJ486976.1), *PR-10c* (AF274852.1) and *Npr1* (DQ450949.1) was not detected at any time points regardless of Y98-16 or transformant inoculating Lijiangxintuanheigu. The expression of *PR-10b* was detected at all selected time points during transformant infecting rice, but it was not detected during Y98-16 infecting rice. The expression of the ethylene synthesase gene was only detected at 0, 24, 48 and 96hpi when Lijiangxintuanheigu was challenged with Y98-16, while the gene expression was detected at all time points when transformant inoculating rice. The expression of *OsPR4* was only detected at 168hpi during transformant inoculating rice. Based on these data, the most tested defense-related genes expression pattern was not any different between Y98-16 and transformant of MgNIP04::GFP, which indicated *MgNIP04* quantitatively influenced pathogenicity of blast fungus.

7. Expression difference of effector-encoding genes from blast isolates with different virulence determined using two-dimensional gel electrophoresis

Fungi maintained their cell living and even growth through material reutilization when they were in nutrition-stress environment. Some research showed that expression quantity of pathogenicity-related genes increased when rice blast strains grew under nitrogen-starvation medium, which enhanced the pathogenicity of blast strains (Talbot et al., 1997).

The two isolates of Y99-63 and Y98-16 were from Yunnnan, China. Virulence test of two isolates of Y98-16 and Y99-63 on rice isogenic lines of IRBL1-24 had been previously performed in our lab, and virulence of Y99-63 was more intensive than Y98-16. To analyze the virulence of extracellularly secreted proteins on rice varieties such as susceptible variety of Lijiangxintuanheigu, resistant variety of Tetep and rice isogenic lines of IRBL1-24, we separated the extracellularly secreted proteins when Y98-16 and Y99-63 grew under nitrogen starvation for 48h, and the wounded rice leaves were inoculated with extracellularly secreted proteins. The result showed that necrosis speck occurred around the wounded leaves and wounded stems of rice when secreted proteins were inoculated on leaves or stems for 48h, and speck diameter of leaves or stems treated with secreted proteins was 2 to 4 fold larger than leaves or stems treated with sterilized water.

We compared difference of extracellularly secreted proteins from Y99-63 and Y98-16 growing under nitrogen-starvation medium for 48h using two-dimensional electrophoresis technology (Figure 5). The result showed that more proteins spots were detected from Y99-63 growing under nitrogen-starvation medium than Y98-16 (Table 3). And pI and molecular weight of secreted proteins had an apparent difference between Y99-63 and Y98-16 (Figure 6 and Figure 7).

Strain	Replicate group	Protein spots	Protein matched spots	Match Rate 1	Match Rate 2
*					
Y98-16	3	253±10	253±10	100%	100%
Y99-63	3	262±10	132±8	50.4%	52.2%

Note: * mean master reference gel; Match rate 1 for the match-point block of gel protein spots representing the ratio; Match rate 2 is the ratio of match point to total master.

Table 3. Comparison of 2-DE maps of two strains in *M.oryzae*.

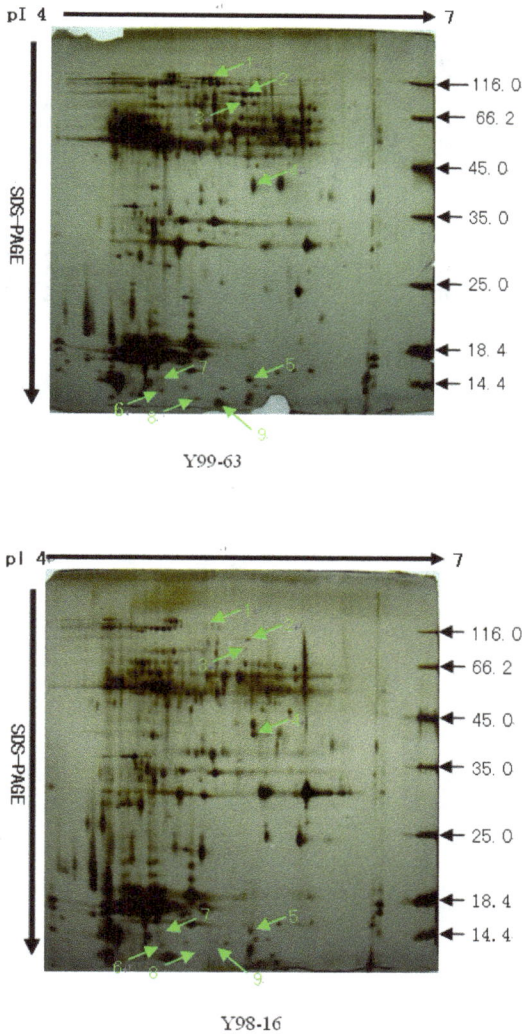

Fig. 5. 2-DE maps of Y99-63 and Y98-16.Protein spots labeled with arrow (1 to 5)are co-expressed in two strains of Y99-63 and Y98-16 with the pI ranging from 5.5 to 6.0, but the expression level of Y99-63 is over five times more than Y98-16. Protein spots indicated by arrow (6 to 9) are proteins which are specifically expressed in the strain Y99-63, with their MW ranging from 10 to 20 kDa and pI from 5.0 to 6.0. MW is indicated on the right side in KD. IEF is abbreviation for isoelectric focusing and SDS-PAGE is for SDS-polyacrylamide gel electrophoresis.

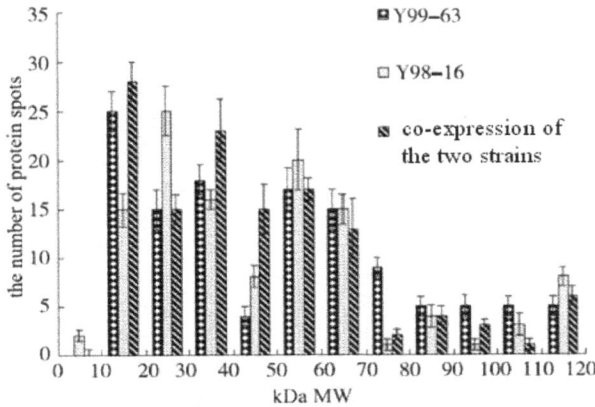

Fig. 6. Comparison of the distribution of the proteins spots based on molecular weight.

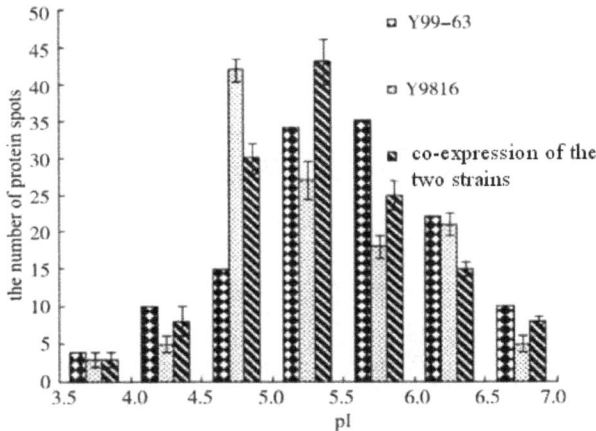

Fig. 7. Comparison of the distribution of the proteins spots based on pI.

8. Acknowledgment

This work was supported by the National Basic Research Program (No. 2011CB100400) from The Ministry of Science and Technology of China and the National Natural Science Funds, China (30860161), respectively.

9. References

Birch, P. R., Rehmany, A. P., Pritchard, L., Kamoun, S. & Beynon, J. L. (2006). Trafficking Arms: Oomycete Effectors Enter Host Plant Cells. Trends Microbiol. 14:8-11

Bohnert, H. U., Fudal, I., Dior, W., Tharreau, D., Notteghem J. L. & Lebrun, M. H.(2004). A Putative Polyketide Synthase/Peptide Synthetase from *Magnaporthe grisea* Signals Pathogen Attack to Resistance Rice. Plant Cell, 16:2499-2513.

Bradford, M. M. (1976). A Rapid and Sensitive Method for the Quantitation of Microgram Quantities of Protein Utilizing the Principle of Protein-dye Binding. Anal Biochem, 72:249–254.

Catanzariti, A.M., Dodds, P.N. & Ellis, J.G. (2007). Avirulence Proteins from Haustoria-Forming Pathogens. FEMS Microbiol. Lett. 269:181-188.

Chisholm, S.T., Coaker, G., Day, B. & Staskawicz, B. J. (2006). Host-Microbe Interactions: Shaping the Evolution of the Plant Immune Response. Cell. 124:803-814.

Clare, L., Salmond & Gorge P.C. (2004). The Nep1-like Proteins – a Growing Family of Microbial Elicitors of Plant Necrosis. Mol Plant Pathol, 5:353–359.

Collemare, J., Pianfetti, M., Houlle, A. E., Morin, D., Camborde, L., Gagey, M.J., Barbisan, C., Fudal, I., Lebrun, M.H. & Böhnert, H.U.(2008). *Magnaporthe grisea* Avirulence Gene *ACE1* Belongs to an Infection-Specific Gene Cluster Involved in Secondary Metabolism. New Phytol. 179:196-208.

Couch, B. C. & Kohn L.M. (2002). A multilocus gene genealogy concordant with host preference indicates segregation of a new species, *Magnaporthe oryzae*, from M. *grisea*. Mycologia 94(4): 683-693.

Farman, M. L. & Leong, S. A.(1998). Chromosome Walking to the AVR1-CO39 Avirulence Gene of Magnaporthe grisea: Discrepancy between the Physical and Genetic Maps. Genetics, 150:1049-1058.

Fudal, I., Bohnert, H. U., Tharreau, D & Lebrun, M. H. (2005). Transposition of MINE, a Composite Retrotransposon, in the Avirulence Gene *ACE1* of the Rice Blast Fungus Magnaporthe grisea. Fungal. Genet. Biol., 42:761-772

Idnurm, A. & Howlett, B.J. (2001).Pathogen city Genes of Phytopathogenic Fungi. Mol. Plant Pathol.2:241-255.

Jia Y, Correll, J.C., Lee, F.N., Eizenga, G.C., Yang, Y. , Gealy, D.R., Valent, B.,& Zhu, Q. 2001. Understanding molecular interaction mechanisms of the *Pi-ta* rice resistance genes and the rice blast pathogen. Phytopathology 91(Suppl. 6):S44(abstr.).

Judelson, H.S., Tooley, P.W. (2000). Enhanced Polymerase Chain Reaction Methods for Detecting and Quantifying Phytophthora infestans in Plants. Phytopathology 90:1112–1119.

Kamoun, S. (2006). A Catalogue of the Effector Secretome of Plant Pathogenic Oomycetes. Annual Review of Phytopathology. 44: 41–60.

Kamoun, S. (2007). Groovy Times: Filamentous Pathogen Effectors Revealed. Curr. Opin. Plant Biol.10:358-365.

Kang, S., Sweigard, J. A., & Valent, B. (1995). The PWL Host Specificity Gene Family in the Blast Fungus Magnaporthe grisea. Mol. Plant-Microbe Interact, 8:939-948

Khang, C.H., Park, S. Y., Lee, Y. H., Valent, B., & Kang, S.(2008). Genome Organization and Evolution of the *AVR-Pita* Avirulence Gene Family in the *Magnaporthe grisea* Species Complex. Mol. Plant-Microbe Interact, 21:658-670.

Mahuku, G.S., Goodwin, P.H. & Hall, R. (1995). A Competitive Polymerase Chain Reaction to Quantify DNA of Leptosphaeria maculans During Blackleg Development in Oilseed Rape. Mol Plant-Microbe Interact 8:761–767.

Nickel, W. (2003). The mystery of nonclassical protein secretion. *Eur. J. Biochem.* 270, 2109-2119.

O'Connell, R.J. & Panstruga, R. (2006). Tete a tete inside a Plant Cell: Establishing Compatibility between Plants and Biotrophic Fungi and Oomycetes. New Phytol.171:699-718.

Orbach, M. J., Farrall, L., Sweigard, J.A., Chumley, F. G. & Valent, B. (2000). A Telomeric Avirulence Gene Determines Efficacy for the Rice Blast Resistance Gene Pi-ta. Plant Cell, 12:2019-2032

Randall, T.A. et al. (2005). Large-Scale Gene Discovery in the Oomycete *Phytophthora infestans* Reveals Likely Components of Phytopathogenicity Shared with True Fungi. Mol. Plant Microbe Interact. 18:229-243.

Sweigard, J. A., Carroll, A. M., Kang, S., Farrall, L., Chumley, F. G. & Valent, B. (1995). Identification, Cloning, and Characterization of *PWL2*, a Gene for Host Species Specificity in the Rice Blast Fungus. Plant Cell, 7:1221-1233.

Talbot NJ,McCafeny HRK, Ma M et al. (1997). Nitrogen Starvation of the Rice Blast Fungus *Magnaporthe grisea* may Act as an Environmental Cue for Disease Symptom Expression. Physiological and Molecular Plant Pathology.50:179-195

Talbot, N. J. (2003). On the Trail of a Cereal Killer: Exploring the Biology of *Magnaporthe grisea*. Annu. Rev. Microbiol. 57:177-202.

Tjalsma, H., Noback, M. A., Bron, S., Venema, G., Yamane, K., & van Dijl J. M. (1997).*Bacillus subtilis* Contains Four Closely Related Type I Signal Peptidases with Overlapping Substrate Specificities: Constitutive and Temporally Controlled Expression of Different *sip* Genes. Journal Biological Chemistry, 272:25983–25992.

Tjalsma, H., Bolhuis, A., van Roosmalen, M. L., Wiegert, T., Schumann, W., Broekhuizen, C. P., Quax, W., Venema, G., Bron, S., & van Dijl, J. M. (1998).Functional analysis of the secretory precursor processing machinery of *Bacillus subtilis*: identification of a eubacterial homolog of archaeal and eukaryotic signal peptidases. *Genes Developement*, 12:2318–2331.

Tjalsma, H., van den Dolder, J., Meijer, W. J. J., Venema, G., Bron, S., & van Dijl, J. M. (1999). The plasmid-encoded type I signal peptidase SipP can functionally replace the major signal peptidases SipS and SipT of *Bacillus subtilis*. *Journal Bacteriology*, 181:2448–2454.

Valent, B., Farrall, L., & Chumley, F. G.(1991). Magnaporthe grisea Genes for Pathogenicity and Virulence Identified Through a Series of Backcrosses. Genetics, 127:87-101.

Yoshida, K., Saitoh, H., Fujisawa, S., Kanzaki, H., Matsumura, H., Yoshida, K., Tosa, Y., Chuma, I., Takano,Y., Win, J., Kamoun, S., & Terauchi, R. (2009). Association Genetics Reveals Three Novel Avirulence Genes from the Rice Blast Fungal Pathogen *Magnaporthe oryzae*. Plant Cell, 21:1573-1591

6

Molecular Tools for Detection of Plant Pathogenic Fungi and Fungicide Resistance

Nieves Capote[1], Ana María Pastrana[1],
Ana Aguado[1] and Paloma Sánchez-Torres[2]
[1]IFAPA Las Torres-Tomejil, Junta de Andalucía, Alcalá del Río, Sevilla
[2]IVIA, Generalitat Valenciana, Moncada, Valencia
Spain

1. Introduction

Plant pathogenic fungi are the causal agents of the most detrimental diseases in plants, including economically important crops, provoking considerable yield losses worldwide. Fungal pathogens can infect a wide range of plant species or be restricted to one or few host species. Some of them are obligate parasites requiring the presence of the living host to grow and reproduce, but most of them are saprophytic and can survive without the presence of the living plant, in the soil, water or air. Isolates of a fungal species can be differentiated by morphological characteristics, host range (*formae speciales*), pathogenic aggressiveness (pathotypes or races) or their ability to form stable vegetative heterokaryons by fusion between genetically different strains (belonging to the same vegetative compatibility group, VCG).

Detection and accurate identification of plant pathogens is one of the most important strategies for controlling plant diseases to initiate preventive or curative measures. Special interest should be taken in the early detection of pathogens in seeds, mother plants and propagative plant material to avoid the introduction and further spreading of new pathogens in a growing area where it is not present yet. For that reason, the availability of fast, sensitive and accurate methods for detection and identification of fungal pathogens is increasingly necessary to improve disease control decision making. Traditionally, the most prevalent techniques used to identify plant pathogens relied upon culture-based morphological approaches. These methods, however, are often time-consuming, laborious, and require extensive knowledge of classical taxonomy. Other limitations include the difficulty of some species to be cultured *in vitro*, and the inability to accurately quantify the pathogen (Goud & Termorshuizen, 2003). These limitations have led to the development of molecular approaches with improved accuracy and reliability. A high variety of molecular methods have been used to detect, identify and quantify a long list of plant pathogenic fungi. Molecular methods have also been applied to the study of the genetic variability of pathogen populations, and even for the description of new fungal species. In general, these methods are much faster, more specific, more sensitive, and more accurate, and can be performed and interpreted by personnel with no specialized taxonomical expertise. Additionally, these techniques allow the detection and identification of non-culturable microorganisms, and due to its high degree of specificity, molecular techniques can distinguish closely related organisms at different taxonomic levels. Here, we review the most important tools for molecular detection of plant pathogenic fungi, their applicability, and their implementation in horticultural and agricultural practices.

On the other hand, once the pathogenic fungus is already established in a given crop growing area, the use of synthetic fungicides constitutes the main strategy to control plant diseases, since these compounds act quickly and effectively. The disadvantages of continued use of fungicides are their limited spectrum and the emergence of resistant fungal isolates. This fact leads to many yield losses as control systems are not longer effective. The development of resistance to fungicides in fungal pathogens and the growing public concern over the health and environmental hazards associated with the high level of pesticide have resulted in a significant interest in knowing more about fungal resistance. The emergence of more stringent regulations regarding pesticide residues means that one of the main priorities is to ensure food security by reducing the use of fungi toxics. For this reason, it is important to identify and characterize the mechanisms involved in the emergence of strains resistant to fungicides used for control diseases and to know the molecular methods currently available to detect them.

2. Molecular methods for detection of plant pathogenic fungi

2.1 Polymerase Chain Reaction (PCR)

The polymerase chain reaction (PCR) is the most important and sensitive technique presently available for the detection of plant pathogens. PCR allows the amplification of millions of copies of specific DNA sequences by repeated cycles of denaturation, polymerisation and elongation at different temperatures using specific oligonucleotides (primers), deoxyribonucleotide triphosphates (dNTPs) and a thermostable *Taq* DNA polymerase in the adequate buffer (Mullis & Faloona, 1987). The amplified DNA fragments are visualized by electrophoresis in agarose gel stained with EtBr, SYBR Green or other safer molecule able to intercalate in the double stranded DNA, or alternatively by colorimetric (Mutasa et al., 1996) or fluorometric assays (Fraaije et al., 1999). The presence of a specific DNA band of the expected size indicates the presence of the target pathogen in the sample. Advances in PCR-based methods, such as real-time PCR, allow fast, accurate detection and quantification of plant pathogens in an automated reaction. Main advantages of PCR techniques include high sensitivity, specificity and reliability. Moreover, it is not necessary to isolate the pathogen from the infected material reducing the diagnosis time from weeks to hours, and allowing the detection and identification of non-culturable pathogens. This characteristic has been especially useful in the analysis of symptomless plants. However, the frequent presence of PCR inhibitors in the plant tissues or soil can reduce considerably the sensitivity of the reaction and even result in false negative detection. Many attempts have been carried out to overcome this issue (see below). Another disadvantage of the PCR methodology is the occurrence of false positive results due to the presence of DNA or PCR products (amplicons) contaminants. For this reason, it is advisable to separate physically and temporally pre- and post-PCR analysis. Another failure of PCR-based methods is the inability to discriminate viable from non-viable fungi or fungal structures, which might inform of the real threat for the plant. Development of a prior PCR step involving enrichment culturing (BIO-PCR) (Ozakman &

Schaad, 2003) or amplification of fungal RNA (RT-PCR) (Lee et al., 1989) may solve this problem to same extent. Like this, many PCR variants have been developed to improve sensitivity, specificity, rapidity and throughput, and to allow the quantification of the fungal pathogen in the plant and the environment.

2.1.1 Starting material

Collection and preparation of samples is a critical step for the detection of plant pathogenic fungi. The starting material may be symptomatic plant tissue (roots, leaves, stems, flowers, fruits or seeds), soil, water or air. Also, latent infections can be detected on symptomless plants.

In the case of infected plants, the first step consists in the cultivation of the fungi. After surface sterilisation of the plant tissue (e.g. with 1% sodium hypochlorite or 50% hydrogen peroxide) small pieces are transferred to Petri dishes containing an appropriate nutrient medium (e.g. Potato Dextrosa Agar, PDA) supplemented with antibiotics to prevent bacterial contaminants (usually streptomycin), and incubated at required temperature (25-30°C) for pathogen development. If the fungal pathogen has infected deep internal tissues, cutting the plant material to expose core tissues may be necessary. Pure fungal colonies must be obtained either by isolating single spore or single hyphal tips. In the first case, a spore suspension of the fungus is prepared and serial dilutions in sterile water are obtained. An optimal dilution is transferred to Petri dishes containing 2% agar-water and incubated at appropriate temperature to induce spore germination. A single germinating spore is then isolated and a pure colony of the fungus is obtained. To get a single hyphal tip, a small segment of fungal growth is transferred to a new Petri dish containing nutrient medium and incubated at optimum temperature to allow growth of the mycellium. A separate hyphal edge is them transferred to new agar plate to obtain a pure colony of the fungal species. Isolation of the fungus from an infected plant tissue can however obviate the presence in the infected plant of other non-culturable pathogens or fungi having special culture conditions. Competition between distinct fungi can also mask the detection by culturing of an infecting pathogen. For DNA extraction, fungal mycelium must be firstly homogenized in a mortar in the presence of liquid nitrogen. Other homogenisation procedures obviate the use of liquid nitrogen, e.g, grinding the mycellium inside centrifuge tubes with the help of sealed tips or plastic pistils that fit perfectly into the tubes coupled to household drills (González et al., 2008).

Alternatively, total DNA from the plant and the fungi can be isolated together from the infected plant tissue. That allows skipping the fungi culture step, although DNA from different fungal species or strains may be obtained. Homogenisation of the infected plant tissues is usually performed in the presence of liquid nitrogen. Then, an extraction buffer is added to obtain a crude extract. Alternatively, plant material may be introduced into individual plastic bags containing a soft net in the case of tender material (Homex, Bioreba; Stomacher, AES Laboratoire) or a heavy net in the case of dry or harder material such as bark tissue or seeds (PlantPrint Diagnostics). Homogenisation can be made by the help of a manual roller or by the use of special apparatus designed to facilitate the homogenisation (e.g. Homex, Bioreba). As in the case of fungal mycellium, small amounts of plant tissue can be homogenised inside centrifuge tubes using pistils coupled to electric drills in the presence of extraction buffers.

When analysing soil samples, the main focus for phytopathologists is the isolation of DNA from different microorganisms and then, the specific detection and monitoring of the fungus(i) of interest. Classical approach consists of cultivating the soil fungi in different media and screen for the desired pathogen. However, many microorganisms from the soil community can not be isolated by this procedure. An alternative method is to isolate DNA directly from the soil sample without prior culturing. Protocols using enzymatic (e.g. protease, chitinase, glucanase) or mechanical lysis (glass-beads beating, freeze-thawing, vigorous shaking, microwave or grinding in liquid nitrogen) have been reported, but a combination of both procedures seems to be the more effective (Anderson I. C. & Parkin, 2007, Jiang et al., 2011). In the same way, improved protocols for an efficient isolation of DNA from water for detection and monitoring of plant pathogens have been reported (Pereira et al., 2010).

During the homogenisation process, polysaccharides and phenolic compounds from plants or humic and fulvic acids from soils can be released that can inhibit the *Taq* DNA polymerase leading to the occurrence of false negatives (Munford et al., 2006; Tebbe & Vahjen, 1993; Wilson, 1997). This problem may be partially overcome by the use in the extraction buffer of some compounds such as polyvinylpyrrolidone (PVP) or cetyltrimethyl ammonium bromide (CTAB) for plant extracts, and bovine sero albumine (BSA) for soil samples (Anderson I.C. & Cairney, 2004), or by removing inhibitors by the use of spin/vacuum columns. Some PCR variants such as Magnetic Capture-Hybridisation PCR have been developed to remove the presence of PCR inhibitors in plant extracts (see below). In addition, it is increasingly common to use an internal positive control of the PCR reaction either by the amplification of a conserved plant gene (e.g. cytochrome oxidase I, *cox* I) in multiplex (Bilodeau, et al., 2009) or in a parallel assay (Garrido et al., 2009), or by the addition of an exogenous DNA and their corresponding primers to each reaction (Cruz-Pérez et al., 2001).

2.1.2 DNA extraction methods

There are no universally validated nucleic-acids extraction protocols for fungi, infected plant material or soil. Many published protocols are available to ensure an efficient and reproducible method for DNA extraction from plants (revised by Demeke and Jenkins, 2010; Biswas and Biswas, 2011); from fungi (Chi et al., 2009; Feng et al., 2010; González-Mendoza et al., 2010; Niu et al., 2008; Zelaya-Molina et al., 2011; Zhang, Y. J. et al., 2010); and from soil (revised by Hirsch et al., 2010).

The use of commercial kits for nucleic acids extraction, either general or specifically designed for plant material, fungi or soil is gaining acceptance because they are easy to use and are enable to efficiently remove inhibitory compounds during the purification process. They are generally based on magnetic beads or spin columns, although quicker protocols are also available (e.g. QuickExtract™ Plant DNA Extraction Solution, Epicentre). Automated or semi-automated systems have also been developed to allow the isolation of nucleic acids from different samples, among others, QIAxtractor, QIAgen; 6700 Automated Nucleic Acid Workstation, Life Technologies; Magna Pure LC extraction system, Roche; Solucion m2000, Abbott.

Purification of nucleic acids is labour intensive, costly, time-consuming and not applicable when a large number of samples need to be analysed. Several attempts have been undertaken to avoid the nucleic acids isolation step. One of them uses few microliters of crude extract loaded and immobilized on small pieces of paper, e.g FTA cards. A subsequent lysis of the cells in appropriate buffer allows the release of nucleic acids that are fixed in the membrane and protected from degradation. DNA can be stored on dry cards for several years in a dry place at room temperature without decreasing the sensitivity of detection (Smith & Burgoyne, 2004). Moreover, membrane immobilised-DNA is suitable of transportation or mail to other laboratories. Suzuki et al. (2006) reported that nucleic acids recovered from FTA cards could be used for the detection of *Aspergillus oryzae*, releasing the DNA from the fungal tissue by treatment in a microwave oven before application to the membranes. Grund et al. (2010) used FTA cards coupled to PCR for the detection of plant pathogens including oomycetes such as *Phytophthora* and filamentous fungi such as *Fusarium*.

2.1.3 Selection of target DNA to amplify

Generally, conserved known genes with enough sequence variation are selected for designing PCR diagnostic assays and performing phylogenetic analysis. The most common region used for these purposes has been the internal transcribed spacer (ITS) region of ribosomal RNA genes. rDNA region consists of multiple copies (up to 200 copies per haploid genome) arranged in tandem repeats comprising the 18S small subunit, the 5.8S, and the 28S large subunit genes separated by internal transcribed spacer regions (ITS1 and ITS2) (Bruns et al., 1991; Liew et al., 1998). This region contains highly conserved areas adequate for genera- o species-consensus primer designing (RNA ribosomal genes), alternate with highly variable areas that allow discrimination over a wide range of taxonomic levels (ITS region) (White et al., 1990). The ITS region is ubiquitous in nature and found in all eukaryotes. In addition, the high copy numbers of rRNA genes in the fungal genome enable a highly sensitive PCR amplification. Furthermore, a large numbers of ribosomal sequences are publicly available in databases, facilitating the validation and the reliability of the detection assays.

Traditionally, molecular identification of plant pathogenic fungi is accomplished by PCR amplification of ITS region followed by either restriction analysis (Durán et al., 2010) or direct sequencing and BLAST searching against GenBank or other databases (White et al., 1990). Identification could be a challenge when using BLAST analysis with ITS sequences because there can be minimal or no differences between some species or, in some cases, intraspecific variation can confuse the boundaries between species (e.g., *P. fragariae* var. *fragariae* and *P. fragariae* var. *rubi* have identical ITS sequences). The sequence analysis of the ITS region has additionally served to proposed new species. Abad et al. (2008) aligned ITS regions of *Phytophthora* spp. associated with root rot from different geographic origins and hosts with GenBank sequences from other *Phytophthora* species and proposed a new isolate *Phytophthora bisheria* sp. nov. Just as, Burgess et al. (2009) also distinguished new and undescribed taxa of *Phytophthora* from natural ecosystems. This alignment analysis has also been used for the identification of new species of *Pythium* (de Cock & Lévesque, 2004; Nechwatal & Mendgen, 2006; Paul, 2006, 2009; Paul et al., 2005; Paul & Bala, 2008).

The ITS region has also been widely used in fungal taxonomy and it is known to show variation between species e.g., between *Phytium ultimum* and *P. helicoides* (Kageyama et al., 2007); *Peronospora arborescens* and *P. cristata* (Landa et al., 2007); *Colletotrichum gloeosporioides* and *C. acutatum* (Kim J. T. et al., 2008), and within species e.g allowing differentiation of

Puccinia striiformis f. sp. *tritici* (Zhao et al., 2007), and distinguishing between weakly and high virulent isolates of *Leptosphaeria maculans* (Xue et al., 1992), or difining anastomosis subgroups of *Rhizoctonia solani* (Budge, 2009; Godoy-Lutz et al., 2008).

The ITS region remains an important locus for molecular identification of fungi. However, as more sequence data is collected from a wider range of fungal isolates, the utility of alternative loci for accurate species identification is increasing. The intergenic spacer sequence (IGS) placed between the 28S and 18S rRNA genes is the region with the greatest amount of sequence variation in rDNA. It is frequently used in PCR-based methods when there are not enough differences available across the ITS. Primers in this region have been designed to detect and identify *Verticillium dahliae* and *V. alboatrum* (Schena et al., 2004) and to distinguish pathogenic and non-pathogenic *Fusarium oxysporum* in tomato (Validov et al., 2011). As another example, Inami et al. (2010) differentiated *Fusarium oxysporum f. sp lycopersici*, and its races using primers and TaqMan-MGB probes based on IGS and avirulent SIX genes.

Other housekeeping genes with higher variability are being more extensively used to develop diagnostics for fungi, including nuclear genes such as β-tubulin (Aroca et al., 2008; Fraaije et al., 2001; Mostert et al., 2006), translation elongation factor 1 alpha (*TEF* 1α) (Geiser et al., 2004; Knutsen et al., 2004, Kristensen et al., 2005), calmodulin (Mulè et al., 2004), avirulence genes (Lievens et al., 2009), and mitochondrial genes such as the multicopy *cox* I and *cox* II and their intergenic region (Martin & Tooley, 2003; Nguyen & Seifert, 2008; Seifert et al., 2007). Mating type genes also show high diversity and fast evolutionary rate and could be used for inter- and intra-species differentiation, e.g. Foster et al. (2002) distinguished between the two mating types of *Pyrenopeziza brassicae*. Moreover, Martínez-Espinoza et al. (2003) used mating type genes to specifically detect *Ustilago maydis* in maize cultivars. To enhance the specificity of a diagnostic assay, a combination of multiple diagnostic regions is recommended. Many authors have followed this multi-locus diagnostic strategy, e.g. Collado-Romero et al. (2008) studied the evolutionary relationships among *Verticillium dahliae* vegetative compatibility groups by AFLP fingerprints and sequence analysis of actin, β-tubulin, calmodulin, and histone 3 genes, the ITS region, and a *V. dahliae*-specific sequence; Dixon et al. (2009) demonstrated the host specialisation and phylogenetic diversity of *Corynespora cassiicola* using the ITS region, actine gene and two random hypervariable loci; Glienke et al. (2011) performed sequence analysis of the ITS region and partial *TEF* 1α, actin and glyceraldehyde-3-phosphate dehydrogenase (GPDH) genes to study the genetic diversity of *Phyllosticta* spp. allowing differentiation of pathogenic and non pathogenic species and describing two new *Phyllosticta* species; Inderbitzin et al. (2010) verified a high species diversity in *Botryosphaeriaceae* species performing phylogenetic analyses based on six loci, including the ITS region and *TEF* 1α, GPDH, a heat shock protein, histone-3 and β–tubulin genes; Shimomoto et al. (2011) used RAPDs and sequences from β-tubulin, *TEF* 1α, calmodulin and actin genes to detect pathogenic and genetic variation among isolates of *Corynespora cassiicola*.

2.1.4 Single nucleotide polymorphisms (SNPs)

Closely related pathogens showing different host ranges or pathogenicity often differ in only a single to a few base pairs in target genes commonly used for identification. Therefore, the ability to discriminate single nucleotide polymorphisms (SNPs) should be pursued in any diagnostic assay. Based on the DNA nucleotide sequence difference in the mitochondrial *cox* I gene, a SNP method has been developed to detect and differentiate isolates of *Phytophthora ramorum* from Europe and those originating in the United States (Kroon et al., 2004). In other experiment, polymorphisms detected in the microsatellite flanking regions of *Phytophthora infestans* allowed the development of a SNP genetic marker system for typing this pathogen (Abbott et al., 2010). Many of the examples described below take advantage of SNPs for designing highly specific detection assays.

2.1.5 Design of primers and probes

PCR methods are based on the use of specific oligonucleotides or primers that specifically hybridise with the DNA target and that are required to initiate the synthesis of the new DNA chain. In some real-time PCR methods additional specific oligonucleotides are used, named probes, that hybridise with the target DNA between the two primers. The design of primers and probes is crucial for PCR to be specific and efficient. Primer specificity relies on its sequence, length and GC content, which determines its melting temperature (T_m, the temperature at which 50% of primer-target duplex are hybridized). DNA amplified fragments (amplicons) size must be short enough to ensure efficiency of the reaction and high sensitivity (Singh and Singh, 1997). Real-time PCR usually requires shorter amplicons than conventional PCR. Balancing of primers concentration is necessary especially in multiplex PCR reactions and in real-time PCR using non-specific fluorescent dyes. Also, formation of hairpin structures or complementarities between primers must be avoided.

The first step for designing primers and probes consists in the aligment of the sequences of interest by the blastn program (http://blast.ncbi.nlm.nih.gov/Blast.cgi) (Altschul et al., 1997) using sequences from the GenBank, EMBL and DDBJ databases. Partial or complete nucleotide sequences of many fungal genes are available at the National Center for Biotechnology Information (NCBI) (http://www.ncbi.nlm.nih.gov/Genbank/) (Bethesda, MD, USA). Consensus sequences are used to design primers for detection of members of a same genus or species. If the consensus is not possible, degenerate primers can be used, although this may severely affect the overall sensitivity of the PCR reaction. On the other hand, variable sequences are useful for the differentiation of pathogens at lower taxonomic levels and for the analysis of the molecular variability of fungal population in phylogenetic studies.

Software packages for primer and probe design are available, among others Primer Express, Applied Biosystems; LightCycler Probe Design, Roche; Primer Explorer, Eiken Chemical Co.; Beacon Designer, Premier Biosoft International; Primer Premier, Premier Biosoft; Primer Analysis Software, OLIGO; Oligo perfect designer, Life Technologies/Invitrogen; or Primer3 (http://frodo.wi.mit.edu/primer3/).

2.1.6 PCR-based methods

2.1.6.1 Conventional PCR

Identification of fungal pathogens by conventional PCR may be achieved at different taxonomic levels (genus, species or strain) depending on the specificity of the primers. As recent examples, PCR methods for identification of *Sclerotium rolfsii* (Jeeva et al., 2010) and *Colletotrichum capsici* (Torres-Calzada et al., 2011) have been developed based in specific sequences of the ITS region. Improved variants of PCR have emerged with higher sensitivity, specificity and throughput and allowing the quantification of fungi in infected plants or environment.

2.1.6.2 Nested-PCR and Cooperational-PCR (Co-PCR)

Nested PCR approach is used when an improvement of the sensitivity and/or specificity of detection is necessary. This method consists in two consecutive rounds of amplification in which two external primers amplify a large amplicon that is then used as a target for a second round of amplification using two internal primers (Porter-Jordan et al., 1990). The two reactions are usually performed in separated tubes involving time and effort and increasing the risks of false positives due to cross contamination. However, some improvements in the relative concentrations of the external and internal primers have permitted to perform the two reactions in a single closed tube supporting high throughput. This method has been widely used for detection and/or further characterisation of numerous fungi (Aroca & Raposo, 2007; Grote et al., 2002; Hong et al., 2010; Ippolito et al., 2002; Langrell et al., 2008; Meng and Wang 2010; Mercado-Blanco et al., 2001; Qin et al., 2011; Wu et al., 2011).

An alternative PCR method that enhances sensitivity and minimise contamination risks is the Co-operational PCR. In Co-PCR a single reaction containing the four primers, one pair internal to other, enhances the production of the longest fragment by the co-operational action of all amplicons (Olmos et al., 2002). Co-PCR is usually coupled with dot blot hybridisation by using a specific probe to enhance the specificity of the detection and provide a sensitivity level similar to nested PCR method. Martos et al. (2011) used this method for sensitive and specific detection of grapevine fungi. In both nested- and Co-PCR methods, the use of external primers can be used for generic amplification and the internal primers for further and more specific characterisation of the amplified product at species or strain level.

2.1.6.3 Multiplex PCR

Multiplex PCR is based on the use of several PCR primers in the same reaction allowing the simultaneous and sensitive detection of different DNA targets, reducing time and cost. This method is useful in plant pathology since plants are usually infected by more than one pathogen. Different fragments specific to the target fungi were simultaneously amplified and identified on the basis of their molecular sizes on agarose gels. Although the efficiency of amplification is strongly influenced by amplicon size (shorter amplicons may be amplified preferentially over longer ones), an accurate and careful design of primers and the optimisation of their relative concentrations are required to overcome this drawback and get an equilibrate detection of all target fungi. Multiplex PCR technique has been used for the simultaneous detection and differentiation of *Podosphaera xanthii* and *Golovinomyces cichoracearum* in sunflower (Chen et al., 2008); for detecting *Phytophthora lateralis* in cedar trees and water samples, including detection of an internal control in the same reaction (Winton & Hansen, 2001); for determining the mating type of the pathogens *Tapesia yallundae* and *T. acuformis* (Dyer et al., 2001); for differentiating two pathotypes of *Verticilliun albo-atrum* infecting hop (Radišek et al., 2004) and for distinguishing among eleven taxons of wood decay fungi infecting hardwood trees (Guglielmo et al., 2007). Due to the complexity of the design this technique has recently been replaced by other multiplexing techniques including multiplex real-time PCR (see below).

Another multiplexing method that allows simultaneous detection and identification of multiple oomycetes and fungi in complex plant or environmental samples is the use of ligation detection (LD) system using padlock probes (PLPs). PLPs are long oligonucleotide probes containing asymmetric target complementary regions at their 5' and 3' ends. Padlock probes also incorporate a desthiobiotin moiety for specific capture and release, an internal endonuclease IV cleavage site for linearization, and a unique sequence identifier, the so-called ZipCode, for standardised microarray hybridisation. DNA samples are PCR amplified and subjected to PLP ligation. Under perfectly hybridisation with the target, the PLPs are circularized by enzymatic ligation. Then, the probes are captured with streptavidin-coupled magnetic beads, cut at the internal cleavage site, allowing only the originally ligated PLPs to be visualized by hybridisation on a universal complementary ZipCode microarray. Padlock probes have been used for the simultaneous detection of *Phytophthora cactorum*, *P. nicotianae*, *Pythium ultimum*, *P. aphanidermatum*, *P. undulatum*, *Rhizoctonia solani*, *Fusarium oxysporum* f. sp. *radicis-lycopersici*, *F. solani*, *Myrothecium roridum*, *M. verrucaria*, *Verticillium dahliae* and *V. alboatrum* in samples collected from horticultural water circulation systems in a single assay (van Doorn et al., 2009).

2.1.6.4 Magnetic Capture-Hybridisation (MCH)-PCR

This technique was developed to circumvent the presence of PCR inhibitors in plant extracts. Magnetic beads are coated with a biotinilated oligonucleotide that is specific of a DNA region of the fungus of interest. Hybridisation takes place between the fungal DNA and magnetic beads-oligomer and the conjugate is recovery separated from inhibitory compounds. After the magnetic capture-hybridisation, PCR amplification was carried out using species-specific primers. Langrell & Barbara, (2001) used this method to detect *Nectria galligena* in apple and pear trees.

2.1.6.5 PCR-ELISA

This serological-based PCR method uses forward and reverse primers carrying at their 5' end biotin and an antigenic group (e.g. fluorescein), respectively (Landgraf et al., 1991). PCR amplified DNA can be immobilized on avidin or streptavidin-coated microtiter plates via the biotin moiety of the forward primer and then can be quantified by an ELISA specific for the antigenic group of the reverse primer (e.g. anti-fluorescein antibody detected by colorimetric reactions). PCR-ELISA method is as sensitive as nested PCR. In addition, it does not require electrophoretic separation and/or hybridisation, and can be easily automated. All reactions can be performed in 96-well microtiter plates for mass screening of PCR products making them very suitable for routine diagnostic purposes. This procedure has been used for detection and differentiation of *Didymella bryoniae* from related *Phoma* species in cucurbits (Somai et al., 2002) and for detection of several species of *Phytophtora* and *Pythium* (Bailey et al., 2002).

2.1.6.6 Reverse Transcription (RT)-PCR

An important limitation of molecular methods is the inability to distinguish living or dead fungi or fungal structures. So, results from detection and identification of fungal plant pathogens should be validated by pathogenicity tests. Since mRNA is degraded rapidly in dead cells, the detection of mRNA by RT–PCR is considered an accurate indicator of cell viability (Sheridan et al., 1998). In RT-PCR the RNA is reverse transcribed using the enzyme reverse transcriptase. The resulting cDNA is then amplified using conventional or any other

PCR-based method. RT-PCR has been used to detect viable populations of *Mycosphaerella graminicola* in wheat (Guo et al., 2005). A RT-nested-PCR method was applied for detection of *Oidium neolycopersici* in tomato (Matsuda et al., 2005). Even so, the most frequent application of this technique in phytopathology is the analysis of plant and fungal gene expression during disease development (Yang et al., 2010).

2.1.6.7 *in situ* PCR

This technique allows the amplification of specific gene sequences within intact cells or tissues combining two technologies: PCR and *in situ* hybridisation (ISH) (Long, 1993; Nuovo, 1992). The improved sensitivity of this technique allows the localization of one target copy per cell (Haase et al., 1990; Nuovo et al., 1991). However, background detection is usually high because nonspecific DNA synthesis during *in situ* PCR on tissue sections may occur (Nuovo et al., 1994). In addition, it is a time-consuming technique due to the need for a hybridisation step and technically demanding procedures such as light microscopy. Bindsley et al. (2002) used *in situ* PCR technique to identify *Blumeria graminis* spores and mycelia on barley leaves.

2.1.6.8 PCR-DGGE

This method is mainly applied for the analysis of the genetic diversity of microbial communities without the need of any prior knowledge of the species (Muyzer, 1999; Gothwal et al., 2007; Portillo et al., 2011). DGGE (Denaturing Gradient Gel Electrophoresis) and its variant TGGE (Temperature Gradient Gel Electrophoresis) use chemical gradient such as urea (DGGE) or temperature (TGGE) to denature and separate DNA samples when they are moving across an acrylamide gel. In PCR-DGGE target DNA from plant or environmental samples are firstly amplified by PCR and then subjected to denaturing electrophoresis. Sequence variants of particular fragments migrate at different positions in the denaturing gradient gel, allowing a very sensitive detection of polymorphisms in DNA sequences. In addition, PCR-DGGE primers contain a GC rich tail in their 5′ end to improve the detection of small variations (Myers et al., 1985). The bands obtained in the gel can be extracted, cloned or reamplified and sequenced for identification, being even possible to identify constituents that represent only 1% of the total microbial community. These techniques are very suitable for the identification of novel or unknown organisms and the most abundant species can be readily detected.

This method is however time-consuming, poorly reproducible and provides relative information about the abundance of detected species. Interpretation of the results may be difficult since the microheterogeneity present in some target genes may appear as multiple bands in the gel for a single species, leading to an overestimation of the community diversity. Furthermore, fragments with different sequences but similar melting behaviour are not always correctly separated. In other cases, the analysis of complex communities of microorganisms may result in blurred gels due to the large number of bands obtained.

A PCR-DGGE detection tool based in the amplification of the ITS region has been recently applied to detect multiple species of *Phytophthora* from plant material and environmental samples (Rytkönen et al., 2011). Other authors have used this technique to compare the structure of fungal communities growing in different conditions or environments, e.g. to study the impact of culture management such as biofumigation, chemifumigation or fertilisation on the relative abundance of soil fungal species (Omirou et al., 2011; Wakelin et al., 2008).

2.1.6.9 Real-time PCR

Real-time PCR is currently considered the gold standard method for detection of plant pathogens. This technique allows the monitoring of the reaction during the amplification process by the use of a fluorescent signal that increases proportionally to the number of amplicons generated and to the number of targets present in the sample (Wittwer et al., 1997).

Many are the advantages of real time PCR over conventional PCR, including that this system does not require the use of post PCR processing (electrophoresis, colorimetric reaction or hybridisation), avoiding the risk of carryover contamination and reducing assay labour and material costs. In addition to its improved sensitivity and specificity, this technique allows the accurate quantification of the target pathogen, by interpolating the quantity measured to a standard curve with known amounts of target copies. This quantification characteristic is very useful in phytopathology in order to correlate the amount of fungus in a biological sample with the disease state, or to monitor the progress of the disease in an infected plant (Garrido et al., 2009). In addition, real-time PCR is a high throughput method for the analysis of a large number of samples due to the use of a plate-based system which permits the analysis of 96 or 384 samples at the same time. This characteristic facilitates the robotisation of the nucleic acids extraction and master mix preparation steps, reducing personal and gaining time. Portable real-time PCR machines are emerging as diagnostic tools for *on site* detection under field conditions providing a realistic option to perform molecular tests at the same place of the collection of samples (e.g. portable Cepheid SmartCycler). This is especially interesting in situations where a rapid diagnostic test is needed. Another advantage of real-time PCR is the capability to perform multiplex detection of two or more pathogens in the same reaction (see below).

Real-time RT-PCR chemistry can be based on the use of doubled-stranded DNA binding dyes, such as SYBR Green, specific fluorescent labelled probes such as TaqMan, Molecular Beacons, or Scorpions, or dye-primer based systems, such as hairpin primers or Plexor system.

a. Non-specific fluorescent dyes

SYBR Green I is a fluorescence intercalating dye with a high affinity for double-stranded DNA. The overall fluorescent signal from a reaction is proportional to the amount of double-stranded DNA (dsDNA) present in the sample, and increase as the target is amplified. The main advantage of the use of intercalator dyes is that no probe is required, which reduces assay setup and running costs. Binding dyes are also attractive because protocols using established primers and PCR conditions can readily be converted to the real-time method. However, intercalating dyes detect accumulation of both specific and non-specific PCR products. So, to assess the specificity of the reaction, it is necessary a further step of melting curve analysis that allows the identification of the PCR product by its T_m. Additionally, a fine optimisation of primers concentration is crucial to avoid formation of also detected primer-dimers. Quantification of targets using SYBR Green is not very accurate, since the amount of fluorescent signal is proportional to the mass of dsDNA produced in the reaction (amplification of a longer product will generate more signal than a shorter one). Generally, small amplicons must be selected (between 50 and 200 bp) for optimal efficiency. SYBR Green real-time PCR with melting curve analysis has been described as a simple, rapid, and reliable technique for the detection and identification of phytopathogenic fungi even in multiplex assays (Tabla 1).

Pathogen	Real-time chemistry	Reference	Host plant
Botrytis cinerea	SYBR Green	Diguta el al., 2010	Grape
Cladosporium fulvum	SYBR Green	Yan et al., 2008	Tomato
Colletotrichum acutatum	SYBR Green	Samuelian et al., 2011	Grape
Fusarium avenaceum	SYBR Green	Moradi et al., 2010	Wheat
Fusarium culmorum	SYBR Green	Moradi et al., 2010	Wheat
	SYBR Green multiplex	Brandfass & Karlovsky, 2006	Cereals
Fusarium graminearum	SYBR Green	Moradi et al., 2010	Wheat
	SYBR Green multiplex	Brandfass & Karlovsky, 2006	Cereals
Fusarium oxysporum	SYBR Green	Jiménez-Fernández et al., 2010	Chickpea
			Melon
			Pea
			Soil
Fusarium oxysporum f. sp. *vasinfectum*	SYBR Green	Abd-Elsalam et al., 2006	Cotton
Fusarium oxysporum f. sp. *ciceris*	SYBR Green	Jiménez-Fernández et al., 2011	Chickpea
			Soil
Fusarium poae	SYBR Green	Moradi et al., 2010	Wheat
Fusarium verticillioides	SYBR Green	Kurtz et al., 2010	Corn
Greeneria uvicola	SYBR Green	Samuelian et al., 2011	Grape
Macrophomina phaseolina	SYBR Green	Babu et al., 2011	Chickpea
			Soybean
			Pigeon pea
Phoma sclerotioides	SYBR Green	Larsen et al., 2007	Alfalfa
			Wheat
Phoma tracheiphila	SYBR Green	Demontis et al., 2008	Citrus
Phytophthora capsici	SYBR Green	Silvar et al., 2005	Pepper
Phytophthora cryptogea	SYBR Green	Minerdi et al., 2008	Gerbera
Plasmodiophora brassicae	SYBR Green	Sundelin et al., 2010	Oilseed rape
Puccinia horiana	SYBR Green	Alaei et al., 2009	Chrysanthemum
Pythium irregular	SYBR Green	Schroeder et al., 2006	Wheat
			Barley
			Soil
Pythium ultimum	SYBR Green	Schroeder et al., 2006	Wheat
			Barley
			Soil
Rhizoctonia oryzae	SYBR Green	Okubara et al., 2008	Cereals
Rhizoctonia solani	SYBR Green	Okubara et al., 2008	Cereals
Rhynchosporium secalis	SYBR Green	Fountaine et al. 2007	Barley
Sclerotinia sclerotiorum	SYBR Green	Yin et al., 2009	Oilseed rape
Verticillium dahliae	SYBR Green, Plexor	Attallah et al., 2007	Potato
	SYBR Green	Gayoso et al., 2007	Hot pepper
	SYBR Green	Markakis et al., 2009	Olive

Table 1. Examples of SYBR Green real-time PCR assays for detection of plant pathogenic fungi (last 6 years).

b. Specific fluorescent labelled probes

b.1 Hydrolysis probes

TaqMan chemistry (Heid et al., 1996) is based on the use of an oligonucleotide probe located between the two PCR primers and labelled with a fluorophore covalently attached to the 5'-end (reporter) and a quencher on the 3'end. When the reporter and the quencher are close, the emission of fluorescence is inhibited. After the PCR denaturation step, primers and probes specifically hybridise to the complementary target. The probe is then cleaved by the 5'-3' exonuclease activity of the *Taq* DNA polymerase causing the separation of the fluorophore and the quencher and allowing the reporter dye to emit fluorescence. The fluorescence detected in the real-time PCR thermal cycler is directly proportional to the fluorophore released and the amount of DNA template present in the PCR.

TaqMan probes were designed to increase the specificity of the reaction because detection and accurate quantification require high complementarity with the target sequence. The number of phytopathogenic fungi detected by this method has exponentially increased in recent years. Some examples are detailed in Table 2.

Modified probes have been designed to improve the TaqMan reaction specificity such as MGB probes that include a Minor Groove Binding group at the 3' end raising the *Tm* of the hybrid. That allows the use of shorter and more specific probes. The high specificity of MGB probes make them very suitable for specific detection of fungal species based on SNPs (Massart et al., 2005). For the same purpose, primers or probes can be synthesized with a lock nucleic acid (LNA), which are modified nucleotides that form methylene bridges after binding to the target DNA (Braasch & Corey, 2001). That provokes the lock of the duplex structure, improving their binding affinity and stability and allowing the use of higher annealing temperatures. LNA primers have been used for the specific detection of *P. ramorum* (Tomlinson et al., 2007) and the multiplex detection of species of *Phytophthora* (Bilodeau et al., 2009). The high specificity of TaqMan probes may compromise however the accuracy of the detection and quantification due to the existence of interstrain variability in the target sequence that may result in failure to detect or underestimation of the amount of DNA targets in the sample.

In addition to its high specificity, one of the advantages of the TaqMan chemistry is that probes can be labeled with different reporter dyes (FAM, VIC, TET, TAMRA, HEX, JOE, ROX, Cy5, Texas Red, etc) which allows multiplexing detection of two or more distinct pathogens in the same reaction increasing throughput (Aroca et al., 2008; Bilodeau et al., 2009). However, for this porpose the synthesis of different probes is required making the multiplex detection analysis more expensive. The types and number of fluorescent labels that can be used depend upon the detection capabilities of the real-time instrument used. Additionally, designing of real-time multiplex assays may be difficult. Special attention must be focus on avoiding primer competition that could strongly drop the levels of specificity and sensitivity. In fact, TaqMan multiplex assays usually show lower detection sensitivity than single reactions.

b.2 Hairpin probes

Molecular Beacons (MB) (Tyagi & Kramer, 1996) are specific oligonucleotide probes (15-40-mer) flanked by two complementary 5-7-mer arms sequences, with a fluorescent dye covalently attached to the 5' end and a quencher dye at the 3' end. When the molecular beacon is in an unbound state the arms form a stem/loop structure in which the fluorophore

Pathogen	Real-time chemistry	Reference	Host plant
Biscogniauxia mediterranea	TaqMan	Luchi et al., 2005	Oak
Botrytis squamosa	TaqMan	Carisse et al., 2009	Onion
Chalara fraxinea	TaqMan	Ioos et al., 2009	Ash trees
Colletotrichum acutatum	TaqMan	Garrido et al., 2009	Strawberry
Colletotrichum gloesporioides	TaqMan	Garrido et al., 2009	Strawberry
Colletotrichum spp.	TaqMan	Garrido et al., 2009	Strawberry
Discula destructiva	TaqMan	Zhang, N. et al., 2011	Dogwood
Fusarium avenaceum	TaqMan MGB	Kulik et al., 2011	Cereals
Fusarium equiseti	Molecular Beacons	Macía-Vicente et al., 2009	Barley
Fusarium foetens	TaqMan	De Weerdt et al., 2006	Begonia
Fusarium graminearum	TaqMan	Demeke et al., 2010	Wheat / Barley
Fusarium oxysporum f. sp. *lycopersici* and its races	TaqMan MGB	Inami et al., 2010	Tomato
Fusarium poae	TaqMan MGB	Kulik et al., 2011	Cereals
Fusarium tricinctum	TaqMan MGB	Kulik et al., 2011	Cereals
Fuscoporia torulosa	Scorpions	Campanile et al., 2008	Oak
Gremmeniella abietina	TaqMan	Børja et al., 2006	Spruce
Guignardia citricarpa	TaqMan	van Gent-Pelzer et al., 2007	Citrus fruit
Macrophomina phaseolina	TaqMan MGB	Babu et al., 2011	Chickpea / Soybean / Pigeon pea
Mycosphaerella graminicola	TaqMan MGB	Bearchell et al., 2005	Wheat
Phaeoacremonium aleophilum	Taqman multiplex	Aroca et al., 2008	Grapevine wood
Phaeoacremonium mortoniae	Taqman multiplex	Aroca et al., 2008	Grapevine wood
Phaeoacremonium parasiticum	Taqman multiplex	Aroca et al., 2008	Grapevine wood
Phaeoacremonium viticola	Taqman multiplex	Aroca et al., 2008	Grapevine wood
Phaeospora nodorum	TaqMan MGB	Bearchell et al., 2005	Wheat
Phialophora gregata	TaqMan	Malvick & Impullitti, 2007	Soybean / Soil
Phoma tracheiphila	TaqMan	Demontis et al., 2008	Citrus
Phoma tracheiphila	TaqMan	Licciardello et al., 2006	Citrus
Phomopsis sp.	TaqMan	Børja et al., 2006	Spruce
Phytophthora citricola	TaqMan multiplex	Schena et al., 2006	Forest trees
Phytophthora erythroseptica	TaqMan	Nanayakkara et al., 2009	Potato
Phytophthora kernoviae	TaqMan multiplex	Schena et al., 2006	Forest trees

Pathogen	Real-time chemistry	Reference	Host plant
Phytophthora pseudosyringae	TaqMan multiplex	Tooley et al., 2006	Rododendron and other host species
Phytophthora quercina	TaqMan multiplex	Schena et al., 2006	Forest trees
Phytophthora ramorum	TaqMan multiplex	Bilodeau et al., 2009	Oak
	TaqMan	Hughes et al., 2006	Parrotia persica
	TaqMan multiplex	Schena et al., 2006	Forest trees
	Scorpion and Molecular Beacon	Tomlinson et al., 2007	
	TaqMan multiplex	Tooley et al., 2006	Rododendron and other host species
Phytophthora spp.	TaqMan	Tomlinson et al., 2005	Rhododendron
	TaqMan multiplex	Bilodeau et al., 2009	Oak
Plasmopara viticola	TaqMan	Valsesia et al., 2005	Grapevine
Pochonia chlamydosporia	Molecular Beacon	Macía-Vicente et al., 2009	Barley
Puccinia coronata	TaqMan	Jackson et al., 2006	Oat
Puccinia graminis	TaqMan	Barnes & Szabo, 2007	Cereals Grasses
Puccinia recondita	TaqMan	Barnes & Szabo, 2007	Cereals Grasses
Puccinia striiformis	TaqMan	Barnes & Szabo, 2007	Cereal Grasses
Puccinia triticina	TaqMan	Barnes & Szabo, 2007	Cereals Grasses
Pyrenophora teres	TaqMan MGB	Leisova et al., 2006	Barley
Pyrenophora teres f. maculata	TaqMan MGB	Leisova et al., 2006	Barley
Pyrenophora teres f. teres	TaqMan MGB	Leisova et al., 2006	Barley
Pythium vexans	TaqMan	Tewoldemedhin et al., 2011	Apple
Rhynchosporium secalis	TaqMan, LNA	Fountaine et al., 2007	Barley
Rosellinia necatrix	Scorpion	Ruano-Rosa et al., 2007	Avocado
Thielaviopsis basicola	TaqMan	Huang & Kang, 2010	Tobacco Soil
Ustilaginoidea virens	TaqMan	Ashizawa et al., 2010	Rice Soil

Table 2. Examples of real-time PCR based on specific fluorescent labelled probes or primers for detection of plant pathogenic fungi (last 6 years).

and the quencher are in close proximity and fluorescence is quenched. When the probe hibridises to the target sequence the complementary arms separate, thus allowing the emission of fluorescence and hence making possible the detection and quantification of the target sequence. Because the stem/loop structure is very thermostable molecular beacons must have a high specificity to hybridise to a target. This makes the chemistry appropriate for the detection of single nucleotide differences in mutation and SNP analyses. Molecular beacons have allowed real-time specific quantification of *Fusarium equiseti* and *Pochonia chlamydosporia* (a nematode parasitic fungus) in barley roots (Macía-Vicente et al., 2009).

Scorpions® are bifunctional molecules in which an upstream hairpin probe is covalently linked to a downstream primer sequence (Whitcombe et al., 1999). The hairpin probe contains a fluorophore at the 5' end and a quencher at the 3' end. The loop portion of the scorpion probe is complementary to the target sequence. During the amplification reaction the probe becomes attached to the target region synthesized in the first PCR cycle. Following the second cycle of denaturation and annealing, the probe and the target hybridise resulting in separation of the fluorophore from the quencher and an increase in the fluorescence emitted. Improvement of Scorpions sensitivity has been achieved by placing the quencher in a separate oligonucleotide (Scorpions bi-probes) allowing greater separation of fluorophore and quencher and giving stronger signals. As with all dye-probe based methods, Scorpion probes follow strict design considerations for secondary structure and primer sequence to ensure that a secondary reaction will not compete with the correct probing event. Scorpion technology has been used for the detection of *Rosellinia necatrix* in roots of different plant host species and soils (Ruano-Rosa et al., 2007; Schena & Ippolito, 2003), and for the detection of *Fuscoporia torulosa* in holm oaks (Campanile et al., 2008).

c. Specific fluorescent labelled primers

Unlike other real-time chemistries, in which the incorporation of the fluorescent dye increases with the increasing number of copies of the DNA target, Plexor system measures the decrease of the fluorescence over time by the quenching of the dye. One of the two primers is labelled with a fluorescent dye and modified with methylisocytosine (isodC) residue at the 5' end, whereas the other primer is not. The dabcyl-iso dGTP (iso-dG) present in the real-time PCR reaction cocktail is incorporated at the position complementary to the iso-dC label acting as a quencher and reducing the fluorescence over time. A quantitative real-time using Plexor primers has been developed for the detection and quantification of *Verticillium dahliae* in potato (Atallah et al., 2007).

2.2 Isothermal amplification methods

An efficient and cost-effective alternative to PCR is the possibility of isothermal amplification that does not require thermocycler aparatus. Loop-Mediated Isothermal Amplification (LAMP) (Notomi et al., 2000) uses a set of six oligonucleotide primers with eight binding sites hybridizing specifically to different regions of a target gene, and a thermophilic DNA polymerase from *Geobacillus stearothermophilus* for DNA amplification. This technique can specifically amplify the DNA target using only a heated block in less than 1 hour. Amplification products can be detected directly by visual inspection in vials using SYBR Green I, or by measuring the increased turbidity (due to the production of large amounts of magnesium pyrophosphate), as well as by electrophoresis on agarose gel. LAMP method is very suitable for field testing and potentially valuable to laboratories without PCR facilities. This isothermal method has been applied for the rapid detection of *Fusarium graminearum* in contaminated wheat seeds (Abd-Elsalam et al., 2011) and for the detection of *Phytophthora ramorum* and *P. kernoviae* in field samples (Tomlinson et al., 2007, 2010).

2.3 Fingerprinting

Fingerprinting approaches allow the screening of random regions of the fungal genome for identifying species-specific sequences when conserved genes have not enough variation to successfully identify species (McCartney et al., 2003). Fingerprinting analyses are generally used to study the phylogenetic structure of fungal populations. However, these techniques have been also useful for identifying specific sequences used for the detection of fungi at very low taxonomic level, and even for differentiate strains of the same species with different host range, virulence, compatibility group or mating type.

2.3.1 Restriction fragment length polymorphism (RFLP)

RFLP involves restriction enzyme digestion of the pathogen DNA, followed by separation of the fragments by electrophoresis in agarose or polyacrilamide gels to detect differences in the size of DNA fragments. Polymorphisms in the restriction enzyme cleavage sites are used to distinguish fungal species. Although DNA restriction profile can be directly observed by staining the gels, Southern blot analysis is usually necessary. DNA must be transferred to adequate membranes and hybridised with an appropriate probe. However, the Southern blot technique is laborious, and requires large amounts of undegraded DNA. RFLPs have been largely used for the study of the diversity of micorrhizal and soil fungal communities (Thies, 2007; Kim Y. T. et al., 2010; Martínez-García et al., 2011). Although used for differentiation of pathogenic fungi (Hyakumachi et al., 2005) this early technique has been progressively supplanted by other fingerprint techniques based in PCR.

PCR-RFLP combines the amplification of a target region with the further digestion of the PCR products obtained. PCR primers specific to the genus *Phytophthota* were used to amplify and further digest the resulting amplicons yielding a specific restriction pattern of 27 different *Phytophthora* species (Drenth et al., 2006). PCR-RFLP analysis of the ITS region demonstrated the presence of different anastomosis group (AG) within isolates of *Rhizoctonia solani* (Pannecoucque & Höfte, 2009); It also allowed the differentiation of pathogenic and non pathogenic strains of *Pythium myriotolum* (Gómez-Alpizar et al., 2011). In other cases, the analysis of the ITS region by this technique failed in differentiating closely related species (e.g., clade 1c species such as *Phytophthora infestans* and *P. mirabilis*) (Grünwald et al., 2011).

2.3.2 Random Amplified Polymorphic DNA (RAPD)

RAPD analyses rely on PCR amplification of the pathogen genome with short arbitrary sequences (usually decamers) that are used as primers. These primers are probably able to find distinct complementary sequences in the genome producing specific banding patterns. The resulting PCR fragments are then separated by electrophoresis to obtain fingerprints that may distinguish fungal species varieties or strains (Welsh & McClelland, 1990; Williams et al., 1990). Some of the specific DNA fragments detected in a profile may be cut out of the gel and sequenced to obtain a SCAR (Sequence-characterized amplified region), into which specific primers can be designed for a more precise PCR detection. SCAR primers have been used for instance to specifically identify *Phytophthora cactorum* (Causin et al., 2005), *Fusarium subglutinans* (Zaccaro et al., 2007) and *Guignardia citricarpa* (Stringari et al., 2009) in infected plant material; to distinguish among several *formae speciales* of *Fusarium oxysporum* (Lievens et al., 2008); to differentiate the bioherbicidal strain of *Sclerotinia minor* from like organisms (Pan et al., 2010); and to establish two different groups in *Gaeumannomyces graminis* var. *tritici* (Daval et al., 2010).

RAPD results are also useful for the analysis of the genetic diversity among populations. Fingerprints are scored for the presence (1) or absence (0) of bands of various molecular weight sizes in the form of binary matrices. Data are analyzed to obtain statistic coefficients among the isolates that are then clustered to generate dendrograms. RAPDs have been used to analyze the genetic diversity among different species and races of *Fusarium* spp. (Lievens et al., 2007; Arici & Koc 2010) and different pathotypes of *Elsinoë* spp. (Hyun et al., 2009). This technique has also been applied to differentiate fungi isolates according to their host plant (Midorikawa et al., 2008), enzyme production profiles (Saldanha et al., 2007) or geographical origin and chemotypes (Zheng et al., 2009).

The RAPD technique is rapid, inexpensive and does not require any prior knowledge of the DNA sequence of the target organism. Results obtained from RAPD profiles are easy to interpret because they are based on amplification or non amplification of specific DNA sequences. In addition, RAPD analyses can be carried out on large numbers of isolates without the need for abundant quantities of high-quality DNA (Nayaka et al., 2011). Disadvantages of this technique include poor reproducibility between laboratories, and the inability to differentiate non-homologous co-migrating bands. In addition, RAPDs are dominant markers so, they cannot measure the genetic diversity affected by the number of alleles at a locus, nor differentiate homozygotes and heterozygotes individuals. This is not an issue with haploid fungi, but it can be a problem with many basidiomycetes and oomycetes that are heterokaryons, diploids or polyploids (Fourie et al., 2011).

2.3.3 Amplified fragment length polymorphism (AFLP)

AFLP analysis (Vos et al., 1995) consists in the use of restriction enzymes to digest total genomic DNA followed by ligation of restriction half-site specific adaptors to all restriction fragments. Then, a selective amplification of these restriction fragments is performed with PCR primers that have in their 3' end the corresponding adaptor sequence and selective bases. The band pattern of the amplified fragments is visualized on denaturing polyacrilamide gels. The AFLP technology has the capability to amplify between 50 and 100 fragments at one time and to detect various polymorphisms in different genomic regions simultaneously. It is also highly sensitive and reproducible. As with other fingerprinting techniques, no prior sequence information is needed for amplification (Meudt & Clarke 2007). The disadvantages of AFLPs are that they require high molecular weight DNA, more technical expertise than RAPDs (ligations, restriction enzyme digestions, and polyacrylamide gels), and that AFLP analyses suffer the same analytical limitations of RAPDs (McDonald et al., 1997).

Depending on the primers used and on the reaction conditions, random amplification of fungal genomes produces genetic polymorphisms specific at the genus, species or strain levels (Liu et al., 2009). As a result, AFLP has been used to differentiate fungal isolates at several taxonomic levels e.g. to distinguish *Cladosporium fulvurn* from *Pyrenopeziuz brassicae* species (Majer et al., 1996), *Aspergillus carbonarius* from *A. ochraceus* (Schmidt et al., 2004), and *Colletotrichum gossypii* from *C. gossypii* var. *cephalosporioides* (Silva et al., 2005); also to differentiate *Monilinia laxa* that infect apple trees from isolates infecting other host plants (Gril et al., 2008); and to separate non-pathogenic strains of *Fusarium oxysporum* from those of *F. commune* (Stewart et al., 2006). AFLP markers have also been used to construct genetic linkage maps e.g. of *Phytophthora infestans* (Van der Lee et al., 1997). Specific AFLP bands may also be used for SCAR markers development used in PCR-based diagnostic tests. Using SCAR markers Cipriani et al. (2009) could distinguish isolates of *Fusarium oxysporum* that specifically infect the weed *Orobanche ramose*. AFLP profiles have also been widely used for the phylogenetic analysis of *Fusarium oxysporum* complex (Baayen et al., 2000; Fourie et al., 2011; Groenewald et al., 2006).

2.3.4 Microsatellites

Microsatellites, also known as simple sequence repeats (SSRs) or short tandem repeats (STRs), are motifs of one to six nucleotides repeated several times in all eukaryotic genomes (generally in non-coding regions). These nucleotide units can differ in repeat number among individuals and their distribution in the genome is almost random. Using primers flanking such variable regions PCR products of different lengths can be obtained. So, the microsatellites are highly versatile genetic markers that have been widely exploited for DNA fingerprinting. The advantages of SSRs are that they are multiallelic, codominant, highly polymorphic and several thousand potentially polymorphic markers are available. Moreover, it is possible the analysis of samples with limited DNA amounts or degraded DNA with high reproducibility. The microsatellites have a high mutation rate and are able to gain and lose repeat units by DNA-replication slippage, a mutation mechanism that is specific to tandemly repeated sequences (Schlöetterer, 2000). This characteristic can create difficulties for populations-genetic analyses. Other drawbacks of the SSRs include the requirement of a prior knowledge of the DNA sequences of the flanking regions and their cost and low throughput because of difficulties for automation and data management. Moreover, a high number of microsatellite loci are necessary for a reliable phylogenetic reconstruction. However, the next-generation sequencing technologies and multiplexing microsatellites solve, in part, these problems.

Microsatellites have been used for the study of the genetic diversity of plant pathogenic fungi within species e.g. *Ascochyta rabiei* (Bayraktar et al., 2007), *Ceratocystis fimbriata* (Rizatto et al., 2010), *Macrophomina phaseolina* (Jana et al., 2005), *Puccinia graminis* and *P. triticina* (Szabo, 2007; Szabo & Kolmer, 2007), *Sclerotinia subarctica* and *S. sclerotiorum* (Winton et al., 2007); and for genetic map construction, e.g. Zheng et al. (2008) constructed a genetic map of *Magnaporthe grisea* consisting of 176 SSR markers. In other experiment, microsatellite markers specific for *Phytophthora ramorum* were employed to distinguish between A1 and A2 mating types isolates of this pathogen from two different geographic origins (Prospero et al., 2004).

To reduce the cost of developing microsatellites a novel technique has emerged based on sequence tagged microsatellites (STMs). Each STM is amplified by PCR using a single primer specific to the conserved DNA sequence flanking the microsatellite repeat in combination with a universal primer that anchors to the 5′-ends of the microsatellites (Hayden et al., 2002). STMs have been developed for the plant pathogens *Rhynchosporium secalis* (Keipfer et al., 2006) and *Pyrenophora teres* (Keipfer et al., 2007).

2.4 DNA hybridisation technology

The use of Southern blot or dot blot hybridisation techniques using selected probes from DNA libraries was a strategy for the identification of plant pathogens prior to the introduction of PCR-based methods with greater sensitivity, simplicity and speed (Takamatsu et al., 1998; Levesque et al., 1998; Xu et al., 1999). Nevertheless, new and revolutionary methods based in hybridisation have been recently developed for detection and differentiation of phytopathogenic fungi:

2.4.1 DNA arrays

A DNA array is a collection of species-specific oligonucleotides or cDNAs (known as probes) immobilized on a solid support that is subjected to hybridisation with a labelled target DNA. Macroarrays are membrane-based arrays containing spotted samples of 300 µm in diameter or more. Microarrays uses higher density chips such as glass or silicon, or microscopic beads in where thousands of sample spots (less than 200 µm in diameter) are immobilised via robotisation. The target DNA is a labelled PCR fragment, amplified with universal primers, spanning a genomic region that includes species-specific sequences. Probe-target hybridisation is usually detected and quantified by detection of fluorophore-, silver-, or chemiluminescence-labeled targets and the relative abundance of nucleic acids sequences in the target can be determined. DNA micro- and macro-arrays are generally used for gene expression profiling but are also powerful tools for identification and differentiation of plant pathogens (Anderson N. et al., 2006; Lievens & Thomma, 2005). Currently, it is one of the most suitable techniques to detect and quantify multiple pathogens present in a sample (plant, soil, or water) in a single assay (Lievens et al., 2005a). The specificity of the DNA array technology allows an accurate SNP detection. This characteristic is crucial for diagnostic application since closely related pathogens may differ in only a single base pair polymorphism for a target gene. Specificity of the assay depends on the number and position of the mismatch(es) in the oligonucleotide probes, the oligonucleotide sequence and the lenght of the amplicon target. For instance, mismatches at the 3′ end of the oligonucleotide must be avoided and center mismatches are usually the most discriminatory sites (Lievens et al., 2006). Furthermore, using longer amplicons as targets increases the sensitivity but decreases the specificity of the array hybridisation. For improving specificity and robustness, the use of multiple oligonucleotides for a single pathogen and the use of multiple diagnostic regions are desirable. One of the main drawbacks of this technique is however, the lack of sensitivity. To reach sensitive detections, PCR amplification before array hybridisation is required, biasing the results through the species that are more represented in the sample.

This technology has been applied for detecting oomycete plant pathogens by using specific oligonucleotides designed on the ITS region (Anderson N. et al., 2006; Izzo & Mazzola, 2009). Another ITS and rRNA genes-based microarrays allowed the multiple detection and quantification of tomato pathogens (*Verticillium*, *Fusarium*, *Pythium* and *Rhizoctonia*) confirmed by real-time PCR analysis (Lievens et al., 2005b), the monitoring of *Phytophthora* species diversity in soil and water samples (Chimento et al., 2005), and the identification and differentiation of toxin producing and non-producing *Fusarium* species in cereal grains (Nicolaisen et al., 2005). Using a *cox* I high density oligonucleotide microarray Chen et al. (2009) could identify *Penicillium* species. Moreover, Lievens et al. (2007) could detect and differentiate *F. oxysporum* f. sp. *cucumerinum* and *F. oxysporum* f. sp. *radicis-cucumerinum* pathogens by a DNA array containing genus-, species- and *forma specialis*-specific oligonucleotides.

Additionally, A DNA array for simultaneous detection of over 40 different plant pathogenic soilborne fungi and 10 bacteria that frequently occur in greenhouse crops has been developed. This array, called DNA Multiscan® (http://www.dnamultiscan.com/), is routinely used worldwide by companies that offer disease diagnostic services and advice to commercial growers.

2.5 Sequencing

As discussed above, morphological characteristics are not always enough to identify a pathogen. One of the most direct approaches to do that consists in the PCR amplification of a target gene with universal primers, followed by sequencing and comparison with the available publicly databases. In addition, new fungal species have been described by using sequencing approaches. However, the use of sequence databases to identify organisms based on DNA similarity may have some pitfalls including erroneous and incomplete sequences, sequences associated with misidentified organisms, the inability to easily change or update data, and problems associated with defining species boundaries, all of them leading to erroneous interpretation of search results. An effort for generating and archiving high quality data by the researchers community should be the remedy of this drawback (Kang et al., 2010). Other limitation of sequencing as diagnostic tool is the need to sequence more than one locus for the robustness of the result, and the impractical of this method in cases when rapid results are needed such as for the control or eradication of serious plant disease outbreaks. Nevertheless, the increase of sequencing capacity and the decrease of costs have allowed the accumulation of a high numbers of fungal sequences in publicly accessible sequence databases, and sequences of selected genes have been widely used for the identification of specific pathogens and the development of sequence-based diagnostic methods.

2.5.1 Massive sequencing techniques

The Sanger sequencing method has been partially supplanted by several "next-generation" sequencing technologies able to produce a high number of short sequences from multiple organisms in short time. Massive sequencing technologies offer dramatic increases in cost-effective sequence throughput, having a tremendous impact on genomic research. They have been used for standard sequencing applications, such as genome sequencing and resequencing, and quantification of sequence variation. The next-generation technologies commercially available today include the 454 GS20 pyrosequencing-based instruments (Roche Applied Science), the Solexa 1G analyzer (Illumina, Inc.), and the SOLiD instrument (Applied Biosystems).

Pyrosequencing is a DNA sequencing technology based on the sequencing-by-synthesis principle. The technique is built on a four enzyme real-time monitoring of DNA synthesis by chemiluminescence and a fifth protein, SSB, which can be included to enhance the quality of the obtained sequences and thereby prolong the read length. The detection system is based on the pyrophosphate released when a nucleotide is introduced in the DNA strand. Thereby, the signal can be quantitatively connected to the number of bases added (Ahmadian et al., 2006). The pyrosequencing principle is used by the 454 platform, the first next-generation sequencing technology released to the market by Roche Applied Science. 454 technology is based in emulsion PCR (Tawfik et al., 1998), which uses fixing adapter-ligated-DNA fragments to streptavidin beads in water-in-oil emulsion droplets. In each droplet the DNA fixed to these beads is then amplified by PCR producing about 10^7 copies of a unique DNA template per bead. Each DNA-bound bead is placed into a ~29 µm well on a PicoTiterPlate, a fiber optic chip, and analyzed using a pyrosequencing reaction. The use of the picotiter plate allows hundreds of thousands of pyrosequencing reactions to be carried out in parallel, massively increasing the sequencing throughput. 454 platform is capable of generating 80–120 Mb of sequence in 200 to 300 bp reads in a 4 h run. This technology enables a rapid and accurate quantification of sequence variation, including mutation detection, SNP genotyping, estimation of allele frequency and gene copy number, allelic imbalance and methylation status. Pyrosequencing can be applied to any DNA source, including degraded or low-quality DNA. Disadvantages are short read lengths, which may be problematic for sequence assembly particularly in areas associated with sequence repeats, the need for expensive biotinylated primers, and the inability to accurately detect variants within long (~ 5 or 6 bp) homopolymer stretches. In addition, multiplexing, while possible, is difficult to design.

The pyrosequencing technology has not been widely applied for the control of fungal plant diseases yet. However, Nunes et al. (2011) applied 454 sequencing technology to elucidate and characterize the small RNA transcriptome (15 - 40 nt) of mycelia and appressoria of *Magnaporthe oryzae*. Thus, they propose that a better understanding of key small RNA players in *M. oryzae* pathogenesis-related processes may illuminate alternative strategies to engineer plants capable of modifying the *M. oryzae* small transcriptome, and suppress disease development in an effective manner. Another application of this new sequencing technology is the rapid generation of genomic information to identify putative single-nucleotide polymorphisms (SNPs) to be used for population genetic, evolutionary, and phylogeographic studies on non-model organisms. Thus, Broders et al. (2011) described the sequencing, assembly and discovery of SNPs from the plant fungal pathogen *Ophiognomonia clavigignenti-juglandacearum*, for which virtually no sequence information was previously available. Moreover, Malausa et al. (2011) described a high-throughput method for isolating microsatellite markers based on coupling multiplex microsatellite enrichment and 454 pyrosequencing in different organisms, such as *Phytophthora alni* subsp. *uniformis*.

The principle of the Illumina/Solexa system is also based on sequencing-by-synthesis chemistry, with novel reversible terminator nucleotides for the four bases each labeled with a different fluorescent dye, and a special DNA polymerase enzyme able to incorporate them (Ansorge, 2010). In the ABI SOLiD (Sequencing by Oligo Ligation and Detection), the sequence extension reaction is not carried out by polymerases but rather by ligases. In the sequencing-by-ligation process, a sequencing primer is hybridized to single-stranded copies of the library molecules to be sequenced. (Kircher & Kelso, 2010). These two above mentioned systems have not been currently used in studies on plant pathogenic fungi.

2.5.2 DNA barcoding

DNA barcoding is a taxonomic method that uses a short genetic marker in the DNA to identify an organism as belonging to a particular species. It has facilitated the description of numerous new species and the characterisation of species complexes. Current fungal species identification platforms are available: *Fusarium*-ID was created as a simple, web-accessible BLAST server that consisted of sequences of the *TEF 1 α* gene from representative species of *Fusarium* (Geiser et al., 2004). Sequences of multiple marker loci from almost all known *Fusarium* species have been progressively included for supporting strain identification and phylogenetic analyses. Two additional platforms have been constructed: the Fusarium Comparative Genomics Platform (FCGP), which keeps five genomes from four species, supports genome browsing and analysis, and shows computed characteristics of multiple gene families and functional groups; and the Fusarium Community Platform (FCP), an online research and education forum. All together, these platforms form the Cyber infrastructure for *Fusarium* (CiF; http://www.fusariumdb.org/) (Park, B. et al., 2011).

For *Phytophthora* identification two web-based databases have been created: (i) *Phytophthora* Database (http://www.phytophthoradb.org/) based in nine loci sequences including the ITS region and the 5′ portion of the large subunit of rRNA genes; nuclear genes encoding 60S ribosomal protein L10, β-tubulin, enolase, heat shock protein 90, TigA fusion protein, and TEF 1α; the mitochondrial gene *cox* II and spacer region between *cox* I and *cox* II genes (Park, J. et al., 2008); and (ii) Phytophthora-ID (http://phytophthora-id.org/) based on sequences of the ITS and the *cox* I and *cox* II spacer regions (Grunwald et al., 2011). Additional web-based databases are available including UNITE (http://unite.ut.ee/index.php), an ITS database supporting the identification of ectomycorrhizal fungi (Koljalg et al., 2005); TrichOKey (http://www.isth.info/tools/molkey/index.php), a database supporting the identification of *Hypocrea* and *Trichoderma* species (Druzhinina et al., 2005); and BOLD (http://www.boldsystems.org/) containing ITS and *cox* I databases from oomycetes (Ratnasingham et al., 2007; Robideau et al., 2011). Consortium for the Barcode of Life (CBOL) is an international collaborative effort which aims to use DNA barcoding to generate a unique genetic barcode for every species of life on earth. The *cox* I mitochondrial gene is emerging as the standard barcode region for eukariotes.

3. Fungicide resistance

Despite extensive fungicide use in the previous 90 years, resistance emerged as a practical problem as recently as 1970. The incidence of resistance has been restricted largely to systemic fungicides that operate to biochemical targets (single-site inhibitors). These included several of major groups of fungicides: sterol demethylation, bezimidazoles, pyrimidines, phenylamines, dicarboximides, carboxanilides and morpholines.

The resistance to toxic compounds is a genetic adaptation of the fungus to one or more fungicides that leads to a reduction in sensitivity to these compounds. This phenomenon, described as genotypic or acquired resistance, is found in numerous fungi that have been sensitive to the fungicide prior to exposure. Fungicide resistance can be acquired by single mutations in genes of the pathogen or by increasing the frequency of subpopulations that are naturally less sensitive. We have to differentiate between resistance and natural or intrinsic insensitivity of species that are not sensitive to the action of these compounds (Delp & Bekker,

1985; Brent, 1995). The term "lower sensitivity" is used in practical situations where there is a decreased sensitivity to a fungicide without an effect on field performance. The term "field resistance" is used when both the level of resistance and frequency of resistant strains are high and coincident, resulting in noticeable decline of field performance (Hewit, 1998).

It is to be expected that the evolutionary and dynamic progress of selection should eventually produce fungi that are resistant to fungicides. Therefore fungicide resistance is variable, with fungi developing resistance to some fungicides more rapidly than to others and in some cases no resistance had been reported after long periods of fungicide use.

In fungicide resistance, we should consider how pathogens have the ability to evolve resistant and how fungicides varies their susceptibility to resistance. Hence, resistance risk is determined by the target pathogen and selected fungicide.

Populations of fungi are so diverse that mechanisms of resistance may be present within the population before the fungicide is applied. The rate of resistance development in a population depends on the resultant mechanism of resistance such as how resistance characters are inherited, the epidemiology of the fungus, the environment, and the persistence of selective pressure.

3.1. Mechanisms of fungicide resistance

Fungicide resistance can be conferred by various mechanisms including: (I) an altered target site, which reduces the binding of the fungicide; (II) the synthesis of an alternative enzyme capable of substituting the target enzyme; (III) the overproduction of the fungicide target; (IV) an active efflux or reduced uptake of the fungicide; and (V) a metabolic breakdown of the fungicide. In addition, some unrecognized mechanisms could also be responsible for fungicide resistance.

3.1.1 The reduction of intracellular concentration of antifungal compound

Currently, the most widespread hypothesis to explain the reduced levels of toxic products in the cell is based on active efflux of these compounds by ABC transporters (ATP binding cassette) and MFS (Major Facilitator Superfamily) (Hayashi et al., 2001, 2002a, 2002b, 2003; Stergiopoulos et al., 2002a, 2002b; Stergiopoulos & de Waard, 2002, Vermeulen et al., 2001). ABC and MFS transporters are the most studied so far and have been described in many fungi such as *Aspergillus nidulans, Botrytis cinerea, Mycosphaerella graminicola, Magnaporthe grisea, Penicillium digitatum*, etc (Andrade et al., 2000a, 2000b; del Sorbo et al., 2000; Nakaune, 2001; Schoonbeek et al., 2003; Stergiopoulos et al., 2002a; Stergiopoulos & de Waard, 2002; Vermeulen et al., 2001; Yoder and Turgeon 2001; Zwiers et al., 2003). Its function is to prevent or reduce the accumulation of compounds and therefore to avoid or minimize their toxic action (Bauer et al., 1999, Pao et al., 1998). ABC transporters comprise a large family of proteins and are located outside the plasma membrane or within the cell in intracellular compartments as vacuoles, endoplasmic reticulum, peroxisomes and mitochondria. They can carry a wide variety of toxic against the gradient (Del Sorbo et al., 2000; Theodoulou, 2000). ABC transporters include systems both capture and removal, generally showing activity on a wide range of substrates (fungicides, drugs, alkaloids, lipids, peptides, sterols, flavonoids, sugars, etc), but there are specific transporters for substrates (Bauer et al., 1999, Del Sorbo et al., 2000).

The overexpression of ABC and MFS genes plays an essential role in the resistance of chemically unrelated phenomenon described as multidrug resistance or MDR drugs (Del Sorbo et al., 2000; White, 1997). This phenomenon has been observed in a wide variety of organisms and can be a real threat to the effective control of fungal pathogens (Fling et al., 1991). In phytopathogenic fungi, these transporters can also be a virulence factor in providing protection against defense compounds produced by the plant or mediating the secretion of host-specific toxins. Also play a major role in determining the baseline sensitivity to fungicides and other antifungal agents. (De Waard, 1997; Stergiopoulos et al., 2003a, 2003b).

In *P. digitatum*, causal agent of citrus green rotten, four ABC transporters have been identified so far. ABC transporters PMR1 (Hamamoto et al., 2001b; Nakaune et al., 1998) and PMR5 (Nakaune, 2001; Nakaune et al., 2002) have been studied previously. Disruption of PMR1 in sensitive and resistant strains results in an increased sensitivity to DMIs and other compounds (Nakaune et al., 1998; Nakaune et al., 2002). However, the introduction of resistant strains from PMR1 restores the resistance while the introduction of PMR1 from sensitive strains does not have the same effect and does not restore the resistance. This suggests that although PMR1 plays an important role in the sensitivity of *P. digitatum* against DMIs alone does not explain the differences between sensitive and resistant strains (Hamamoto et al., 2001). Another of the genes studied is PMR5. This gene has highly homologous to PMR1, and also to *atrB* from *Aspergillus nidulans* and *BcAtrB* from *Botrytis cinerea* (Schoonbeek et al., 2003), however, is strongly induced by benzimidazoles, resveratrol and other compounds, but not for DMIs. This shows the different substrate specificity of both proteins and may play an important role in providing protection against natural or synthetic toxic compounds (Nakaune et al., 2002).

Sequence analysis in all four ABC transporter genes in several sensitive and resistant strains revealed no mutations in PMR1, PMR3 and PMR4, and point mutations only were observed in both the promoter and coding regions of PMR5 in multiple resistant strains (TBZ- and DMI-resistant) (Sánchez-Torres & Tuset, 2011). But no explanation was ascertained for the absence of sequence changes relating to fungicide resistance in the other ABC transporters, particularly in the PMR1 gene given that transcription of PMR1 has proven to be strongly activated in the presence of different fungicides (Hamamoto et al., 2001b; Nakaune, 2001; Nakaune et al., 1998).

To date, MFS transporters have been described in several fungi, e.g. *Botrytis cinerea* shows a broad spectrum of resistance to different fungicides and their expression has been induced by many of them, particularly noteworthy *Bcmfs4* induction in the presence of strobilurin (trifloxiestrobin) (Hayashi et al., 2002a, 2002b, 2003; Schoonbeek et al., 2003; Vermeulen et al., 2001). All this means that these transporters are potential candidates for the study of factors involved in resistance based on active efflux of these toxic compounds since they have remarkably broad substrate specificity although they can also transport specific compounds.

Recently, five different MFS transporters have been identified and characterized in the postharvest phytopathogenic fungus *Penicillium digitatum* (PdMFS1-PdMFS5). Sequence analysis of these five genes revealed different genomic structure and although all genes seem to be implicated in pathogenicity, only 2 out of five MFS transporters confirmed to be involved in fungicide resistance (Sánchez-Torres et al., submitted). Therefore, the most recent thought for fungicide resistance based on active efflux of these toxic compounds is now discussed. These results suggest that many genes could be involved in the mechanisms conferring fungicide resistance to phytopathogenic fungi and some are fungicide-dependent.

From a practical standpoint, the fact that ABC and MFS transporters determine the baseline sensitivity to fungicides, are responsible for MDR and can act as virulence factors implies that these carriers are an attractive target for chemical control. In this context, inhibitors of these transporters could improve the effectiveness of control and reduce the virulence of fungal pathogens.

3.1.2 Changes in binding target that causes a reduced affinity of the compound fungicide

The most extent mechanism to confer DMI resistance involved mutations of *CYP51* gene, the target enzyme of DMIs fungicides and has been described for a large number of pathogens such as *Botrytis cinerea* (Albertini et al., 2002, Albertini & Leroux, 2004), a substitution of Phe for Tyr at position 136 (Y136F) was found in *Uncinula necator* (Délye et al., 1997) and also in *Erisiphe graminis* f. sp. *hordei* (Délye et al., 1998). Two single nucleotide mutations of *CYP51* resulting in amino acid substitutions Y136F and K147Q in *Blumeria graminis* were also found (Wyand & Brown, 2005). Different mutations were also found in *Tapesia* sp (Albertini et al., 2003), *Penicillium italicum* (Joseph-Horne & Hollomon, 1997), *Ustilago maydis* (Butters et al., 2000) and *Blumeriella jaapii* (Ma et al., 2006)

Similarly, mutations have been described in the cytochrome b gene that lead to change of its corresponding protein G143A, conferring resistance to strobilurin (Avila-Adame & Koller, 2003a, 2003b; Gisi et al., 2000, Zheng et al., 2000, Zhang, Z. et al., 2009) or the amino acid substitution in the β-tubulin target protein involved in development of resistance to benzimidazoles in *Botrytis cinerea* (Banno, 2008), *Venturia inaequalis* and *Penicillium italicum* (Koenraadt et al., 1992), *Monilinia fructicola* (Ma et al., 2003a), *M. laxa* (Ma & Michailides, 2005), *P. expansum* (Baraldi et al., 2003) and *P. digitatum* (Sánchez-Torres & Tuset, 2011).

3.1.3 Over-expression of the target of union of fungicide

P45014DM increased by over-expression of the *CYP51* gene has been described as a mechanism of resistance to azoles. In *Penicillium digitatum* a unique 126-bp sequence in the promoter region of *CYP51* was tandem repeated five times in resistant isolates and was present only one in sensitive isolates. This provided a quick and easy method to detect DMI-resistant strains of *P. digitatum* (Hamamoto et al., 2001). Insertions in the promoter were also found in *Blumeriella jaapii* (Ma et al., 2006), *Venturia inaequalis* (Schnabel & Jones 2001) and in *Monilinia fructicola* (Luo et al., 2008) and recently another insertion of 199-bp was found in *P. digitatum CYP51A* gene (Ghosoph et al., 2007) and *PdCYP51B* gene (Sun et al., 2011; Sánchez-Torres et al., submitted) leading to resistant phenotypes.

3.1.4 Compensation of the toxic effects of the fungicide by altered biosynthetic or metabolic pathway

This phenomenon has been described both in the case of DMIs because they exert their toxic effect by depletion of ergosterol and the accumulation of C14 methylated precursors and in the case of strobilurins since the presence of AOX (alternative oxidase) allows the use an alternative route in mitochondrial respiration (Gisi et al., 2000; Schnabel et al., 2001; Wood & Hollomon, 2003).

3.2 Molecular detection of fungicide resistance in phytopathogenic fungi

The procedures for detecting fungicide resistance using conventional methods are labor-intensive and time-consuming if large numbers of isolates have to be tested. Advances in molecular biology have provided new opportunities for rapidly detecting fungicide resistant genotype once the mechanisms of resistance have been elucidated at a molecular level. Several molecular techniques, such as PCR, PCR-restriction fragment length polymorphism (PCR-RFLP), allele specific PCR, allele-specific real-time PCR and massive sequencing techniques (all above described) have been used successfully to detect fungicide-resistant genotypes of several plant pathogens. Many examples of these techniques have been reported.

3.2.1 PCR detection of DMI-resistant isolates

PCR has revolutionized molecular biology and diagnostics and has become a fast tool for detecting fungicide-resistant pathogens. In *Penicillium digitatum*, the DMI resistance resulted from over-expression of the *PdCYP51* genes driven by a tandem repeat of five copies of a 126 transcriptional enhancer (Hamamoto et al., 2000), or insertion of 199-bp in the promoter region of *PdCYP51A* (Gosoph et al., 2007) or by the presence of 199-bp enhancer in the promoter region of *PdCYP51B* (Sun et al., 2011). Based on the DNA sequence of the *PdCYP51*-genes, PCR primers have been developed which are able to distinguish not only between sensitive and resistant DMI strains but allow identification of molecular mechanism that takes place (Hamamoto et al., 2001; Gosoph et al., 2007; Sánchez-Torres & Tuset, 2011; Sun et al., 2011).

3.2.2 PCR-RFLP and primer-introduced restriction analysis PCR (PIRA-PCR)

PCR amplification followed by restriction enzyme analysis (PCR-RFLP) is a general technique to detect a point mutation that alters a restriction enzyme site. This method has been used to rapidly detect benzimidazole-resistant isolates of *Monilinia laxa* from stone fruit and almond crops in California (Ma & Michailides, 2005) and can also detect azoxystrobin-resistant isolates of *Alternaria alternata, A. tenuissima,* and *A. arborescens* within a few hours (Ma et al., 2003b). Since the PCR-RFLP is an easy and rapid technique, these assays have been developed for detecting benzimidazole resistance in *Botrytis cinerea* (Luck & Gillings, 1995), *Cladobotryum dendroides* (McKay et al., 1998), *Helminthosporium solani* (Cunha & Rizzo, 2003; McKay & Cooke, 1997), and for detecting strobilurin resistance in *Blumeria graminis* f.sp. *hordei* (Baumler et al., 2003), *Erysiphe graminis* f. sp. *tritici* (Sierotzki et al., 2000), and *Podosphaera fusca* (Ishii et al., 2001).

Although PCR-RFLP is a simple method for detection of point mutations, some target DNA fragments may not contain a restriction endonuclease recognition sequence at the site of point mutation. Thus, primer-introduced restriction analysis PCR (PIRA-PCR) has become a useful method to create diagnostic artificial RFLPs (Haliassos et al., 1989). A PIRA-PCR assay was developed for detecting carpropamid resistant *Magnaporthe grisea* (Kaku et al., 2003).

3.2.3 Allele-specific PCR and quantitative allele-specific real-time PCR

Allele-specific PCR is another simple and rapid method for detecting point mutations. Usually, one of two PCR primers used in an allele specific amplification is designed to amplify preferentially one allele by matching the desired allele and mismatching the other allele at the 3' end of primer.

Allele-specific PCR assays have been developed for detecting benzimidazole-low resistant isolates of *Monilinia laxa* (Ma & Michailides, 2005), and azoxystrobin-resistant isolates of *Alternaria alternata*, *A. tenuissima*, and *A. arborascens* (Ma & Michailides, 2004). Additionally, allele-specific PCR assays have been developed for the rapid detection of strobilurin-resistant isolates of *Blumeria graminis* f. sp. *tritici* (Fraaije et al., 2002) and *Mycosphaerella fijiensis* (Gisi et al., 2002), and DMI-resistant isolates of *Erysiphe graminis* f. sp. *hordei* (Dèlye et al., 1997).

Real-time PCR technique can be used to quantitatively determine the amount of target DNA in a sample. An allele-specific real-time PCR assay has been used to follow the dynamics of *QoI* resistant allele *A143* in field populations of *Blumeria graminis* f. sp. *tritici* before and after fungicide application (Fraaije et al., 2002). A real-time PCR assay to rapidly detect azoxystrobin-resistant *Alternaria* has been developed in California pistachio orchards (Ma & Michailides, 2005). Additionally, a real-time PCR using the *Alternaria* specific PCR primer pair can quantify both resistant and sensitive alleles in the same tested samples, thereby enabling a rapid determination of frequencies of the azoxystrobin-resistant allele in *Alternaria* populations.

4. Conclusion

Advances in the development of molecular methods, especially PCR technology have provided diagnostic laboratories with powerful tools for detection and identification of phytopathogenic fungi. Molecular techniques have also contributed to elucidate the phenotypic and genetic structure within species and the complexity of plant and environment fungal populations. New technologies and improved methods with reduced cost and improved speed, throughput, multiplexing, accuracy and sensitivity have emerged as an essential strategy for the control of plant fungal diseases. These advances have been complemented by the development of new nucleic acids extraction methods, increased automation, reliable internal controls, multiplexing assays, *online* information and *on site* molecular diagnostics. Nevertheless, molecular diagnostic tools should be complemented with other techniques, either traditional culture-based methods or the newly emerged proteomic, a promising tool for providing information about pathogenicity and virulence factors that will open up new possibilities for crop disease diagnosis and crop protection.

On the other hand, fungicides continue to play a key role in strategies for the control of diseases in crops, and the development of resistance in the target pathogens is a continuing risk. This fact leads to many losses as control systems are not longer effective. Therefore, the better understanding on mechanisms developed during fungicide resistance is essential for a better management of chemical control, environment and human health.

Great advances have been made in the development of molecular methods to identify and monitor resistance of plant pathogens to fungicides. The highly sensitive methods can improve our ability of studying the evolution of fungicide resistance at the population level. Molecular techniques can be also developed based on the different fungicide mechanisms to rapidly detect resistant isolates. Furthermore, a timely detection of resistance levels in populations of phytopathogenic fungi in a field would help growers make proper decisions on resistance management programs to control plant diseases.

5. Acknowledgements

Authors want to thank Dr. J.M. Colmenero for critical reading of the manuscript. A.M. Pastrana is recipient of an IFAPA fellowship from the Consejería de Agricultura y Pesca, Junta de Andalucía, Spain.

6. References

Abad, Z.G.; Abad, J.A.; Coffey, M.D.; Oudemans, P.V.; Man, W.A.; de Gruyter, H.; Cunnington, J. & Louws, F.J. (2008). *Phytophthora bishetia* sp nov., a new species identified in isolates from the Rosaceous raspberry, rose and strawberry in three continents. *Mycologia*, Vol.100, No.1, pp. 99-110, ISSN 0027-5514

Abbott, C.L.; Gilmore, S.R.; Lewis, C.T.; Chapados, J.T.; Peters, R.D.; Platt, H.W.; Coffey, M.D. & Lévesque, C.A. (2010). Development of a SNP genetic marker system based on variation in microsatellite flanking regions of *Phytophthora infestans*. *Canadian Journal of Plant Pathology*, Vol.4, No.32, pp. 440-457, ISSN 0706-0661

Abd-Elsalam, K.; Bahkali, A.; Moslem, M.; Amin, O.E. & Niessen, L. (2011). An optimized protocol for DNA extraction from wheat seeds and Loop-Mediated Isothermal Amplification (LAMP) to detect *Fusarium graminearum* contamination of wheat grain. *International Journal of Molecular Sciences*, Vol.12, No.6, pp. 3459-3472, ISSN 1422-0067

Abd-Elsalam, K.A.; Asran-Amal, A.; Schnieder, F.; Migheli, Q. & Verreet, J.-A. (2006). Molecular detection of *Fusarium oxysporum* f. sp. *vasinfectum* in cotton roots by PCR and real-time PCR assay. *Journal of Plant Diseases and Protection*, Vol.113, No.1, pp. 14–19, ISSN 1861-3829

Ahmadian, A.; Ehn, M. & Hober, S. (2006). Pyrosequencing: History, biochemistry and future. *Clinica Chimica Acta*, Vol.363, No.1-2, pp. 83-94, ISSN 0009-8981

Alaei, H.; Baeyen, S.; Maes, M.; Hofte, M. & Heungens, K. (2009). Molecular detection of *Puccinia horiana* in *Chrysanthemum* x *morifolium* through conventional and real-time PCR. *Journal of Microbiological Methods*, Vol.76, No.2, pp. 136-145, ISSN 0167-7012

Albertini, C.; Gredt, M. & Leroux, P. (2003). Polymorphism of the 14_a demethylase gene (CYP51) in the cereal eyespot fungi *Tapesia acuformis* and *Tapesia yallundae*. *European Journal of Plant Pathology*, Vol.109, pp. 117–128, ISSN 0929-1873

Albertini, C. & Leroux, P. (2004). A *Botrytis cinerea* putative 3-keto reductase gene (ERG27) that is homologous to the mammalian 17 beta-hydroxysteroid dehydrogenase type 7 gene (17 beta-HSD7). European Journal of Plant Pathology Vol.110, No.7, pp.723-733, ISSN: 0929-1873

Albertini, C.; Thebaud G.; Fournier, E. & Leroux P. (2002). Eburicol 14*a*-demethylase gene (CYP51) polymorphism and speciation in *Botrytis cinerea*. *Mycological Research*, Vol.106, pp. 1171-1178, ISSN 0953-7562

Altschul, S.F.; Madden, T.L.; Schaffer, A.A.; Zhang, J.H.; Zhang, Z.; Miller, W. & Lipman, D.J. (1997). Gapped BLAST and PSI-BLAST: a new generation of protein database search programs. *Nucleic Acids Research*, Vol.25, pp. 3389-3402

Anderson, I.C. & Cairney, J.W.G. (2004). Diversity and ecology of soil fungal communities: increased understanding through the application of molecular techniques. *Environmental Microbiology*, Vol.6, No. 8, pp. 769–779

Anderson, I.C. & Parkin, P.I. (2007). Detection of active soil fungi by RT-PCR amplification of precursor rRNA molecules. *Journal of Microbiological Methods*, Vol.68, pp. 248-253

Anderson, N.; Szemes, M.; O'Brien, P.; De Weerdt, M.; Schoen, C.; Boender, P. & Bonants, P. (2006). Use of hybridization melting kinetics for detecting *Phytophthora* species using three-dimensional microarrays: demonstration of a novel concept for the differentiation of detection targets. *Mycological Research*, Vol.110, pp. 664-671

Andrade, A.C.; Del Sorbo, G.; Van Nistelrooy, J.G.M. & de Waard, M.A. (2000a). The ABC transporter AtrB from *Aspergillus nidulans* is involved in resistance to all major classes of fungicides and natural toxic compouns. *Microbiology U.K.*, Vol.146, pp. 1987-1997, ISSN 1350-0872

Andrade, A.C.; Van Nistelrooy, J.G.M.; Peery, R.B.; Skatrud, P.L. & de Waard M.A. (2000b). ABC transporters from *Aspergillus nidulans* are involved in protection against cytotoxic agents and antibiotic production. *Molecular General Genetics*, Vol.263, pp. 966-977, ISSN 0026-8925

Ansorge, W.J. (2010). Novel Next-Generation DNA Sequencing Techniques for Ultra High-Throughput Applications in Bio-Medicine, In: *Molecular Diagnostics* (Second Edition), G.P. Patrinos and W.J. Ansorge, pp. 365-378, Elsevier, ISBN 978-0-12-374537-8

Arici, S.E. & Koc N.K. (2010). RAPD–PCR analysis of genetic variation among isolates of *Fusarium graminearum* and *Fusarium culmorum* from wheat in Adana Turkey. *Pakistan Journal of Biological Sciences*, Vol.13, No. 3, pp. 138–142, ISSN 1028-8880

Aroca, A. & Raposo, R. (2007). PCR-based strategy to detect and identify species of *Phaeoacremonium* causing grapevine diseases. *Applied and Environmental Microbiology*, Vol.73, No.9, pp. 2911–2918, ISSN 0099-2240

Aroca, A.; Raposo, R. & Lunello, P. (2008). A biomarker for the identification of four *Phaeoacremonium* species using the beta-tubulin gene as the target sequence. *Applied Microbiology and Biotechnology*, Vol.80, No.6, pp. 1131-1140

Ashizawa, T.; Takahashi, M.; Moriwaki, J. & Hirayae, K. (2010). Quantification of the rice false smut pathogen *Ustilaginoidea virens* from soil in Japan using real-time PCR. *European Journal of Plant Pathology*, Vol.128, No.2, pp. 221-232, ISSN 0929-1873

Attallah, Z.K.; Bae, J.; Jansky, S.H.; Rouse, D.I. & Stevenson, W.R. (2007). Multiplex real-time quantitative PCR to detect and quantify *Verticillium dahliae* colonization in potato lines that differ in response to *Verticillium* wilt. *Phytopathology*, Vol.97, No.7, pp. 865–872

Avila-Adame, C. & Koller, W. (2003a). Characterization of spontaneous mutants of *Magnaporthe grisea* expressing stable resistance to the Qo-inhibiting fungicide azoxystrobin. *Current Genetics*, Vol.42, pp. 332-338, ISSN 0172-8083

Avila-Adame, C. & Koller, W. (2003b). Impact of alternative respiration and target-site mutations on responses of germinating conidia of *Magnaporthe grisea* to Qo-inhibiting fungicides. *Pest Managment Science*, Vol.59, pp. 303-309, ISSN 1526-498X

Baayen, R.P.; O'Donnell, K.; Bonants, P.J.M.; Cigelnik, E.; Kroon, L.P.N.M.; Roebroeck, E.J.A. & Waalwijk, C. (2000). Gene genealogies and AFLP analyses in the *Fusarium oxysporum* complex identify monophyletic and nonmonophyletic formae speciales causing wilt and rot disease. *Phytopathology*, Vol.90, No.8, pp. 891-900, ISSN 0031-949X

Babu, B.K..; Mesapogu, S.; Sharma A.; Somasani, S.R. & Arora, D.K. (2011). Quantitative real-time PCR assay for rapid detection of plant and human pathogenic *Macrophomina phaseolina* from field and environmental samples. *Mycologia* Vol.103, No.3, pp.466-473, ISSN: 0027-5514

Bailey, A.M.; Mitchell, D.J.; Manjunath, K.L.; Nolasco, G. & Niblett, C.L. (2002). Identification to the species level of the plant pathogens *Phytophthora* and *Pythium* by using unique sequences of the ITS1 region of ribosomal DNA as capture probes for PCR Elisa. *Fems Microbiology Letters*, Vol.207, No.2, pp. 153-158, ISSN 0378-1097

Banno, S.; Fukumori, F.; Ichiishi, A.; Okada, K.; Uekusa, H.; Kimura, M. & Fujimura, M. (2008). Genotyping of benzimidazole-resistant and dicarboximide-resistantmutations in *Botrytis cinerea* using real-time polymerase chain reaction assays. *Phytopathology*, Vol.98, pp. 397–404, ISSN 0031-949X

Baraldi, E.; Mari, M.; Chierici, E.; Pondrelli, M.; Bertollini, P. & Pratella, G.C. (2003). Studies of thiabendazole resistance of *Penicillium expansum* of pears: pathogenic fitness and genetic characterization. *Plant Pathology*, Vol.52, pp. 362-370, ISSN 0032-0862

Barnes, C.W. & Szabo, L.J. (2007). Detection and identification of four common rust pathogens of cereals and grasses using real-time polymerase chain reaction. *Phytopathology*, Vol.97, No.6, pp. 717-727, ISSN 0031-949X

Bauer, B.E.; Wolfger, H. & Kuchler, K. (1999). Inventory and function of yeast ABC proteins: about sex, stress, pleiotropic drug and heavy metal resistance. *Biochimestry Biophysic Acta*, Vol.1461, No.2, pp. 217-236, ISSN 0005-2736

Baumler, S.; Felsenstein, F.G. & Schwarz, G. (2003). CAPS and DHPLC analysis of a single nucleotide polymorphism in the cytochrome b gene conferring resistance to strobilurin in field isolates of *Blumeria graminis* f. sp. *hordei. Journal of Phytopathology*, Vol.151, pp. 149–152, ISSN 0931-1785

Bayraktar, H.; Dolar, F.S. & Tor, M. (2007). Determination of genetic diversity within *Ascochyta rabiei* (pass.) labr., the cause of ascochyta blight of chickpea in Turkey. *Journal of Plant Pathology*, Vol.89, No.3, pp. 341-347, ISSN 1125-4653

Bearchell, S.J.; Fraaije, B.A.; Shaw, M.W. & Fitt, B.D.L. (2005). Wheat archive links long-term fungal pathogen population dynamics to air pollution. *Proceedings of the National Academy of sciences of the United States of America*, Vol.102, No.15, pp. 5438-5442, ISSN 0027-8424

Bilodeau, G.J.; Pelletier, G.; Pelletier, F.; Lévesque, C.A. & Hamelin, R.C. (2009). Multiplex real-time polymerase chain reaction (PCR) for detection of *Phytophthora ramorum*, the causal agent of sudden oak death. *Canadian Journal of Plant Pathology*, Vol.31, No.2, pp. 195-210, ISSN 0706-0661

Bindslev, L.; Oliver, R.P. & Johansen, B. (2002). *In situ* PCR for detection and identification of fungal species. *Mycological Research*, Vol.106, No.3, pp. 277–279, ISSN 0953-7562

Biswas, K. & Biswas, R.A. (2011). Modified method to isolate genomic DNA from plants without liquid nitrogen. *Current Science*, Vol.100, No.11, pp. 1622-1624 , ISSN 0011-3891

BOLD, Barcode of Life Database Systems: (http://www.boldsystems.org)

Børja, I.; Solheim, H.; Hietala, A.M. & Fossdal, C.G. (2006). Etiology and real-time polymerase chain reaction-based detection of *Gremmeniella-* and *Phomosis-* associated disease in Norway spruce seedlings. *Phytopathology*, Vol.96, No.12, pp. 1305-1314, ISSN 0031-949X

Braasch, D.A. & Corey, D.R. (2001) Locked nucleic acid (LNA): fine-tuning the recognition of DNA and RNA. *Chemistry & Biology*, Vol.8, No.1, pp. 1-7, ISSN 1074-5521

Brandfass, C. & Karlovsky, P. (2006). Simultaneous detection of *Fusarium culmorum* and *F-graminearum* in plant material by duplex PCR with melting curve analysis. *BMC Microbiology*, Vol.6, No.4, ISSN 1471-2180

Brent, K.J. (1995). *Fungicide resistance in crop pathogen: How can it be managed?*. FRAC Monograph N° 1 Global Crop Protection Federation, ISBN 90-72398-07-6, Brussels

Broders, K.D.; Woeste, K.E.; SanMiguel, P.J.; Westerman, R.P. & Boland, G.J. (2011). Discovery of single-nucleotide polymorphisms (SNPs) in the uncharacterized genome of the ascomycete *Ophiognomonia clavigignenti-juglandacearum* from 454 sequence data. *Molecular Ecology Resources*, Vol.11, No.4, pp. 693-702, ISSN 1755-098X

Bruns T.D., White T.J. & Taylor J.W. (1991). Fungal molecular systematics. *Annual Review of Ecology and Systematics*, Vol.22, pp. 525-564.

Budge, G.E.; Shaw, M. W.; Colyer, A.; Pietravalle, S. & Boonham, N. Molecular tools to investigate *Rhizoctonia solani* distribution in soil. *Plant Pathology*, Vol.58, No.6, pp. 1071-1080, ISSN 0032-0862

Burgess, T.I.; Webster, J.L.; Ciampini, J.A.; White, D.; Hardy, G.E.S. & Stukely, M.J.C. (2009). Re-evaluation of *Phytophthora* species isolated during 30 years of vegetation health surveys in western Australia using molecular techniques. *Plant Disease*, Vol.93, No.3, pp. 215-223, ISSN 0191-2917

Butters, J.A.; Zhou, M. & Hollomon, D.W. (2000). The mechanism of resistance tosterol 14_-demethylation inhibitors in amutant (Erg40) of *Ustilago maydis*. *Pest Managment Science*, Vol.56, pp. 257–263, ISSN 1526-498X

Campanile, G.; Schena L. & Luisi, N. (2008). Real-time PCR identification and detection of *Fuscoporia torulosa* in *Quercus ilex*. *Plant Pathology*, Vol.57, No.1, pp. 76-83, ISSN 0032-0862

Carisse, O.; Tremblay, D.M.; Lévesque, C.A.; Gindro, K.; Ward, P. & Houde, A. (2009). Development of a TaqMan real-time PCR assay for quantification of airborne conidia of *Botrytis squamosa* and management of Botrytis leaf blight of onion. *Phytopathology*, Vol.99, No.11, pp. 1273-1280, ISSN 0031-949X

Causin, R.; Scopel, C.; Grendene, A. & Montecchio L. (2005). An improved method for the detection of *Phytophthora cactorum* (L.C.) Schröeter in infected plant tissues using SCAR markers. *Journal of Plant Pathology*, Vol.87, No.1, pp. 25-35, ISSN 1125-4653

Chen, R.S.; Chu, C.; Cheng, C.W.; Chen W.Y. & Tsay J.G. (2008). Differentiation of two powdery mildews of sunflower (Helianthus annuus) by a PCR-mediated method based on ITS sequences. *European Journal of Plant Pathology*, Vol.121, No.1, pp. 1-8, ISSN 0929-1873

Chen, W.; Seifert, K.A. & Lévesque, C.A. (2009). A high density COX1 barcode oligonucleotide array for identification and detection of species of *Penicillium subgenus Penicillium. Molecular Ecology Resources*, Vol.9 (Suppl. 1), pp. 114-129, ISSN 1755-098X

Chi, M.-H.; Park, S.-Y. & Lee, Y.-H. (2009). A quick and safe method for fungal DNA extraction. *Plant Pathology Journal*, Vol.25, No.1, pp. 108-111, ISSN 1598-2254

Chimento, A.; Scibetta, S.; Schena, L.; Cacciola, S.O. & Cooke, D.E.L. (2005). A new method for the monitoring of *Phytophthora* diversity in soil and water. *Journal of Plant Pathology*, Vol.87, pp. 290

CiF, Cyber infrastructure for *Fusarium* : (http://www.fusariumdb.org)

Cipriani, M.G.; Stea, G.; Moretti, A.; Altomare, C.; Mulè, G. & Vurro, M. (2009). Development of a PCR-based assay for the detection of *Fusarium oxysporum* strain FT2, a potential mycoherbicide of *Orobanche ramosa*. *Biological Control*, Vol.50, No.1, pp. 78–84, ISSN 1049-9644

Collado-Romero, M.; Mercado-Blanco, J.; Olivares-Garcia, C. & Jimenez-Diaz, R. M. (2008). Phylogenetic analysis of *Verticillium dahliae* vegetative compatibility groups. *Phytopathology*, Vol.98, No.9, pp. 1019-1028, ISSN 0031-949X

Cruz-Perez, P.; Buttner, M.P. & Stetzenbach, L. D. 2001. Detection and quantitation of *Aspergillus fumigatus* in pure culture using polymerase chain reaction. *Molecular and Cellular Probes*, Vol.15, No.2, pp. 81-88, ISSN 0890-8508

Cunha, M.G. & Rizzo, D.M. (2003). Development of fungicide cross resistance in *Helminthosporium solani* population from California. *Plant Disease*, Vol.87, No.7, pp. 798–803, ISSN 0191-2917

Daval, S.; Lebreton L.; Gazengel K.; Guillerm-Erckelboudt A.-Y. & Sarniguet A. (2010). Genetic evidence for differentiation of *Gaeumannomyces graminis* var. *tritici* into two major groups. *Plant Pathology*, Vol.59, No. 1, pp. 165–178, ISSN 0032-0862

de Cock, A.W.A.M. & Lévesque, A. (2004). New species of *Pythium* and *Phytophthora*. *Studies in Mycology*, Vol.50, Special Issue, pp. 481–487, ISSN 0166-0616

De Waard, M.A. (1997). Significance of ABC transporters in fungicide sensitivity and resistance. *Pesticide Science*, Vol.51, pp. 271-275, ISSN 0031-613X

de Weerdt, M.; Zijlstra, C.; van Brouwershaven, I.R.; van Leeuwen, G.C.M.; de Gruyter, J. & Kox, L.F.F. (2006). Molecular detection of *Fusarium foetens* in Begonia. *Journal of Phytopathology*, Vol.154, No.11-12, pp. 694-700, ISSN 0931-1785

Del Sorbo, G.; Schoonbeek, H. & de Waard, M.A. (2000). Fungal transporters involved in efflux of natural toxic compounds and fungicides. *Fungal Genetics Biology*, Vol.30, pp. 1-15, ISSN 1087-1845

Delp, C.J. & Dekker, J. (1985). Fungicide resistance: definitions and use of terms. *EPPO Bulletin*, Vol.15, pp. 333-335

Dèlye, C.; Laigret, F. & Coster, M.F. (1997). A mutation in 14 α demethylase gene of *Uncinula necator* that correlates with resistance to a sterol biosynthesis inhibitor. *Applied Environmental Microbiology*, Vol.63, pp. 2966-2970, ISSN 0099-2240

Demeke, T. & Jenkins, G.R. (2010). Influence of DNA extraction methods, PCR inhibitors and quantification methods on real-time PCR assay of biotechnology-derived traits. *Analytical and Bioanalytical Chemistry*, Vol.396, No.6, pp. 1977–1990, ISSN 1618-2642

Demeke, T.; Gräfenhan, T.; Clear, R.M.; Phan, A.; Ratnayaka, I.; Chapados, J.T.; Patrick, S.K.; Gaba, D.; Seifert, K.A. & Lévesque, C.A. (2010). Development of a specific TaqMan® real-time PCR assay for quantification of *Fusarium graminearum* clade 7 and comparison of fungal biomass determined by PCR with deoxynivalenol content in wheat and barley. *International Journal of Food Microbiology*, Vol.141, No.1-2, pp. 45-50, ISSN 0168-1605

Demontis, M.A.; Cacciola, S.O.; Orru, M.; Balmas, V.; Chessa, V.; Maserti, B.E.; Mascia, L.; Raudino, F.; Quirico, G.M.D.S.L. & Migheli, Q. (2008). Development of real-time PCR systems based on SYBR (R) Green I and TaqMan (R) technologies for specific quantitative detection of *Phoma tracheiphila* in infected Citrus. *European Journal of Plant Pathology*, Vol.120, No.4, pp. 339-351, ISSN 0929-1873

DNA Multiscan®: (http://www.DNAmultiscan.com)

Diguta, C. F.; Rousseaux, S.; Weidmann, S.; Bretin, N.; Vincent, B.; Guilloux-Benatier, M. & Alexandre, H. (2010). Development of a qPCR assay for specific quantification of *Botrytis cinerea* on grapes. *Fems Microbiology Letters*, Vol.313, No.1, pp. 81-87, ISSN 0378-1097

Dixon, L.J.; Schlub, R.L.; Pernezny, K. & Datnoff, L.E. (2009). Host specialization and phylogenetic diversity of *Corynespora cassiicola*. *Phytopathology*, Vol.99, No.9, pp. 1015-1027, ISSN 0031-949X

Drenth, A.; Wagals, G.; Smith, B.; Sendall, B.; O'Dwyer, C.; Irvine, G. & Irwin, J.A.G. (2006). Development of a DNA-based method for the detection and identification of *Phytophthora species*. *Australasian Plant Pathology* , Vol.35, No.2, pp. 147–159, ISSN 0815-3191

Druzhinina, I. S.; Kopchinskiya, A. G.; Komoja, M.; Bissettb, J.; Szakacs, G. & Kubiceka, C.P. (2005). An oligonucleotide barcode for species identification in *Trichoderma* and *Hypocrea*. *Fungal Genetics and Biology*, Vol.42, No.10, pp. 813-828, ISSN 1087-1845

Durán, A.; Gryzenhout, M; Drenth, A; Slippers, B; Ahumada, R; Wingfield, B.D. & Wingfield, M.J. (2010). AFLP analysis reveals a clonal population of *Phytophthora pinifolia* in Chile. *Fungal Biology*, Vol.114, No.9, pp. 746-752, ISSN 1878-6146

Dyer, P.S.; Furneaux, P.A.; Douhan, G. & Murray, T.D. (2001). A multiplex PCR test for determination of mating type applied to the plant pathogens *Tapesia yallundae* and *Tapesia acuformis*. *Fungal Genetics and Biology*, Vol.33, No.3, pp. 173-180, ISSN 1087-1845

Feng, J.; Hwang, R.; Chang, K. F.; Hwang, S. F.; Strelkov, S. E.; Gossen, B. D. & Zhou, Q. (2010). An inexpensive method for extraction of genomic DNA from fungal mycelia. *Canadian Journal of Plant Pathology*, Vol.32, No.3, pp. 396-401, ISSN 0706-0661

Fling, M.E.; Kopf, J.; Tamarkin, A.; Gorman, J.A.; Smith, H.A. & Koltin, Y. (1991). Analysis of a *Candida albicans* gene encodes a novel mechanism for resistance to benomyl and methotrexate. *Molecular General Genetics*, Vol.227, pp. 318-329, ISSN 0026-8925

Foster, S.J.; Ashby, A.M. & Fitt, B.D.L. (2002). Improved PCR-based assays for pre-symptomatic diagnosis of light leaf spot and determination of mating type of *Pyrenopeziza brassicae* on winter oilseed rape. *European Journal of Plant Pathology*, Vol.108, No.4, pp. 379-383, ISSN 0929-1873

Fountaine, J.A.; Shaw, M.W.; Napier, B.; Ward, E. & Fraaije, B. A. (2007). Application of real-time and multiplex polymerase chain reaction assays to study leaf blotch epidemics in barley. *Phytopathology*, Vol.97, No.3, pp. 297-303, ISSN 0031-949X

Fourie, G.; Steenkamp, E.T.; Ploetz, R.C.; Gordon, T.R. & Viljoen, A. (2011). Current status of the taxonomic position of *Fusarium oxysporum formae specialis cubense* within the

Fusarium oxysporum complex. *Infection Genetics and Evolution*, Vol.11, No.3, pp. 533-542, ISSN 1567-1348

Fraaije, B.A.; Butters, J.A.; Coelho, J.M.; Johes, D.R. & Hollomon, D.W. (2002). Following the dynamics of strobilurin resistance in *Blumeria graminis f. sp. tritici* using quantitative allele-specific realtime PCR measurements with the fluorescent dye SYBR green I. *Plant Pathology*, Vol.51, pp. 45–54, ISSN 0032-0862

Fraaije, B.A.; Lovell, D.J.; Coelho, J.M.; Baldwin, S. & Hollomon, D.W. (2001). PCR-based assays to assess wheat varietal resistance to blotch (*Septoria tritici* and *Stagonospora nodorum*) and rust (*Puccinia striiformis* and *Puccinia recondita*) diseases. *European Journal of Plant Pathology*, Vol.107, No.9, pp. 905-917, ISSN 0929-1873

Fraaije, B.A., Lovell, D.J., Rohel, E.A. & Hollomon, D.W. (1999). Rapid detection and diagnosis of *Septoria tritici* epidemics in wheat using a polymerase chain reaction PicoGreen assay. *Journal of Applied Microbiology*, Vol.86, No.4, pp. 701-708, ISSN 1364-5072

Garrido, C.; Carbú, M.; Fernández-Acero, F.J.; Boonham, N.; Colyer, A.; Cantoral, J.M. & Budge G. (2009). Development of protocols for detection of *Colletotrichum acutatum* and monitoring of strawberry anthracnose using real-time PCR. *Plant Pathology*, Vol.58, No.1, pp. 43–51, ISSN 0032-0862

Gayoso, C.; de Ilárduya, O.M.; Pomar, F. & de Cáceres, F.M. (2007). Assessment of real-time PCR as a method for determining the presence of *Verticillium dahliae* in different Solanaceae cultivars. *European Journal of Plant Pathology*, Vol.118, No.3, pp. 199–209, ISSN 0929-1873

Geiser, D. M.; Jiménez-Gasco, M.; Kang, S.; Makalowska, I.; Veeraraghavan, N.; Ward, T.J.; Zhang, N.; Kuldau, G.A. & O'Donnell, K. (2004). *Fusarium*-ID v.1.0: A DNA sequence database for identifying *Fusarium*. *European Journal of Plant Pathology*, Vol.110, No.5-6, pp. 473-479, ISSN 0929-1873

Ghosoph, J.M.; Schmidt, L.S.; Margosan, D.A. & Smilanick, J.L. (2007). Imazalil resistance linked to a unique insertion sequence in the PdCYP51 promoter region of *Penicillium digitatum*. *Postharvest Biology Technology*, Vol.44, pp. 9–18, ISSN 0925-5214

Gisi, U.; Chin, K.M.; Knapova, G.; Küng Färber, R.; Mohr, U.; Parisi, S.; Sierotzki, H, & Steinfeld, U. (2000). Recent development in elucidating modes of resistance to phenylamide, DMI and strobilurin fungicides. *Crop Protection*, Vol.19, pp. 863-872, ISSN 0261-2194

Gisi, U.; Sierotzki, H.; Cook, A. & McCaffery, A. (2002). Mechanisms influencing the evolution of resistance to Qo inhibitor fungicides. *Pest Managment Science*, Vol.58, pp. 859–867, ISSN 1526-498X

Glienke, C.; Pereira, O.L.; Stringari, D.; Fabris, J.; Kava-Cordeiro, V.; Galli-Terasawa, L.; Cunnington, J.; Shivas, R.G.; Groenewald, J.Z. & Crous, P.W. (2011). Endophytic and pathogenic *Phyllosticta* species, with reference to those associated with Citrus Black Spot. *Persoonia*, Vol.26, pp. 47-56, ISSN 0031-5850

Godoy-Lutz, G.; Kuninaga, S.; Steadman, J. R. & Powers, K. (2008). Phylogenetic analysis of *Rhizoctonia solani* subgroups associated with web blight symptoms on common bean based on ITS-5.8S rDNA. *Journal of General Plant Pathology*, Vol.74, No.1, pp. 32-40, ISSN 1345-2630

Gómez-Alpizar, L.; Saalau, E.; Picado I.; Tambong, J.T. & Saborio, F. (2011). A PCR-RFLP assay for identification and detection of *Pythium myriotylum*, causal agent of the cocoyam root rot disease. *Letters in Applied Microbiology*, Vol.52, No.3, pp. 185-192, ISSN 0266-8254

González, C.; Noda, J.; Espino, J.J & Brito, N. (2008). Drill-assisted genomic DNA extraction from *Botrytis cinerea*. *Biotechnology Letters*, Vol.30, No.11, pp. 1989-1992, ISSN 0141-5492

González-Mendoza, D.; Argumedo-Delira, R.; Morales-Trejo, A.; Pulido-Herrera, A.; Cervantes-Díaz, L.; Grimaldo-Juarez, O. & Alarcon, A. (2010). A rapid method for isolation of total DNA from pathogenic filamentous plant fungi. *Genetics and Molecular Research*, Vol.9, No.1, pp. 162-166, ISSN 1676-5680

Goud, J.C. & Termorshuizen A.J. (2003) Quality of methods to quantify microsclerotia of *Verticillium dahliae* in soil. *European Journal of Plant Pathology*, Vol.109, No.6, pp. 523-534, ISSN 0929-1873

Gril, T.; Celar, F.; Munda, A.; Javornik, B. & Jakse, J. (2008). AFLP analysis of intraspecific variation between *Monilinia laxa* isolates from different hosts. *Plant Disease*, Vol.92, No.12, pp.1616-1624, ISSN 0191-2917

Groenewald, S.; Van Den Berg, N.; Marasas, W.F.O. & Viljoen, A. (2006). The application of high-throughput AFLPs in assessing genetic diversity in *Fusarium oxysporum* f.sp. *cubense*. *Mycological Research*, Vol.110, pp. 297-305, ISSN 0953-7562

Grote, D.; Olmos, A.; Kofoet, J.J.; Tuset, E.; Bertolini, E. & Cambra, M. (2002). Specific and sensitive detection of *Phytophthora nicotianae* by simple and nested-PCR. *European Journal of Plant Pathology*, Vol.108, No.3, pp. 197-207, ISSN 0929-1873

Grund, E.; Darissa, O. & Adam, G. (2010). Application of FTA (R) cards to sample microbial plant pathogens for PCR and RT-PCR. *Journal of Phytopathology*, Vol.158, No.11-12, pp. 750-757, ISSN 0931-1785

Grünwald, N.J.; Martin, F.N.; Larsen, M.M.; Sullivan, C.M.; Press, C.M.; Coffey, M.D.; Hansen, E.M. & Parke, JL. (2011). *Phytophthora*-ID.org: A sequence-based *Phytophthora* identification tool. *Plant Disease*, Vol.95, No.3, pp. 337-342, ISSN 0191-2917

Guglielmo, F.; Bergemann, S.E.; Gonthier, P.; Nicolotti, G. & Garbelotto, M. (2007). A multiplex PCR-based method for the detection and early identification of wood rotting fungi in standing trees. *Journal of Applied Microbiology*, Vol.103, pp. 1490-1507, ISSN 1364-5072

Guo, J.R.; Schnieder, F.; Beyer, M. & Verreet, J.A. (2005). Rapid detection of *Mycosphaerella graminicola* in wheat using reverse transcription-PCR assay. *Journal of Phytopathology*, Vol.153, No.11-12, pp. 674-679, ISSN 0931-1785

Haase, A.T.; Retzel, E.F. & Staskus, K.A. (1990). Amplification and detection of lentiviral DNA inside cells. *Proceedings of the National Academy of Sciences of the United States of America*, Vol.87, No.13, pp. 4971-4975, ISSN 0027-8424

Hamamoto, H.; Hasegawa, K.; Nakaune, R.; Lee, Y.J.; Makizumi, Y.; Akutsu, K. & Hibi T. (2000). Tandem repeat of a transcription enhancer upstream of the sterol 14 α demethylase gene (*CYP51*) in *Penicillium digitatum*. *Applied Environmental Microbiology*, Vol.66, pp. 3421-3426, ISSN 099-2240

Hamamoto, H.; Nawata, O.; Hasegawa, K.; Nakaune, R.; Lee, Y.J.; Makizumi, Y.; Akutsu, K. & Hibi T. (2001). The role of the ABC transporter gene *PMR1* in demethylation inhibitor resistance in *Penicillium digitatum*. *Pest Biochemical Physiology*, Vol.70, pp. 19-26, ISSN 0048-3575

Hayashi, K.; Schoonbek, H.J. & de Waard, M.A. (2002a). Bcmfs1. A novel major facilitator superfamily transporter from *Botrytis cinerea* provides tolerance towards the natural toxic compounds camptothecin and cercosporin and towards fungicides. *Applied Environmental Microbiology*, Vol.68, pp. 4996-5004, ISSN 0099-2240

Hayashi, K.; Schoonbeek, H.J. & de Waard, M.A. (2002b). Expression of the ABC transporter *BcatrD* from *Botrytis cinerea* reduces sensitivity to sterol demethylation inhibitor fungicides. *Pest Biochemical Physiology*, Vol.73, pp. 110-121, ISSN 0048-3575

Hayashi, K.; Schoonbeek, H.J. & de Waard MA. (2003). Modulators of membrane drug transporters potentiate the activity of the DMI fungicide oxpoconazole against *Botrytis cinerea*. *Pest Managment Science*, Vol.59, pp. 294-302, ISSN 1526-498X

Hayashi, K.; Schoonbeek, H.J.; Sugiura, H. & de Waard, M.A. (2001). Multidrug resistance in *Botrytis cinerea* associated with decreased accumulation of the azole fungicide oxpoconazole and increased transcription of the ABC transporter gene *BcatrD*. *Pest Biochemical Physiology*, Vol.70, pp. 168-179, ISSN 0048-3575

Heid, C.A.; Stevens, J.; Livak, K.J. & Williams, P.M. (1996). Real time quantitative PCR. *Genome Research*, Vol.6, pp. 986–994

Hewit, H. G. (1998). *Fungicides in crop protection*. CAB Intenational. University Press, ISBN 0-85199-201-3, Cambridge. UK.

Hirsch, P.R.; Mauchline, T.H. & Clark, I.M. (2010). Culture-independent molecular techniques for soil microbial ecology. *Soil Biology & Biochemistry*, Vol.42, pp. 878-887, ISSN 0038-0717

Hong, S.Y.; Kang, M.R.; Cho, E.J.; Kim, H.K. & Yun, S.H. (2010). Specific PCR detection of four quarantine *Fusarium Species* in Korea. *Plant Pathology Journal*, Vol.26, No.4, pp. 409-416, ISSN 1598-2254

Huang, J. & Kang, Z. (2010). Detection of *Thielaviopsis basicola* in soil with real-time quantitative PCR assays. *MicrobiologicaL Research* , Vol.165, No.5, pp. 411-417, ISSN 0944-5013

Hughes, K.J.D.; Giltrap, P.M.; Barton, V.C.; Hobden E.; Tomilson, J.A. & Barber, P. (2006). On-site real-time PCR detection of *Phytophthora ramorum* causing dieback of *Parrotia persica* in the UK. *Plant Pathology*, Vol.55, No.6, pp. 813-813, ISSN 0032-0862

Hyakumachi, M.; Priyatmojo, A.; Kubota, M. & Fukui, H. (2005). New anastomosis groups, AG-T and AG-U, of binucleate *Rhizoctonia* spp. causing root and stem rot of cut-flower and miniature roses. *Phytopathology*, Vol.95, No.7, pp. 784-792, ISSN 0031-949X

Hyun, J. W.; Yi, S. H.; MacKenzie, S. J.; Timmer, L. W.; Kim, K. S.; Kang, S. K.; Kwon, H. M. & Lim, H. C. (2009). Pathotypes and genetic relationship of worldwide collections of *Elsinoe* spp. causing scab diseases of citrus. *Phytopathology*, Vol.99, No.6, pp. 721-728, ISSN 0031-949X

Inami, K.; Yoshioka, C.; Hirano, Y.; Kawabe, M.; Tsushima, S.; Teraoka, T. & Arie, T. (2010). Real-time PCR for differential determination of the tomato wilt fungus, *Fusarium oxysporum f. sp lycopersici*, and its races. *Journal of General Plant Pathology*, Vol.76, No.2, pp.116-121, ISSN 1345-2630

Ioos, R.; Kowalski, T.; Husson, C. & Holdenrieder, O. (2009). Rapid in planta detection of *Chalara fraxinea* by a real-time PCR assay using a dual-labelled probe. *European Journal of Plant Pathology*, Vol.125, No.2, pp. 329-335, ISSN 0929-1873

Ippolito, A.; Schena, L. & Nigro, F. (2002). Detection of *Phytophthora nicotianae* and *P. citrophthora* in citrus roots and soils by nested PCR. *European Journal of Plant Pathology*, Vol.108, No.9, pp. 855-868, ISSN 0929-1873

Ishii, H.; Fraaije, B.A.; Sugiyama, T.; Noguchi, K.; Nishimura, K.; Takeda, T.; Amano, T. & Hollomon, D.W. (2001). Occurrence and molecular characterization of strobilurin resistance in cucumber powdery mildew and downy mildew. *Phytopathology*, Vol.91, pp. 1166–1171, ISSN 0031-949

Izzo, A.D. & Mazzola, M. (2009). Hybridization of an ITS-based macroarray with ITS community probes for characterization of complex communities of fungi and fungal-like protists. *Mycological Research*, Vol.113, pp. 802-812, ISSN 0953-7562

Jackson, E.W.; Avant, J.B.; Overturf, K.E. & Bonman, J.M. (2006). A quantitative assay of *Puccinia coronata* f. sp *avenae* DNA in *Avena sativa*. *Plant Disease*, Vol.90, No.5, pp. 629-636, ISSN 0191-2917

Jana, T.; Sharma, T.R. & Singh, N.K. (2005). SSR-based detection of genetic variability in the charcoal root rot pathogen *Macrophomina phaseolina*. *Mycological Research*, Vol.109, Part 1, pp. 81-86, ISSN 0953-7562

Jeeva, M.L.; Mishra, A.K.; Vidyadharan, P.; Misra, R.S. & Hegde, V. (2010). A species-specific polymerase chain reaction assay for rapid and sensitive detection of *Sclerotium rolfsii*. *Australasian Plant Pathology*, Vol.39, No.6, pp. 517-523, ISSN 0815-3191

Jiang, Y.X.; Wu, J.G.; Yu, K.Q.; Ai, CX; Zou, F. & Zhou, H.W. (2011) Integrated lysis procedures reduce extraction biases of microbial DNA from mangrove sediment. *Journal of Bioscience and Bioengineering*, Vol.111, No.2, pp. 153-157, ISSN 1389-1723

Jiménez-Fernández, D.; Montes-Borrego, M.; Navas-Cortés, J.A.; Jiménez-Díaz, R.M. & Landa, B.B. (2010). Identification and quantification of *Fusarium oxysporum* in planta and soil by means of an improved specific and quantitative PCR assay. *Applied Soil Ecology*, Vol.46, No.3, pp. 372–382, ISSN 0929-1393

Jiménez-Fernández D.; Montes-Borrego M.; Jimenez-Diaz R.M.; Navas-Cortes, J.A. & Landa, B.B. (2011). In planta and soil quantification of *Fusarium oxysporum* f. sp. *ciceris* and evaluation of Fusarium wilt resistance in chickpea with a newly developed quantitative polymerase chain reaction assay. *Phytopathology* Vol. 101, No.2, pp. 250-262, ISSN: 0031-949X

Joseph-Horne, T. & Hollomon, D.W. (1997). Molecular mechanisms of azole resistance. *FEMS Microbiology Letters*, Vol.149, pp. 141–149

Justesen, A.F.; Hansen, H.J.; Pinnschmidt, H.O. (2008). Quantification of *Pyrenophora graminea* in barley seed using real-time PCR. *European Journal of Plant Pathology*, Vol.122, No.2, pp. 253-263, ISSN 0929-1873

Kageyama, K.; Senda, M.; Asano, T.; Suga, H. & Ishiguro, K. (2007). Intra-isolate heterogeneity of the ITS region of rDNA in *Pythium helicoides*. *Mycological Research*, Vol.111, pp. 416-423, ISSN 0953-7562

Kaku, K.; Takagaki, M.; Shimizu, T. & Nagayama, K. (2003). Diagnosis of dehydratase inhibitors in melanin biosynthesis inhibitor (MBID) resistance by primer-introduced restriction enzyme analysis in scytalonedehydratase gene of *Magnaporthe grisea*. *Pest Managment Science*, Vol.59, pp. 843–846, ISSN 1526-498X

Kang, S.; Mansfield, M.A.; Park, B.; Geiser, D.M.; Ivors, K.L.; Coffey, M.D.; Grünwald, N.J.; Martin, F.N.; Lévesque, C.A. & Blair, J.E. (2010). The promise and pitfalls of sequence-based identification of plant-pathogenic fungi and oomycetes. *Phytopathology*, Vol.100, No.8, pp. 732-737, ISSN 0031-949X

Keiper, F.J.; Capio, E.; Grcic, M. & Wallwork, H. (2007). Development of sequence tagged microsatellites for the barley net blotch pathogen *Pyrenophora teres*. *Molecular Ecology Notes*, Vol.7, No.4, pp. 664–666, ISSN 1471-8278

Keiper, F.J.; Hayden, M.J. & Wallwork, H. (2006) Development of sequence tagged microsatellites for the barley scald pathogen *Rhynchosporium secalis*. *Molecular Ecology Notes*, Vol.6, No.4, pp. 543–546, ISSN 1471-8278

Kim, J.T.; Park, S.Y.; Choi, W.; Lee, Y.H. & Kim H.T. (2008). Characterization of *Colletotrichum* isolates causing anthracnose of pepper in Korea. *Plant Pathology Journal*, Vol.24, No.1, pp. 17-23, ISSN 1598-2254

Kim, Y.T.; Cho, M.; Jeong, J.Y.; Lee, H.B. & Kim, S.B. (2010). Application of Terminal Restriction Fragment Length Polymorphism (T-RFLP) analysis to monitor effect of biocontrol agents on rhizosphere microbial community of hot pepper (*Capsicum annuum* L.). *Journal of Microbiology*, Vol.48, No.5, pp. 566-572, ISSN 1225-8873

Kircher, M. & Kelso, J. (2010). High-throughput DNA sequencing - concepts and limitations. *Bioessays*, Vol.32, No. 6, pp. 524-536

Knutsen, A.K.; Torp, M. & Holst-Jensen, A. (2004). Phylogenetic analyses of the *Fusarium poae*, *Fusarium sporotrichioides* and *Fusarium langsethiae* species complex based on partial sequences of the translation elongation factor-1 alpha gene. *International Journal of Food Microbiology*, Vol.95, No.3, pp. 287– 295, ISSN 0168-1605

Koenraadt, H.; Somerville, S.C.; Jones, A.L. (1992). Characterization of mutations in the beta-tubulin gene of benomyl-resistant field strains of *Venturia inaequalis* and other plant pathogenic fungi. *Phytopathology*, Vol.82, pp. 1348-1354, ISSN 0031-949X

Koljalg, U.; Larsson, K.-H.; Abarenkov, K.; Nilsson, R.H.; Alexander, I.J.; Eberhardt, U.; Erland, S.; Hoiland, K.; Kjoller, R.; Larsson, E.; Pennanen, T.; Sen, R.; Taylor, A.F.S., Tedersoo, L., Vralstad, T., & Ursing, B.M. (2005). UNITE: A database providing web-based methods for the molecular identification of ectomycorrhizal fungi. *New Phytologist*. Vol.166, No.3, pp. 1063-1068, ISSN 0028-646X

Kristensen, R.; Torp, M.; Kosiak, B. & Holst-Jensen, A. (2005). Phylogeny and toxigenic potential is correlated in *Fusarium* species as revealed by partial translation elongation factor 1 alpha gene sequences. *Mycological Research*, Vol.109, No.2, pp. 173–186, ISSN 0953-7562

Kroon, L.P.N.M.; Verstappen, E.C.P.; Kox, L.F.F.; Flier, W.G. & Bonants, P. (2004). A rapid diagnostic test to distinguish between American and European populations of *Phytophthora ramorum Phytopathology*, Vol.94, No.6, pp. 613-620, ISSN 0031-949X

Kulik, T.; Jestoi, M. & Okorski, A. (2011). Development of TaqMan assays for the quantitative detection of *Fusarium avenaceum/Fusarium tricinctum* and *Fusarium poae* esyn1 genotypes from cereal grain. *FEMS Microbiology Letters*, Vol.314, No.1, pp. 49-56, ISSN 0378-1097

Kurtz, B.; Karlovsky, P. & Vidal, S. (2010). Interaction between western corn rootworm (Coleoptera: Chrysomelidae) larvae and root-infecting *Fusarium verticillioides*. *Environmental Entomology*, Vol.39, No.5, pp. 1532-1538, ISSN 0046-225X

Landa, B.B.; Montes-Borrego, M.; Munoz-Ledesma, F.J. & Jimenez-Diaz, R.M. (2007). Phylogenetic analysis of downy mildew pathogens of opium poppy and PCR-Based in planta and seed detection of *Peronospora arborescens*. *Phytopathology*, Vol.97, No.11, pp. 1380-1390, ISSN 0031-949X

Landgraf, A., Reckmann, B. & Pingoud, A. (1991). Direct analysis of polymerase chain-reaction products using enzyme-linked-immunosorbent-assay techniques. *Analytical Biochemistry*, Vol.198, pp. 86-91, ISSN 0003-2697

Langrell, S.R.H. & Barbara, D.J. (2001). Magnetic capture hybridization for improved PCR detection of *Nectria galligena* from lignified apple extracts. *Plant Molecular Biology Reporter*, Vol.19, No.1, pp. 5–11, ISSN 0735-9640

Langrell, S.R.H.; Glen, M. & Alfenas, A.C. (2008). Molecular diagnosis of *Puccinia psidii* (guava rust) - a quarantine threat to Australian eucalypt and *Myrtaceae* biodiversity. *Plant Pathology*, Vol.57, No.4, pp. 687-701, ISSN 0032-0862

Larsen, J.E.; Hollingsworth, C.R.; Flor, J.; Dornbusch, M.R.; Simpson, N.L. & Samac, D.A. (2007). Distribution of *Phoma sclerotioides* on alfalfa and winter wheat crops in the north central United States. *Plant Disease*, Vol.91, No.5, pp. 551-558, ISSN 0191-2917

Lee, M.S.; LeMaistre, A.; Kantarjian, H.M.; Talpaz, M.; Freireich, E.J.; Trujillo, J.M. & Stass, S.A. (1989). Detection of two alternative bcr/abl mRNA junctions and minimal residual disease in Philadelphia chromosome positive chronic myelogenous leukemia by polymerase chain reaction. *Blood*, Vol.73, No.8, pp. 2165–2170, ISSN 0006-4971

Leisova, L.; Minarikova, V.; Kucera, L. & Ovesna, J. (2006). Quantification of *Pyrenophora teres* in infected barley leaves using real-time PCR. *Journal of Microbiological Methods*, Vol.67, No.3, pp. 446-455, ISSN 0167-7012

Lévesque, C.A.; Harlton, C.E. & de Cock, A.W.A.M. (1998). Identification of some oomycetes by reverse dot-blot hybridization. *Phytopatholgy*, Vol.88, No.3, pp. 213-222, ISSN 0031-949X

Licciardello, G.; Grasso, F. M.; Bella, P.; Cirvilleri, G.; Grimaldi, V. & Catara, V. (2006). Identification and detection of *Phoma tracheiphila*, causal agent of citrus mal secco disease, by real-time polymerase chain reaction. *Plant Disease*, Vol.90, No.12, pp. 1523-1530, ISSN 0191-2917

Lievens, B. & Thomma B.P.H.J., (2005). Recent developments in pathogen detection arrays: implications for fungal plant pathogens and use in practice. *Phytopathology*, Vol.95, No.12, pp. 1374-1380, ISSN 0031-949X

Lievens, B.; Brouwer, M.; Vanachter, A.C.R.C.; Levesque, C.A.; Cammue, B.P.A. & Thomma B.P.H.J. (2005b). Quantitative assessment of phytopathogenic fungi in various substrates using a DNA macroarray. *Environmental Microbiology*, Vol.7, No.11, pp. 1698-1710, ISSN 1462-2912

Lievens, B.; Claes, L.; Vakalounakis, D.J., Vanachter, A.C.R.C. & Thomma, B.P.H.J. (2007). A robust identification and detection assay to discriminate the cucumber pathogens *Fusarium oxysporum* f. sp *cucumerinum* and f. sp *radicis-cucumerinum*. *Environmental Microbiology*, Vol.9, No.9, pp. 2145-2161, ISSN 1462-2912

Lievens, B.; Claes, L.; Vanachter, A.C.R.C.; Cammue, B.P.A. & Thomma, B.P.H.J. (2006). Detecting single nucleotide polymorphisms using DNA arrays for plant pathogen diagnosis. *FEMS Microbiology Letters* Vol.255, No.1, pp.129-139, ISSN 0378-1097

Lievens, B.; Grauwet, T.J.M.A.; Cammue, B.P.A. & Thomma B.P.H.J. (2005a). Recent developments in diagnostics of plant pathogens: a review. *Recent Research Developments in Microbiology*, Vol.9, pp. 57-79

Lievens, B.; Rep, M. & Thomma, B.P.H.J. (2008). Recent developments in the molecular discrimination of *formae speciales* of *Fusarium oxysporum*. *Pest Management Science*, Vol.64, No.8, pp. 781–788, ISSN 1526-498X

Lievens, B.; Van Baarlen P.; Verreth C.; Van Kerckhove, S.; Rep, M. & Thomma, B.P.H.J. (2009). Evolutionary relationships between *Fusarium oxysporum* f. sp *lycopersici* and *F. oxysporum* f. sp *radicis-lycopersici* isolates inferred from mating type, elongation factor-1 alpha and exopolygalacturonase sequences. *Mycological Research*, Vol.113, pp. 1181-1191, Part: 10, ISSN 0953-7562

Liew, E.C.Y.; MacLean, D.J. & Irwin, J.A.G. (1998). Specific PCR based detection of *Phytophthora medicaginis* using the intergenic spacer region of the ribosomal DNA. *Mycological Research*, Vol.102, pp. 73-80, ISSN 0953-7562

Liu, J.H.; Gao L.; Liu T.G. & Chen W.Q. (2009). Development of a sequence-characterized amplified region marker for diagnosis of dwarf bunt of wheat and detection of *Tilletia controversa* Kühn. *Letters in Applied Microbiology*, Vol.49, No.2, pp. 235–240, ISSN 0266-8254

Long, A.A. (1998). In-situ polymerase chain reaction: foundation of the technology and today's options. *European Journal of Histochemistry*, Vol.42, pp. 101–109, ISSN 1121-760X

Luchi, N.; Capretti, P.; Pinzani, P.; Orlando, C. & Pazzagli, M. (2005). Real-time PCR detection of *Biscogniauxia mediterranea* in symptomless oak tissue. *Letters in Applied Microbiology*, Vol.41, No.1, pp. 61-68, ISSN 0266-8254

Luck, J.E. & Gillings, M.R. (1995). Rapid identification of benomyl resistant strains of *Botrytis cinerea* using the polymerase chain reaction. *Mycological Research*, Vol.99, pp. 1483–1488, ISSN 0953-7562

Luo, C.; Cox, K.D.; Amiri, A. & Schnabel, G. (2008). Occurrence and detection of the DMIresistance-associated genetic element 'Mona' in *Monilinia fructicola*. *Plant Disease*, Vol.92, pp. 1099–1103, ISSN 0191-2917

Ma, Z. & Michailides, T.J. (2004). An allele-specific PCR assay fordetecting strobilurin-resistant *Alternaria* isolates from pistachio in California. *Journal of Phytopathology*, Vol.152, pp. 118–121, ISSN 0931-1785

Ma, Z. & Michialides, T.J. (2005). Advances in undertanding molecular mechanisms of fungicide resistance and molecular detection of resistant genotypes in phytopathogenic fungi. *Crop Protection*, Vol.24, pp. 853-863, ISSN 0261-2194

Ma, Z.; Felts, D. & Michailides, T.J. (2003b). Resistance to azoxystrobin in *Alternaria* isolates from pistachio in California. *Pesticide Biochemical Physiology*, Vol.77, No.2, pp. 66–74, ISSN 0048-3575

Ma, Z.; Proffer, T.J.; Jacobs J.L. & Sundin G.W. (2006). Overexpression of the 14α-demethylase target gene (CYP51) mediates fungicide resistance in *Blumeriella jaapii*. *Applied Environmental Microbiology*, Vol.72, pp. 2581–2585, ISSN 0099-2240

Ma, Z.; Yoshimura, M. & Michailides, T.J. (2003a). Identification and characterization of benzimidazole resistance in *Monilinia fructicola* from stone fruit orchards in California. *Applied Environmental Microbiology*, Vol.69, pp. 7145–7152, ISSN 0099-2240

Macia-Vicente, J.G.; Jansson, H.B.; Talbot, N.J. & Lopez-Llorca, L.V. (2009). Real-time PCR quantification and live-cell imaging of endophytic colonization of barley (*Hordeum vulgare*) roots by *Fusarium equiseti* and *Pochonia chlamydosporia*. *New Phytologist*, Vol.182, No.1, pp. 213-228, ISSN 0028-646X

Malausa, T.; Andre, G.; Meglecz, E.; Blanquart, H.; Duthoy, S.; Costedoat, C.; Dubut, V.; Pech, N.; Castagnone-Sereno, P.; Delye, C.; Feau, N.; Frey, P.; Gauthier, P.; Guillemaud, T.; Hazard, L.; Le Corre, V.; Lung-Escarmant, B.; Male, P.J.G.; Ferreira, S.; Martin, J.F. (2011). High-throughput microsatellite isolation through 454 GS-FLX Titanium pyrosequencing of enriched DNA libraries. *Molecular Ecology Resources*, Vol.11, No.4, pp. 638-644, ISSN 1755-098X

Malvick, D.K. & Impullitti, A.E. (2007). Detection and quantification of *Phialophora gregata* in soybean and soil samples with a quantitative, real-time PCR assay. *Plant Disease*, Vol.91, No.6, pp. 736-742, ISSN 0191-2917

Markakis, E.A.; Tjamos, S.E.; Antoniou, P.P.; Paplomatas, E.J. & Tjamos, E.C. (2009). Symptom development, pathogen isolation and real-time QPCR quantification as factors for evaluating the resistance of olive cultivars to *Verticillium* pathotypes. *European Journal of Plant Pathology*, Vol.124, No.4, pp. 603–611, ISSN 0929-1873

Martin, F.N. & Tooley, P.W. (2003). Phylogenetic relationships among *Phytophthora* species inferred from sequence analysis of mitochondrially encoded cytochrome oxidase I and II genes. *Mycologia*, Vol.95, No.2, pp. 269-284, ISSN 0027-5514

Martínez-Espinoza, A.D.; León-Ramírez, C.G.; Singh, N. & Ruiz-Herrera, J. (2003). Use of PCR to detect infection of differentially susceptible maize cultivars using *Ustilago maydis* strains of variable virulence. *International Microbiology*, Vol.6, No.2, pp. 117-120, ISSN 1139-6709

Martínez-García, L.B.; Armas, C.; Miranda, J.D.; Padilla, F.M. & Pugnaire, F.I. (2011) .Shrubs influence arbuscular mycorrhizal fungi communities in a semi-arid environment. *Soil Biology & Biochemistry*, Vol.43, No.3, pp. 682-689, ISSN 0038-0717

Martos, S.; Torres, E.; El Bakali, M.A.; Raposo, R.; Gramaje, D.; Armengol, J. & Luque, J. (2011) Co-operational PCR coupled with Dot Blot Hybridization for the detection of *Phaeomoniella chlamydospora* on infected grapevine wood. *Journal of Phytopathology*, Vol.159, No.4, pp. 247-254, ISSN 0931-1785

Massart, S.; De Clercq, D.; Salmon, M.; Dickburt, C. & Jijakli, M.H. (2005). Development of real-time PCR using Minor Groove Binding probe to monitor the biological control agent *Candida oleophila* (strain O). *Journal of Microbiological Methods*, Vol.60, No.1, pp. 73-82, ISSN 0167-7012

Matsuda, Y.; Sameshima, T.; Moriura, N.; Inoue, K.; Nonomura, T.; Kakutani, K.; Nishimura, H.; Kusakari, S.; Tamamatsu, S. & Toyoda, H. (2005). Identification of individual powdery mildew fungi infecting leaves and direct detection of gene expression of single conidium by polymerase chain reaction. *Phytopathology*, Vol.95, No.10, pp. 1137–1143, ISSN 0031-949X

McCartney, H.A.; Foster, S.J.; Fraaije, B.A. & Ward, E. (2003). Molecular diagnostics for fungal plant pathogens. *Pest Management Science*, Vol.59, No.2, pp. 129–142, ISSN 1526-498X

McDonald, B.A. (1997). The population genetics of fungi: tools and techniques. *Phytopathology*, Vol.87, No. pp. 448–453, ISSN 0031-949X

McKay, G.J. & Cooke, L.R. (1997). A PCR-based method to characterize and identify benzimidazole resistance in *Helminthosporium solani*. *FEMS Microbiology Letters*, Vol.152, pp. 371–378, ISSN 0378-1097

McKay, G.J.; Egan, D.; Morris, E. & Brown, A.E. (1998). Identification of benzimidazole resistance in *Cladobotryum dendroides* using a PCR-based method. *Mycological Research*, Vol.102, pp. 671–676, ISSN 0953-7562

Meng, J. & Wang, Y. (2010). Rapid detection of *Phytophthora nicotianae* in infected tobacco tissue and soil samples based on its *Ypt1* gene. *Journal of Phytopathology*, Vol.158, No.1, pp. 1–7, ISSN 0931-1785

Mercado-Blanco, J.; Rodríguez-Jurado, D.; Pérez-Artés, E. & Jiménez-Díaz, R.M. (2001). Detection of the nondefoliating pathotype of *Verticillium dahliae* in infected olive plants by nested PCR. *Plant Pathology*, Vol.50, No. 5, pp. 609-619

Meudt, H.M. & Clarke, A.C. (2007). Almost forgotten or latest practice? AFLP applications, analyses and advances. *Trends in Plant Science*. Vol.12, No.3, pp. 106–117, ISSN 1360-1385

Midorikawa, G.E.O.; Pinheiro, M.R.R.; Vidigal, B.S.; Arruda, M.C.; Costa, F.F.; Pappas, G.J. Jr.; Ribeiro, S. G.; Freire, F. & Miller, R.N.G. (2008). Characterization of *Aspergillus flavus* strains from Brazilian Brazil nuts and cashew by RAPD and ribosomal DNA analysis. *Letters in Applied Microbiology*, Vol.47, No.1, pp. 12-18, ISSN 0266-8254

Minerdi, D.; Moretti M.; Li Y.; Gaggero, L.; Garibaldi, A. & Gullino, M.L. (2008). Conventional PCR and real time quantitative PCR detection of *Phytophthora cryptogea* on *Gerbera jamesonii*. *European Journal of Plant Pathology*, Vol.122, No.2, pp. 227-237, ISSN 0929-1873

Moradi, M.; Oerke, E.-C.; Steiner, U.; Tesfaye, D.; Schellander, K. & Dehne, H.-W. (2010). Microbiological and Sybr®Green Real-Time PCR detection of major *Fusarium* Head Blight pathogens on wheat ears. *Microbiology*, Vol.79, No.5, pp. 646–654, ISSN 0026-2617

Mostert, L.; Groenewald, J.Z.; Summerbell, R.C.; Gams, W. & Crous, P.W. (2006). Taxonomy and pathology of *Togninia* (Diaporthales) and its *Phaeoacremonium* anamorphs. *Studies in Mycology*, Vol.54, pp. 1–113, ISSN 0166-0616

Mulè, G.; Susca, A.; Stea, G. & Moretti, A. (2004). Specific detection of the toxigenic species *Fusarium proliferatum* and *F. oxysporum* from asparagus plants using primers based on calmodulin gene sequences. *FEMS Microbiology Letters*. Vol.230, No.2, pp. 235-240

Mullis, K.B. & Faloona, F.A. (1987). Specific synthesis of DNA in vitro via a polymerase-catalyzed chain-reaction. *Methods in Enzymology*, Vol.155, pp. 335-350

Mumford, R.; Boonham, N.; Tomlinson, J. & Barker, I. (2006). Advances in molecular phytodiagnostics-new solutions for old problems. *European Journal of Plant Pathology*, Vol.116, pp. 1-19, ISSN 0929-1873

Nakaune, R. (2001). ABC transporter genes involved in multidrug resistance in *Pencillium digitatum*. *Journal of General Plant Pathology*, Vol.67, No.3, pp. 251

Nakaune, R.; Adachi, K.; Nawata, O.; Tomiyama, M.; Akutsu, K. & Hibi, T. (1998). A novel ATP-binding cassette transporter involved in multidrug resistance in phytopathogenic fungus *Pencillium digitatum*. *Applied Environmental Microbiology*, Vol.64, pp. 3983-3988, ISSN 0099-2240

Nakaune, R.; Hamamoto, H.; Imada, J. & Akutsu, K. (2002). A novel ABC transporter gene, *PMR5*, is involved in multidrug resistance in the phytopathogenic fungus *Pencillium digitatum*. *Molecular Genetics and Genomics*, Vol.267, pp. 179-185, ISSN 1617-4615

Nanayakkara, U.N.; Singh, M.; Al-Mugharabi, K.I. & Peters, R.D. (2009). Detection of *Phytophthora erythroseptica* in above-ground potato tissues, progeny tubers, stolons and crop debris using PCR techniques. *American Journal of Potato Research*, Vol.86, No.3, pp. 239-245, ISSN 1099-209X

Nayaka, S.C.; Wulff, E.G.; Udayashankar, A.C.; Nandini, B.P.; Niranjana, S.R.; Mortensen, C.N. & Prakash, H.S. (2011). Prospects of molecular markers in *Fusarium* species diversity. *Applied Microbiology and Biotechnology*. Vol.90, No.5, pp. 1625-1639, ISSN 0175-7598

NCBI, National Center for Biotechnology Information: (http://www.ncbi.nlm.nih.gov/Genbank/)

Nechwatal, J. & Mendgen, K. (2006). *Pythium litorale* sp nov., a new species from the littoral of Lake Constance, Germany *Fems Microbiology Letters*, Vol.255, No.1, pp. 96-101, ISSN 0378-1097

Nguyen, H.D.T. & Seifert, K.A. (2008). Description and DNA barcoding of three new species of *Leohumicola* from South Africa and the United States. *Persoonia*, Vol.21, pp. 57–69, ISSN 0031-5850

Nicolaisen, M.; Justesen, A.F.; Thrane, U.; Skouboe, P. & Holmstrom, K. (2005). An oligonucleotide microarray for the identification and differentiation of trichothecene producing and nonproducing *Fusarium* species occurring on cereal grain. *Journal of Microbiological Methods*, Vol.62, No. 1, pp. 57–69, ISSN 0167-7012

Niu, C.; Kebede, H.; Auld, D.L.; Woodward, J.E.; Burow, G. & Wright R.J. (2008). A safe inexpensive method to isolate high quality plant and fungal DNA in an open laboratory environment. *African Journal of Biotechnology*, Vol.7, No.16, pp. 2818-2822, ISSN 1684-5315

Notomi, T.; Okayama, H.; Masubuchi, H.; Yonekawa, T.; Watanabe, K.; Amino, N. & Hase, T. (2000). Loop-mediated isothermal amplification of DNA. *Nucleic Acids Research*, Vol.28, No. 12, Article Number: e63, ISSN 0305-1048

Nunes, C.C.; Gowda, M.; Sailsbery, J.; Xue, M.F.; Chen, F.; Brown, D.E.; Oh, Y.; Mitchell, T.K. & Dean, R.A. (2011). Diverse and tissue-enriched small RNAs in the plant pathogenic fungus, *Magnaporthe oryzae*. *BMC Genomics*, Vol.12, Article Number: 288, ISSN 1471-2164

Nuovo, G.J. (1992). *PCR in situ hybridization: Protocols and applications*. Raven Press, ISBN 0881679402, New York

Nuovo, G.J.; MacConnell, P. & Gallery, F. (1994). Analysis of nonspecific DNA synthesis during in situ PCR and solution-phase PCR. *Genome Research*, Vol.4, pp. 89-96

Nuovo, G.J., MacConnell, P.; Forde, A. & Delvenne, P. (1991). Detection of human papillomavirus DNA in formalin fixed tissues by in situ hybridization after amplification by the polymerase chain reaction. *American Journal of Pathology*, Vol.139, No.4, pp. 847-854, ISSN 0002-9440

Okubara, P.A.; Schroeder, K.L. & Paulitz, T.C. (2008). Identification and quantification of *Rhizoctonia solani* and *R. oryzae* using real-time polymerase chain reaction. *Phytopathology*, Vol.98, No.7, pp. 837-847, ISSN 0031-949X

Olmos, A.; Bertolini, E. & Cambra, M. (2002). Simultaneous and co-operational amplification (Co-PCR): a new concept for detection of plant viruses. *Journal of Virological Methods*, Vol.106, No.1, pp. 51-59, ISSN 0166-0934

Omirou, M.; Rousidou, C.; Bekris, F.; Papadopoulou, K.K.; Menkissoglou-Spiroudi, U.; Ehaliotis, C. & Karpouzas, D.G. (2011). The impact of biofumigation and chemical fumigation methods on the structure and function of the soil microbial community. *Microbial Ecology*, Vol.61, No.1, pp. 201–213, ISSN 0095-3628

Ozakman, M. & Schaad, N.W. (2003). A real-time BIO-PCR assay for detection of *Ralstonia solanacearum* race 3, biovar 2, in asymptomatic potato tubers. *Canadian Journal of Plant Pathology*, Vol.25, No.3, pp. 232-239, ISSN 0706-0661

Pan, L.; Ash, G.J.; Ahn, B. & Watson, A.K. (2010). Development of strain specific molecular markers for the *Sclerotinia minor* bioherbicide strain IMI 344141. *Biocontrol Science and Technology*, Vol.20, No.9, pp. 939-959, ISSN 0958-3157 print/ISSN 1360-0478 online

Pannecoucque, J. & Hofte M. (2009). Detection of rDNA ITS polymorphism in *Rhizoctonia solani* AG 2-1 isolates. *Mycologia*, Vol.101, No. 1, pp. 26-33, ISSN 0027-5514

Pao, S.S.; Paulsen, I.T. & Saier MH Jr. (1998). Major facilitator superfamily. *Microbiology Molecular Biology Review*, Vol.62, pp. 1-34, ISSN 1092-2172

Park, B.; Park, J.; Cheon,g K-C.; Choi, J.; Jung, K.; Kim, D.; Lee, Y.H.; Ward, T.J.; O'Donnell, K.; Geiser, D.M. & Kang, S. (2011). Cyber infrastructure for Fusarium: three integrated platforms supporting strain identification, phylogenetics, comparative genomics and knowledge sharing. *Nucleic Acids Research* Vol.39, Supplement. 1, pp. D640-D646, ISSN 0305-1048

Park, J.; Park, B.; Veeraraghavan, N.; Blair, J.E.; Geiser, D.M.; Isard, S.; Mansfield, M.A.; Nikolaeva, E.; Park, S.-Y.; Russo, J.; Kim, S.H.; Greene, M.; Ivors, K.L.; Balci, Y.; Peiman, M.; Erwin, D.C.; Coffey, M.D.; Jung, K.; Lee, Y.-H.; Rossman, A.; Farr, D.; Cline, E.; Grünwald, N.J.; Luster, D.G.; Schrandt, J.; Martin, F.; Ribeiro, O.K.; Makalowska, I. & Kang, S. (2008). *Phytophthora* database: forensic database supporting the identification and monitoring of *Phytophthora*. *Plant Disease*, Vol.92, No.6, pp. 966-972, ISSN 0191-2917

Paul, B. (2006). *Pythium apiculatum* sp. nov. isolated from Burgundian vineyards: morphology, taxonomy, ITS region of its rRNA, and comparison with related species. *FEMS Microbiology Letters*, Vol.263, No.2, pp. 194–199, ISSN 0378-1097

Paul, B. (2009). *Pythium burgundicum* sp. nov. isolated from soil samples taken in French vineyards. *FEMS Microbiology Letters*, Vol.301, No. , pp. 109-114, ISSN 0378-1097

Paul, B. & Bala, K. (2008). A new species of *Pythium* with inflated sporangia and coiled antheridia, isolated from India. *FEMS Microbiology Letters*, Vol.282, No.2, pp. 251–257, ISSN 0378-1097

Paul, B., Bala, K., Gognies, S. & Belarbi, A. (2005). Morphological and molecular taxonomy of *Pythium longisporangium* sp. nov. isolated from the Burgundian region of France. *FEMS Microbiology Letters*, Vol.246, No.2, pp. 207–212, ISSN 0378-1097

Pereira, V.J.; Fernandes, D.; Carvalho, G.; Benoliel, M.J.; Romao, M.V.S. & Crespo, M.T.B. (2010). Assessment of the presence and dynamics of fungi in drinking water sources using cultural and molecular methods. *Water Research*, Vol.44, No.17, Special Issue: SI, pp. 4850-4859, ISSN 0043-1354

Phytophthora Database: (http://www.phytophthoradb.org)

Phytophthora-ID: (http://phytophthora-id.org)

Porterjordan, K.; Rosenberg, E.I.; Keiser, J.F.; Gross, J.D.; Ross, A.M.; Nasim, S. & Garrett, C.T. (1990). Nested polymerase chain reaction assay for the detection of cytomegalovirus overcomes false positives caused by contamination with fragmented DNA. *Journal of Medical Virology*, Vol.30, No.4, pp. 85-91, ISSN 0146-6615

Portillo, M.C.; Villahermosa, D.; Corzo, A. & González, J.M. (2011). Microbial community fingerprinting by differential display-denaturing gradient gel electrophoresis. *Applied and Environmental Microbiology*, Vol.77, No.1, pp. 351–354, ISSN 0099-2240

Primer3: (http://frodo.wi.mit.edu/primer3)

Prospero, S.; Black, J.A. & Winton, L.M. (2004). Isolation and characterization of microsatellite markers in *Phytophthora ramorum*, the causal agent of sudden oak death. *Molecular Ecology Notes*, Vol.4, No.4, pp. 672-674, ISSN 1471-8278

Qin, L.; Fu, Y.; Xie, J.; Cheng, J.; Jiang, D.; Li, G. & Huang, J. (2011). A nested-PCR method for rapid detection of *Sclerotinia sclerotiorum* on petals of oilseed rape (*Brassica napus*). *Plant Pathology*, Vol.60, No.2, pp. 271-277, ISSN 0032-0862

Saldanha, R.L.; García, J.E.; Dekker, R.F.H.; Vilas-Boas, L.A. & Barbosa, A.M. (2007). Genetic diversity among *Botryosphaeria* isolates and their correlation with cell wall-lytic enzyme production. *Brazilian Journal of Microbiology*, Vol.38, No.2, pp. 259-264, ISSN 1517-8382259

Samuelian, S.K.; Greer, L.A.; Savocchia, S. & Steel, C.C. (2011). Detection and monitoring of *Greeneria uvicola* and *Colletotrichum acutatum* development on grapevines by real-time PCR. *Plant Disease*, Vol.95, No.3, pp. 298-303, ISSN 0191-2917

Sánchez-Torres, P. & Tuset, J.J. (2011). Molecular insights into fungicide resistance in sensitive and resistant *Penicillium digitatum* strains infecting citrus. *Postharvest Biology and Technology*, Vol.59, No.2, pp. 159–165, ISSN 0925-5214

Schena, L. & Ippolito, A. (2003). Rapid and sensitive detection of *Rosellinia necatrix* in roots and soils by real time Scorpion-PCR. *Journal of Plant Pathology*, Vol.85, No.1, pp. 15-25, ISSN 1125-4653

Schena, L.; Hughes K.J.D. & Cooke D.E.L. (2006). Detection and quantification of *P. ramorum*, *P. kernoviae*, *P. citricola* and *P. quercina* in symptomatic leaves by multiplex realtime PCR. *Molecular Plant Pathology*, Vol.7, No.5, pp. 365-379, ISSN 1464-6722

Schena, L., Nigro, F., Ippolito, A. & Gallitelli, D. (2004). Real-time quantitative PCR: a new technology to detect and study phytopathogenic and antagonistic fungi. *European Journal of Plant Pathology*, Vol.110, No.9, pp. 893-908, ISSN 0929-1873

Schmidt, H.; Taniwaki, M.H.; Vogel, R.F. & Niessen, L. (2004) Utilization of AFLP markers for PCR-based identification of *Aspergillus carbonarius* and indication of its presence in green coffee samples. Journal of Applied Microbiology Vol.97, No.5, pp.899-909, ISSN 1364-5072

Schnabel, G. & Jones, A.L. (2001). The 14α-demethylase (*CYP51A1*) gene is overexpressed in *Venturia inaequalis* strains resistant to myclobutanil. *Phytopathology*, Vol.91, pp. 102–110, ISSN 0031-949X

Schnabel, G.; Dait, Q. & Paradkar, M.R. (2001). Cloning and expression analysis of the ATP-binding cassette transporter gene *MFABC1* and the alternative oxidase gene *MfAOX1* from *Monilinia fructicola*. *Pest Managment Scence*, Vol.59, pp. 1143-1151, ISSN 1526-498X

Schoonbeek, H-J.; van Nistelrooy, J.G.M. & de Waard, M.A. (2003). Functional analysis of ABC transporter genes from *Botrytis cinerea* identifies *BcatrB* as a transporter of eugenol. *European Journal of Plant Pathology*, Vol.109, pp. 1003-1011, ISSN 0929-1873

Schroeder, K.L.; Okubara, P.A.; Tambong, J.T.; Levesque, C.A. & Paulitz, T.C. (2006). Identification and quantification of pathogenic *Pythium* spp. from soils in eastern Washington using real-time polymerase chain reaction. *Phytopathology*, Vol.96, No.6, pp. 637-647, ISSN 0031-949X

Seifert, K.A.; Samson, R.A.; deWaard, J.R.; Houbraken, J.; Lévesque C.A.; Moncalvo, J-M.; Louis-Seize, G. & Hebert P.D.N. (2007). Prospects for fungus identification using CO1 DNA barcodes, with Penicillium as a test case. *Proceedings of the National Academy of Sciences of the United States of America*, Vol.104, No.10, pp. 3901-3906, ISSN 0027-8424

Sheridan, G.E.C.; Masters, C.I.; Shallcross, J.A. & Mackey, B.M. (1998). Detection of mRNA by reverse transcription-PCR as an indicator of viability in *Escherichia coli* cells. *Applied Environmental Microbiology*, Vol.64, No.4, pp. 1313–1318, ISSN 0099-2240

Shimomoto, Y.; Sato, T.; Hojo, H.; Morita, Y.; Takeuchi, S.; Mizumoto, H.; Kiba, A. & Hikichi, Y. (2011). Pathogenic and genetic variation among isolates of *Corynespora cassiicola* in Japan. *Plant Pathology*, Vol.60, No.2, pp. 253-260, ISSN 0032-0862

Sierotzki, H.; Wullschleger, J. & Gisi, U. (2000). Point mutation in cytochrome b gene conferring resistance to strobilurin fungicides in *Erysiphe graminis* f. sp. *tritici* field isolates. *Pesticide Biochemical Physiology*, Vol.68, pp. 107–112, ISSN 0048-3575

Silvar, C.; Díaz, J. & Merino, F. (2005). Real-time polymerase chain reaction quantification of *Phytophthora capsici* in different pepper genotypes. *Phytopathology*, Vol.95, No.12, pp. 1423–1429, ISSN 0031-949X

Singh, M. & Singh, R.P. (1997). Potato virus Y detection: Sensitivity of RT-PCR depends on the size of fragment amplified. *Canadian Journal of Plant Pathology-Revue Canadienne De Phytopathologie*, Vol.19, pp. 149-155, ISSN 0706-0661

Smith, L.M. & Burgoyne, L.A. (2004). Collecting, archiving and processing DNA from wildlife samples using FTA® databasing paper. *BMC Ecology*, 4:4

Somai, B.M.; Keinath, A.P. & Dean, R.A. (2002). Development of PCR-ELISA detection and differentiation of *Didymella bryoniae* from related *Phoma* species. *Plant Disease*, Vol.86, No.7, pp. 710–716, ISSN 0191-2917

Stergiopoulos, I. & de Waard, M.A. (2002). Activity of azole fungicides and ABC transporters modulators on *Mycosphaerella graminicola*. *Journal of Phytopathology*, Vol.150, pp. 313-320, ISSN 0931-1785

Stergiopoulos, I.; Gielkens, M.M.C.; Goodall, S.D.; Venema, K. & de Waard, M.A. (2002b). Molecular cloning and charaterization of three new ATP-binding cassette transporter genes from the wheat pathogen *Mycosphaerella graminicola*. *Gene*, Vol.289, pp. 141-149, ISSN 0378-1119

Stergiopoulos, I.; van Nistelrooy, J.G.; Kema, G.H. & de Waard, M.A. (2003b). Multiple mechanisms account for variation in base-line sensitivity to azole fungicides in field isolates of *Mycosphaerella graminicola*. *Pest Managment Science*, Vol.59, pp. 1333-1343, ISSN 1526-498X

Stergiopoulos, I.; Zwiers, L-H. & de Waard, M.A. (2002a). Secretion of natural and synthetic toxic compounds from filamentous fungi by membrane transporters of the ATP-binding cassette and major facilitator superfamily. *European Journal of Plant Pathology*, Vol.108, pp. 719-734, ISSN 0929-1873

Stergiopoulos, I.; Zwiers, L-H. & de Waard, M.A. (2003a). The ABC transporter MgAtr4 is a virulence factor of *Mycosphaerella graminicola* that effects colonisation of substomatal cavities in wheat leaves. *Molecular Plant-Microbe Interaction*, Vol.16, pp. 689-698, ISSN 0894-0282

Stewart, J.E.; Kim, M.-S.; James, R.L.; Dumroese, R.K. & Klopfenstein, N.B. (2006). Molecular characterization of *Fusarium oxysporum* and *Fusarium commune* isolates from a conifer nursery. *Phytophathology*, Vol.96, pp. 1124-1133, ISSN 0031-949X

Stringari, D.; Glienke, C.; de Christo, D.; Maccheroni, W. & de Azevedo, J.L. (2009). High molecular diversity of the fungus *Guignardia citricarpa* and *Guignardia mangiferae* and new primers for the diagnosis of the Citrus Black Spot. *Brazilian archives of biology and technology*, Vol.52, No.5, pp. 1063-1073, ISSN 1516-8913

Sun, X.; Wang, J.; Feng, D.; Ma, Z. & Li, H. (2011). PdCYP51B, a new putative sterol 14α-demethylase gene of *Penicillium digitatum* involved in resistance to imazalil and other fungicides inhibiting ergosterol synthesis. *Applied Microbioly and Biotechnology*, Vol.91, No.4, pp. 1107-1119, ISSN 0175-7598

Sundelin, T.; Christensen, C.B.; Larsen, J.; Møller, K.; Lübeck, M.; Bødker, L. & Jensen, B. (2010). In planta quantification of *Plasmodiophora brassicae* using signature fatty acids and real-time PCR. *Plant Disease*, Vol.94, No.4, pp. 432-438, ISSN 0191-2917

Suzuki, S.; Taketani, H.; Kusumoto, K.I. & Kashiwagi, Y. (2006). High-throughput genotyping of filamentous fungus *Aspergillus oryzae* based on colony direct polymerase chain reaction. *Journal of Bioscience and Bioengineering*, Vol.102, No.6, pp. 572–574, ISSN 1389-1723

Takamatsu, S.; Nakano, M.; Yokota, H. & Kunoh, H. (1998). Detection of *Rhizoctonia solani* AG-2-2-IV, the causal agent of large patch of Zoysiagrass, using plasmid DNA as probe. *Annals of the Phytopathological Society of Japan*, Vol.64, No.5, pp. 451–457

Tawfik, D.S. & Griffiths, A.D. (1998). Man-made cell-like compartments for molecular evolution, *Nature Biotechnology*, Vol.16, No.7, pp. 652–656, ISSN 1087-0156

Tebbe, C.C. & Vahjen, W. (1993). Interference of humic acids and dna extracted directly from soil in detection and transformation of recombinant-DNA from bacteria and a yeast. *Applied and Environmental Microbiology* Vol.59, No.8, pp. 2657-2665, ISSN 0099-2240

Tewoldemedhin, Y.T.; Mazzola, M.; Botha W.J.; Spies, C.F.J. & McLeod, A. (2011). Characterization of fungi (*Fusarium* and *Rhizoctonia*) and oomycetes (*Phytophthora* and *Pythium*) associated with apple orchards in South Africa. *European Journal of Plant Pathology*, Vol.130, No 2, pp. 215-229, ISSN 0929-1873

Theodoulou, F.L. (2000). Plant ABC transporters. *Biochimica Biophysica Acta*, Vol.1465, pp. 79-103, ISSN 0005-2736

Thies, J.E. (2007). Soil microbial community analysis using terminal restriction fragment length polymorphisms. *Soil Science Society of America Journal*, Vol.71, No.2, pp. 579-591, ISSN 0361-5995

Tomlinson J. A.; Barker I. & Boonham N. (2007). Faster, simpler, more-specific methods for improved molecular detection of *Phytophthora ramorum* in the field. *Applied and Environmental Microbiology* Vol.73, No.12, pp. 4040-4047, ISSN 0099-2240

Tomlinson, J.A.; Boonham, N.; Hughes, K.J.D.; Griffen, R.L. & Barker, I. (2005). On-site DNA extraction and real-time PCR for detection of *Phytophthora ramorum* in the field. *Applied and Environmental Microbiology* Vol.71, No.11, pp. 6702-6710, ISSN 0099-2240

Tomlinson J.A.; Dickinson M.J. & Boonham N. (2010). Rapid detection of *Phytophthora ramorum* and *P. kernoviae* by two-minute DNA extraction followed by isothermal amplification and amplicon detection by generic lateral flow device. *Phytopathology* Vol.100, No.2, pp, 143-149, ISSN 0031-949X

Tooley, P.W.; Martin, F.N.; Carras, M.M. & Frederick, R.D. (2006). Real-time fluorescent polymerase chain reaction detection of *Phytophthora ramorum* and *Phytophthora pseudosyringae* using mitochondrial gene regions. *Phytopathology*, Vol.96, No.4, pp. 336–345, ISSN 0031-949X

Torres-Calzada, C.; Tapia-Tussell, R.; Quijano-Ramayo, A.; Martin-Mex, R.; Rojas-Herrera, R.; Higuera-Ciapara, I.; Perez-Brito, D. (2011). A species-specific polymerase chain reaction assay for rapid and sensitive detection of *Colletotrichum capsici*. *Molecular Biotechnology*, Vol.49, No.1, pp. 48-55, ISSN 1073-6085

TrichOKey, Molecular Barcode: (http://www.isth.info/tools/molkey/index.php)

Tyagi, S. & Kramer, F.R. (1996). Molecular beacons: Probes that fluoresce upon hybridization. *Nature Biotechnology*, Vol.14, No.3, pp. 303-308, ISSN 1087-0156

UNITE, A molecular database for identification of fungi: (http://unite.ut.ee/index.php)

Validov, S.Z.; Kamilova F.D. & Lugtenberg B.J.J. (2011). Monitoring of pathogenic and non-pathogenic *Fusarium oxysporum* strains during tomato plant infection. *Microbial Biotechnology*, Vol.4, No.1, pp. 82-88, ISSN 1751-7907

Valsesia, G.; Gobbin, D.; Patocchi, A.; Vecchione, A.; Pertot, I. & Gessler, C. (2005). Development of a high-throughput method for quantification of *Plasmopara viticola* DNA in grapevine leaves by means of quantitative real-time polymerase chain reaction. *Phytopathology*, Vol.95, No.6, pp. 672-678, ISSN 0031-949X

van Doorn, R.; Slawiak, M.; Szemes, M.; Dullemans, A.M.; Bonants, P.; Kowalchuk, G.A. & Schoen, C.D. (2009). Robust detection and identification of multiple oomycetes and fungi in environmental samples by using a novel cleavable padlock probe-based ligation detection assay. *Applied and Environmental Microbiology*, Vol.75, No.12, pp. 4185-4193, ISSN 0099-2240

van Gent-Pelzer, M.P.E.; van Brouwershaven, I.R.; Kox, L.F.F. & Bonants, P.J.M. (2007). A TaqMan PCR method for routine diagnosis of the quarantine fungus *Guignardia citricarpa* on citrus fruit. *Journal of Phytopathology*, Vol.155, No.6, pp. 357-363, ISSN 0931-1785

VanderLee, T.; DeWitte, I.; Drenth, A.; Alfonso, C. & Govers, F. (1997). AFLP linkage map of the oomycete *Phytophthora infestans*. *Fungal Genetics and Biology*, Vol.21, No.3, pp. 278–291, ISSN 1087-1845

Vermeulen, T.; Schoonbeek, H. & de Waard, M.A. (2001). The ABC transporter BcatrB from *Botrytis cinerea* is a determinant of the activity of the phenylpyrrole fungicide fludiioxonil. *Pest Managment Science*, Vol.57, pp. 393-402, ISSN 1526-498X

Vos, P.; Hogers, R.; Bleeker, M.; Reijans, M.; Vandelee, T.; Hornes, M.; Frijters, A.; Pot, J.; Peleman, J.; Kuiper, M. & Zabeau, M. (1995). AFLP: a new technique for DNA fingerprinting. *Nucleic Acids Research*, Vol.23, No.21, pp. 4407–14, ISSN 0305-1048

Wakelin S.A.; Warren, R.A.; Kong, L. & Harvey, P.R. (2008). Management factors affecting size and structure of soil *Fusarium* communities under irrigated maize in Australia. *Applied Soil Ecology*, Vol.39, pp. 201-209, ISSN 0929-1393

Welsh, J. & McClelland, M. (1990). Fingerprinting genomes using PCR with arbitrary primers. *Nucleic Acids Research*, Vol.18, No.24, pp. 7213–7218, ISSN 0305-1048

Whitcombe, D.; Kelly, S.; Mann, J.; Theaker, J.; Jones, C. & Little, S. (1999). Scorpions (TM) primers - a novel method for use in single-tube genotyping. *American Journal of Human Genetics*, Vol.65, No.4, pp. A412-A412, Meeting Abstract: 2333, ISSN 0002-9297

White, T. (1997). Increased mRNA levels of *ERG16*, *CDR*, and *MDR1* correlate with increases in azole resistance in *Candida albicans* isolates from a patient infected with human immunodeficiency virus. *Antimicrobe Agents Chemotherapy*, Vol.41, pp. 1482-1487

White, T.J.; Bruns, T.; Lee, S. & Taylor, J.W. (1990). Amplification and direct sequencing of fungal ribosomal RNA genes for phylogenetics. In: *PCR Protocols: A Guide to Methods and Applications*, Innis, M.A.; Gelfand, D.H.; Sninsky, J.J. & White, T.J. Academic Press, Inc., New York. pp. 315-322

Williams, J.G.K.; Kubelik, A.R.; Livak, K.J.; Rafalski, J.A. & Tingey, S.V. (1990). DNA polymorphisms amplified by arbitrary primers are useful as genetic markers. *Nucleic Acids Research*, Vol.18, No.22, pp. 6531–6535, ISSN 0305-1048

Wilson, I.G. (1997) Inhibition and facilitation of nucleic acid amplification. Applied and Environmental Microbiology Vol.63, No.10, pp. 3741–3751, ISSN 0099-2240

Winton, L.M. & Hansen, E.M. (2001). Molecular diagnosis of *Phytophthora lateralis* in trees, water, and foliage baits using multiplex polymerase chain reaction. *Forest Pathology*, Vol.31, No.5, pp. 275-283, ISSN 1437-4781

Winton, L.M.; Krohn, A.L. & Leiner, R.H. (2007). Microsatellite markers for *Sclerotinia subarctica* nom. prov., a new vegetable pathogen of the High North. *Molecular Ecology Notes*, Vol.7, No.6, pp. 1077-1079, ISSN 1471-8278

Wittwer, C.T.; Herrmann, M.G.; Moss, A.A. & Rasmussen, R.P. (1997). Continuous fluorescence monitoring of rapid cycle DNA amplification. *Biotechniques*, Vol.22, No.1, pp. 130–138, ISSN 0736-6205

Wood, P.M. & Hollomon, D.W. (2003). A critical evaluation of the role of alternative oxidase in the performance of strobilurin and related fungicides acting at the Qo site of complex III. *Pest Managment Science*, Vol.59, pp. 499-511, ISSN 1526-498X

Wu, C.P.; Chen, G.Y.; Li, B.; Su, H.; An, Y.L.; Zhen, S.Z. & Ye, J.R. (2011). Rapid and accurate detection of *Ceratocystis fagacearum* from stained wood and soil by nested and real-time PCR. *Forest Pathology*, Vol.41, No.1, pp. 15-21, ISSN 1437-4781

Wyand, R.A. & Brown, J.K.M. (2005). Sequence variation in the *CYP51* gene of *Blumeria graminis* associated with resistance to sterol demethylase inhibiting fungicides. *Fungal Genetics Biology*, Vol.42, pp. 726–765, ISSN 1087-1845

Xu, M.L.; Melchinger, A.E. & Lübberstedt, T. (1999). Species-specific detection of the maize pathogens *Sporisorium reiliana* and *Ustilago maydis* by dot-blot hybridization and PCR-based assays. *Plant Disease*, Vol.83, No.4, pp. 390–395, ISSN 0191-2917

Xue, B; Goodwin, P.H. & Annis, S.L. (1992). Pathotype identification of *Leptosphaeria maculans* with PCR and oligonucleotide primers from ribosomal internal transcribed spacer sequences. *Physiological and Molecular Plant Pathology*, Vol.41, No.3, pp. 179-188, ISSN 0885-5765

Yan, L.; Zhang, C.; Ding, L. & Ma, Z. (2008). Development of a real-time PCR assay for the detection of *Cladosporium fulvum* in tomato leaves. *Journal of Applied Microbiology,* Vol.104, No.5, pp. 1417-1424, ISSN 1364-5072

Yang F.; Jensen J. D.; Svensson B.; Jorgensen, H. J. L.; Collinge, D.B. & Finnie, C. (2010). Analysis of early events in the interaction between *Fusarium graminearum* and the susceptible barley (*Hordeum vulgare*) cultivar Scarlett. *Proteomics,* Vol.10, No.21, pp. 3748-3755, ISSN 1615-9853

Yin, Y.; Ding, L.; Liu, X.; Yang, J. & Ma, Z. (2009). Detection of *Sclerotinia sclerotiorum* in planta by a real-time PCR assay. *Journal of Phytopathology,* Vol.157, No.7-8, pp. 465–469, ISSN 0931-1785

Yoder, O.C. & Turgeon, B. (2001). Fungal genomics and pathogenicity. *Current Opinion in Plant Biology,* Vol.4, pp. 315-321, ISSN 1369-5266

Zaccaro, R. P.; Carareto-Alves, L.M.; Travensolo, R.F.; Wickert, E. & Lemos, E.G.M. (2007). Use of molecular marker SCAR in the identification of *Fusarium subglutinans,* causal agent of mango malformation. *Revista brasileira de fruticultura,* Vol.29, No.3, pp. 563-570, ISSN 0100-2945

Zelaya-Molina L.X.; Ortega M.A. & Dorrance A.E. (2011). Easy and efficient protocol for oomycete DNA extraction suitable for population genetic analysis. *Biotechnology Letters,* Vol.33, No.4, pp. 715-720, ISSN 0141-5492

Zhang, N.; Tantardini, A.; Miller, S.; Eng, A. & Salvatore, N. (2011). TaqMan real-time PCR method for detection of *Discula destructiva* that causes dogwood anthracnose in Europe and North America. *European Journal of Plant Pathology,* Vol.130, No.4, pp. 551-558, ISSN 0929-1873

Zhang, Y.J.; Zhang, S.; Liu, X.Z.; Wen, H.A. & Wang, M. (2010). A simple method of genomic DNA extraction suitable for analysis of bulk fungal strains. *Letters in Applied Microbiology,* Vol.51, No.1, pp. 114-118, ISSN 0266-8254

Zhang, Z.; Zhu, Z.; Ma, Z., & Li, H. (2009). A molecular mechanism of azoxystrobin resistancein *Penicillium digitatum* UV mutants and a PCR-based assay for detection of azoxystrobin-resistant strains in packing- or store-house isolates. *Interantional Journal of Food Microbiology,* Vol.131, pp. 157–161, ISSN 0168-1605

Zhao, J.; Wang, X. J.; Chen, C. Q.; Huang, L. L. & Kang, Z.S. (2007). A PCR-Based assay for detection of *Puccinia striiformis* f. sp *tritici* in wheat. *Plant Disease,* Vol.91, No.12, pp. 1969-1674, ISSN 0191-2917

Zheng, D.; Olaya, G. & Koller, W. (2000). Characterization of laboratory mutants of *Venturia inaequalis* resistant to the strobilurin-related fungicide kresoxim-methyl. *Current Genetics,* Vol.38, pp. 148-155, ISSN 0172-8083

Zheng, P.H.; Meng, X.Y.; Yu, Z.X.; Wu, Y; Bao, Y.L.; Yu, C.L. & Li, Y.X. (2009). Genetic and phytochemical analysis of *Armillaria mellea* by RAPD, ISSR, and HPLC. *Analytical Letters,* Vol.42, No.10, pp. 1479-1494, ISSN 0003-2719

Zheng, Y.; Zhang, G.; Lin, F.C.; Wang, Z.H.; Jin, G.L.; Yang, L.; Wang, Y.; Chen, X.; Xu, Z.H.; Zhao, X.Q.; Wang, H.K.; Lu, J.P.; Lu, G.D. & Wu, W.R. (2008). Development of microsatellite markers and construction of genetic map in rice blast pathogen *Magnaporthe grisea. Fungal Genetics and Biology,* Vol.45, No.10, pp. 1340-1347, ISSN 1087-1845

Zwiers, L-H.; Stergiopoulos, I.; Gielkens, M.M.C.; Goodall, S.D. & de Waard, M.A. (2003). ABC transporters of the wheat pathogen *Mycosphaerella graminicola* function as protectants against biotic and xenobiotic toxic compounds. *Molecular Genetics and Genomics,* Vol.269, pp. 499-507, ISSN 1617-4615

7

The Role of the Extracellular Matrix (ECM) in Phytopathogenic Fungi: A Potential Target for Disease Control

Kenichi Ikeda, Kanako Inoue, Hiroko Kitagawa,
Hiroko Meguro, Saki Shimoi and Pyoyun Park
Kobe University
Japan

1. Introduction

Crop yield loss as a result of disease has an economic impact on many people. To protect against disease, plant pathologists have developed various fungicides and disease resistant cultivars. In parallel, pathogens have evolved to escape from disease protection measures through, for example, the emergence of fungicide resistant isolates and the breakdown of disease resistant cultivars. To resolve these issues, we should develop various disease protection strategies for different types of target sites. Therefore, we must improve our understanding of the infection mechanism of pathogens. Pathogens possess several classes of genes that are essential for causing disease, i.e., pathogenicity genes or virulence genes, the products of which are called pathogenicity factors. Phytopathogenic fungi have developed various pathogenic factors, e.g., adhesion molecules to the host cells, sensor machineries against host plants, invasion machineries into the host cells, adaptation ability on the host cells, and so on. These factors comprise potential targets for disease control. Fungal adhesion to host cells is an initial important step to establish infection, which is considered to be a universal mechanism across plant pathogenic fungi. In this chapter, we provide a review of the components required for fungal adhesion to the host cell, and propose fungal adhesion as a potential target for disease control.

2. Importance of fungal adhesion to the host cell

Pathogenic fungi have various strategies for propagation. Some fungi produce asexual spores, sexual spores, sclerotia, budding progenies, or just extending mycelia. These propagules comprise minimal compartmental units to transmit genetic information, and are highly replicable. These small propagules are able to travel great distances and colonize novel niches. Dispersing propagules may land on various types of environment. Each pathogenic fungus is capable of adapting to establish individuals on a specific environment. The mechanism of pathogenic fungal adhesion is considered to occur in a number of ways. For example, the hydrophobic interaction is a universal system of adhesion, i.e., most fungal surfaces are covered with hydrophobic components, such as hydrophobin. In addition, either protein-protein or protein-carbohydrate interaction also occurs.

The plant surface is coated with a cuticle (wax) to prevent desiccation, particularly when the environment is highly hydrophobic. When the spores of pathogens land on the plant surface, the spores elongate into germ tubes and have elaborate infection machineries, such as appressorium to infect plant cells. However, the spores must first settle on the plant surface to accomplish infection. If the spores are not able to adhere to the plant surface, they are easily dislodged by the wind or rain fall. Moreover, the spore germlings encounter the counterforce of infection pressure from the plant surface. Therefore, firm adhesion enables spore germlings to anchor tightly onto the host surface, and is important for the differentiation of specialized infection structures. Adhesion is required for infection in most plant pathogens, and this mechanism is also developing in phytopathogenic fungi. We showed that adhesion ability was variable depending on the living strategies of different fungal species (Fig. 1). For example, the saprotrophic fungus *Neurospora crassa* has weak adhesion ability on plastic surface. In contrast, phytopathogenic fungi, such as *Alternaria alternata* Japanese pear pathotype and *Magnaporthe oryzae*, adhered strongly to plastic surfaces. Since plastic surfaces are highly hydrophobic, they mimic the plant surface. Hence, phytopathogenic fungi might have evolved to settle on the surfaces of plants. However, the precise molecular mechanism of fungal adhesion on plant surfaces remains undetermined.

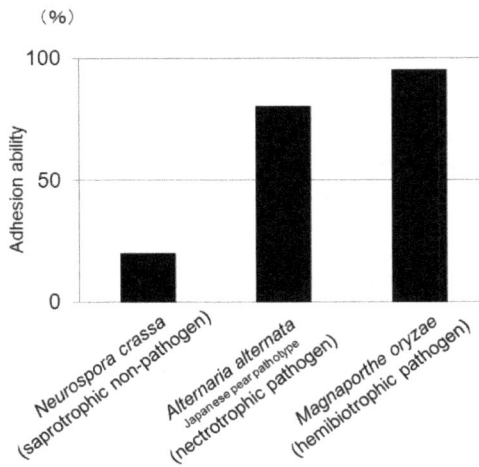

Fig. 1. Variation in the ability of fungal adhesion to plastic surfaces. The saprotrophic fungus *Neurospora crassa* has a weak adhesion ability. The phytopathogenic fungi, *Alternaria alternata* Japanese pear pathotype and *Magnaporthe oryzae*, have high adhesion ability. Adhesion ability was determined by total number of germlings was first counted under the microscope and then washed by dipping in distilled water 100 times vertically to remove the detached germlings.

In animal pathogenic fungi, adhesion to host cells is also important to manifest pathogenicity. However, because the animal cell has no cell wall and no wax layer, adhesion mechanisms to the host cell are supposed to be different to that of phytopathogenic fungi. Moreover, the environment within the animal body is different to that of the atmosphere, i.e., low oxygen concentrations, high carbon dioxide concentrations, as well as being filled with serum. Animal pathogenic fungi have evolved various adhesive proteins called adhesins (Sundstrom, 1999). In the next section, we describe various components of adhesion in animal and plant pathogenic fungi.

3. Components of adhesion strategies

3.1 Animal pathogenic fungal adhesion

Candida albicans causes severe oropharyngeal and esophageal mucositis in patients that have human immunodeficiency virus (HIV), and is the most intensively studied animal pathogenic fungi. *C. albicans* normally propagates budding yeast cells (blastospores). When the blastospores attach to the animal cells, the blastospores transform into a filamentous form to invade underlying tissues. The cell wall components between the blastospore and filamentous hypha are different. Therefore, the adhesion of *C. albicans* may be divided into two different steps, depending on morphological switching.

Blastospore adhesion is the first step of adhesion to the host epithelial and endothelial cells, especially to the extracellular matrix, and is important for the morphological switching of yeast cells from blastospore to filamentous forms (Gale *et al.*, 1998; Klotz *et al.*, 1993). This adhesion step is dependent on the presence of calcium ions (Klotz *et al.*, 1993). Ultrastructure analysis has revealed that the cell wall of blastospores is surrounded by a fibrillar reticulated layer (fimbriae) that contains mannose sugar, suggesting the importance of glycoprotein in this process (Bobichon *et al.*, 1994). The molecular approach elucidated the presence of several adhesive proteins, called adhesins. Since the blastospore adheres to the extracellular matrix of the animal cell, the involvement of the integrin-like component was suspected to be a component of adhesion. The homologous gene of vertebrate leukocyte integrins was cloned as *INT1* and characterized (Gale *et al.*, 1996; Gale *et al.*, 1998). When the *INT1* gene was introduced into *Saccharomyces cerevisiae*, the transformants producing INT1 exhibited enhanced aggregation (Gale *et al.*, 1996). Furthermore, the disruption of *INT1* suppressed blastospore adhesion to epithelial cells, hyphal growth, and virulence in mice (Gale *et al.*, 1998). In addition to INT1, agglutinin family proteins, ALA1 and ALS1, and the fibronectin binding protein, FLS5, were also involved in blastospore adhesion (Fu *et al.*, 1998; Rauceo *et al.*, 2006).

Once the blastospores settle on the host cells, morphological switching occurs to form filamentous growth (germ tubes, pseudohyphae, and hyphae). Proline and glutamine rich protein-encoding genes have been abundantly expressed in hyphae but not yeast forms, and were designated as *HWP1* (Staab *et al.*, 1996). The disruption of the *HWP1* gene reduced the stable attachment of blastospores to human buccal epithelial cells, and reduced their capacity to cause systemic candidiasis in mice (Staab *et al.*, 1999). The HWP1 protein also served as a substrate for mammalian transglutaminases, suggesting that the blastospore adherence mechanism may be involved in the cross-linking of HWP1 to unidentified proteins through transglutaminase activity (Staab *et al.*, 1999). Moreover, adhesin family proteins, ALS3 and HYR1, were also found to be involved in filamentous growth adhesion (Sundstrom, 1999).

Other adhesion components of animal pathogenic fungi have also been also studied. In *Aspergillus fumigatus*, a hydrophobin RODA was involved in adhesion (Thau *et al.*, 1994). In *Rhizopups oryzae*, the spores adhered to laminin and type IV collagen, but not to fibronectin, with adhesion decreasing during the elongation of germ tubes (Bouchara *et al.*, 1996).

3.2 Plant pathogenic fungal adhesion

Plant pathogenic fungi have several adhesion strategies. This review mainly focuses on the spore dispersal of plant pathogenic fungi. When the spores land on the plant surface, they germinate and develop appressoria to enter into the plant cell (Fig. 2). Appressoria adhere tightly to the plant surface. This process is divided into two steps, i.e., spore adhesion and germ tube adhesion. During germ tube elongation, the fungal surface is covered with *de novo* synthesized compounds collectively called the extracellular matrix (ECM), which might be involved in adhesion (Beckett *et al.*, 1990). The term ECM is confusing because the outer surface components of animal cells are also called the extracellular matrix (ECM). In this chapter, ECM corresponds to the secreted products from plant pathogenic fungi only. The adhesion strength and the adhesion components in each step are different for each plant pathogenic fungus. Here, we provide an overview of the typical adhesion mechanisms in plant pathogenic fungi.

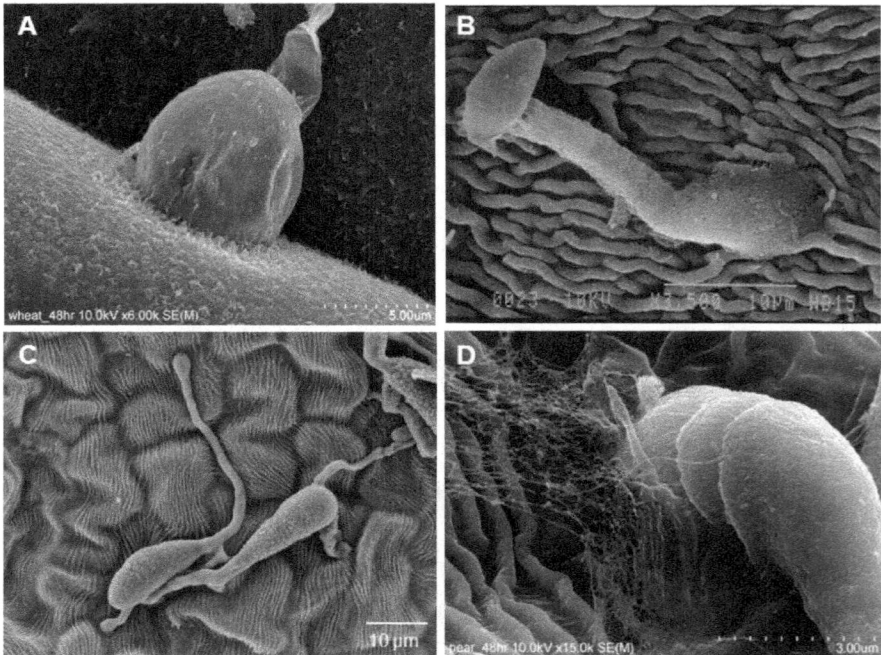

Fig. 2. Scanning electron microscope image of appressoria of phytopathogenic fungi on the host plant surface. (A) Appressorium of *Magnaporthe oryzae* on a wheat leaf. (B) Appressorium of *Venturia nashicola* on a Japanese pear leaf. Infection structure (C) and appressorium (D) of *Alternaria alternata* Japanese pear pathotype on a Japanese pear leaf.

3.2.1 *Magnaporthe oryzae*

The blast fungus *M. oryzae* is one of the most destructive fungal diseases in gramineous crop plants, especially rice, barley, and wheat. *M. oryzae* produces millions of conidia, asexual spores, and disperses to distant areas. Successful infection by this fungus requires the following steps: (1) the spore to land on the host surface and elongate the germ tube from the apical spore cell and (2) the tip of the germ tube to differentiate into the appressorium, which is the infection machinery. Subsequently, the fungi penetrate into the host cuticle and generate a penetration peg from the bottom of the appressorium (Howard et al., 1991). Although the conidia itself has low adhesion ability, the hydrated conidia attains strong adhesion ability within 30 minutes of incubation (Hamer et al., 1988). Ultrastructural analysis has revealed that the spore tip mucilage (STM) is preserved at the apical spore and released after hydration (Braun and Howard, 1994). The attached spores on the surface start to germinate and secrete ECM. The secreted ECM supports the enhanced adhesion ability of the fungi on the plant surface. It is not known whether the STM and the ECM are the same component. During the infection process, some essential environmental signals have been identified as appressorium (infection)-inducing factors, such as the hardness (Xiao et al., 1994) and the hydrophobicity of the attachment surface (Jelitto et al., 1994; Lee & Dean, 1994), and the chemical components from the plant surface (Gilbert et al., 1996). In addition, a number of up-regulating genes have also been characterized (Talbot, 2003). However, the available information was not sufficient to identify the principal components and regulation mechanisms of fungal adhesion in *M. oryzae*.

While the components of STM remain unclear, it was found that STM was reactive with lectin concanabalin A (ConA; α-D-glucose and α-(1,3)-/α-(1,6)-D-mannose binding), suggesting that STM consists of mannose containing glycoprotein (Hamer et al., 1988). The ECM was also ConA-positive (Xiao et al., 1994). The effect of three different specific sugar-binding lectins, namely ConA, PSA (α-D-glucose and α-(1,6)-D-mannose binding), and WGA (chitin binding), on appressorium formation and the adhesion ability of the *M. oryzae* germlings was evaluated (Fig. 3). High concentrations (50 µg ml⁻¹) of lectins, ConA, and WGA were found to inhibit appressorium formation and adhesion. These results indicate that these lectin treatments covered the cell wall surface, and inhibited the development of the cell wall architecture (data not shown). In contrast, low concentrations (10 µg ml⁻¹) of these lectins only affected adhesion, suggesting that low concentrations of lectins are preferentially bound to the adhesion components (Fig. 3). Moreover, although PSA was observed to bind to the cell wall surface, it did not affect the adhesion. ConA and PSA commonly bind to the α-D-glucose and the α-(1,6)-D-mannose moieties; however, only ConA binds to α-(1,3)-D-mannose, suggesting that α-(1,3)-D-mannose is a potential adhesion component (Fig. 3). Our unpublished study revealed that the incubation of *M. oryzae* spores in modified nutrient conditions produced less adherent spore germlings. Furthermore, the less adherent spore germlings were ConA-negative (K. Inoue, unpublished data), suggesting that mannose sugar may be important for adhesion.

Protein secretion via the Golgi pathway might be responsible for ECM production. In the treatment with cycloheximide at the early step of germination, the germ tube and the appressorium were not formed, and the germlings were easily removed (Fig. 4). In the treatments with monensin and tunicamycin, the adhesion rate was significantly lower than for the water control (Fig. 4). Hence, N-glycosylation of the mannose sugar moiety might be important for the maturation of the adhesive protein(s). This result also agreed with the conclusion of the lectin treatment experiments.

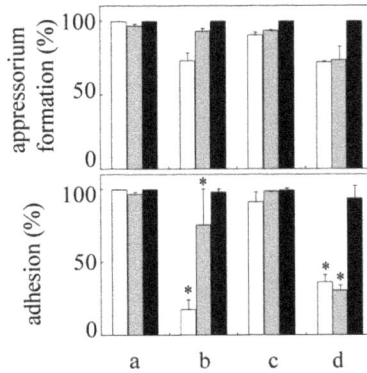

Fig. 3. Effects of lectins on appressorium formation and adhesion of *M. oryzae* germlings on the hydrophobic surface. The rate of appressorium formation (upper) and adhesion (lower) in *M. oryzae* germlings treated with the water control (a), 10 µg ml⁻¹ ConA (b), 10 µg ml⁻¹ PSA (c), and 10 µg ml⁻¹ WGA (d) at 0 hour post-inoculation (hpi; white bars), 1 hpi (grey bars), and 6 hpi (black bars). Bars indicate standard deviation. * indicates a significant difference from the control ($p < 0.05$).

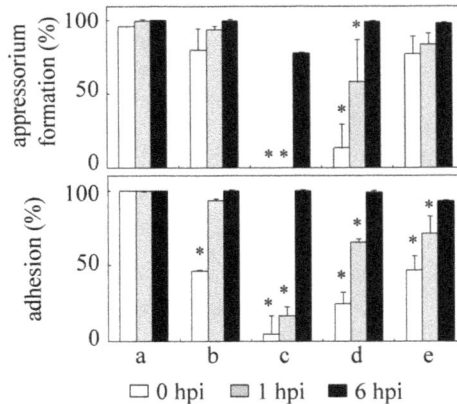

Fig. 4. The effects of inhibitors on appressorium formation and the adhesion of *M. oryzae* germlings on the hydrophobic surface. The rate of appressorium formation (upper) and adhesion (lower) of *M. oryzae* on the hydrophobic surface by inhibitor treatments with the water control (a), 50 µg ml⁻¹ colchicine (b), 0.1 µg ml⁻¹ cycloheximide (c), 1 µg ml⁻¹ monensin (d), and 10 µg ml⁻¹ tunicamycin (e) at 0 hpi (white bars), 1 hpi (grey bars), and 6 hpi (black bars). Bars indicate standard deviation. *, indicates a significant difference from the control ($p < 0.05$).

One of the candidate ECM components in *M. oryzae*, *Emp1*, was isolated from the cDNA library, and was homologous to the extracellular matrix protein *Fem1* in *Fusarium oxysporum* (Ahn *et al.*, 2004). The *emp1* disruption mutants showed a slight reduction in adhesion ability, but were largely involved in appressorium formation, suggesting that EMP1 was involved in surface recognition (Ahn *et al.*, 2004).

Ultrastrucutural analysis was used to elucidate the ECM components, and revealed that the ECM was an electron dense and fibrous structure (Fig. 5; K. Inoue *et al.*, 2007). This fibrous structure was supposed to be similar to the components of the cell adhesion factors (extracellular matrix in animal cells), such as collagen, fibronectin, and laminin. We then performed immunohistochemical analysis, using specific antibodies against mammal cell adhesion factors. Positive signals were detected from the treatments with antibodies against collagen, vitronectin, fibronectin, laminin, and integrin (K. Inoue *et al.*, 2007). Similar results have been reported by other research groups (Bae *et al.*, 2007; Dean *et al.*, 1994), and for other plant pathogenic fungi (Celerin *et al.*, 1996; Corrêa *et al.*, 1996; Gale *et al.*, 1998; Hyon *et al.*, 2009; Jian *et al.*, 2007; Kaminskyj and Heath, 1995; Manning *et al.*, 2004; Sarma *et al.*, 2005). Cell to cell adhesion mediated by integrin was involved in Arg-Gly-Asp (RGD) motifs (Giancotti & Ruoslahti, 1999). The germlings treated with RGD peptide reduced adhesion ability and appressorium formation (Bae *et al.*, 2007). The suppression effect by the RGD treatment was reversed by supplementing with cAMP, suggesting that RGD recognition may be involved in the signaling cascade for appressorium formation (Bae *et al.*, 2007). Based on this evidence, the functions and the components of fungal ECM resemble animal ECM. However, a similar nucleotide sequence to that of the animal ECM component was not detected from the fungal genome sequences. This suggests that only the restricted tertiary protein structure had a resemblance, with the amino acids sequences being highly diverged.

Melanin is also one of the cell surface components. The fungicide treatments, chlobenthiazone and tricyclazole, that inhibit melanin synthesis, have been shown to reduce germling adhesion (S. Inoue *et al.*, 1987).

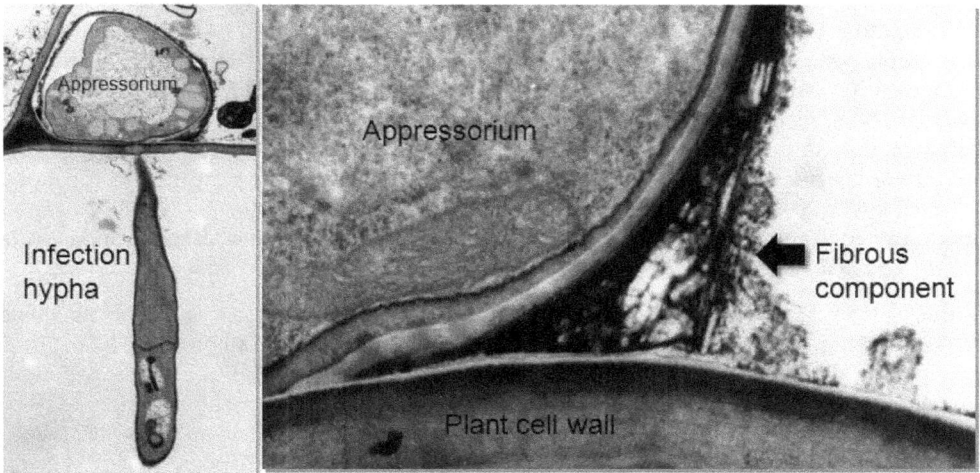

Fig. 5. Transmission electron microscopy images of *M. oryzae* appressoria on the leaf surface of wheat. Note: fibrous components were observed at the interface between the plant and fungal appressorium. Adapted from Inoue *et al.* (2007), with permission.

Hydrophobicity is one of the important factors for appressorium formation in *M. oryzae* (Jelitto *et al.*, 1994; Lee & Dean, 1994). To determine whether substrate hydrophobicity is essential for fungal adhesion and appressorium formation, fungal differentiation and adhesion were compared on hydrophilic and hydrophobic surfaces. On the hydrophobic surface, most of germlings produced appressoria, with ca. 100% adherence over a six hour period (Fig. 6). On the hydrophilic surface, appressoria were not formed; however, over a six hour period ca. 100% of germlings adhered to the surface (Fig. 6). This result suggests that adhesion ability is dispensable for appressorium formation. The spore germlings may sense hydrophobicity, hardness, or chemical signals to form appressoria.

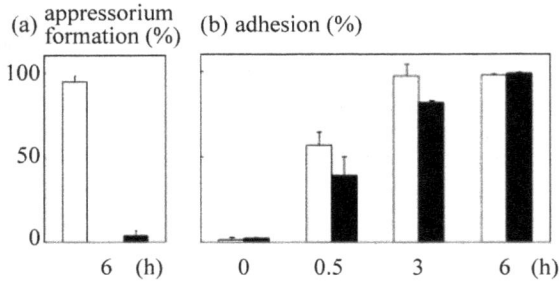

(a) appressorium formation (%)   (b) adhesion (%)

6 (h)   0   0.5   3   6   (h)

Fig. 6. The effect of surface characteristics on appressorium formation and the adhesion of M. oryzae germlings. (a) The rates of appressorium formation of M. oryzae germlings on hydrophobic (white bar) and hydrophilic (black bar) surfaces at 6 hpi. (b) The rates of adhesion to hydrophobic (white bars) and hydrophilic (black bars) surfaces at 0, 0.5, 3, and 6 hpi. Bars indicate standard deviation.

Fungal hydrophobins are cell-surface hydrophobic proteins with four disulphide bonds that are ubiquitous in filamentous fungi (Kershaw et al., 2005; Wösten, 2001). Hydrophobins exist in multiple copies, and these proteins are divided into two groups based on their physical properties (Wessels, 1994). Hydrophobins seem to form part of the ECM. Class I hydrophobins form highly insoluble polymers, whereas class II hydrophobins form polymers that are soluble in some organic solvents (Sunde et al., 2008). These hydrophobins are differentially expressed during the growth stage, and seem to have different functions (Nielsen et al., 2001; Segers et al., 1999; Wessels et al., 1991; Whiteford et al., 2004). In M. oryzae, two hydrophobins are well characterized: class I Mpg1 (Talbot et al., 1993) and class II Mhp1 (Kim et al., 2005).

In the mpg1 null mutant, appressorium formation was reduced, suggesting the involvement of host-sensing (Beckerman and Ebbole, 1996; Talbot et al., 1993; Talbot et al., 1996). In a previous study, adhesion in the mpg1 null mutant was not significantly affected in a conventional wash experiment, but was affected in a boiling-SDS wash experiment (Talbot et al., 1996). In contrast, an RNA silencing experiment of Mpg1 reduced adhesion (K. Ikeda, unpublished data). The Mpg1 silencing mutants reduced ConA-positives (K. Ikeda, unpublished data), suggesting that Mpg1 is a ConA-positive glycoprotein, or that Mpg1 recruits ConA-positive glycoproteins to the cell surface. The complex formations involved in Mpg1 may strengthen adhesion and contribute to pathogenicity. The hydrophobin RolA in Aspergillus oryzae was associated with cutinase and allowed the substrate to degrade on the hydrophobic surface (Takahashi et al., 2005). One of the cutinase genes Cut2 in M. oryzae was involved in appressorium formation, but not involved in adhesion (Skamnioti & Gurr, 2007). Other cutinase genes may also be involved in adhesion.

The class II hydrophobin *Mhp1* disruption mutants affected pleiotropically, hydrophobicity, pathogenicity, and spore viability (Kim *et al.*, 2005). In our unpublished data (K. Ikeda), *mhp1* null mutants showed no alteration, except for reducing hydrophobicity. The class II hydrophobin cerato-ulmin in *Ophiostoma novo-ulmi* not only functioned in adhesion, but was also toxic to the host plant, as well as providing tolerance against desiccation (Temple *et al.*, 1997). Paradoxically, the *Cladosporium fulvum* hydrophobin HCf-6 suppressed adhesion ability (Lacroix *et al.*, 2008). Further analysis of the relationship between hydrophobin and adhesion is required.

3.2.2 *Colletotrichum* species

The genus *Colletotrichum* is one of the most important genera of plant pathogens. The genus *Colletotrichum* encompasses numerous species, and the key criterion for their identification is mainly based on plant host determination. The ungerminated conidium of *Colletotrichum* species possesses adhesion ability. Adhesion is more effective on hydrophobic than hydrophilic surfaces, and is required for *de novo* protein synthesis in *C. graminicola* and *C. musae* (Mercure *et al.*, 1994a; Sela-Buurlage *et al.*, 1991). However, the adhesion strength of *C. graminicola* conidia was not strong, with up to 30% of conidia adhering only, and was influenced by corn leaf age (Mercure *et al.*, 1994a). Attached conidia were ConA positive and were effectively detached by pronase E treatment (Mercure *et al.*, 1994b; Sela-Buurlge *et al.*, 1991). Ultrastructural analysis revealed that the ungerminated conidia of *C. lindemuthianum* and *C. truncatum* were coated with fibrillar ECM, suggesting that the conidia of *Colletotrichum* species may be covered with adhesive ECM that is dissimilar to *M. oryzae* (Hamer *et al.*, 1988; O'Connell *et al.*, 1996; Van Dyke and Mims, 1991). To study the ECM molecular components in *C. lindemuthianum*, monoclonal antibodies (MAbs) were raised against conidia (Pain *et al.*, 1992). A MAb UB20 specifically recognized the conidium surface and UB20 treatment inhibited conidium adhesion (Hughes *et al.*, 1999). However, UB20 treatment did not affect appressorium formation and pathogenicity. The conidium adhesion deficiency was compensated after germination, suggesting that germ tube adhesion was involved in appressorium formation and pathogenicity (Rawlings *et al.*, 2007).

MAbs were also raised against conidium germlings of *C. lindemuthianum* (Pain *et al.*, 1994a,b). MAbs UB26 and UB31 were observed to bind to the ECM surrounding germ tubes and appressoria, but not to the conidia, suggesting that ECM compounds were different for conidia and germ tubes (Hutchison *et al.*, 2002). Moreover, the distribution of UB26 and UB31 positive signals was similar, but spatial distribution was different, i.e., UB31 antigens were located close to the cell wall, while UB26 antigens extended further from the cell wall (Hutchison *et al.*, 2002). These results suggest that multiple components exist in the ECM, and are involved in germling adhesion.

3.2.3 *Botrytis cinerea*

B. cinerea is an important fungal pathogen of a number of food and ornamental crops. Conidium adhesion to host surfaces is an important early event in the infection process. Adhesion ability is correlated with the water contact angle of the substrate, suggesting that hydrophobicity is important for conidium adhesion (Doss *et al.*, 1993). However, the percent adhesion of germinated conidia is larger than that of ungerminated conidia (Doss *et al.*,

1993). The adhesion ability of *B. cinerea* seems to be strengthened with the differentiation of germlings. Germling attachment was resistant to removal by boiling or by treatment with hydrolytic enzymes, periodic acid, or sulfuric acid, but was readily removed by a strong base, sodium hydrate (Doss *et al.*, 1995). The ECM components of germ tubes detected the presence of enzymatic activities such as polygalacturonase, laccase, and cutinase (Doss, 1999). Moreover, the ECM of germlings contained melanin (Doss *et al.*, 2003). These components were similar to *M. oryzae* and are assumed to be involved in adhesion. The class I hydrophobin *BcHpb1* was also partially involved in adhesion, although the disruption mutant of *BcHpb1* retained pathogenicity (Izumitsu *et al.*, 2010).

3.2.4 *Ustilago violacea* (*Microbotryum violacea*)

The anther smut fungus *Ustilago* (*Microbotryum*) *violacea* propagates yeast-like sporidial cells. The cell walls of smut contained a hair-like appendage (fungal fimbriae) that was similar to the Gram-negative bacteria cell wall appendage, fimbriae (Poon & Day, 1974). The fungal fimbriae might be essential for the later stages of conjugation (Poon & Day, 1975). The component of fungal fimbriae contained protein with motifs similar to collagen (Celerin *et al.*, 1996). Collagen is characteristic of the glycine-rich repeat motif, and functions in the cell adhesion of animal connective tissues. Collagen family proteins seem to have a conserved conformation between animal and fungi, although amino acid sequences are divergent. The relationship between the fungal fimbriae of smut and the fibrillar ECM of spore coat in other fungal species remains unclear.

3.2.5 Other fungi

Conidium adhesion was analyzed, and differences were found in various plant pathogenic fungi. In *Bipolaris sorokiniana* and *Venturia inaequalis*, conidium adhesion to the hydrophobic surface is weak, due to the hydrophobic interaction (Apoga *et al.*, 2001; Schumacher *et al.*, 2008). In *Nectria haematococca*, 90 kDa glycoprotein containing mannose sugar was associated with the development of the adhesiveness of macroconidia (Kwon & Epstein, 1997). In *V. inaequalis*, conidia released adhesive material that was termed spore tip glue (STG) (Schumacher *et al.*, 2008). The STG was ConA-negative, suggesting that it was different to STM found in *M. oryzae* (Hamer *et al.*, 1988; Schumacher *et al.*, 2008). These adhesion molecules of ungerminated conidia were pre-synthesized and preserved.

After the germination of conidia in most plant pathogenic fungi, the adhesion ability is increased by the production of ECM from germ tubes. In *Uromyces vicia-fabae* and *Blumeria graminis*, cutinase and esterase in the extracellular matrix (ECM) appeared to play a function in adhesion, either on the hydrophobic or hydrophilic surface, even during pre-germination (Deising *et al.*, 1992; Tucker & Talbot, 2001). Treatment with serine-esterase inhibitor diisopropyl fluorophosphate prevented adhesion (Deising *et al.*, 1992). This evidence suggests that adhesion may be partially involved in the combination of hydrophobin with cutinase as a molecular mechanism in *A. oryzae* (Takahashi *et al.*, 2005). However, the components of other ECM remain undetermined.

In the insect pathogenic fungus *Metarhizium anisopliae*, two proteins, MAD1 and MAD2, were involved in adhesion (Wang & Legar, 2007). MAD1 was involved in adhesion to the insect surface, blastospore production, and virulence to caterpillars. In comparison, MAD2 was involved in adhesion to the plant surface, but showed no effect on fungal differentiation and entomopathogenicity (Wang & Legar, 2007).

4. Detachment of germlings from the host surface

Fungal adhesion to the host cells is expected to be one of the pathogenicity factors (K. Inoue *et al.*, 2007). The regulation of fungal adhesion is expected to lead to disease control. The adhesion mechanisms of pathogenic fungi are common, with specific features existing in each fungus. We attempted to detach pathogenic fungi from the host or artificial substrate by using enzymatic activity.

4.1 Effects of hydrolytic enzymes on appressorium formation and adhesion of *M. oryzae* germlings

Various types of hydrolytic enzymes were tested for detachment activity. In enzyme treatments of 1 hpi, germlings were detached without affecting appressorium formation, when using α-mannosidase, β-mannosidase, pronase E, collagenase N-2, collagenase S-1, and gelatinase B (K. Inoue *et al.*, 2011). In the enzyme treatments of 6 hpi, most germlings produced appressoria, with the inhibition of ECM production being difficult. In this situation, pronase E and all types of collagenase and gelatinase caused a significant detachment of germlings (K. Inoue *et al.*, 2011). Pronase E is known to be a mucoprotein-degrading enzyme, and had the ability of moderate removal effect in *B. sorokiniana* (Apoga *et al.*, 2001). In particular, collagenase type S-1 and gelatinase B seemed to be the most effective ECM target-specific enzymes, and had a minimal effect on appressorium formation, even at the early stages of application. Mannose moiety was also a target for ECM degradation. However, there are some discrepancies with a previous study; Xiao *et al.* (1994) reported that α-glucosidase and α-mannosidase are effective in removing germlings; however, we found that β-mannosidase is more effective (K. Inoue *et al.*, 2011). These time-lapse experiments clearly show that the timing of adding enzymes influenced the results, which might explain the discrepancies between the two studies. The mannose-degrading enzymes were effective at the early stage, but became ineffective with time. Most of the other substrate-specific enzymes, such as glycan-, protein-, and lipid-degrading enzymes, were difficult to degrade (K. Inoue *et al.*, 2011). These results suggest that the adhesive compounds of ECM consist of glycoproteins with mannose sugars, which gradually accumulate with time. In the pathogenicity test on the host plant, lesion formation was remarkably suppressed on the treatment with crude collagenase, collagenase S-1, and gelatinase B (K. Inoue *et al.*, 2011). A similar detachment effect by crude collagenase was also reported in the *Alternaria alternata* Japanese pear pathotype (Hyon *et al.*, 2009), suggesting universality in filamentous fungi.

4.2 SEM analysis of enzyme effects on the adhesion of *M. oryzae* germlings to the host leaf surface

To examine the effect of the enzymes on germling detachment to the leaf surface, each enzyme was applied to *M. oryzae* germlings at the appropriate time, and incubated for up to 25 hpi. The specimens were then observed by using SEM. It was difficult to ascertain whether the absence of spores was the result of the enzyme treatments or the lack of spores at the onset of the experiment. *M. grisea* reportedly produces cutinases (Skamnioti & Gurr, 2007; Sweigard *et al.*, 1992). Therefore, this pathogen can degrade the wax of plant surfaces. The detached infection structure would be recognizable as vestiges of degraded wax on the wheat surface. In the treatment with cellulase or protease, the infection structures tightly adhered onto the surface (K. Inoue *et al.*, 2011). In contrast, the treatment with crude collagenase or gelatinase B (matrix metalloproteinases; MMPs) resulted in the detachment of germlings, and vestiges of the presence of germlings were observed (K. Inoue *et al.*, 2011). The detachment effect caused by the treatment with crude collagenase was validated at 18 hpi. In contrast, the detachment effect was cancelled out at 24 hpi, at which time fungal penetration into the host was established.

The SEM observation clearly shows the effects of the enzymes on the degradation of the interface between host plant and germlings. We demonstrated that the reduction in pathogenicity is attributable to the detachment of germlings based on treatment with effective enzymes. In this study, MMPs were confirmed to be useful for protecting wheat from *M. oryzae*.

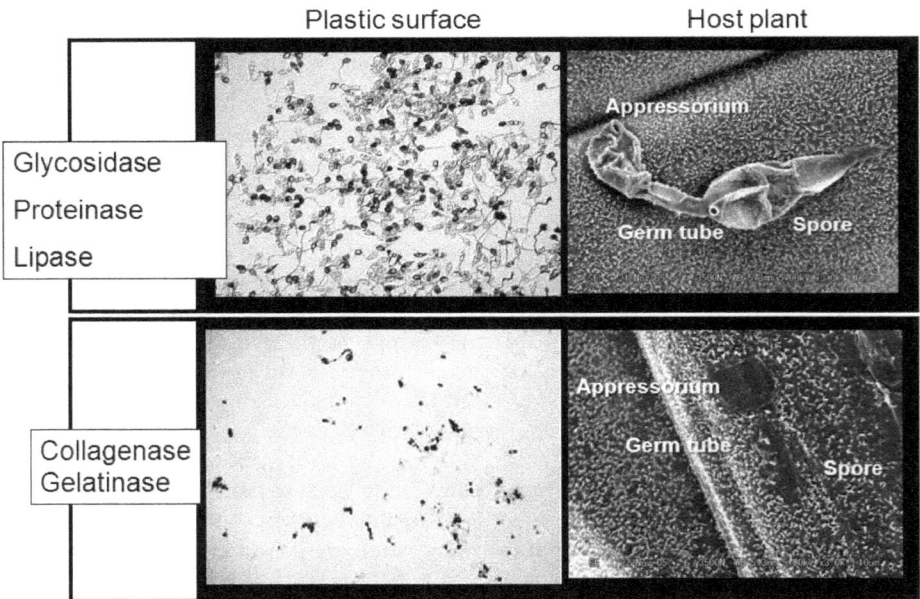

Fig. 7. Detachment effects of *M. oryzae* germlings following treatment with hydrolytic enzymes. Left photos: light microscopy; right photos: scanning electron microscopy. The treatments with collagenase were detached, and only the vestiges of germlings were observed (lower photos). Other hydrolytic enzymes, such as glycosidase were, not affected by germling detachment (upper photos).

4.3 Screening of gelatinolytic bacteria for biological control

To screen collagenolytic/gelatinolytic bacteria, we examined two different screening methods, i.e., (1) utilization of a leaf-associating bacteria library and (2) direct screening from the field, and incubation of samples with collagen (Fig. 8). Collagen is fibrous and difficult to catabolize for nutrient acquisition; therefore, we expected that direct screening with collagen incubation would allow the enrichment of high gelatinolytic genera (Shimoi *et al.*, 2010).

The leaf-associated library was screened for gelatinolytic activity on a 96-microtiter plate (first screening) and Petri dish (second screening). In the case of direct screening samples from the field, five bacterial isolates, which showed especially high gelatinolytic activity, were further characterized as *Acidovorax* sp. (Ac1), *Chryseobacterium* sp. (Ch1 and Ch2), *Sphingomonas* sp. (Sp1), and *Pseudomonas* sp. (Ps1) (Shimoi *et al.*, 2010). The effects of the proteinase inhibitors, EDTA, antipain, aprotinin, and PMSF on bacterial gelatinolytic activity were evaluated. The gelatinolytic activities of Ac1, Ch1, Ch2, and Sp1 were inhibited by EDTA in a dose-dependent manner (Shimoi *et al.*, 2010). In contrast, gelatinolytic activity in Ps1 was not inhibited by EDTA (Shimoi *et al.*, 2010). Bacterial gelatinolytic activities were not inhibited by antipain, aprotinin, or PMSF. These enzymatic characters suggested that all isolates, except Ps1, produced metalloproteinases. Therefore, we examined the effects of each metal cation on gelatinolytic activity in the presence of 5 mM EDTA. The gelatinolytic activities of Sp1 and Ac1 were restored by supplementation with 10 mM Ca^{2+} (Shimoi *et al.*, 2010). In contrast, the gelatinolytic activity of Ch1 and Ch2 were restored by supplementation with either 10 mM Ca^{2+} or 1 mM Zn^{2+}, and to a lesser extent with 2 mM Mn^{2+} (Shimoi *et al.*, 2010).

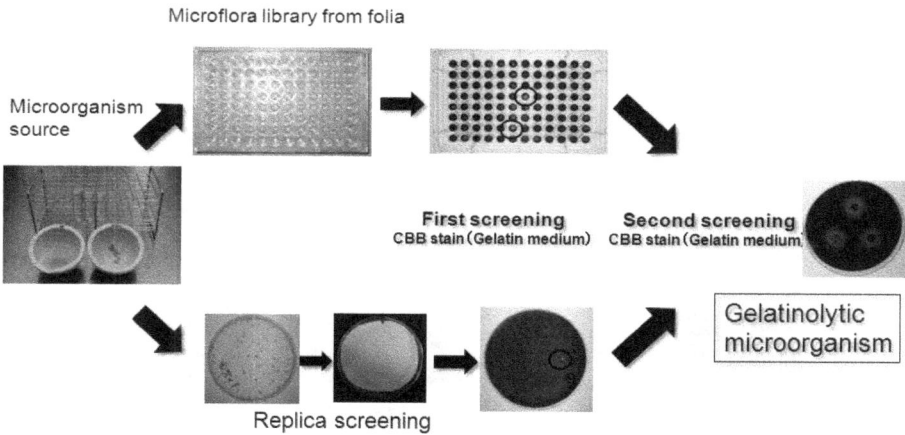

Fig. 8. Screening method to isolate gelatinolytic bacteria from nature. Adapted from Shimoi *et al.* (2010), with permission.

The effects of bacterial gelatinolytic activity on the germling adhesion of *M. oryzae* were evaluated. Treatment with each of the five bacterial isolates significantly decreased the percentage of germling adhesion (Fig. 9). In particular, the effects of Ac1 and Ch1 on germling detachment were comparable to the effects of commercial collagenase placed simultaneously on spore suspensions (Fig. 9). The detachment of germlings decreased when bacterial cultures were added 6 h after the inoculation of the spore suspension. Germling adhesion was restored when EDTA was added to the mixtures of bacteria and spore suspension, except in the case of Ps1.

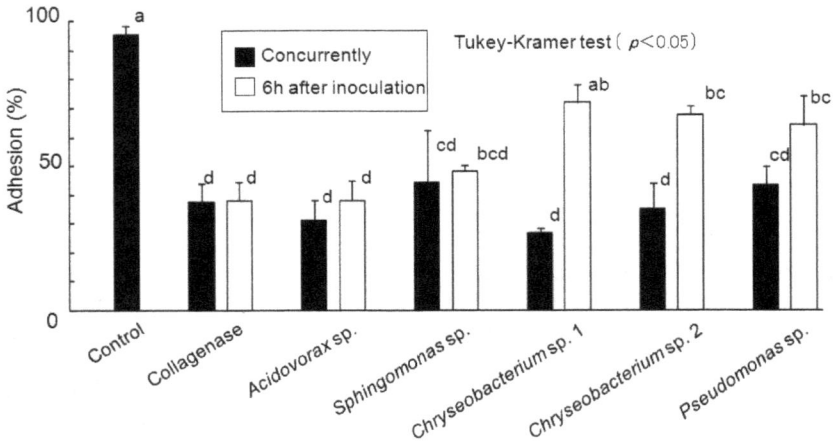

Fig. 9. Detachment effect of *M. oryzae* germlings treated with gelatinolytic bacteria or commercial collagenase. Gelatinolytic bacteria and collagenase were treated concurrently black bars) or 6 hours after inoculation (white bars) with *M. oryzae* spores. Adapted from Shimoi *et al.* (2010), with permission.

We evaluated the protective effects of the selected bacterial isolates on rice blast disease. Spore suspensions of *M. oryzae* and each bacterial culture were inoculated onto rice leaves (cv. Lijiangxintuanheigu; LTH). The treatment with all the bacterial isolates significantly decreased disease indices compared to the control (Fig. 10, Shimoi *et al.*, 2010).

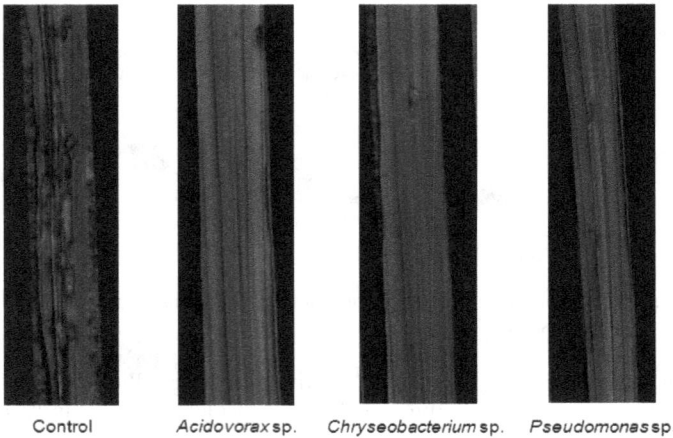

Control   *Acidovorax* sp.   *Chryseobacterium* sp.   *Pseudomonas* sp.

Fig. 10. Disease symptoms of rice blast inoculated with *M. oryzae* and gelatinolytic bacteria. Control: *M. oryzae* spore suspension alone. Adapted from Shimoi *et al.* (2010), with permission.

5. Conclusion

Germling adhesion is a general feature of plant pathogenic fungi, but its components are complicated. This adhesion ability is considered to be a promising a target for protection against disease. We found that glycoprotein degrading enzymes effectively detach germlings, while gelatinolytic bacteria showed a protective effect against disease. If these gelatinolytic bacteria can produce cell lytic enzymes, such as chitinase or glucanase, efficiency against disease protection would increase synergistically. Our future goals are to screen specific bacteria that produce multiple enzymes and provide a stable habitat in the foliar environment.

6. Acknowledgments

This research was supported by Grants-in-Aid for Scientific Research B (No. 18380033), Grants-in-Aid for Young Scientists B (No. 19780036), and Grants-in Aid for Young Scientists A (No. 23688006) from the Japan Society for the Promotion of Science.

7. References

Ahn, N., Kim, S., Choi, W., Im, H. & Lee, Y. H. (2004). Extracellular matrix protein gene, *EMP1*, is required for appressorium formation and pathogenicity of the rice blast fungus, *Magnaporthe grisea*. *Molecules and cells*, 17:166-173.

Apoga, D., Jansson, H. B. & Tunlid, A. (2001). Adhesion of conidia and germlings of the plant pathogenic fungus *Bipolaris sorokiniana* to solid surfaces. *Mycological Research*, 105: 1251-1260.

Bae, C.-Y., Kim, S., Choi, W. B. & Lee, Y.-H. (2007). Involvement of extracellular matrix and integrin-like proteins on conidial adhesion and appressorium differentiation in *Magnaporthe oryzae*. *Journal of Microbiology and Biotechnology*, 17:1198-1203.

Beckerman, J. L. & Ebbole, D. J. (1996). MPG1, a gene encoding a fungal hydrophobin of *Magnaporthe grisea*, is involved in surface recognition. *Molecular Plant-Microbe Interaction*, 9:450-456.

Beckett, A., Tatnell, J. A. & Taylor, N. (1990). Adhesion and pre-invasion behaviour of urediniospores of *Uromyces viciae-fabae* during germination on host and synthetic surfaces. *Mycological Research*, 94:865-875.

Bobichon, H., Gache, D. & Bouchet, P. (1994). Ultrarapid cryofixation of *Candida albicans*: evidence for a fibrillar reticulated external layer and mannan channels within the cell wall. *Cryo-Letters*, 15:161-172.

Bouchara, J.-P., Oumeziane, N. A., Lissitzky, J.-C., Larcher, G., Tronchin, G. & Ghabasse, D. (1996). Attachment of spores of the human pathogenic fungus *Rhizopus oryzae* to extracellular matrix components. *European Journal of Cell Biology*, 70:76-83.

Braun, E. J. & Howard, R. J. (1994). Adhesion of fungal spores and germlings to host plant surfaces. *Protoplasma*, 181:202-212.

Celerin, M., Ray, J. M., Schisler, N. J., Day, A. W., Steter-Stevenson, W. G. & Laudenbach, D. E. (1996). Fungal fimbriae are composed of collagen. *EMBO Journal*, 15:4445-4453.

Corrêa, A., Staples, R. C. & Hoch, H. C. (1996). Inhibition of thigmostimulated cell differentiation with RGD-peptides in *Uromyces* germlings. *Protoplasma*, 194:91-102.

Dean, R. A., Lee, Y. H., Mitchell, T. K. & Whitehead, D. S. (1994). Signalling systems and gene expression regulating appressorium formation in *Magnaporthe grisea*. In *Rice blast disease*, R. S. Zeigler et al. (eds), CAB International, Wallingford, pp. 23-34.

Deising, H., Nicholson, R. L., Haug, M., Howard, R. J. & Mendgen, K. (1992). Adhesion pad formation and the involvement of cutinase and esterases in the attachment of uredospores to the host cuticle. *The Plant Cell*, 4:1101-1111.

Doss, R. P. (1999). Composition and enzymatic activity of the extracellular matrix secreted by germlings of *Botrytis cinerea*. *Applied and Environmental Microbiology*, 65:404-408.

Doss, R. P., Deisenhofer, J., von Nidda, H.-A. K., Soeldner, A. H. & McGuire, R. P. (2003). Melanin in the extracellular matrix of germlings of *Botrytis cinerea*. *Phytochemistry*, 63:687-691.

Doss, R. P., Potter, S. W., Chastagner, G. A. & Christian, J. K. (1993). Adhesion of nongerminated *Botrytis cinerea* conidia to several substrata. *Applied and Environmental Microbiology*, 59:1786-1791.

Doss, R. P., Potter, S. W., Soeldner, A. H., Christian, J. K. & Fukunaga, L. E. (1995). Adhesion of germlings of *Botrytis cinerea*. *Applied and Environmental Microbiology*, 61:260-265.

Fu, Y., Rieg, G., Fonzi, W. A., Belanger, P. H., Edwards Jr, J. E. & Filler, S. G. (1998). Expression of the *Candida albicans* gene ALS1 in *Saccharomyces cerevisiae* induces adherence to endothelial and epithelial cells. *Infection and Immunity*, 66:1783-1786.

Gale, C. A., Bendel, C. M., McClellan, M., Hauser, M., Becker, J. M., Berman, J. & Hostetter, M. K. (1998). Linkage of adhesion, filamentous growth, and virulence in *Candida albicans* to a single gene, *INT1*. *Science*, 279:1355-1358.

Gale, C. A., Finkel, D., Tao, N., Meinke, M., McClellan, M., Olson, J., Kendrick, K. & Hostetter, M. (1996). Cloning and expression of a gene encoding an integrin-like protein in *Candida albicans*. *Proceedings of the National Academy of Sciences of the United States of America*, 93:357-361.

Giancotti, F. G. & Ruoslahti, E. (1999). Integrin signaling. *Science*, 285:1028-1032.

Gilbert, R. D., Johnson, A. M. & Dean, R. A. (1996). Chemical signals responsible for appressorium formation in the rice blast fungus *Magnaporthe grisea*. *Physiological and Molecular Plant Pathology*, 48:335-346.

Hamer, J. E., Howard, R. J., Chumley, F. G. & Valent, B. (1988). A mechanism for surface attachment in spores of a plant pathogenic fungus. *Science*, 239: 288–290.

Howard, R. J., Ferrari, M. A., Roach, D. H. & Money, N. P. (1991). Penetration of hard substrates by a fungus employing enormous turgor pressures. *Proceedings of the National Academy of Sciences of the United States of America*, 88:11281-11284.

Hughes, H. B., Carzaniga, R., Rawlings, S. L., Green, J. R. & O'Connell, R. J. (1999). Spore surface glycoproteins of *Colletotrichum lindemuthianum* are recognized by a monoclonal antibody which inhibits adhesion to polystyrene. *Microbiology*, 145:1927-1936.

Hutchison, K. A., Green, J. R., Wharton, P. S. & O'Connell, R. J. (2002). Identification and localisation of glycoproteins in the extracellular matrices around germ-tubes and appressoria of *Colletotrichum* species. *Mycological Research*, 106:729-736.

Hyon, G.-S., Muranaka, Y., Ikeda, K., Inoue, K., Hosogi, N., Meguro, H., Yamada, T., Hida, S., Suzuki, T. & Park, P. (2009). The extracellular matrix produced from *Alternaria alternata* Japanese pear pathotype plays a possible role of adhesion on the surfaces of host leaves during plant infection. *Journal of Electron Microscopy Technology for Medicine and Biology*, 23:1-8.

Inoue, K., Suzuki, T., Ikeda, K., Jiang, S., Hosogi, N., Hyon, G.-S., Hida, S., Yamada, T. & Park, P. (2007). Extracellular matrix of *Magnaporthe oryzae* may have a role in host adhesion during fungal penetration and is digested by matrix metalloproteinases. *Journal of General Plant Pathology*, 73:388-398.

Inoue, K., Onoe, T., Park, P. & Ikeda, K. (2011). Enzymatic detachment of spore germlings in *Magnaporthe oryzae*. *FEMS Microbiology Letters*, 323:13-19 (doi: 10.1111/j.1574-6968.2011.02353.x)

Inoue, S. Kato, T., Jordan, V. W. L. & Brent, K. J. (1987). Inhibition of appressorial adhesion of *Pyricularia oryzae* to barley leaves by fungicides. *Pesticide Science*, 19:145-152.

Izumitsu, K., Kimura, S., Kobayashi, H., Morita, A., Saitoh, Y. & Tanaka, C. (2010). Class I hydrophobin BcHpb1 is important for adhesion but not for later infection of *Botrytis cinerea*. *Journal of General Plant Pathology*, 76:254-260.

Jelito, T. C., Page, H. A. & Read, N. D. (1994). Role of external signals in regulating the pre-penetration phase of infection by the rice blast fungus, *Magnaporthe grisea*. *Planta*, 194:471-477.

Jian, S., Hyon, G.-S., Inoue, K., Park, P. & Ishii, H. (2007). Immunohistochemical and cytochemical analysis of extracellular matrix produced from *Venturia nashicola*, scab fungus, on the surfaces of susceptible Japanese pear leaves. *Journal of Electron Microscopy Technology for Medicine and Biology*, 21:7-11.

Kaminskyj, S. G. W. & Heath, I. B. (1995). Integrin and spectrin homologues, and cytoplasm-wall adhesion in tip growth. *Journal of Cell Science*, 108:849-856.

Kershaw, M. J., Thornton, C. R., Wakley, G. E. & Talbot, N. J. (2005). Four conserved intramolecular disulphide linkages are required for secretion and cell wall localization of a hydrophobin during fungal morphogenesis. *Molecular Microbiology*, 56:117-125.

Kim, S., Ahn, I. P., Rho, H. S. & Lee, Y. H. (2005). *MHP1*, a *Magnaporthe grisea* hydrophobin gene, is required for fungal development and plant colonization. *Molecular Microbiology*, 57:1224-1237.

Klotz, S. A., Rutten, M. J., Smith, R. L., Babcock, S. R. & Cunningham, M. D. (1993). Adherence of *Candida albicans* to immobilized extracellular matrix proteins is mediated by calcium-dependent surface glycoproteins. *Microbial Pathogenesis*, 14:133-147.

Kwon, Y. H. & Epstein, L. (1997). Isolation and composition of the 90 kDa glycoprotein associated with adhesion of *Nectria haematococca* macroconidia. *Physiological and Molecular Plant Pathology*, 51:63-74.

Lacroix, H., Whiteford, J. R. & Spanu, P. D. (2008). Localization of *Cladosporium fulvum* hydrophobins reveals a role for HCf-6 in adhesion. *FEMS Microbiology Letters*, 286:136-144.

Lee, Y. H. & Dean, R. A. (1994). Hydrophobicity of contact surface induces appressorium formation in *Magnaporthe grisea*. *FEMS Microbiology Letters*, 115:71-75.

Manning, V. A., Andrie, R. M., Trippe, A. F. & Ciuffetti, L. M. (2004). Ptr ToxA requires multiple motifs for complete activity. *Molecular Plant-Microbe Interaction*, 17:491-501.

Mercure, E. W., Leite, B. & Nicholson, R. L. (1994a). Adhesion of ungerminated conidia of *Colletotrichum graminicola* to artificial hydrophobic surfaces. *Physiological and Molecular Plant Pathology*, 45:407-420.

Mercure, E. W., Kunoh, H. & Nicholson, R. L. (1994b). Adhesion of *Colletotrichum graminicola* conidia to corn leaves: a requirement for disease development. *Physiological and Molecular Plant Pathology*, 45:421-440.

Nielsen, P. S., Clark, A. J., Oliver, R. P., Huber, M. & Spanu, P. D. (2001). HCf-6, a novel class II hydrophobin from *Cladosporium fulvum*. *Microbiological Research*, 156:59-63.

O'Connell, R. J., Pain, N. A., Hutchison, K. A., Jones, G. L. & Green, J. R. (1996). Ultrastructure and composition of the cell surfaces of infection structures formed by the fungal plant pathogen *Colletotrichum lindemuthianum*. *Journal of Microscopy*, 181:204-212.

Pain, N. A., Green, J. R., Gammie, F. & O'Connell, R. J. (1994a) Immunomagnetic isolation of viable intracellular hyphae of *Colletotrichum lindemuthianum* (Sacc. & Magn.) Briosi & Cav. from infected bean leaves using a monoclonal antibody. *New Phytologist*, 127:223-232.

Pain, N. A., O'Connell, R. J., Mendgen, K. & Green, J. R. (1994b) Identification of glycoproteins specific to biotrophic intracellular hyphae formed in the *Colletotrichum lindemuthianum*-bean interaction. *New Phytologist*, 127:233-242.

Pain, N. A., O'Connell, R. J., Bailey, J. A. & Green, J. R. (1992). Monoclonal antibodies which show restricted binding to four *Colletotrichum* species: *C. lindemuthianum, C. malvarum, C. orbiculare* and *C. trifolii*. *Physiological and Molecular Plant Pathology*, 40:111-126.

Poon, N. H. & Day, A. W. (1974). Fimbriae in the fungus, *Ustilago violacea*. *Nature*, 250:648-649.

Poon, N. H. & Day, A. W. (1975). Fungal fimbriae. I. Structure, origin, and synthesis. *Canadian Journal of Microbiology*, 21:537-546.

Rauceo, J. M., De Armond, R., Otoo, H., Kahn, P. C., Klotz, S. A., Gaur, N. K. & Lipke, P. N. (2006). Threonine-rich repeats increase fibronectin binding in the *Candida albicans* adhesin Als5p. *Eukaryotic Cell*, 5:1664-1673.

Rawlings, S. L., O'Connell, R. J. & Green, J. R. (2007). The spore coat of the bean anthracnose fungus Colletotrichum lindemuthianum is required for adhesion, appressorium development and pathogenicity. *Physiological and Molecular Plant Pathology*, 70:110-119.

Sarma, G. N., Manning, V. A., Ciuffetti, L. M. & Karplus, P. A. (2005). Structure of Ptr ToxA: an RGD-containing host-selective toxin from *Pyrenophora tritici-repentis*. *Plant Cell*, 17:3190-3202.

Schumacher, C. F. A., Steiner, U., Dehne, H. W. & Oerke, E. C. (2008). Localized adhesion of nongerminated *Venturia inaequalis* conidia to leaves and artificial surfaces. *Phytopathology*, 98:760-768.

Segers, G. C., Hamada, W., Oliver, R. P. & Spanu, P. D. (1999). Isolation and characterisation of five different hydrophobin-encoding cDNAs from the fungal tomato pathogen *Cladosporium fulvum*. *Molecular and General Genetics*, 261:644-652.

Sela-Buurlage, M. B., Epstein, L. & Rodriguez, R. J. (1991). Adhesion of ungerminated *Colletotrichum musae* conidia. *Physiological and Molecular Plant Pathology*, 39:345-352.

Shimoi, S., Inoue, K., Kitagawa, H., Yamasaki, M., Tsushima, S., Park, P. & Ikeda, K. (2010). Biological control for rice blast disease by employing detachment action with gelatinolytic bacteria. *Biological Control,* 55:85-91.

Skamnioti, P. & Gurr, S. J. (2007). *Magnaporthe grisea* cutinase2 mediates appressorium differentiation and host penetration and is required for full virulence. *Plant Cell,* 19:2674-2689.

Staab, J. F., Bradway, S. D., Fidel, P. L. & Sundstrom, P. (1999). Adhesive and mammalian transglutaminase substrate properties of *Candida albicans* Hwp1. *Science,* 283:1535-1538.

Staab, J. F., Ferrer, C. A. & Sundstrom, P. (1996). Developmental expression of a tandemly repeated, proline- and glutamine-rich amino acid motif on hyphal surfaces of *Candida albicans. The Journal of Biological Chemistry,* 271:6298-6305.

Sunde, M., Kwan, A. H., Templeton, M. D., Beever, R. E. & Mackay, J.P. (2008). Structural analysis of hydrophobins. *Micron,* 39:773-784.

Sundstrom P. (1999). Adhesins in *Candida albicans. Current Opinion in Microbiology,* 2:353-357.

Sweigard, J. A., Chumley, F. G. & Valent, B. (1992). Cloning and analysis of *CUT1,* a cutinase gene from *Magnaporthe grisea. Molecular and General Genetics,* 232:174-182.

Takahashi, T., Maeda, H., Yoneda, S., Ohtaki, S., Yamagata, Y., Hasegawa, F., Gomi, K., Nakajima, T. & Abe, K. (2005). The fungal hydrophobin RolA recruits polyesterase and laterally moves on hydrophobic surfaces. *Molecular Microbiology,* 57:1780-1796.

Talbot, N. J. (2003). On the trail of a cereal killer: exploring the biology of *Magnaporthe grisea. Annual Review of Microbiology,* 57:177-202.

Talbot, N. J., Ebbole, D. J. & Hamer, J. E. (1993). Identification and characterization of *MPG1,* a gene involved in pathogenicity from the rice blast fungus *Magnaporthe grisea. Plant Cell,* 5:1575-1590.

Talbot, N. J., Kershaw, M. J., Weakley, G. E., de Vries, O. M. H., Wessels, J. G. H. & Hamer, J. E. (1996). MPG1 encodes a fungal hydrophobin involved in surface interactions during infection-related development of *Magnaporthe grisea. Plant Cell,* 8:985-999.

Temple, B., Horgen, P. A., Bernier, L. & Hintz, W. E. (1997). Cerato-ulmin, a hydrophobin secreted by the causal agents of Dutch elm disease, is a parasitic fitness factor. *Fungal Genetics and Biology,* 22:39-53.

Thau, N., Monod, M., Crestani, B., Rolland, C., Tronchin, G., Latgé, J.-P. & Paris, S. (1994). *rodletless* mutants of *Aspergillus fumigatus. Infection and Immunity,* 62: 4380-4388.

Tucker, S. L. & Talbot, N. J. (2001). Surface attachment and pre-penetration state development by plant pathogenic fungi. *Annual Reviews in Phytopathology,* 39:385-417.

Van Dyke, C. G. & Mims, C. W. (1991). Ultrastructure of conidia, conidium germination, and appressorium development in the plant pathogenic fungus *Colletotrichum truncatum. Canadian Journal of Botany,* 69:2355-2467.

Wang, C. & Leger, R. J. S. (2007). The MAD1 adhesin of *Metarhizium anisopliae* links adhesion with blastospore production and virulence to insects, and the MAD2 adhesin enables attachment to plants. *Eukaryotic Cell,* 6:808-816.

Wessels, J.G.H. (1994). Developmental regulation of fungal cell formation. *Annual Reviews of Phytopathology,* 32:413-437.

Wessels, J. G. H., de Vries, O. M. H., Ásgeirsdóttir, S. A. & Schuren, F. H. J. (1991). Hydrophobin genes involved in formation of aerial hyphae and fruit bodies in *Schizophyllum*. *Plant Cell*, 3:793-799.

Whiteford, J. R., Lacroix, H., Talbot, N. J. & Spanu, P. D. (2004). Stage-specific cellular localisation of two hydrophobins during plant infection by the pathogenic fungus, *Cladosporium fulvum*. *Fungal Genetics and Biology*, 41:624-634.

Wösten, H. A. B. (2001). Hydrophobins: multipurpose proteins. *Annual Reviews in Microbiology*, 55:625-646.

Xiao, J.-Z., Ohshima, A., Kamakura, T., Ishiyama, T. & Yamaguchi, I. (1994). Extracellular glycoprotein(s) associated with cellular differentiation in *Magnaporthe grisea*. *Molecular and Plant-Microbe Interaction*, 7:639-644.

8

Novel Methods for the Quantification of Pathogenic Fungi in Crop Plants: Quantitative PCR and ELISA Accurately Determine *Fusarium* Biomass

Kurt Brunner, Andreas Farnleitner and Robert L. Mach

Vienna University of Technology, Institute of Chemical Engineering, Vienna
Austria

1. Introduction

Fungi of the genus *Fusarium* are worldwide occurring plant pathogens which cause severe damages to numerous cultivable plants (Weiland et al., 2000; Mirete et al., 2004; Youssef et al., 2007; Li et al., 2008) with the highest economical losses upon infection of maize, wheat and barley (Windels et al., 2000; Nganje et al., 2004). *Fusarium* caused diseases can destroy crops within several weeks and the infection leads to quality losses in two different aspects: besides the reduced yield due to reduced kernel size, the fungus produces various toxic metabolites while colonizing the plant. These mycotoxins heavily impair the quality of the harvest (McMullen et al., 1997). The acute or chronic toxicity of *Fusarium* released compounds led to the introduction of national or international limits to regulate mycotoxin levels in food and feed (e.g. limits of the European Community since 2006).

Most *Fusarium* species are widely distributed in substrates such as soil, on subterranean and aerial plant parts, plant debris, and on dead organic matter. Many *Fusarium* species have active or passive means to disperse spores or conidia in the atmosphere. The ability to grow on a broad range of substrates combined with their efficient dispersal mechanisms enables a widespread distribution of these fungi (Burgess, 1981). *Fusarium* species are well adapted to grass hosts (Leonard & Bushnell, 2003) and can colonize on many agricultural commodities such as rice, bean and soybean. According to several studies *F. graminearum* and *F. culmorum* are among the most aggressive plant pathogenic fungi known. *F. graminearum* mainly occurs in temperate and warmer regions of the USA, China and the southern hemisphere (Osborne & Stein, 2007). It is regarded as one of the most vigorous toxin producers and has therefore become the most intensively studied plant pathogen (Goswami & Kistler, 2004). In contrast, *F. culmorum* predominates in the cooler regions including the U.K., Northern Europe and Canada (Osborne & Stein, 2007) and is also associated with the occurrence of many mycotoxins (Desjardins, 2006).

2. Toxigenicity and impacts

A worldwide problem of agricultural industry producing wheat, barley and maize is the fungal disease *Fusarium* head blight (FHB), with maize and wheat as the economically most important host plants. The disease is associated with *F. graminearum, F. culmorum, F. poae, F. avenaceum, F. sporotrichoides* and *Microdochium nivale*. Each fungus contributing to FHB has particular biological and environmental requirements, which in part explains the frequency of occurrence in specific locations.

This destructive disease was initially described in 1884 in England and was considered a major threat during the early years of the twentieth century (Goswami & Kistler, 2004). Nowadays, *Fusarium* head blight is a disease of massive economic impact worldwide and has been ranked by the United States Department of Agriculture (USDA) as the worst plant disease to hit the nation (Windels, 2000). The losses due to direct and secondary economic impact of FHB on wheat and barley were estimated at $2.7 billion for the period from 1998 – 2000 in the US (Nganje et al., 2002). Epidemics of FHB are strongly influenced by various factors such as local and regional environment, the physiological state and genetic background of the host, as well as pathogen related factors including host adaptation and virulence.

Infected plant debris serves as primary substrate for common FHB pathogens, which are able to survive the winter period as saprophytic mycelium or thick walled chlamydospores. Warm, moist conditions in spring are favorable for the development and maturation of conidia and perithecia, which produce ascospores (Markell & Frankl, 2002). These sticky spores are discharged from the surface of crop debris and dispersed by wind, rain or even insects to the host plants. Infected wheat or maize presents brown, dark purple to black lesions on the exterior. The head of cereals has a characteristically bleached appearance, hence the name *Fusarium* head blight. During prolonged wet periods infected spikelets, glumes and kernels present pink to salmon-orange spore masses of the fungus. Maize infections occur at the ear apex following colonization of the kernels with white mycelium, which turns pink to red with time.

Fungal colonization is affected by numerous variables such as spike morphology, canopy density, plant height, rainfall, relative humidity, temperature and host plant resistance (Kolb et al., 2001; Rudd et al., 2001). *Fusarium* infections are increased due to the lack of completely resistant plant material and a frequent application of unsuitable cropping systems. Pathogen survival is favored in reduced tillage systems as residue burial speeds up the decomposition and reduces pathogen reproduction and survival (Khonga et al., 1988; Pereyra et al., 2004). Petcu et al., (1998) investigated the influence of crop rotations on the severity of FHB. Studies show that the economically most lucrative cultivation of alternating maize and wheat cultures turned out to be problematic. The increased production of these *Fusarium* favorable crops and the dramatic increase in their residues remaining on the soil surface provide a large increase in the amount of niche available to the pathogen. *Fusarium* non-host plants used as preceding crops or intertillage of wheat and maize are often less profitable and therefore of low interest for farmers. In North and South America erosion causes a dramatic loss of fertile topsoil. To avoid this loss, no-till systems have been established to overcome the drawback of conventional farming. However, *Fusarium* inoculum density increases in soil in reduced tillage systems compared to plough treated fields (Steinkellner & Langer, 2004).

3. Plant resistance to *Fusarium*

Plant resistance can be defined as the relative amount of heritable qualities possessed by a plant that reduces the degree of damage to the plant caused by pathogens. Schroeder and

Christensen (Schroeder & Christensen, 1963) described in their work two phenotypic measures of disease resistance; type I resistance operates against initial infection and type II resistance describes the resistance to spreading within infected tissues. A third type of *Fusarium* resistance (type III) was described by Wang and Miller (1988) as the insensitivity of wheat lines to toxins or the ability of the resistant cultivar to degrade mycotoxins.

Integrated control strategies are essential to prevent *Fusarium* diseases in modern agriculture. The use of resistant cultivars is considered necessary for numerous reasons. Firstly, modern agronomic practices such as reduced tillage and cereal and maize rich crop rotations have the tendency to increase the amount of *Fusarium* related diseases. In addition to this, chemical control measures are only partly effective as only few fungicides are sufficiently active against *Fusarium* and their application must be performed within a time frame of a few days around the flowering stage of the plants. For the protection of maize cultures no fungicides are commercially available up to now. Understanding *Fusarium* life-cycle and infection pathways is the first step in preventing disease. Nevertheless, uncontrollable factors such as weather conditions during as well as economic or ecologic interests prevent sustainable success in the reduction of these pathogens. Plant breeders have made significant progress in the development of *Fusarium* resistant maize and wheat varieties by identifying genetic regions, which are linked to the resistance of plants (Buerstmayer et al., 2003; Draeger et al., 2007). Resistance against *Fusarium* is a quantitative trait, which is governed by many independent genes distributed on several distinct genetic regions (quantitative trait loci, QTLs) in the plant genome. Therefore an effective approach to investigate FHB resistance is the identification of the QTLs and their mapping (Buerstmayer et al., 2009). The mapping of QTLs is the basis for further efforts towards a straightforward marker assisted resistance breeding strategy. Interestingly, often the best regionally adapted and highly productive crop varieties are susceptible to FHB. For this reason, breeders are faced with the difficult task to combine adaption to certain locations with high yield and *Fusarium* resistance.

The production of mycotoxins has been identified as a crucial factor for successful infection . Deoxynivalenol is considered the most common occurring mycotoxin involving *Fusarium* infection. However, DON levels in infected grain vary significantly amongst wheat cultivars (Bai et al., 2001) infected with the same *Fusarium* isolate. The evaluation of *Fusarium* resistance of either the parents in current breeding programs or the control of resistance levels of novel lines can be carried out by chemical analysis of the DON accumulation in the seeds. In general DON levels in resistant plants are much lower than those found in more susceptible cultivars. Artificial infection of wheat-ears with *Fusarium* conidia followed by determination of DON content and visual scoring of the head blight symptoms are the currently established methods for resistance assessment. Nevertheless, the determinations of disease symptoms in combination with the mycotoxin content of grain are both indirect methods to evaluate plant resistance. Furthermore, both methods observe the same effect, as visual disease symptoms are mainly caused by mycotoxin intoxication and not by the growth of mycelium. Studies have shown that the amount of fungal mycelia formed during infection not always correlates well with these parameters (Waalwijk et al, 2004; Hill et al., 2006; Brunner et al., 2009; Kulik et al. 2011). Asymptomatic kernels may contain significant amounts of mycotoxins while symptomatic kernels within the same sample may not.

4. Methods for the identification of *Fusarium* and quantification of the infection

Quantification and identification of *Fusarium* species in agriculture or plant pathology is traditionally carried out using culture based methods. Usually single Kernels are spread on petri dishes with appropriate medium and the number of infected kernels is counted after several days. For a further classification of the species a morphological investigation of the grown mycelium is necessary. This kind of species determination is time consuming and requires experience and specific expertise. Furthermore, culture based methods rely on living propagules and the obtained results not always reflects the biological situation. The pathogen detection is only reliable at late stages of the infection when a spread of the disease can no longer be controlled by fungicides (McCArtney et al., 2003). Besides these drawbacks the results are only semi-quantitative as only the number of infected kernels can be determined but not the grade of infection of each kernel. Other conventional methods include instrumental analysis such as ergosterol quantification. Ergosterol is a characteristic compound of fungal cell walls. However this approach is not specific to pathogens or a certain fungal species and the ergosterol amount relative to the fungal mycelia varies within *Fusarium* species (Pasanen et al., 1999). Finally, the analytical determination of ergosterol is not easier than that of DON, resulting in an alternative method, which is not commercially applicable to determine the *Fusarium* resistance of plants.

Recent studies aim at developing more direct techniques to quantify *Fusarium* related diseases which can i) reduce the number of analyses to accurately assess *Fusarium* infection, ii) eliminate pleiotropy between disease symptomology and mycotoxin production, iii) reduce the error associated with environmental effects and iv) reduce errors related with asymptomatic expression of FHB (Hill et al., 2008). In general, two distinct approaches have become accepted within the last two decades: immunoassays and (quantitative) PCR. The enzyme linked immunosorbent assay (ELISA) has been developed as a direct measure of *Fusarium* spp. biomass in infested grain samples or plant tissue. ELISA provides at least genus specificity through specific fungal antigens and its ease of sample preparation and low costs make it an interesting novel method. Another approach to overcome the drawbacks of conventional identifications is the development of screening techniques based on DNA identification which are nowadays well established for *Fusarium* species and a broad range of available literature guarantees to find a suitable PCR assay for many applications. Unlike conventional detection methods, samples can be tested directly with ELISA or PCR – no elaborate isolation and cultivation steps are necessary for a suitable detection or quantification.

4.1 Enzyme linked immunosorbent assays (ELISAs)

Enzyme linked immunosorbent assays are based on the specific recognition capabilities of antibodies. These antibodies are usually derived from the immunization of animals (usually rabbits, mice, chicken or goat) with certain immunogens such as culture filtrates or mycelial compounds. After repeated injection of the immunogen blood samples are taken and the serum is used either as a whole or it is applied after certain clean-up steps for the ELISA tests. For the production of the well-defined monoclonal antibodies lymphocytes are isolated from immunized mice and are fused to myeloma cells. The resultant clonal hybridoma cells can be maintained in culture. Soon after the development of the ELISA method in the early 1970s scientists working on agricultural and plant pathology related topics recognized the potential of this analytical tool. Within only a few years the number of annual ELISA based publications in the AGRICOLA database rose from zero to more than 150 and numerous antibodies were developed to detect plant pathogens belonging to different genera.

4.1.1 ELISAs based methods for the detection and quantification of *Fusarium*

The first immuno assay for *Fusarium* was developed at the University of Göttingen in 1989 by J.G. Unger and published in his PhD thesis. The author succeeded in the recognition of *F. culmorum* exoantigens (compounds secreted by *Fusarium* during growth) by an antibody. However, after this first success of the ELISA in *Fusarium* detection almost 15 years passed by until this *F. culmorum* ELISA was applied to practical applications (Chala et al., 2003). During these field tests the authors found out that the developed antibody is not specific to *F. culmorum* but binds to antigens from various *Fusarium* species. Over the years numerous studies on the development of anti-*Fusarium* antibodies have been published. Although most authors used *F. graminearum* and/or *F. culmorum* for immunization, no group succeeded in the production of highly specific antibodies. Gan et al., (1997) immunized chicken with soluble exoantigens and with soluble frations from homogenized mycelia of F. graminarum, *F. poae* and *F. sporotrichioides*. The *F. graminearum* and *F. sporotrichioides* exoantigen antibodies showed cross-reactivities with almost all common *Fusarium* species. The antibody against *F. poae* exoantigens showed absolutely no cross-reactivity against other *Fusarium* species or against other filamentous fungi. Up to now this is the only species-specific antibody for a *Fusarium* spp. Hill et al., (2006) probably produced the best-characterized antibody so far by isolating monoclonal antibodies produced by IF8 cell lines for the detection of *Fusarium*. These antibodies detected species from different phylogenetic clades and the authors also demonstrated that no cross-reaction – even with closely related ascomycetes – could be observed. Unfortunately, no publication describes the essential analytical parameters like limit of detection, limit of quantification or a linear range of the developed tests.

In general, antibodies detect many different *Fusarium* species but the specificity of PCR based assays has not yet been obtained by immunoassays. Neither the detection of particular species – except one example for *F. poae* – nor the detection of a particular toxigenic group (e.g. trichothecene producers or fumonisin producers) is possible with ELISA methods. Furthermore, most antibodies were raised against unknown compounds secreted by *Fusarium* during growth in liquid medium (exoantigens). As usually the secretion is not a constant attribute but is subjected to complex regulation, the amount of exoantigens secreted during the infection of grains might not always be constant. For this reason, the amount of *Fusarium* antigen detected in a sample may not always correlate with the fungal biomass.

4.1.2 Applications of *Fusarium* ELISAs for agriculture and plant pathology

Many of the developed PCR based tests for the determination of *Fusarium* focus on the screening of harvested grain and frequently the correlation of *Fusarium* DNA content and toxin accumulation is the central topic. Only few (references) of the many *Fusarium* PCR publications deal with plant pathology related questions whereas all developed ELISA methods were extensively used in field applications to monitor the efficacy of fungicide application (Chala et al., 2003) or to assess the *Fusarium* resistance of crop lines (Miedaner et al., 1994, Miedaner et al., 2004, Hill et al., 2006, Slikova 2009, Rohde & Rabenstein 2005).

4.1.2.1 Evaluation of fungicide efficacy with a *Fusarium* ELISA

Chala et al., (2003) presented a comprehensive study on the evaluation of fungicide application to reduce *Fusarium* infection and DON content in wheat. Winter wheat was planted in four field replications and artificially infected with *F. culmorum* conidia at flowering stage by spray inoculation. Five different fungicides were tested at different application times alone and in various combinations. The efficiency of the treatment was determined by visual scoring of disease symptoms, *Fusarium* ELISA, yield a 1000-grain weight. The combination of these different methods allows the evaluation of the results obtained by the immunoassay. The results of the field trials reveal clearly that only one fungicide showed a high efficiency in the reduction of fungal biomass and DON content. All applied methods led to the same conclusion. Interestingly, the application of some fungicides at a later growth stage after flowering did not affect the fungal biomass but slightly increased the production of DON. In general the study demonstrates clearly the potential of the developed *Fusarium* ELISA to obtain a deeper insight into biomass formation. The main advantage of the ELISA based determination of infection is the enhanced failure-safety.

4.1.2.2 Evaluation of *Fusarium* resistance with ELISA tests

The comparison of the resistance of different plant varieties is often a challenge for breeders. The increase in *Fusarium* resistance of novel crop lines is just marginal and for some plants like durum or triticale disease symptoms are usually not obvious. Furthermore, other fungal diseases like *Microdochium nivale* or *Septoria nodurum* also lead to similar symptoms as *Fusarium* head blight and the result of the visual assessment might be misinterpreted. Besides the PCR based methods described above the ELISA tests represent the only available tool to determine fungal biomass and so the application for resistance tests is self-evident.

The first field test for *Fusarium* resistance based on ELISA analysis was performed in the early 1990s (Miedaner et al., 1994). Winter rye was infected with *F. culmorum* colonized wheat flour in November when the plants have reached the three-leaf status. The colonization of the host plant started early after inoculation and increased continuously till full maturity. Interestingly, different genotypes showed the highest variance in *Fusarium* protein content during medium growth states. In adult states almost no difference in *Fusarium* infection between resistant and susceptible plants was found. The disappearance of the discrepancy results in saprophytic growth of *F. culmorum* on rye during ripening and therefore the medium growth stages are the optimal date to discriminate resistant from susceptible varieties. The same group also investigated the *F. culmorum* resistance of wheat and triticale cultivars in six different environments and demonstrated the performance of ELISA biomass determination for two more cereal crops (Miedaner et al., 2004). However, the correlation of the ELISA based *Fusarium* protein content to the DON content in a sample was not better than the correlation between DON and visual disease symptom assessment. Another ELISA test based on monoclonal antibodies was optimized for the quantification of *Fusarium* in barley (Hill et al., 2006) and was used to study the influence of the applied analytical method on the results of resistance tests (Hill et al., 2008). A mapping population was grown in two environments and breeding lines were grown at four different locations and disease data were collected by visual disease assessment, determination of the DON accumulation and the *Fusarium* ELISA. The obtained data were subjected to statistical analyses and the authors calculated a model for the prediction of different replications and different environments necessary to certainly identify differences in *Fusarium* resistances. Interestingly, the determination of DON was identified as the factor that is least feasible for resistance studies. The quantification of the *Fusarium* biomass by ELISA required less different locations and fewer field replicates per location to classify barley varieties according to their *Fusarium* resistance than visual disease assessment or toxin measurements.

Often the correlation between visual disease scoring and DON content is better than between ELISA results and DON content. This fact has also been demonstrated in a Slovakian study (Slikova et al., 2009) for the infection of winter wheat with *F. culmorum*. As visual assessment just records the damage caused by the fungal toxins the correlation between symptoms and the DON content is obvious. The fact that visual symptoms and DON content are not independent features makes the combined use of only these two factors for resistance determination questionable. Therefore, Slikova et al. (2009) recommend the application of the *Fusarium* ELSIA to get more reliable results for plant resistance than with the two other methods.

4.2 Detection and quantification of *Fusarium* by PCR methods

The extensive application of DNA based identification technologies has increased the knowledge on suitable diagnostic DNA fragments of *Fusarium* species such as ITS (internal transcribes spacer) or IGS (intergenic spacer) sequences (Gagkaeva & Yli-Mattila, 2004, Jurado et al., 2006, Konstantinova & Yli-Mattila, 2004, Kulik 2008, Yli-Mattila, 2004), mitochondrial DNA (Laday et al., 2004), the β-tubulin encoding gene (Yli-Mattila, 2004, Mach et al., 2004, Reischer et al., 2004), the translation elongation factor gene (Knutsen et al., 2004) and the calmodulin gene (Mule et al., 2004) which were sequenced from numerous *Fusarium* spp. As a result highly specific PCR primers could be developed for *Fusarium* detection. In order to distinguish between producers and non-producers of certain toxins the genes from biosynthesis pathways for mycotoxins were accurately studied. The sequence information gained throughout numerous molecular taxonomic investigations allows not only qualitative applications to detect particular species but also quantitative measurement of fungal biomass. Different techniques like DGGE, RFLP and AFLP can be applied for the identification of adequate PCR targets for a review see Brunner & Mach, (2010), for quantitative detection methods however, only the real-time PCR has been shown to be applicable. Molecular diagnosis of plant pathogenic fungi has been proven to be highly specific, very sensitive and fast and relatively insensitive to microbial backgrounds and non-target organisms (McCartney et al., 2003).

Within the last decade numerous PCR based assays have been published for the detection of most agriculturally important *Fusarium* species. An intensive literature study reveals that all such publications can be separated into two distinct groups. Either the authors focus on highly specific test systems to allow the differentiation of *Fusarium* species or the focus is the comprehensive detection of a whole group sharing a common feature. The latter assays usually use a key-gene for a biosynthetic pathway – e.g. for mycotoxin synthesis – as a target for PCR. This system allows the detection or even the quantification of all isolates, which are able to produce a certain class of toxins even if they belong to different species or to different genera.

4.2.1 Species-specific PCR assays

Species-specific assays allow the determination of one particular *Fusarium* species. This method focuses on high selectivity in order to quantify a target species even in a background of highly similar isolates. Several TaqMan based PCR assays have been developed for the quantification of some of the predominant species associated with head blight in Europe including *F. graminearum*, *F. poae*, *F. culmorum*, *F. sporotrichioides*, *F. verticillioides* or *F. avenaceum* . The authors clearly demonstrated the advantage of quantitative real time PCR systems combined with a high-throughput DNA extraction protocol over morphologic based methods. Unlike conventional agar-plating techniques, the quantification of different fungal species is possible directly from infested material (grain or plant tissue) and the entire procedure is less time consuming than microbiological methods (for a review see Brunner & Mach, 2010).

4.2.2 Group specific assays

Group specific assays were designed for the simultaneous quantification of all strains, which produce certain toxins such as trichothecenes, fumonisins or enniatin, based on key-genes involved in the mycotoxin biosynthesis. In contrast to the species-specific assays, a whole group of *Fusarium* spp., regardless of taxonomic origin, that is able to produce a certain class of mycotoxins can be quantified in a single run. A correlation between qPCR determined fungal biomass and DON, fumonisin or enniatin content in cereals could be shown using these group specific assays.

Detectable species	PCR target	Type of assay	Reference
F. graminearum, F. culmorum, F. avenaceum	RAPD derived	qualitative	Schilling et al., 1996
trichothecene producers	*tri5*	qualitative	Niessen et al., 1998
F. graminearum, F. culmorum	RAPD derived	qualitative	Nicholson et al., 1998
trichothecene producers	*tri5*	quantitative	Schnerr et al., 2001
trichothecene producers	*tri5*	qualitative	Edwards et al., 2001
F. graminearum, F. poae	IGS	quantitative	Yli-Mattila et al., 2008
fumonisin producers	*fum1*	quantitative	Bluhm et al., 2004
F. graminearum	*tub1*	quantitative	Reischer et al., 2004
F. graminearum, F. culmorum, F. avenaceum	RAPD derived	quantitative	Waalwijk et al. 2004
F. solani	rRNA, small subunit	quantitative	Li et al., 2008
fumonisin producers	*fum1*	quantitative	Waalwijk et al. 2008
enniatin producers	*esyn1*	quantitative	Kulik et al. 2011

Table 1. Overview of frequently applied PCR assays. This table cites the original method development papers (qualitative and quantitative) but not potentially following publications demonstrating the practical application of these assays.

4.2.3 Applications of the *Fusarium* PCR in agriculture and plant pathology

In contrast to the practical applications of the developed ELISA tests, the PCR assays are only rarely used to address typical questions related to plant pathology.

Some methods were developed to identify the chemotype of different *Fusarium* strains on a molecular basis. These assays are designed to differentiate between nivalenol, 3- and 15-acetyl-deoxynivalenol chemotypes of *F. graminearum*, *F. culmorum*, and F. cerealis. The biosynthetic pathway of trichothecene production is highly conserved between different *Fusarium* species and the genes within the trichothecene gene cluster are well investigated. Based on this knowledge three genes were chosen in all studies to identify the chemotypes: tri3, tri5, tri7 and tri13 or regions between these genes. The developed assays were applied to *Fusarium* monitoring programs in numerous countries of Europe and Nortehrn America (Chandler et al., 2003; Jennings et al. 2004; Lia et al., 2005; Quarta et al., 2006; Stepien et al., 2008;).

A tri5 based assay was used to evaluate the efficiency of fungicides against *F. graminearum* and *F. culmorum* (Edwards et al., 2001. Three chemical fungicides were applied alone in various concentrations and in combinations. Interestingly, the authors could demonstrate that different pathogenic fungi require different fungicides for the optimal control of the disease. The study revealed a good correlation between the DON content of samples and the *Fusarium* DNA. Furthermore, the ration between DON and the fungal DNA was not altered by the application of fungicides. This fact is of particular interest as previous in vitro studies postulated the enhanced toxin production of *Fusarium* after the treatment with antifungal compounds.

Most of the published studies focus on correlations between accumulated toxins and the PCR determined fungal DNA concentration in harvested grain and the PCR is postulated as a screening method for applications in food and feed safety. For some of these studies species specific assays were used to assess the biomass of various *Fusarium* species (Schnerr et al., 2000; Waalwijk et al., 2004; Brandfass & Karlovsky, 2006; Yli-Mattila et al. 2008) or group specific assays were used to identify of quantify certain toxin producers (Schnerr et al. 2001, 2002; Bluhm et al. 2004; Waalwijk et al., 2008; Kulik et al., 2011).

Although the sensitive PCR methods represent an optimal tool to monitor even minor amounts of *Fusarium* during colonialization of a host these applications are rare. Reischer et al. (2004) developed a TaqMan based quantitative PCR assay to quantify *F. graminearum* directly from infected plant material. A central spikelet of wheat heads was artificially infected with conidia and the increase of fungal biomass was monitored by real-time PCR and the disease symptoms were also assessed visually. Due to the superior sensitivity of the PCR assay the disease could be detected and quantified the first day after inoculation, whereas the first visual symptoms became obvious five days later. Nicholson et al. (1998) used PCR tests for *F. culmorum* and *F. graminearum* to address the fungal spread in wheat. The authors found out that the severity of *Fusarium* stem rot caused by *F. culmorum* correlates not only with the inoculums load but also with the time point the inoculums is applied. The earlier in the season the conidia are applied the more fungal DNA could be measured in the stem. The influence of trichothecene production on successful infection was tested in the same study with toxigenic and atoxigenic *F. graminearum* isolates. The toxin producers colonized grain better than non-producers, supporting the theory of trichothecens as a virulence factor. However, non-producing strains could also infect the wheat heads but the amount of detected fungal DNA was only 1 to 10% of the DNA amount found in plants infected with toxin producing strains. The infected wheat heads were also subjected to visual assessment of disease symptoms. Although the conservative method led to the same results the difference between toxigenic and atoxigenis strains was less pronounced. The above described assay for *F. culmorum* was also used to monitor natural stem rot in wheat in the U.K. (Nicholson et a., 2002), The authors found out that usually other pathogens than *F. culmorum* were the causal agent of this disease as almost no quantifiable DNA of *Fusarium* was present in the samples. The same *F. culmorum* assay was again used in combination with a *F. poae* assay to evaluate fungicide efficiency for *Fusarium* head blight control in a greenhouse experiment (Doohan et al., 1999). Under all tested conditions the applied fungicides reduced the disease severity between 20% and 80% and in general the results could be confirmed by visual monitoring of the disease symptoms.

A Canadian *Fusarium* monitoring program integrated PCR tests for *F. avenaceum, F. culmorum,* F. crookwellense, *F. poae, F. sporotrichioides, F. equiseti, F. pseudograminearum,* and *F. graminearum* to compare results with the traditional agar plating method (Demeke et al., 2005). For 83% of the tested grain samples the two methods led to the same results. However, the agar plate method not always gave positive results when DNA was found in samples. This leads to the assumption that PCR based methods provide enhanced sensitivity. Additionally, *F. graminearum, F. pseudograminearum* and *F. crookwellense* could not be distinguished by morphological analysis whereas PCR could clearly differentiate these species. This comprehensive study clearly demonstrated the power of DNA based methods if integrated into *Fusarium* monitoring programs. PCR allows the quantification (if applied as real-time PCR) and classification of numerous strains within a few hours of analysis. Classical methods need some days to weeks to give the same results and quantification is difficult as only the number of infected kernels can be determined.

An LNA-TaqMan based assays was used to investigate the differences in *Fusarium* resistance of twenty novel wheat lines ranging from highly susceptible to highly resistant (Brunner et al., 2009). The wheat-ears were inoculated with *F. graminearum* and *F. culmorum* in two consecutive years and the formation of mycotoxins, the accumulation of *Fusarium* DNA and the visual disease symptoms were recorded. This study demonstrated a certain discrepancy between visual scoring and quantitative PCR results. In accordance to other studies the visible symptoms matched perfectly the DNA content in medium to low resistant lines. In contrast, highly resistant lines with low *Fusarium* biomass – but nevertheless high amount of toxins – did sometimes not show any disease symptoms. The authors addressed this effect to the detoxification mechanism of high resistant plants as some toxins can be "masked" by the plant. So they are converted to less plant-toxic metabololites which do not damage the wheat-ears.

5. Conclusions

Visual assessment of disease caused by *Fusarium* species is frequently insufficient to identify the causal agent or to quantify the plant pathogen. In natural systems pathogenic fungi often occur in a combination of species, which can induce similar symptoms and a visual discrimination might be impossible. Other methods like agar plate assays rely on the fact that viable propagules are present in a sample. Especially in harvested grain this is not always the case as long dry periods during ripening in combination with intense UV radiation can reduce the survival of fungal mycelium or even of conidia. Furthermore, the morphologic discrimination of grown mycelium is difficult and experienced personal is indispensible. Novel methods for the detection and quantification of fungal biomass based on marker molecules (e.g. antigens or DNA) have the potential to revolutionize the field of pathogen monitoring in plant pathology. Within the last twenty years immunoassays and PCR methods found their way into this scientific field. ELISA test for different *Fusarium* species have been developed intensely in the 1990s and were applied in numerous field studies. However, the specificity of the produced antibodies leaded a lot to desire. Although particular species were used for the immunization of animals the resulting antibodies did detect many different *Fusarium* species. Only one group succeeded in the production of polyclonal antibodies which were specific to a single species, namely to *F. poae* . Furthermore, the re-production of antibodies is difficult as immunization is a complex procedure. The resulting immunoglobulins can vary in their features and therefore a careful characterization must be performed each production cycle. Interestingly with the availability of affordable real-time PCR cyclers on the market published studies using *Fusarium* ELISA almost disappeared. Most probably this tendency can be ascribed to two crucial advantages of PCR methods: i) these methods can be performed in high throughput and are easily transferred to other laboratories. Primers are synthesized commercially, are cheap and they are available within a few days. ii) PCR assays are highly specific and can be used to detect either genera or species and even distinctive isolates can be differentiated. Although quantitative PCR provides numerous advantages to study *Fusarium* in its environment only a few studies have been published to address biological problems. This might be due to the fact that nowadays food safety related questions are of high relevance and so the comparison between fungal DNA and mycotoxins is estimated as more important than other applications. On the other hand, the barrier for traditional plant pathologist might still be high to enter a technical field like quantitative PCR.

An investigation of the NCBI pubmed database reveals that the published studies applying *Fusarium* PCR methods in field-trials increase continuously. Although less novel assays are developed, more and more previously published methods are included in current studies, which might indicate the changeover from PCR based method development to PCR applications.

6. Acknowledgements

The authors thank the Federal Country Lower Austria and the European Regional Development Fund (ERDF) of the European Union for financial support.

7. References

Bai, G.H., Plattner, R., Desjardins, A. & Kolb, F. (2001). Resistance to *Fusarium* head blight and deoxynivalenol accumulation in wheat. *Plant Breeding*, 120(1), pp 1-6.
Baird, R., Abbas, H.K., Windham, G., Williams, P., Baird, S., Ma , P., Kelley, R., Hawkins L. & Scruggs M. (2008). Identification of select fumonisin forming *Fusarium* species using PCR applications of the polyketide synthase gene and its relationship to fumonisin production in vitro. *International Journal of Molecular Sciences*, 9: 554-570.

Bluhm, B.H., Cousin, M.A. & Woloshuk, C.P. (2004). Multiplex real-time PCR detection of fumonisin-producing and trichothecene-producing groups of *Fusarium* species. *Journal of Food Protection*, 67(3): 536-543.

Brandfass, C. & Karlovsky, P. (2006). Simultaneous detection of *F. culmorum* and *F. graminearum* in plant material by duplex PCR with melting curve analysis. *BMC Microbiology*, 6:4

Brunner, K., Kovalsky Paris, M.P., Paolino, G., Bürstmayr, H., Lemmens, M., Berthiller, F., Schuhmacher, R., Krska, R. & Mach R.L. (2009). A reference-gene-based quantitative PCR method as a tool to determine *Fusarium* resistance in wheat. *Analytical and Bioanalytical Chemistry*, 395(5), pp 1385 - 1394.

Brunner, K. & Mach R.L. (2010). Quantitative Detection of Fungi by Molecular Methods: A Case Study on *Fusarium*. In: *Molecular Identification of Fungi*. Gherbawy, Y., Springer-Verlag, Berlin Heidelberg, pp 93-105.

Buerstmayr, H., Steiner, B., Hartl, L., Griesser, M., Angerer, N., Lengauer, D., Miedaner, T., Schneider, B. & Lemmens, M. (2003). Molecular mapping of QTLs for *Fusarium* head blight resistance in spring wheat. II. Resistance to fungal penetration and spread. *Theoretical and Applied Genetics*, 107(3), pp 503-508.

Buerstmayr, H., Ban, T. & Anderson, J.A. (2009). QTL mapping and marker-assisted selection for *Fusarium* head blight resistance in wheat: a review. *Plant Breeding*, 128(1), pp 1-26.

Burgess, L.W. (1981). General ecology of the Fusaria. P. E. Nelson, T. A. Toussoun, & R. J. Cook, In: *Fusarium: diseases, biology, and taxonomy*. Pennsylvania State University Press, University Park, 1981: p. 225-235.

Chala, A., Weinert J., Wolf G.A. (2003). An Integrated Approach to the Evaluation of the Efficacy of Fungicides Against *Fusarium culmorum*, the Cause of Head Blight of Wheat. *Journal of Phytopathology*, 151, 673-678.

Chandler, E.A., Simpson, D.R., Thomsett,M.A. & Nicholson, P. (2003). Development of PCR assays to Tri7 and Tri13 trichothecene biosynthetic genes, and characterisation of chemotypes of *Fusarium graminearum*, *Fusarium culmorum* and *Fusarium cerealis*. *Physiological and Molecular Plant Pathology* , 62 (2003) 355–367.

Demeke, T., Clear, R.M., Patrick, S.K. & Gaba, D. (2005). Species-specific PCR-based assays for the detection of *Fusarium* species and comparison with the whole seed agar plate method and trichothecene analysis. *International Journal of Food Microbiology*, 103, pp 271-284.

Desjardins, A.E. (2006). *Fusarium* mycotoxins. Chemistry, Genetics, and Biology. APS Press, St. Paul, MN, 2006.

Doohan, F.M., Parry, D.W. & Nicholson, P. (1999). *Fusarium* ear blight of wheat: the use of quantitative PCR and visual disease assessment in studies of disease control. *Plant Pathology*, 48, pp 209-217

Draeger, R., Gosman, N., Steed, A., Chandler, E., Thomsett, M., Srinivasachary, J. Schondelmaier, H., Buerstmayr, H., Lemmens, M. & Schmolke, M.. (2007). Identification of QTLs for resistance to *Fusarium* head blight, DON accumulation and associated traits in the winter wheat variety Arina. *Theoretical and Applied Genetics*, 115(5), pp 617-625.

Edwards, S.G., Progozliev, S.R., Hare, M.C. & Jenkinson, P. (2001). Quantification of trichothecene producing *Fusarium* species in harvested grain by competitive PCR to determine efficiacies of fungicides against *Fusarium* head blight of winter wheat. *Applied and Environmental Microbiology*, 67(4): 1575-1580.

Gagkaeva, T.Y. & Yli-Mattila, T. (2004). Genetic diversity of *Fusarium* graminearum in Europe and Asia. *European Journal of Plant Pathology*, 110: 551-562.

Gan, Z., Marquardt, R.R., Abramson, D. & Clear R.M. (1997). The characterization of chicken antibodies raised against *Fusarium* spp. by enzyme-linked immunosorbent assay and immunoblotting. *International Journal of Food Microbiology*, 16, 38(2-3), pp 191-200.

Goswami, R.S. & Kistler, H.C. 2004). Heading for disaster: *Fusarium* graminearum on cereal crops. *Molecular Plant Pathology*, 2004. 5(6): p. 515-525.

Hill, N.S., Hiatt, E.E. & Chanh, T.C. (2006). ELISA analysis for *Fusarium* in barley. *Crop Science*, 46: p. 2636 - 2642.

Hill, N.S., Neate, S.M., Cooper, B., Horsley, R., Schwarz, P., Dahleen, L.S., Smith, K.P., O'Donnell, K. & Reeves, J. (2008). Comparison of ELISA for *Fusarium*, Visual Screening, and Deoxynivalenol Analysis of *Fusarium* Head Blight for Barley Field Nurseries. *Crop Science*, 48, pp 1389-1398.

Jennings, P., Coates, M.E., Turner, J.A., Chandler, E.A. & Nicholson P. (2004). Determination of deoxynivalenol and nivalenol chemotypes of *Fusarium* culmorum isolates from England and Wales by PCR assay. *Plant Pathology*, 53, pp 182–190.

Jurado, M., Vázquez, C., Sanchis, V. & González-Jaén, M.T. (2006). PCR-based strategy to detect contamination with mycotoxigenic *Fusarium* species in maize. *Systematic and Applied Microbiology*, 29: 681-689.

Khonga, E.B. & Sutton, J.C. (1988). Inoculum production and survival of Gibberella zeae in maize and wheat residues. Canadian Journal of Plant Pathology 10: p. 232-240.

Knutsen, A.K. & Holst-Jensen, A. (2004). Phylogenetic analyses of the *Fusarium* poae, *Fusarium* sporotrichioides and *Fusarium* langsethiae species complex passed on partial sequences of the translation elongation factor-1 alpha gene. *International Journal of Food Protection*, 95: 287-295.

Kolb, F.L., Baib, G-H., Muehlbauer, G.J., Andersonc, J.A., Smithc K.P. & Fedak G. (2001). Host Plant Resistance Genes for *Fusarium* Head Blight: Mapping and Manipulation with Molecular Markers. *Crop Science*, 41(3), pp 611-619.

Konstantinova, P. & Yli-Mattila, T. (2004). IGS-RFLPanalysis and development of molecular markers for identification of *Fusarium* poae, *Fusarium* langsethiae, *Fusarium* sporotrichioides and *Fusarium* kyushuense. *International Journal of Food Microbiology*, 95: 321-331.

Kulik, T. (2008). Detection of *Fusarium* tricinctum from cereal grain using PCR assay. *Journal of Applied Genetics*, 49(3): 305–311.

Kulik, T., Jestoi, M. & Okorski, A. (2011). Development of TaqMan assays fort he quantitative detection of *Fusarium* avenaceum/*Fusarium* tritinctum and *Fusarium* poae esyn1 genotypes from cereal grain. *FEMS Microbiology Letters*, 314, pp49-56.

Laday, M., Mule, G., Moretti, A., Hamari, Z., Juhasz, A., Szecsi, A., Logrieco, A., (2004). Mitochondrial DNA variability in *Fusarium* proliferatum (Gibberella intermedia). *European Journal of Plant Pathology*, 110, pp 563–571

Leonard, K.J. & Bushnell, W. (2003). *Fusarium* head blight of wheat and barley. St.Paul, Minn, USA: APS Press, 2003.

Li, S., Hartman, G. L., Domier L.L. & Boykin, D. (2008). Quantification of *Fusarium solani* f. sp. glycines isolates in soybean roots by colony-forming unit assays and real-time quantitative PCR. *Theoretical and Applied Genetics*, 117: 343-352.

Lia, H., Wua, A., Zhaoa, C., Scholten, O., Löffler H., & Yu-Cai Liaoa (2005).Development of a generic PCR detection of deoxynivalenol- and nivalenol-chemotypes of *Fusarium* graminearum. *FEMS Microbiology Letters*, 243(2), pp 505-511

Mach, R.L., Kullnig-Gradinger, C.M., Farnleitner, A.H., Reischer, G. Adler, A. & Kubicek, C.P. (2004). Specific detection of *Fusarium* langsethiae and related species by DGGE and ARMS-PCR of a β-tubulin (*tub1*) gene fragment. *International Journal of Food Protection*, 95, pp 333-339.

Markell, S.G. & Francl, L.J. (2003). *Fusarium* head blight inoculum: species prevalence and *Gibberella zeae* spore type. *Plant Disease*, 87, pp 814-820.

McCartney, H.A., Foster, S.J., Fraaije, B.A. & Ward E. (2003). Molecular diagnostics for fungal plant pathogens. *Pest Management Sciences*, 59, pp129-142

McMullen, M., Jones, R., & Gallenberg, D. (1997). Scab of Wheat and Barley: A Re-emerging Disease of Devastating Impact. *Plant Disease* 81(12): 1340-1348.

Miedaner, T., Beyer, W., Höxter, H. & Geiger, H.H. (1994). Growth Stage Specific Resistance of Winter Rye to *Microdochium nivale* and *Fusarium* spp. In the Field Assessed by Immunological Methods. *Phytopathology*, 85, pp 416-421.

Miedaner, T., Heinrich, N., Schneider, B., Oettler, G., Rohde, S. & Rabenstein, F. (2004). Estimation of deoxynivalenol (DON) grain content by symptom rating and exoantigen content for resistance selection in wheat and triticale. *Euphytica*, 139(2), pp 123-132

Mirete, S., Vázquez, C., Mulè, G., Jurado M. & González-Jaén M.T. (2004). Differentiation of *Fusarium verticillioides* from Banana Fruits by IGS and EF-1α Sequence Analyses. *European Journal of Plant Pathology*, 110(5-6), pp 515-523.

Mulè, G., Susca, A. Stea, G. & Moretti, A. (2004). A species specific PCR assay based on the calmodulin partial gene for identification of *Fusarium verticillioides*, *F. roliferatum* and *F. subglutinans*. *European Journal of Plant Pathology*, 110: 495-502.

Mulfinger, S., Niessen, L. & Vogel, R. (2000). PCR based quality control of toxigenic *Fusarium* spp.in brewing malt using ultrasonication for rapid sample preparation. *Advances in Food Science*, 22(1/2): p. 38 - 46.

Nganje, W.A., Bangsund, D.A., Leistritz, F.L., Wilson,W.W., & Tiapo, N.M. (2004). Regional Economic Impacts of *Fusarium* Head Blight in Wheat and Barley. *Review of Agricultural Economics*, 26(3), pp 332-337.

Nicholson, P. Simpson, D.R., Wilson, A.H., Chandler, E. & Thomsett A. (2004). Detection and differentiation of trichothecene and enniatin-producing *Fusarium* species on small-grain cereals. *European Journal of Plant Pathology*. 110: 503-514.

Nicholson, P., Simpson, D.R., Weston, G., Rezanoor, H.N., Lees, A.K., Parry, D.W., & Joyce, D. (1998). *Physiological and Molecular Plant Pathology*, 53: 17-37

Osborne, L.E. and Stein, J.M. (2007). Epidemiology of *Fusarium* head blight on small-grain cereals. *International Journal of Food Microbiology*, 119(1-2): p. 103-8.

Pasanen A.L., Yli-Pietilä, K., Pasanen, P., Kalliokoski, P. & Tarhanen J. (1999). Ergosterol Content in Various Fungal Species and Biocontaminated Building Materials. *Applied and Environmental Microbiology*, 65(1), pp 138-142.

Pereyra, S.A., Dill-Macky, R. & A.L. (2004). Sims, Survival and inoculum production of *Gibberella zeae* in wheat residue. *Plant Disease*, 88, pp 724-730.

Petcu, G. & Ioniþã, S. (1998). Influence of crop rotation on weed infestion and *Fusarium* spp. attack, yield and quality of winter wheat. Romanian Agricultural Research, 9-10: 83-91.

Quarta, A., Mita, G., Haidukowski, M., Logrieco, A., Mulè, G., & Visconti, A. (2006). Multiplex PCR assay for the identification of nivalenol, 3- and 15-acetyl-deoxynivalenol chemotypes in *Fusarium*. FEMS Microbiology Letters, 259(1), pp 7-13

Reischer G.H., Lemmens, M., Farnleitner, A.H., Adler, A. & Mach, R.L. (2004). Quantification of *Fusarium graminearum* in infected wheat by species specific real-time PCR applying a TaqMan probe. *Journal of Microbiology Methods*, 59, pp 141-146.

Rhode, S. & Rabenstein, F. (2005). Standardization of an indirect PTA-ELISA for detection of *Fusarium* spp. In infected grains. *Mycotoxin Research,* 21(2), pp 100-104.

Rudd, J.C., Horsley, R.D., McKendry, A.L. & Elias, E.M.. (2001). Host Plant Resistance Genes for *Fusarium* Head Blight: Sources, Mechanisms, and Utility in Conventional Breeding Systems. *Crop Science*, 41(3), pp 620-627.

Schilling, A.G., Möller, E.M. & Geiger H.H. (1996). Polymerase Chain Reaction-Based Assay for Species-Specific Detection of *F. culmorum, F. graminearum* , and *F. avenaceum*. *Molecular Plant Pathololgy*, 86(5):515-522

Schnerr, H., Niessen, L. & Vogel, R.F. (2001). Real-time detection of the tri5 gene in *Fusarium* by LightCycler-PCR using SYBR Green I for continous fluorescence monitoring. Int. *Journal of Food Microbiology*, 71, pp 53-61.

Šliková, S., Šudyová, V., Martinek, P., Polišenská, I., Gregová, E. & Mihálik D. (2009). Assessment of infection in wheat by *Fusarium* protein equivalent levels. *European Journal of Plant Patholohy*, 124, pp 163-170.

Schroeder, H.W. & Christensen, J.J. (1963). Factors affecting resistance of wheat to scab caused by *Gibberella zeae*. *Phyto*pathology 53: pp 831-839.

Steinkellner, S. & Langer, I. (2004). Impact of tillage on the incidence of *Fusarium* spp. in soil. *Plant and Soil,* 267(1-2): pp 13-22.

Stepień, L., Popiel, D., Koczyk, G. & Chełkowski, J. (2008). Wheat-infecting *Fusarium* species in Poland--their chemotypes and frequencies revealed by PCR assay. *Journal of Applied Genetics*, 49(4), pp 433-441

Waalwijk, C., Koch, S.H., Ncube, E., Allwood, J., Flett, B., de Vries, I. & Kema, G.H.J. (2008). Quantitative detection of *Fusarium* spp. and its correlation with fumonisin content in maize from South African subsistence farmers. *World Mycotoxin Journal*, 1(1), pp 39-47.

Waalwijk, C., van der Heide, R., de Vries, I., van der Lee, T., Schoen, C., Corainville, G.C., Häusler-Hahn, I., Kastelein, P., Köhl, J., Lonnet, P., Demarquet, T. & Kema, G.H.J. (2004). Quantitative detection of *Fusarium* in wheat using TaqMan. *European Journal of Plant Pathology*. 110: 481-494.

Wang, Y.Z. & Miller, J.D. (1988). Effects of *Fusarium* graminearum metabolites on wheat tissue in relation to *Fusarium* head blight resistance. *Journal of Phytopathology*, 122, pp 118-125.

Weiland, J.J. & Sundsbak, J.L. (2000). Differentiation and detection of sugar beet fungal pathogens using PCR amplification of actin coding sequences and the ITS region oft he rRNA gene. *Plant Disease*, 84(4), pp 475-482.

Windels, C.E. (2000). Economic and Social Impacts of *Fusarium* Head Blight: Changing Farms and Rural Communities in the Northern Great Plains. *Phytopathology*, 90(1), pp 17-21.

Yli-Mattila, T., Mach, R.L., Alekhina, I.A., Bulat, S.A., Koskinen, S., Kullnig-Gradinger, C.M., Kubicek, C.P. & Klemsdal, S.S. (2004). Phylogenetic relationship of *Fusarium langsethiae* to *Fusarium* poae and *Fusarium* sporotrichioides as inferred by IGS, ITS, β-tubulin sequences and UP-PCR analysis. *International Journal of Food Microbiology*, 95, pp 267-285.

Yli-Mattila, T., Paavanen-Huhtala, S., Jestoi, M., Parikka, P., Hietanieme, V., Gagkaeva, T., Sarlin, T., Haikara, A., Laaksonen, S. & Rizzo, A. (2008). Real-time PCR detection and quantification of *Fusarium poae, F. graminearum, F. sporotrichioides* and *F. langsethiae* in cereal grains in Finland and Russia. Archives of Phytopathology and Plant Protection, 41(4), pp 243-260.

Youssef, S.A., Maymon, M., Zveibil, A., Klein-Gueta, D., Sztejnberg, A., Shalaby, A.A. & Freeman S. (2007). Epidemiological aspects of mango malformation disease caused by *Fusarium mangiferae* and source of infection in seedlings cultivated in orchards in Egypt. *Plant Pathology*, 56, pp 257–263.

9

Gray Mold of Castor: A Review

Dartanhã José Soares

Empresa Brasileira de Pesquisa Agropecuária,
Embrapa Algodão, Campina Grande
Brazil

1. Introduction

Castor plant (*Ricinus communis* L.) is a non-edible oilseed crop with unique oil features for the chemistry industry. The crop was very important in the mid and late nineteenth century and also during WWI. After that the crop lost its importance in developed countries (Godfrey, 1923), but in India and Brazil it has remained as the most important non-edible oilseed crop of the arid and semi-arid regions (Dange et al., 2005; Santos et al., 2007). Nowadays, due the constant pressure for renewable fuels, castor has been investigated as a potential source of biofuel, mainly in Brazil due to governmental stimulus, and this has raised the crop importance once again. Regardless of the lack of a well established crop system, castor hosts several pests and diseases which cause heavy losses in the crop yield. One of the most destructive diseases of castor is gray mold, caused by the fungus *Botryotinia ricini* (Godfrey) Whetzel. Actually, it is the anamorphic phase of *B. ricini*, known as *Amphobotrys ricini* (N.F. Buchw.) Hennebert, that is responsible for disease epidemics and heavy yield losses frequently observed in castor crops. The first epidemic outbreak caused by this fungus was reported by H.E. Stevens of the Florida Experiment Station, Gainesville, Florida (Godfrey, 1919, 1923). At that time, a meticulous study was conducted and much of our knowledge regarding the disease and its causal agent was published in the classic work of Godfrey (1923). Subsequently, only sporadic works were conducted by other scientists around the world, consequently few advances have been made on management of gray mold. Breeding programs have failed in developing varieties with satisfactory resistance levels (Kolte, 1995), and chemical control is still ineffective and economically prohibitive, mainly due to the lack of basal information about the causal organism and its biology. In this chapter, the major aspects of castor gray mold will be reviewed.

2. Gray mold of castor

2.1 Historic and economic importance

Castor gray mold was first reported in the USA in 1918, following pioneering investigations by H.E. Stevens and F. W. Patterson, who promptly suggested that the causal organism of castor gray mold was an unknown *Botrytis* species (Godfrey, 1919, 1923). This fungus had caused serious losses of castor crop in the summer of 1918 mainly in Florida and others southern States, where it was responsible for losses up to 100% of castor yield (Godfrey,

1923). Later, the disease was reported in almost all countries where castor has been cultivated (Kolte, 1995), having nowadays a worldwide distribution.

The first occurrence of this disease in USA was directly linked to seeds imported from Bombay (now Mumbai), India, even though until that time, such disease had not been described in that country (Godfrey, 1923). In his work, Godfrey (1923) did a detailed account of the destructive potential of the gray mold of castor under favourable condition. By attacking mainly reproductive organs of the castor plant, gray mold disease is implicated in direct losses of yield whatever the level of infection.

In India, today the major castor producer, gray mold is found in few states and is regarded as troublesome only in Andhra Pradesh and Tamil Nadu, in the South, where the weather conditions are more favourable for disease development where in 1987, an epidemic outbreak of gray mold occurred (Dange et al., 2005).

In Brazil, the disease was first reported in the São Paulo state in 1932. However, it was only in 1936 that any attention was given to the disease due to the serious losses which occurred that year (Gonçalves, 1936). Currently, gray mold is present in almost all Brazilian states and its importance has grown at the same time that the crop cultivation has been intensified, mainly in those regions where the weather conditions are favourable for disease development, including the Southern and South-eastern Brazilian states (Araújo et al., 2007; Freire et al., 2007). In the region of the "Brejo Paraibano", where the recommended sowing period is between mid-April to early-May (Amorim Neto et al., 2000), the flowering period (mid-June to early-August) usually coincides with highly favourable conditions for disease development (Moraes et al., 2009). Yield losses of up to 100% are quite frequent when highly susceptible cultivars are planted. Conversely, in Bahia the major castor producer in Brazil gray mold is not a problem because the weather conditions are usually not favourable for disease development.

2.2 Etiology, taxonomy and population structure

The causal agent of gray mold of castor was originally described by Godfrey (1919) as *Sclerotinia ricini* Godfrey, based on the holomorph. Later, Whetzel (1945) transferred the species *S. ricini* to the genus *Botryotinia*, which since then has been known as *Botryotinia ricini* (Godfrey) Whetzel. Subsequently, the anamorphic state of *Botryotinia ricini* was named as *Botrytis ricini* N.F. Buchw. (Buchwald, 1949). This led to general confusion between the non-mycologist communities, which adopted the name *Botrytis ricini* N.F. Buchw., instead of *Botryotinia ricini* (Godfrey) Whetzel. In 1973, Hennebert erected the genus *Amphobotrys* to accommodate the anamorphic state of *B. ricini*, based mainly on the distinctive pattern of conidiophore ramification, and since then the anamorphic state became known as *Amphobotrys ricini* (N.F. Buchw.) Hennebert (Hennebert, 1973). Even so, several authors used, and still use, the erroneous name "*Botrytis ricini*" attributing its authority to Godfrey (Barreto & Evans, 1998; Batista et al., 1998; Dange et al., 2005; Lima & Soares, 1990).

Although the correct name to be applied to the causal agent of gray mold of castor is *Botryotinia ricini*, only the anamorphic state is observed in the field by most authors, and thus the name of the anamorph, at the expense of the name of the holomorph, is preferred (Holcomb et al., 1989; Lima et al., 2008).

Botryotinia ricini belongs to Sclerotiniaceae (Helotiales, Ascomycota) and is characterized by its dark, plane-convex, elongated sclerotia (Fig.1), which give rise to cinnamon brown to chestnut brown, long stipitate apothecia, with cylindrical to cylindro-clavate asci, apex slightly thickened, 8-spored; ascopores ellipsoidal, often sub-fusoid, one-celled, bi-guttulate and hyaline; paraphyses hyaline, filiform, septate (Godfrey, 1919). Its anamorphic phase is characterized by cylindrical, straight, dichotomously branched, pale brown conidiophores, with conidiogeneous cell not inflated, thin-walled; conidia globose, maturing synchronically, on short denticles, smooth, one-celled, sub-hyaline to pale brown (Fig.1) (Godfrey, 1919; Hennebert, 1973; Lima et al., 2008). A synanamorph (*Myrioconium* sp. – spermatial state) may sometimes be present on culture media (Godfrey, 1923; Hennebert, 1973; Seifert et al., 2011). According to Kirk et al. (2008), the genus *Amphobotrys* remains monotypic.

Botryotinia ricini is regarded as a homothallic species (Beveer & Weeds, 2007), so sexual reproduction will readily take place. If sexual reproduction had taken place, a high degree of diversity would be expected within the *B. ricini* population; however, Bezerra (2007) shows evidence that, in the state of Paraíba (Northeast Brazil) populations of *B. ricini* are clonal, which means that sexual reproduction had not taken place within those populations. Nonetheless, this conclusion must be viewed with caution because the population sampled was relatively small. Unfortunately, there is no other work on this subject and, therefore, the population structure of *B. ricini* remains unknown.

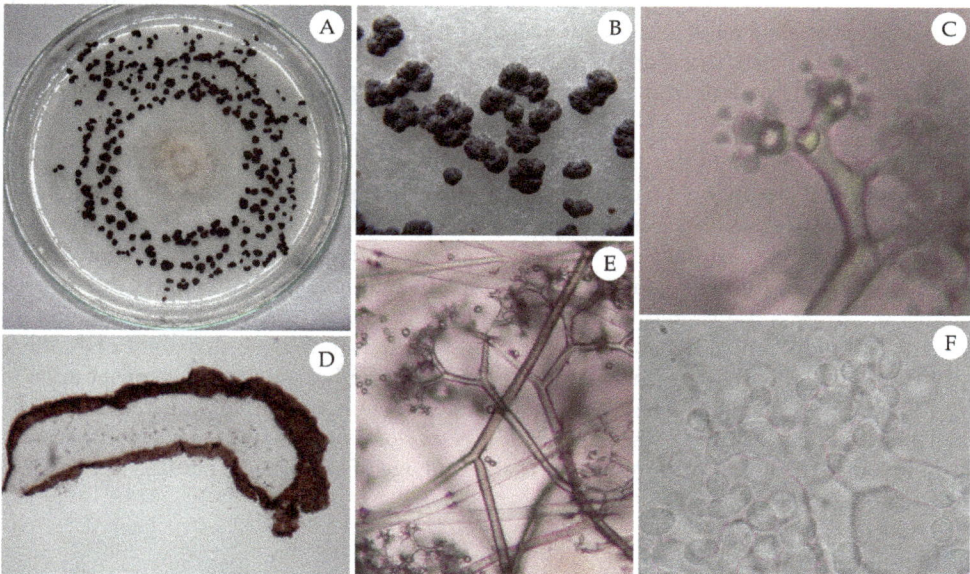

Fig. 1. Dark sclerotia on culture medium (A); close-up view of the sclerotia (B); transversal section through a sclerotium to show its plane-convex form (D); dichotomous branch of the conidiophores (E); conidiogenous cells showing the synchronic conidiogenesis (C); and close-up the conidiogenous cell to show the denticles and globose conidia (F). Photos: D.J. Soares.

2.3 Host penetration and colonization

In his classical work, G.H. Godfrey also investigated the infection process of *B. ricini* on leaves of the castor plant and concluded that penetration occurs directly through the host cuticle, in a process similar to *Botrytis cinerea* (Godfrey, 1923). After penetrating the cuticle, the fungus quickly spreads over the host tissues leading to a complete disorganization and breakdown. Although Godfrey had made mention as to the possible role of an enzymatic action in the penetration process, his conclusion pointed out to a mechanical penetration of the germ-tube, without tissues dissolving, prior to infection. On the other hand, Thomas & Orellana (1963b) found that it was not possible to verify the direct germ-tube penetration of *B. ricini*, through the cuticle or stomata, on castor capsules, before tissue maceration by pectic enzymes action, suggesting that the fungus first degraded the cuticle and later penetrated the host tissues (Orellana & Thomas, 1962). Probably *B. ricini* uses both mechanical and chemical processes to penetrate undamaged host tissue, however, no further studies have been done to clarify these questions.

Although the infection process of *B. ricini* needs to be better understood, it is likely that enzymes, such as lipases and cutinases, play an important role in the infection process similar to several other *Botrytis*-host interactions (Kars & van Kan, 2007). Additionally, *Botrytis* species can also affect the 'redox' process in the host plants, during the colonization of host tissues, through the production of enzymes like superoxide dismutase (Lyon et al., 2007). Due the great biological similarity of *Botrytis* spp. and *B. ricini*, probably, such enzymes also have an important role in the infection process of the causal agent of castor gray mold, as already evidenced by Orellana & Thomas (1962). Hoffmann et al. (2004) had, for example, extracted alpha and beta esterase and superoxide dismutase from *B. ricini*, however they did not perform a study to determine the role of such enzymes in the infection process of this fungus.

It is important to highlight the great distinction between the penetration process of a fungus under controlled and highly favourable condition in contrast with the natural process in the field. In the latter case, all aerial parts of the host are potential targets for deposition and penetration of *B. ricini*, because not only the conidia, regarded as the major propagative unit, usually responsible for the epidemic outbreak, but also the ascospores, sclerotia and mycelia fragments can give rise to infection, as observed in *Botrytis* spp. (Jarvis, 1978). So, besides direct penetration, probably natural openings and wounds also serve as a point of entrance for the fungus. Growth of the fungus on the host surface and consequently its penetration in the host tissues will depend on factors such as inoculum type, free water and nutrient availability, cuticle features, presence of exudates on floral organs and other glands, besides the abundance of natural openings and the size and age of wounds, as pointed out by Holz et al., (2007) for *Botrytis* species.

2.4 Symptoms

The primary targets of the fungus are the inflorescence and the capsules, in any development stage (Fig.2) (Araújo et al., 2007; Dange et al., 2005; Gonçalves, 1936; Lima et al., 2001). Some authors (Drumond & Coelho, 1981; Batista et al., 1996) claim that the male flowers are the first to be infected, but it is not always the case because any part of the inflorescence can be infected, the female flowers being the preferential target. That claim came from the fact that the male flowers are the first to be exposed, at the earlier stage of inflorescence formation; consequently such flowers are exposed longer to the infection units of the fungus. However, as soon as the male flowers suffer anthesis they are no longer a target and are hardly infected, contradicting the statement of Drumond & Coelho (1981) that "the fungus attacks first the male flowers because the anthers, being soaked with the rain water or dew, easily retain the fungus spores carried by the wind".

Fig. 2. Symptoms of gray mold attack on castor inflorescence and raceme. A-B. Symptoms on young inflorescences, before fertilization of female flowers. C-H. Symptoms on capsules at distinct development stages. Photos: D.J. Soares.

The first symptoms are visible as bluish spots on the inflorescences, on both female and male (before anthesis) flowers, and on developing fruits. On fruits, the symptoms can evolve to circular or elliptic, sunken, dark coloured spots that can result in rupture of the capsule (Fig.3 A-B) (Araújo et al., 2007). These symptoms are usually more frequent when a period of low relative humidity unfavourable to fungal sporulation occurs soon after the fungus penetrates the host tissues.

Depending on weather conditions (e.g. long periods with high relative humidity soon after the fungus penetrates the host), the occurrence of yellow ooze at the point of infection is frequent (Fig. 3 C-D) (Batista et al., 1996; Dange et al., 2005) as a result of the rapid enzymatic tissue degradation. The symptoms on the male flowers, before anthesis, are small, pale brown, necrotic spots, which can evolve to larger brown spots with a darker edge (Fig.3 E). The infected flowers and young capsules became softened due the fungal colonization and mycelial growth is, at first, pale gray and later dark olivaceous. A profuse sporulation is usually observed in such stage (Fig.3 F). When the infection starts on immature capsules, they become rotten; if the infection starts later, with fully developed capsules, the seeds usually became hollow, with coat discoloration and weight loss (Dange et al., 2005). On the inflorescence, the male flowers can be infected first, but the fungus has a clear preference for the female flowers (Fig.3 G). Infection can lead to complete destruction of the raceme (Fig.3 H), particularly if it reaches the main stem and the weather conditions are favourable for the disease. Several other plant parts, e.g. leaves, petioles and stem can also be infected, mainly due to the deposition or fall of infected material from the inflorescence or racemes. On leaves, the lesions are usually irregular, but can assume an elliptic or circular pattern, the size is very variable, sometimes coalescing and resulting in a foliar blight (Fig.3 I-J). On petioles and stems, necrotic, sunken lesions usually are formed which can cause the strangulation and consequently death of the parts above the infection point (Fig. K-N) (Batista et al., 1996; Dange et al., 2005).

Fig. 3. Symptoms of gray mold on castor plants. Dark-bluish spot (A) and capsule rupture (B) under unfavourable conditions; Gray-bluish spot with yellowish-brown ooze (C-D) under favourable conditions; Dark brown spot on male flowers before anthesis (E); Profuse sporulation on infected capsule (F); Infected inflorescence showing the fungus preference for female flowers (G); Completely destroyed inflorescence (H); Leaf spot (I) and blight (J); Apical infection (K); Petiole infection (L); Stem (M) and rachis infection (N). Photos: D.J. Soares.

2.5 Epidemiology

There are few studies of the epidemiological aspects of gray mold on castor. Godfrey (1923) mentioned that temperatures around 25°C and high relative humidity are highly favourable to disease development. Such statements have been exhaustively repeated in almost all publications about the subject over the last decades (Araújo et al., 2007; Batista et al., 1998; Gonçalves, 1936; Kimati, 1980; Lima & Soares, 1990; Massola JR & Bedendo, 1997; Melhorança & Staut, 2005;). The minimum and maximum temperature for mycelial growth was established by Godfrey as 12 and 35°C, respectively (Godfrey, 1923).

Some complementary studies have confirmed that temperatures around 25°C are favourable to fungal growth and disease development (Araújo et al., 2003; Suassuna et al., 2003; Sussel, 2008). At temperatures below 20°C, the disease is little expressed and highly dependent on long periods of high relative humidity (Sussel, 2008). According to Sussel et al. (2011), there is a high correlation between the temperature and duration of leaf wetness with the disease incidence and severity. The same authors also concluded that the disease was more intense with a temperature of 28°C and 72 hours of leaf wetness, and that for temperatures below 15° the fungus needs more than 6 hours of leaf wetness, otherwise the disease does not occur (Sussel et al., 2011). Even when under optimal temperatures (near 25°C), the fungus appears to be highly dependent on periods of high humidity, since in a study developed by Esuruoso (1966) with 39 castor varieties in two distinct places, Ibadan and Ilora (West Nigeria), that share similar temperatures (24.4 to 27.9 and 24.2 to 27.2°C, respectively), but that rainfall was twofold higher in Ibadan than Ilora, the disease level in Ibadan was 1.5 fold higher than in Ilora. This dependence on high rainfall, and not only on high relative humidity, was also observed in a study conducted in the Paraiba state (Northeast Brazil) where a high correlation between the disease progress and the accumulated rainfall between the seventh and fifth day before the evaluation was observed (D.J. Soares, unpublished data). The correlation, however, was inversely proportional to the rainfall intensity which means that long periods of low-intensity rains are more favourable than short periods of high-intensity rains.

Sussel (2008) concluded that the disease shows a random distribution pattern, matching its airborne nature. However, under high rainfall, the disease assumes an aggregate pattern, typical of those dispersed by water splash. With reference to aerobiology, Sussel (2008) obtained a positive correlation between the average number of conidia collected daily in air and the weather variables: minimum temperature, mean relative humidity, mean precipitation and leaf wetness. There was an increase in the number of conidia collected with the elevation of the minimum temperature from 14.6 to 18.1°C; the same occurred when the mean relative humidity was raised from 42 to 95% and when the daily precipitation increased from 1 to 20 hours (Sussel, 2008, 2009a).

Under controlled conditions, at 25°C and relative humidity near saturation, Soares et al. (2010) had determined that the incubation period of B. ricini can vary from 44 to 88 hours (average 72 h) and the latent period from 72 to 144 hours (average 96 h), depending on the genotype.

Although in the recent years, there have been some advances in knowledge about the epidemiology of gray mold of castor, several issues remains unresolved. Further studies are needed to understand the role of each potential dispersal unit; how the fungus survives between the growth seasons, for example, as sclerotia, on secondary hosts, on spontaneous (volunteer) castor plants. Additionally, it is also crucial to determine whether it is possible to predict disease development based on environmental variables, like precipitation or surface wetness.

2.5.1 Live cycle and host range

The disease starts with spore deposition on the host surface, followed by penetration and colonization of the host tissues. Soon after colonization, the fungus, under favourable conditions, sporulates profusely on the dead tissues, and then the conidia became the main inoculum source for new infection sites. Although most authors recognized the major role of conidia after the disease had been established in the field, there is much speculation about the primary inoculum source of *B. ricini*.

Godfrey (1923) claims that the fungus survives on soil or crop residues as sclerotia and, under the right conditions, these can produce sexual structures, which will be responsible for the initial infection. However, there is no report of sexual reproduction under natural conditions, other than the original reports of Godfrey (1919, 1923), so the purported role of ascospores as the initial inoculum source remains unclear, despite the fact that apothecia are easily overlooked in the field.

Under tropical climate, the initial inoculum source are probably the conidia from wild castor plants which grows spontaneously near the crop areas all year (Gonçalves, 1936). Wild castor plants can produce flowers throughout the year and consequently new susceptible tissue will be available for the fungus to self perpetuate in its anamorphic state through the year. By infecting the first inflorescence under favourable conditions, the fungus produces abundant sporulation, thus allowing multiple rounds of re-infection, since this pathogen is easily spread by wind, rain splash and, probably, by insects (Fig. 4 A-C) (Dange et al., 2005).

It was also mentioned by Godfrey (1923), that the fungus is seed-borne, the seeds being regarded as the primary inoculum source (Fig. 4 D-E). However, the role of seeds as a primary inoculum source requires further study. Probably the seeds have no essential role at the beginning of the epidemic because there is usually an interval of almost two months between sowing and flowering, so the inoculum originating from the seeds will not be available to infect the flowers. This means that, although *B. ricini* is a seed-borne fungus, it is probably not seed transmitted, because it could hardly infect the crops which grow from them, as conceived by Maude (1996). Thus, the most important role of the seeds is to carry the pathogen to new areas, rather than to act as a primary inoculum source for epidemic on the crop season.

Initially, it was suspected that *B. ricini* was host-specific and that it has a very narrow host range (Godfrey, 1923). Through artificial inoculation, Godfrey (1923) showed that the fungus is extremely dependent on high humidity and it was not able to cause disease, at same levels observed on castor plant, when inoculated on several other hosts, including members of Euphorbiaceae. In field inspections, the same author did not detect any other plant species showing symptoms of natural infection in the surroundings of castor-growing areas which were severely affected by the disease. Nevertheless, since the 1980s several reports of natural infection of *B. ricini* on members of Euphorbiaceae have been made, including both weeds and ornamentals: such as *Caperonia palustris* (Whitney & Taber,1986), *Euphorbia supina*

(Holcomb et al., 1989; Russo & Rossman, 1991), *Euphorbia milli* (Sanoamuang, 1990), *Euphorbia pulcherrima* (Holcomb & Brown, 1990), *Euphorbia heterophylla* (Barreto & Evans, 1998), *Euphorbia inarticulata* (Alwadie & Baka, 2003), *Acalypha hispida* and *Jatropha podagrica* (Lima et al., 2008). Besides these reports of natural infection, artificial inoculations tests have shown that this pathogen has a wide host range within the Euphorbiaceae, including species of economic interest, e.g. cassava (*Manihot utilissima*) (Holcomb et al., 1989; Kumar et al., 2007; Lima et al., 2008). The sole report of natural infection by *B. ricini* outside the Euphorbiaceae family was made by Hansen & Bega (1955) on *Caladium bicolor* (Araceae) and, it is quite likely a case of fungus misidentification.

Fig. 4. Visitor insects (flies and stingless bees) on castor inflorescence infected by *B. ricini* (A-C). Seeds with profuse gray mold growth and sporulation (D-E). Photos: D.J. Soares.

Hence, it is possible that the fungus can survive in the field on several other plant species belonging to Euphorbiaceae, and these can be an inoculum reservoir of the fungus in such a way that it would be ready to infect its preferential host, the castor plant, as soon as susceptible tissues become available.

2.6 Host resistance

The search for resistance to gray mold has been investigated ever since the disease was described. In his classic work, Godfrey (1923) pointed out general conclusions that clearly have been overlooked over the decades by plant breeders and plant pathologists alike. Among these, three are worth mentioning: "(1) Plants of more ornamental type, with stalk, foliage, and sometimes pods in different shades of red or reddish green were more resistant; (2) All smaller, many branched plants, which by their yield indicated commercial possibilities, showed high susceptibility to the disease; and (3) Cross pollination in castor-bean fields probably occurs very extensively. It would require years of work to develop pure strains and then to select and breed for desirable qualities combined with resistance before permanent results could be secured" (Godfrey, 1923). Based on this information, it is clear that only minor progress has been achieved toward the development of resistant cultivars during the last century. As a result, breeding programs have failed to develop a resistant cultivar or hybrid until today.

Gonçalves (1936) reported that "spontaneous varieties" are highly resistant to the disease, since the fungus attacks only few capsules, while most of the capsules of the same raceme or from others raceme remain healthy. Although this behaviour is frequently observed in wild types, and even in some commercial cultivars, this statement must be viewed with caution since it has no scientific basis and probably such behaviour is a result of the wide genetic variation within castor plants.

According to several authors, varieties with more compact racemes, shorter internodes, and male flowers distributed all long the inflorescence are considered to be more susceptible to the pathogen (Batista et al., 1998; Costa et al., 2004; Dange et al., 2005; Milani et al., 2005; Thomas & Orellana, 1963b; Ueno et al., 2006; Zarzycka, 1958): a fact already noted by Godfrey (1923), and also by Esuruoso (1966), who concluded that the disease severity was more intense on short-stalked varieties than on those with longer stalks. Another relevant aspect is that, apparently, the presence of spines in the capsule predisposes them to pathogen attack (Alcântara et al., 2008; Cook, 1981; Lima & Soares, 1990).

Plants with capsules containing high soluble sugar concentrations are more susceptible to fungal development than plants with low soluble sugar concentrations (Orellana & Thomas, 1962). According to these authors, capsule resistance is intimately associated with its capacity of inactivation of pectic, cellulolytic and others hydrolytic enzymes through the products of the phenol oxidation (Thomas & Orellana, 1963a), as well as lower content of water-soluble pectin, higher content of calcium and magnesium and lower sodium and potassium contents (Thomas & Orellana, 1964).

Although several studies have been conducted to assess the resistance of castor bean genotypes, none of them deals with the inheritance of resistance to *B. ricini*, nor how it is governed. Probably, the resistance to gray mold is quantitative and possibly governed by several minor genes.

Quantitative resistance, also called horizontal resistance, is, according to Robinson (1976), universal and occurs in all plants against all parasites; it is also permanent and permits cumulative plant breeding. Horizontal resistance can be either passive or active, and usually is conferred by numerous and complex mechanisms and can include traits such as tolerance and disease-escape which are not strictly resistance mechanisms *per se* (Robinson, 1976). It is also important to highlight the fact that horizontal resistance usually is an efficient way to achieve disease control, and that in the recent years selection for horizontal resistance has become easier with the use of marker-assisted breeding (Keane, 2012).

In Brazil, several studies have been carried out aiming to select for a resistance source to gray mold. However, what has become clear is that whilst there are differences in susceptibility among the assayed genotypes, none has the desired resistance level (Batista et al., 1998; Costa et al., 2004; Lima & Soares, 1990; Milani et al., 2005; Rego Filho et al., 2007). Among the genotypes assessed in different countries, thus far, several distinct levels of susceptibility have been observed, but not immunity (Anjani et al. 2004; Esuruoso, 1996; Zarzycka, 1958). The way that the breeding programs are being conducted, using only information generated by field investigations, and which are usually affected by uncontrollable factors, is much more likely to select for "field resistance" rather than for genetic resistance, i.e., the plant and raceme architecture, together with the weather conditions, play a bigger role in the disease development, by inducing a micro-climate formation, which might be more or less favourable to the pathogen. This means that unless we redirect our thinking to incorporate horizontal resistance, through a careful marker-assisted breeding program, whether it be genetic or morphological, it will probably never be possible to obtain a variety with satisfactory levels of resistance.

Some castor plant varieties have a thick outer wax layer which can act as a constitutive barrier to pathogen infection. However, there is no information about the role of this wax layer in the infection process of *B. ricini* and, although the adhesion of the conidia and subsequently their penetration into the host tissue could be difficult, the wax could also act as an elicitor; being responsible for the initial recognition of the host-pathogen interaction. Thus, the wax layer could actually favour the pathogen rather than inhibit its development. Field observations lead us to hypothesize that the latter, is the most probable scenario in the present pathosystem.

2.6.1 Host resistance assessment

As commented previously, most of the research involving the assessment of genotypes resistance has been conducted under field conditions and without any standard to quantify the disease and, worst still, sometimes using very subjective assessment methods, like that adopted by Lima & Soares (1990). This has been reflected by the fact that it is almost impossible perform a comparison among the different already studies undertaken. However, a diagrammatic scale to assess gray mold severity in castor- was published recently (Fig.5) (Sussel et al., 2009; Sussel, 2009b). This was developed in order to standardize disease assessment in field experiments, and was constructed based on the Weber-Fechner law and divided into 10 levels. Although useful, such a scale faces two crucial factors which might constitute an impediment to its wider adoption: first, there is great variation in raceme architecture among the castor genotypes, and; secondly, it only considers the fully developed raceme. In the first case, as the diagrammatic scale was drawn based on long conical racemes, its use to assess disease severity in genotypes with more or less globose raceme will be difficult. In the second case, the scale does not take into account disease severity in inflorescences or even in immature racemes, where the disease is usually more severe and more difficult to estimate. Therefore, it is possible that the scale will underestimate disease severity. Nonetheless, the simple fact that such a diagrammatic scale is now available is a huge advance towards more reliable disease assessment and the possibility of comparing assays conducted by different researchers in different localities. Chagas et al. (2010) also developed a diagrammatic scale, with six levels, to assess the disease severity of gray mold on castor, but similar to the scale developed by Sussel et al. (2009), this scale was also developed based on long conical, full developed, racemes, and thus, will face the same problem mentioned above.

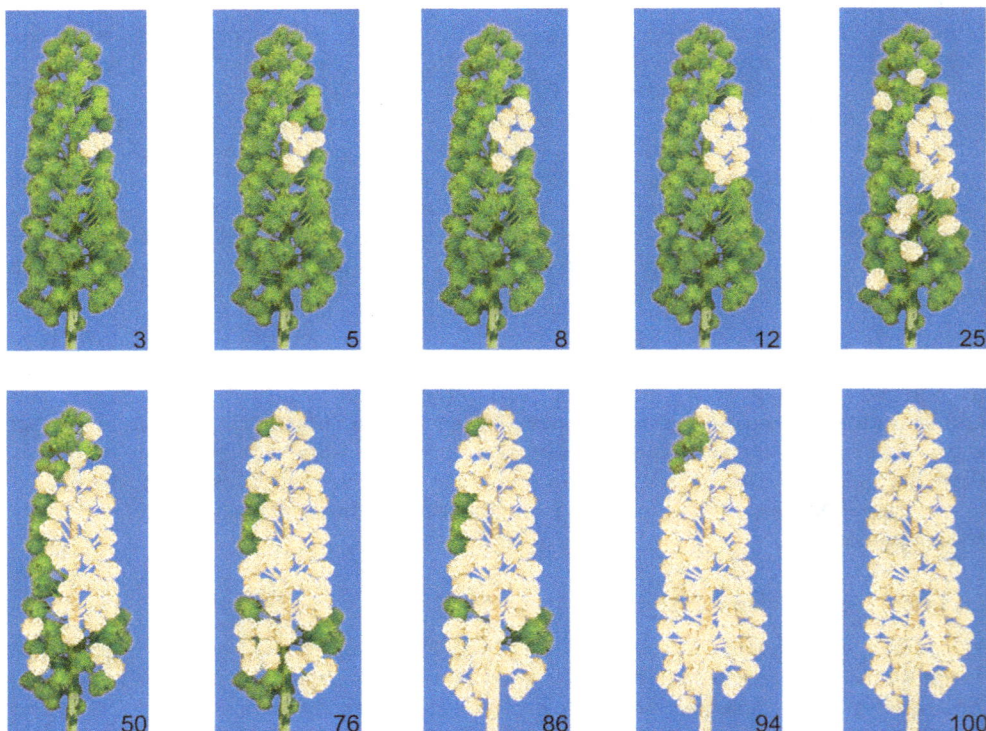

Fig. 5. Diagrammatic scale to assess the gray mold severity on castor (Reprinted from: Sussel et al. *Tropical Plant Pathology*, Vol. 34, No.3, pp.186-191, 2009, by permission). Numbers represents the percent area affected by the disease.

The first attempt to evaluate disease resistance under controlled conditions was made by C.A. Thomas & R.G. Orellana in 1963 using a biochemical test (Thomas & Orellana, 1963b). This methodology was reproduced by O.F. Esuruoso in Nigeria, a few years later (Esuruoso, 1969). This last author concluded that the laboratory results were similar to previous observations from field tests, and none of the tested varieties were resistant to the disease (Esuruoso 1969). At the Embrapa Algodão Station - a branch of the Brazilian Agricultural Research Agency, responsible for developing research on castor - we have developed a controlled method to assess castor resistance to gray mold. The principal advantage of this method is to eliminate the unpredictable changes usually present under field evaluation, and then to select the castor plants based solely on their genetic background rather than select a phenotype. Based on such studies, it has been possible to obtain a clear difference among the so-called susceptible and resistant genotypes, which had been previously screened under field conditions. It has also been possible to stratify the genotypes within at least three ranks: highly susceptible, moderately susceptible and less susceptible (Silva et al., 2008; Soares et al., 2010), although none of the screened genotypes were regarded as resistant.

2.7 Disease management

Without doubt, protection of the inflorescences and immature capsules is crucial to avoid heavy yield losses when castor is cultivated under favourable disease conditions. Fully developed capsules are less susceptible to pathogen attack and the severity levels are usually lower when compared with infection in young capsules or on the inflorescence. However, it is important to note that under highly favourable conditions with high inoculum pressure, losses of 100% are relatively frequent (Anjani et al., 2004). There is no single measure to keep the disease under acceptable levels; as well as there is no knowledge about any acceptable disease level. As the pathogen has a very short incubation period, and is easily wind-dispersed, its destructive potential is very high and usually the growers do not want to take their chances and wait passively, they usually prefer to act before or at the first signs of the disease.

2.7.1 Cultural

Cultural practices are usually applied at aiming to prevent the introduction of inoculum into the field, reducing its survival, spread or build-up, or rendering the host less prone to disease attack (Palti & Rotem, 1983; Termorshuizen, 2001). The use of varietal resistance is regarded as the better method for disease management. However, as highlighted previously, there are no varieties with satisfactory resistance levels to gray mold (Anjani et al., 2003; Araújo et al., 2007; Cook, 1981; Dange et al., 2005; Kolte, 1995; Milani et al., 2005).

Several authors have recommended the use of healthy seeds, removal of plant debris, adequate choice of planting area and growing season, and use of less susceptible cultivars (Galli et al., 1968; Massola Jr. & Bedendo, 2005; Sussel, 2009a). It is also recommended to use plant spacing adjusted for maximum aeration (Kolte, 1995; Lima et al., 2005). The use of healthy seeds, including seed treatment with fungicides, is always a desirable practice, however, its practical benefits for the management gray mold are questionable, because as mentioned previously, it is unlikely that such seeds will serve as an inoculum source for that crop season, so this practice has much more value in promoting vigorous plant growth and avoiding the introduction of the pathogen into new areas. Elimination of alternate and reservoir hosts (euphorbiaceous hosts), as well as removal and destruction of inoculum persisting in plant residues, are welcome practices for management of gray mold, and usually result in lower disease levels. Nevertheless, this practice must be followed by rigorous field inspection since the fungus is easily wind-dispersed, and once established in an area, such practice will become unfeasible. Perhaps, among the recognized cultural practices, the choice of growing season, or sowing time in such a way that spike development and maturity occur during the dry season, consequently avoiding the long, wet periods favourable to disease development, should be the most efficient one (Dange et al., 2005; Kolte, 1995); as evidenced by the fact that in locations where dry weather prevails, the disease does not occur (Godfrey, 1923; Kolte, 1995). This situation is typically observed in the Bahia state of Brazil, as mentioned previously.

2.7.2 Chemical

Seed treatment has been the management strategy most frequently recommended (Araújo et al., 2007; Batista et al., 1996; Godfrey, 1923; Gonçalves, 1936; Massola Jr. & Bedendo, 2005; Milani et al., 2005; Sussel, 2009), mainly to avoid the introduction of the pathogen into new areas. However, as the pathogen is air-borne and since it is already reported in almost all countries where castor is cultivated, the efficacy of such measures needs to be corroborated, because doubts about the role of the seeds in this pathosystem, as a primary inoculum source, still need to be clarified.

After disease establishment, fungicide spraying is usually the only way to stop or reduce the disease progress. However, there are few studies on chemical control of gray mold, and, of the fungicides tested, none is registered for use on castor crops in Brazil. According to Araújo et al. (2007), spraying of systemic fungicides soon after the appearance of the first symptoms delayed the epidemic and reduced disease progress.

In India Dange et al. (2005) recommended two prophylactic sprays with carbendazim (0.05%): the first at 50% of flowering; and the second, when the first disease symptoms appear. Although, Anjani et al. (2004) considered that non-genetic management measures have failed to control gray mold.

In the last decades, significant progress has been achieved regarding the use of fungicides to control plant diseases. Several new active fungicides with distinct modes of action, and usually with high specificity, have provided satisfactory levels of control of many plant diseases. However, for gray mold of castor, the main issue is not the ineffectiveness of fungicidal products, but the lack of research on the most appropriate timing of fungicide application and optimum dose, including also cost-benefit analyses.

Preliminary studies under controlled conditions have shown that carbendazim and azoxystrobin are effective against the gray mold pathogen (Bezerra, 2007). According to Chagas (2009), however, azoxystrobin was ineffective against *B. ricini*, while carbendazin and several others fungicides, including tebuconazole, iprodione and procymidone, were highly effective.

A field study, conducted under highly favourable conditions for disease development and using susceptible varieties, has confirmed that procymidone and iprodione are effective in disease control, but only if applied at the beginning of the epidemic and at weekly intervals. Where the timing of the first spraying was lost, and the application intervals were longer (15 days), the use of the same fungicides could not stop the disease and the losses reached 100% (D.J. Soares unpublished data).

Despite the limited studies dealing with chemical control of gray mold of castor, there are several fungicides known as "botryoticides" which are effective in protecting crops against *Botrytis* spp. (Leroux, 2007), and probably we can assume that these fungicides also are effective against *B. ricini*. In several *Botrytis* pathosystems, there is a high concern about the use of fungicides just before, or even after the harvest because of the toxicological risks of their residues (Leroux, 2007). In contrast, in castor crops, such a restriction is not a concern and fungicides can be applied just before harvest. However, if used indiscriminately, there must be concern about resistance phenomena associated with several major botryticide families, including benzimidazoles, phenylcarbamates and dicarboxymides (Leroux, 2007). Hence, control of gray mold of castor must consider several, rather than just one, management practices, as is currently recommended for the management of other *Botrytis* diseases (Leroux, 2007).

2.7.3 Biological

There are several studies dealing with the use of biological control agents, mainly *Trichoderma* spp. and *Clonostachys rosea* to control diseases caused by *Botrytis* spp. (Elad & Stewart, 2007). If we consider the fact that the genera *Botrytis* and *Amphobotrys* are biologically similar and, probably, phylogenetically related, we could expect that the use of such biological control agents could be applied as an effective strategy in the pathosystem *B. ricini* x *R. communis*. Actually, there are some studies conducted with *Trichoderma* and *Clonostachys rosea* for the control of gray mold of castor and promising results have been obtained (Bhattiprolu & Bhattiprolu, 2006; Chagas, 2009; Demant et al., 2006; Raoof et al., 2003; Tirupathi et al., 2006). However, it is clear that, although promising, these results are still experimental and much work needs to be done before permanent recommendations regarding the use of biological control agents can be secured for gray mold management.

3. Future challenges

The major problem about castor gray mold remains the lack of basic knowledge about its causal agent, how the disease develops, which factors are conducive to epidemics, and how we can manage it. It is necessary to elucidate, for example, what is the role of the climatic variables over the monocyclic components of the disease and how they affect the development of epidemics in the field. A better understanding of such relationships will determine which areas are suitable to grow castor. It is also imperative to know whether sexual reproduction is occurring within *B. ricini* populations in order to prevent fungicide resistance developing, perhaps by recommending permutations of fungicides with distinct active molecules and to determine the role of ascospores during the beginning of epidemics. However, we must first determine which fungicides are effective against gray mold, as well as their timing and frequency of application. We must also determine how resistance to the disease is inherited and if there are any phenotypic or genetic markers associated with it, so that breeders can more effectively generate resistant varieties using marker-assisted programs. Additionally, economic and cost-benefit analyses should be conducted to determine which practices for disease management are worthwhile recommending. In others words, there is still much work to be done before we can better define the best strategies to avoid economic losses due to gray mold in castor crops.

4. Conclusion

Despite being one of the most important diseases of castor worldwide, causing severe losses on castor yield for almost a century, since its first report, gray mold is still poorly studied. The recent concern about renewable energy sources has proportioned a unique opportunity to draw attention back to this pathosystem. So, researchers involved with castor cultivation must now take this opportunity to try to elucidate the many still to be answered questions about this disease in order to mitigate the constant menace of gray mold to castor crops.

5. Acknowledgment

The author wishes to thank the CNPq (Proc. 472953/2009-5) and Petrobrás (TC 0050.0064181.10.9) for the financial support on research of castor gray mold. I also wish to thanks Dr. Harry C. Evans by his priceless suggestions to improve the text.

6. References

Alcântara, M.J.; Nassur, R.C.M.R.; Sá, G.A.; Fraga, A.C. & Neto, P.C. (2008). Avaliação de materiais de mamoneira visando obtenção de resistência ao mofo cinzento, *Proccedings of 5th Congresso Brasileiro de Plantas Oleaginosas, Óleos, Gorduras e Biodiesel*. 04.06.2010. Available from
http://oleo.ufla.br/anais_05/artigos/a5__464.pdf

Alwadie, H.M. & Baka, Z.A.M. (2003). New records of fungal pathogens of *Euphorbia inarticulate* from Aseer region. *Archives of Phytopathology and Plant Protection*, Vol.36, No.3/4, pp.195-209, ISSN 0323-5408

Amorim Neto, M. S.; Araújo, A.E.; Beltrão, N. E.M.; Silva, L.C. & Gomes, D.C. (2000) Zoneamento e época de plantio para a mamoneira – estado da Paraíba. Embrapa Algodão, [Comunicado Técnico 108], ISSN 0102-0099, Campina Grande, Brazil

Anjani, K.; Raoof, M.A.; Ashoka Vardhana Reddy, V. & Hanumanata Rao, C. (2004). Sources of resistance to major castor (*Ricinus communis* L.) diseases. *Plant Genetic Resources Newsletter*. Vol.137, (March 2004) pp.46-48, ISSN 0048-4334

Araújo, A.E.; Suassuna, N.D.; Bandeira, C.M. & Agra, K.N. (2003). Efeito da temperatura na germinação de esporos de *Amphobotrys ricini* (=*Botrytis ricini*). *Fitopatologia Brasileira*, Vol.28, pp.S200 [Abstract], ISSN 0100-4158

Araújo, A.E.; Suassuna, N.D. & Coutinho, W.M. (2007). *Doenças e seu Manejo*. In: *O Agronegócio da Mamona no Brasil*, D.M.P. Azevedo & N.E. de M. Beltrão, (Eds.), 283-303, Embrapa Informação Tecnológica, ISBN 978-85-7383-381-2, Brasília, Brazil

Barreto, R.W. & Evans H.C. (1998). Fungal pathogens of *Euphorbia heterophylla* and *E. hirta* in Brazil and their potential as weed biocontrol agents. *Mycopathologia*, Vol.141, pp.21-36, ISSN 0301-486X

Batista, F.A.S.; Lima, E.F.; Moreira, J.A.N.; Azevedo, D.M.P.; Pires, V.A.; Vieira, R.M. & Santos, J.W. (1998). Avaliação da resistência de genótipos de mamoneira, *Ricinus communis* L., ao mofo cinzento causado por *Botrytis ricini* Godfrey. Embrapa Algodão. [Comunicado Técnico 73], ISSN 0102-0099, Campina Grande, Brazil

Batista, F.A.S.; Lima, E.F.; Soares, J.J. & Azevedo, D.M.P. (1996). Doenças e pragas da mamoneira (*Ricinus communis* L.) e seu controle. Embrapa Algodão, [Circular Técnica 21], ISSN 0100-6460, Campina Grande, Brazil

Beever, R.E. & Weeds, P.L. (2007) *Taxonomy and Genetic Variation of* Botrytis *and* Botryotinia. In: Botrytis: *Biology, Pathology and Control*, Y. Elad; B. Williamson; P. Tudzynski & N. Delen (Eds), 29-52, Springer, ISBN 978-1-4020-6586-6, Dordrecht, The Netherlands

Bezerra, C.S. (2007). Estrutura genética e sensibilidade a fungicidas de *Amphobotrys ricini* agente causal do mofo cinzento da mamoneira. MSc Dissertation (Genetic and Molecular Biology), Universidade Federal do Rio Grande do Norte, Natal, Brazil

Bhattiprolu, S.L. & Bhattiprolu, G.R. (2006). Management of castor grey rot disease using botanical and biological agents. *Indian Journal of Plant Protection*, Vol.34, No.1, pp.101-104, ISSN 0253-4355

Buchwald, N. F. (1949). Studies in the Scletoriniaceae: I. Taxonomy of the Sclerotiniaceae. *Kongelige Veterinær- og Landbohøjskole, Aarsskrift*, Vol.32, pp.1-116. ISSN 03687171

Chagas, H.A. (2009). Controle de mofo-cinzento (*Amphobotrys ricini*) da mamoneira (*Ricinus communis* L.) por métodos químico, biológico e com óleos essenciais. MSc Dissertation, (Agronomy), Universidade Estadual Paulista "Júlio de Mesquita Filho", Botucatu, Brazil

Chagas, H.A.; Basseto, M.A.; Rosa, D.D.; Zanotto, M.D.; & Furtado E.L. (2010). Escala diagramática para avaliação de mofo cinzento (*Amphobotrys ricini*) da mamoneira (*Ricinus communis* L.). *Summa Phytopathologica*, Vol.36, No.2, pp.164-167, ISSN 0100-5405

Cook, A.A. (1981). *Disease of Tropical and Subtropical Field, Fiber and Oil Plants*, Macmillan, ISBN 0-02-949300-5, New York, United States

Costa, R. S.; Suassuna, T. M .F.; Milani, M.; Costa, M. N. & Suassuna, N. D. (2004). Avaliação de resistência de genótipos de mamoneira ao mofo cinzento (*Amphobotrys ricini*), *Proccedings of 1st Congresso Brasileiro de Mamona*. 04.06.2010. Available from: http://www.cnpa.embrapa.br/produtos/mamona/publicacoes/trabalhos_cbm2/049.pdf

Dange, S.R.S; Desal, A.G. & Patel, S.I. (2005). *Diseases of castor*. In: G.S. Saharan; N. Mehta & M.S. Sangwan (Eds), 211-234, *Diseases of Oilseed Crops*, Indus Publishing Co, ISBN 81-7387-176-0, New Delhi, India

Demant, C.A.R.; Furtado, E.L.; Zanotto, M. & Chagas, A.A. Controle do mofo cinzento com o uso de *Trichoderma*, *Proccedings of 2nd Congresso Brasileiro de Mamona*. 04.06.2010. Available from: http://www.cnpa.embrapa.br/produtos/mamona/publicacoes/trabalhos_cbm2/043.pdf

Drumond, O. A. & Coelho S. J. (1981). Doenças da mamoneira. *Informe Agropecuário* Vol.7, No.82, pp.38-42, ISSN 0100-3364

Elad, Y & Stewart, A. (2007). *Microbial Control* of Botrytis *spp.* In: Botrytis: *Biology, Pathology and Control,* Y. Elad; B. Williamson; P. Tudzynski & N. Delen (Eds), 223-236, Springer, ISBN 978-1-4020-6586-6, Dordrecht, The Netherlands

Esuruoso, O.F. (1966). A preliminary study on the susceptibility of certain varieties of castor (*Ricinus communis* L.) to inflorescence blight disease caused by *Sclerotinia* (*Botrytis*) *ricini* (Godfrey) Whet. *The Nigerian Agricultural Journal,* Vol.3, No.1, pp.15-17, ISSN 0300-368X

Esuruoso, O.F. (1969). The reaction of the capsules of certain varieties of castor to a biochemical test for susceptibility to the inflorescence blight diseases. *The Nigerian Agricultural Journal,* Vol.6, No.1, pp.15-17, ISSN 0300-368X

Freire, E.C.; Lima, E.F.; Andrade, F.P.; Milani, M. & Nóbrega, M.B.M. (2007). *Melhoramento Genético.* In: *O Agronegócio da Mamona no Brasil,* D.M.P. Azevedo & N.E. de M. Beltrão, (Eds.), 171-194, Embrapa Informação Tecnológica, ISBN 978-85-7383-381-2, Brasília, Brazil

Galli, F.; Tokeshi, H.; Carvalho, P.C.T.; Balmer, E.; Cardoso, C.O.N. & Salgado, C.L. (1968). *Doenças da Mamoneira.* In: *Manual de Fitopatologia: Doenças das Plantas e seu Controle,* Galli, F.; Tokeshi, H.; Carvalho, P.C.T.; Balmer, E.; Cardoso, C.O.N. & Salgado, C.L., 292-197, Editora Agronomica Ceres, São Paulo, Brazil

Godfrey, G.H. (1919). *Sclerotinia ricini* n. sp. on the castor bean (*Ricinus communis*). *Phytopathology,* Vol.9, pp.565-567, ISSN 0031-949X

Godfrey, G. H. (1923). Gray mold of castor bean. *Journal* of *Agricultural Research,* Vol.23, No. 9, pp.679-715 + 13 plates, ISSN 0095-9758

Gonçalves, R.D. (1936). Mofo cinzento da mamoneira. *O Biológico,* Vol.2, No.7, pp.232-235, ISSN 0366-0567

Hansen H.N. & Bega, R.V. (1955). *Botrytis* rot of *Caladium* tubers. *Plant Disease Reporter,* Vol.39, No.3, pp.283, ISSN 0032-0811

Hennebert, G.L. (1973). *Botrytis* and *Botrytis*-like genera. *Persoonia,* Vol.7, No.2, pp.183-204, ISSN 0031-5850

Hoffmann, L.V.; Coutinho, T.C.; Duarte, E.A.A.; Bandeira, C.M. & Suassuna, N. D. (2004) Cultivo de *Amphobotrys ricini* e detecção das enzimas málica, superóxido dismutase e esterase, *Proccedings of 1st Congresso Brasileiro de Mamona,* 14.06.2010. Available from: http://www.cnpa.embrapa.br/produtos/mamona/publicacoes/trabalhos_cbm1/126.PDF

Holcomb, G.E. & Brown, W.L. (1990). Basal stem rot of cultivated *Poinsettia* caused by *Amphobotrys ricini. Plant Disease,* Vol.74, pp.828, ISSN 0191-2917

Holcomb, G.E.; Jones, J.P. & Wells, D.W. (1989). Blight of prostate spurge and cultivated *Poinsettia* caused by *Amphobotrys ricini. Plant Disease,* Vol.73, pp.74-75, ISSN 0191-2917

Holz, G.; Coertze, S. & Williamson, B. (2007). *The Ecology of* Botrytis *on Plant Surfaces.* In: Botrytis: *Biology, Pathology and Control,* Y. Elad; B. Williamson; P. Tudzynski & N. Delen (Eds), 9-28, Springer, ISBN 978-1-4020-6586-6, Dordrecht, The Netherlands

Jarvis, W.R. (1978). Epidemiology. In: *The Biology of* Botrytis, J.R. Coley-Smith, K. Verhoeff & W.R. Jarvis (Eds), 219-250, Academic Press, ISBN 0-12-179850-X, London, UK

Kars, I. & van Kan, J.A.L. (2007) *Extracellular Enzymes and Metabolites Involved in Pathogenesis of Botrytis*. In: Botrytis: *Biology, Pathology and Control*, Y. Elad; B. Williamson; P. Tudzynski & N. Delen (Eds), 99-118, Springer, ISBN 978-1-4020-6586-6, Dordrecht, The Netherlands

Keane, P.J. (2012) *Horizontal or Generalized Resistance to Pathogens in Plants*. In: *Plant Pathology*, C.J.R. Cumagun (Ed), 317-352, InTech, ISBN 978-953-307-933-2, Viena, Austria

Kimati, H. (1980). *Doenças da Mamoneira*. In: *Manual de Fitopatologia: Doenças das Plantas Cultivadas*, F. Galli, (Ed), 347-351, Editora Agronomica Ceres, São Paulo, Brazil

Kirk, P.M.; Cannon, P.F.; Minter, D.W. & Stalpers, J.A. (2008). *Dictionary of the Fungi*, CABI Publishing, ISBN 978-0-85199-826-8, Wallingford, UK

Kolte, J.S. (1995). *Castor: Diseases and Crop Improvement*. Shipra Publications, ISBN 81-85402-54-X, Delhi, India

Kumar, A.; Reddy, P.N. & Rao, T.G.N. (2007). Host range studies of *Botrytis ricini*, the causal agent of castor grey mold. *Indian Journal of Plant Protection*, Vol.35, No.1, pp.140-141, ISSN 0253-4355

Leroux, P. (2007) *Chemical Control of Botrytis and its Resistance to Chemical Fungicides*. In: Botrytis: *Biology, Pathology and Control*, Y. Elad; B. Williamson; P. Tudzynski & N. Delen (Eds), 195-222, Springer, ISBN 978-1-4020-6586-6, Dordrecht, The Netherlands

Lima, B.V.; Soares, D.J.; Perreira, O.L. & Barreto, R.W. (2008). Natural infection of *Acalypha hispida* and *Jatropra podagrica* inflorescences by *Amphobotrys ricini* in brazil. *Australasian Plant Disease Notes* Vol.3, pp.5-7, ISSN 1833-928X

Lima, E.F.; Araújo, A.E. & Batista, F.A.S. (2001) *Doenças e seu Controle*. In: *O Agronegócio da Mamona no Brasil*, D.M.P. Azevedo & E.F.Lima (Eds), 192-212, Embrapa Informação Tecnológica, ISBN85-7383-116-2, Brasília, Brazil

Lima, E.F. & Soares, J.J. (1990). Resistência de cultivares de mamoneira ao mofo cinzento, causado por *Botrytis ricini*. *Fitopatologia Brasileira*, Vol.15, No.1, pp.96-98, ISSN 0100-4158

Lima, V.P.T.; Graça Leite, E.A.; Botrel, E.P.; Fraga, A.C. &Castro Neto, P. (2005). Avaliação de ataque de mofo cinzento da mamoneira, variedade Al Guarany 2002, em diferentes espaçamentos. *Proccedings of 2nd Congresso Brasileiro de Plantas Oleaginosas, Óleos, Gorduras e Biodiesel*. 12.06.2010. Available from: http://oleo.ufla.br/anais_02/artigos/t183.pdf

Lyon, G.D.; Goodman, B.A. & Williamson, B. (2007). Botrytis cinerea *Perturbs Redox Processes as an Attack Strategy in Plants*. In: Botrytis: *Biology, Pathology and Control*, Y. Elad; B. Williamson; P. Tudzynski & N. Delen (Eds), 119-142, Springer, ISBN 978-1-4020-6586-6, Dordrecht, The Netherlands

Massola JR, N. S. & Bedendo, I. P. (2005). *Doenças da Mamoneira* (Ricinus communis L.). In: *Manual de Fitopatologia: Doenças das Plantas Cultivadas*, H. Kimati; L. Amorim; A. Bergamin Filho; L.E.A. Camargo & J.A.M. Rezende (Eds), 497-500,. Agronomica Ceres, ISBN 85-318-0008-0, São Paulo, Brazil

Maude, R.B. (1996). *Seedborne Diseases and their Control: Principles and Practice*. 978-0851989228, CABI Publishing, Kew, UK

Melhorança, A. L. & Staut, L.A. (2005). Informações técnicas para a cultura da mamona em Mato Grosso do Sul. Embrapa Agropecuária Oeste [Sistemas de Produção 8], ISSN 1679-1320, Dourados, Brazil

Milani M.; Nóbrega, M.B.M.; Suassuna, N.D. & Coutinho, W.M. (2005). Resistência da mamoneira (*Ricinus communis* L.) ao mofo cinzento causado por *Amphobotrys ricini*. Embrapa Algodão [Documentos 137], ISSN 0103-0205, Campina Grande, Brazil.

Moraes, W.B.; Souza, A.F.; Tomas, M.A.; Cecilio, R.A. & Jesus Junior, W.C. (2009). Zoneamento das áreas de risco da ocorrência de mofo cinzento da mamona no Brasil. *Proccedings of 13th Encontro Latino Americano de Iniciação Científica and 9th Encontro Latino Americano de Pós-Graduação*, 01.06.2010. Available from: http://www.inicepg.univap.br/cd/inic_2009/anais/arquivos/0821_1434_04.pdf

Orellana, R.G. & Thomas, C.A. (1962). Nature of predisposition of castor beans to *Botrytis*. I. Relation of leachable sugar and certain other biochemical constituents of the capsule to varietal susceptibility. *Phytopathology*, Vol.52, No.6, pp.533-538, ISSN 0031-949X

Palti, J. & Rotem, J. (1983). *Cultural Practices for the Control of Crop Diseases*. In: *Plant Pathologist's Pocketbook*, A. Johnston & C. Booth (Eds), 186-195, CMI, ISBN 0-85-198-460-6, Kew, England

Raoof, M.A.; Yasmeen, M. & Kausar, R. (2003). Potential of biocontrol agents for the management of castor grey mold, *Botrytis ricini* godfrey. *Indian Journal of Plant Protection*, Vol.31, No.2, pp.124-126, ISSN 0253-4355

Rego Filho, L.M.; Bezerra Neto, F.V. & Santos, Z.M. (2007). Avaliação da incidência de mofo-cinzento em genótipos de mamoneira no período de outono-inverno em campos dos Goytacazes-RJ. *Proccedings of the 2nd Congresso da Rede Brasileira de Tecnologia de Biodiesel 2006*. 27.07.2011. Available from: http://www.biodiesel.gov.br/docs/congresso2007/agricultura/3.pdf

Robinson, R.A. (1976). *Horizontal Pathosystem Analysis*. In: *Plant Pathosysytem (Advanced Series in Agricultural Sciences 3)*, R.A. Robinson, 74-89, Springer-Verlag, ISBN 3-540-07712-X, Berlin, Germany

Russo, V.M. & Rossman, A.Y. (1991). Occurrence of *Amphobotrys ricini* on prostate spurge in Oklahoma. *Plant Disease*, Vol. 75, pp. 750, ISSN 0191-2917

Santos. R.F.; Kouri, J.; Barros, M.A.L.; Firmino, P.T. & Requião. L.E.G. (2007). *Aspectos Econômicos do Agronegócio da Mamoneira*. In: *O Agronegócio da Mamona no Brasil*, D.M.P. Azevedo & N.E. de M. Beltrão, (Eds.), 22-41, Embrapa Informação Tecnológica, ISBN 978-85-7383-381-2, Brasília, Brazil

Sanoamuang, N. (1996). First report of gray mold blight caused by *Amphobotrys ricini* on crown of thorns in Thailand. *Plant Disease*, Vol.80, pp.223, ISSN 0191-2917

Seifert, K.; Morgan-Jones, G.; Gams, W. & Kendrick, B. (2011). *The Genera of Hyphomycetes*, CBS-KNAW Fungal Biodiversity Centre, ISBN 978-90-70351-85-4, Utrecht, The Netherlands

Silva, J.A.; Suassuna, N.D.; Coutinho, W.M. & Milani, M. (2008). Esporulação de *Amphobotrys ricini* em frutos de mamoneira como componente de resistência ao mofo cinzento, *Proccedings of the 3th Congresso Brasileiro de Mamona*. 04.06.2010. Available from: http://www.cnpa.embrapa.br/produtos/mamona/publicacoes/cbm3/trabalhos/FITOSSANIDADE/F%2012.pdf

Soares, D.J.; Fernandes J.N. & Araújo, A.E. (2010). Componentes monociclicos do mofo cinzento (*Amphobotrys ricini*) em diferentes genótipos de mamoneira, *Proceedings of 4th Congresso Brasileiro de Mamona,* 12.07.2011. Available from http://www.cbmamona.com.br/pdfs/FIT-01.pdf

Suassuna, N.D.; Araújo, A.E.; Bandeira, C.M.; Agra, K.N. (2003). Efeito de temperatura no crescimento e esporulação de *Amphobotrys ricini* (=*Botrytis ricini*). *Fitopatologia Brasileira* Vol.28, pp.S232 [Abstract], ISSN0100-4158

Sussel, A.A.B. (2008). Epidemiologia do mofo-cinzento (*Amphobotrys ricini* Buchw.) da mamoneira. PhD Thesis (Phytopathology), (June 2008)Universidade Federal de Lavras, Lavras, Brazil

Sussel, A.A.B. (2009a) Epidemiologia e Manejo do Mofo-cinzento-da-mamoneira. Embrapa Cerrados, [Documentos 241], ISSN 1517-5111, Brasília, Brazil

Sussel, A.A.B. (2009b) Escala Diagramática para Avaliação do Mofo-cinzento-da-mamoneira. Embrapa Cerrados, [Documentos 247], ISSN1517-5111, Brasília, Brazil

Sussel, A.A.B.; Pozza, E.A. & Castro, H.A. (2009). Elaboração e validação de escala diagramática para avaliação da severidade do mofo cinzento da mamoneira. *Tropical Plant Pathology,* Vol.34, No.3, pp.186-191, ISSN 1982-5676

Sussel, A.A.B.; Pozza, E.A. & Castro, H.A. & Lasmar, E.B.C. (2011). Incidência e severidade do mofo-cinzento-da-mamoneira sob diferentes temperaturas, períodos de molhamento e concentração de conídios. *Summa Phytopatologica* Vol.37, No.1, pp.30-34, ISSN0100-5405

Termorshuizen, A.J. (2001). *Cutural Control.* In: *Plant Pathologist's Pokcketbook,* J.M. Waller; J.M. Lenné & S.J. Waller (Eds), 318-327, CABI Publishing, ISBN 0851994598, Wallingford, UK

Thomas, C.A. & Orellana, R.G. (1963a) Nature of predisposition of castor beans to *Botrytis*. II. Raceme compactness, internode lenghs, position of staminate flowers and bloom in raltion to capsule susceptibility. *Phytopathology,* Vol.53, No.2, pp.249-251, ISSN 0031-949X

Thomas, C.A. & Orellana, R.G. (1963b) Biochemical tests indicative of reaction of castor bean to *Botrytis*. *Science,* Vol.139, pp.334-335, ISSN 0036-8075

Thomas, C.A. & Orellana, R.G. (1964) Phenols and pectins in relation to browning and maceration of castorbean capsules by *Botrytis*. *Phytopathologische Zeitschrift,* Vol.50, No.4, pp.359-366, ISSN 0931-1785

Tirupathi, J.; Kumar, C.P.C. & Reddy, D.R.R. (2006). *Trichoderma* as potential biocontrol agents for the management of grey mold of castor. *Journal of Research Angrau,* Vol.34, No.2, pp.31-36

Ueno, B.; Hellwig, T.C.; Nickel, G.; Silva, S.D.A. (2006). Resistência ao mofo cinzento em 15 genótipos de mamoneira cultivadas na região de Pelotas, RS, na safra 2004/2005, *Proceedings of 2nd Congresso Brasileiro de Mamona.* 12.06.2010. Available from: http://www.cnpa.embrapa.br/produtos/mamona/publicacoes/trabalhos_cbm2/049.pdf

Whetzel, H.H. (1945). A synopsis of the genera and species of the Sclerotiniaceae, a family of stromatic inoperculate discomycetes. *Mycologia,* Vol.37, pp.648-714, ISSN 0027-5514

Whitney, N.G. & Taber, R.A. (1986). First report of *Amphobotrys ricini* infecting *Caperonia palustris* in the United States. *Plant Disease*, Vol.70, pp.892, ISSN 0191-2917

Zarzycka, H. (1958). Resistance of some varieties of castor bean (*Ricinus communis*) to the fungus *Botrytis cinerea* Pers. *Acta Agrobotanica*, Vol.7, pp.117-124, ISSN 0065-0951

10

Current Advances in the Fusarium Wilt Disease Management in Banana with Emphasis on Biological Control

R. Thangavelu and M.M. Mustaffa
National Research Centre for Banana, Trichirapalli
India

1. Introduction

Banana (*Musa* spp.) is the fourth most important global food commodity after rice, wheat and maize in terms of gross value production. At present, it is grown in more than 120 countries throughout tropical and subtropical regions and it is the staple food for more than 400 million people (Molina and Valmayor, 1999). Among the production constraints, Fusarium wilt caused by the fungus *Fusarium oxysporum* f.sp *cubense* (Foc) is the most devastating disease affecting commercial and subsistence of banana production through out the banana producing areas of the world (Ploetz, 2005). The disease is ranked as one of the top 6 important plant diseases in the world (Ploetz & Pegg, 1997). In terms of crop destruction, it ranks with the few most devastating diseases such as wheat rust and potato blight (Carefoot and sprott, 1969). The disease almost destroyed the banana export industry, built on the Gros Michel variety, in Central America during the 1950's (Stover, 1962). In addition, the widely grown clones in the ABB 'Bluggoe' and AAA 'Gros Michel and Cavendish' sub groups are also highly susceptible to this disease worldwide. Presently, Fusarium wilt has been reported in all banana growing regions of the world (Asia, Africa, Australia and the tropical Americas) except some islands in the South Pacific, the Mediterranean, Melanesia, and Somalia (Stover, 1962; Anonymous, 1977; Ploetz and Pegg, 2000).

The fungus Foc is the soilborne hyphomycete and is one of more than 100 formae speciales of *F. oxysporum* that causes vascular wilts of flowering plants (Domsch et al. 1980; Nelson et al. 1983). Although Fusarium wilt probably originated in Southeast Asia, (Ploetz and Pegg, 1997), the disease was first discovered at Eagle Farm, Brisbane, Queensland, Australia in 1876 in banana plants var. Sugar (Silk AAB) (Bancroft, 1876). The fungus infects the roots of banana plants, colonizing the vascular system of the rhizome and pseudostem, and inducing characteristic wilting symptoms mostly after 5-6 months of planting and the symptoms are expressed both externally and internally (Wardlaw, 1961; Stover, 1962). Generally, infected plants produce no bunches and if produced, the fruits are very small and only few fingers develop. Fruits ripen irregularly and the flesh is pithy and acidic. The fungus survives in soil for up to 30 years as chlamydospores in infested plant material or in the roots of alternative hosts (Ploetz, 2000).

Since the discovery of Fusarium wilt of banana, though various control strategies like soil fumigation (Herbert and Marx, 1990); fungicides (Lakshmanan et al., 1987); crop rotation

(Hwang, 1985; Su et al., 1986), flood –fallowing (Wardlaw, 1961; Stover, 1962) and organic amendments (Stover, 1962) have been evolved and attempted, yet, the disease could not be controlled effectively except by planting of resistant cultivars (Moore et al., 1999). Planting of resistant varieties also cannot be implemented because of consumer preference (Viljoen, 2002). Under these circumstances, use of antagonistic microbes which protect and promote plant growth by colonizing and multiplying in both rhizosphere and plant system could be a potential alternative approach for the management of Fusarium wilt of banana.

Besides, biological control of Fusarium wilt disease has become an increasingly popular disease management consideration because of its environmental friendly nature which offers a potential alternative to the use of resistant banana varieties and the discovery of novel mechanisms of plant protection associated with certain microorganisms (Weller et al., 2002; Fravel et al., 2003). Biological control of soil borne diseases caused especially by *Fusarium oxysporum* is well documented (Marois et al., 1981; Sivan and Chet, 1986; Larkin and Fravel, 1998; Thangavelu et al., 2004). Several reports have previously demonstrated the successful use different species of *Trichoderma, Pseudomonas, Streptomyces*, non pathogenic *Fusarium* (npFo) of both rhizospheric and endophytic in nature against Fusarium wilt disease under both glass house and field conditions (Lemanceau & Alabouvette, 1991; Alabouvette et al.1993; Larkin & Fravel, 1998; Weller et al. 2002; Sivamani and Gnanamanickam, 1988; Thangavelu et al. 2001; Rajappan et al. 2002; Getha et al. 2005). The details on the effect of these biocontrol agents in controlling Fusarium wilt disease of banana are discussed in detail hereunder.

2. *Trichoderma* spp.

Trichoderma spp., are free-living fungi that are common in soil and root ecosystems. They are highly interactive in root, soil and foliar environments. They produce or release a variety of compounds that induce localized or systemic resistance responses in plants. This fungal bio-control agent has long been recognized as biological agents, for the control of plant disease and for their ability to increase root growth and development, crop productivity, resistance to abiotic stresses, and uptake and use of nutrients. It can be efficiently used as spores (especially, conidia), which are more tolerant to adverse environmental conditions during product formulation and field use, in contrast to their mycelial and chlamydospore forms as microbial propagules (Amsellem et al. 1999). However, the presence of a mycelial mass is also a key component for the production of antagonistic metabolites (Benhamou and Chet 1993; Yedidia et al. 2000). Several reports indicate that *Trichoderma* species can effectively suppress Fusarium wilt pathogens (Sivan and Chet, 1986; Thangavelu et al. 2004). Thangavelu (2002) reported that application of T. *harzianum* Th-10, as dried banana leaf formulation @ 10 g/plant containing 4×10^{31} cfu/g in basal + top dressing on 2, 4 and 6 months after planting in cv. Rasthali recorded the highest reduction of disease incidence (51.16%) followed by *Bacillus subtilis* or *Pseudomonas fluorescens* (41.17%) applications as talc based formulation under both glass house and field conditions. The talc based formulation of T. *harzianum* Th-10 and fungicide treatment recorded only 40.1% and 18.1% reduction of the disease respectively compared to control. In the Fusarium wilt-nematode interaction system also, soil application of biocontrol agents reduced significantly the wilt incidence and also the root lesion and root knot index. In addition to this, 50 to 82% of reduction in nematode population *viz., Pratylenchus coffeae* and *Meloidogyne incognita* was also noted due to application of bioagents and the maximum reduction was due to T. *harzianum* treatment (Thangavelu, 2002). Raghuchander et al. (1997)

reported that *T. viride* and *P. fluorescens* were equally effective in reducing the wilt incidence. Inoculation of potted abaca plants with *Trichoderma viride* and yeast showed 81.76% and 82.52% reduction of wilt disease severity respectively in the antagonist treated plants. (Bastasa and Baliad, 2005).

Similarly, soil application of *T. viride* NRCB1 as chaffy grain formulation significantly reduced the external (up to 78%) and internal symptoms (up to 80 %) of Fusarium wilt disease in tissue cultured as well as sucker derived plants of banana cv. Rasthali (Silk-AAB) and increased the plant growth parameters significantly as compared to the talc powder formulation under pot culture and field conditions (Thangavelu and Mustaffa, 2010).

The possible mechanisms involved in the reduction of Fusarium wilt severity due to *Trichoderma* spp. treatment might be the mycoparasitism, spatial and nutrient competition, antibiosis by enzymes and secondary metabolites, and induction of plant defence system. The mycoparasitism involves in coiling, disorganization of host cell contents and penetration of the host (Papavisas, 1985; University of Sydney, 2003). During the mycoparasitism, *Trichoderma* spp. parasitizes the hyphae of the pathogen and produce extracellular enzymes such as proteolytic enzymes, β-1, 3- glucanolytic enzymes and chitinase etc., which cause lysis of the pathogen. The toxic metabolites such as extracellular enzymes, volatiles and antibiotics like gliotoxin and viridin which are highly fungistatic substances (Weindling, 1941) are considered as elements involved in antibiosis. In addition, *Trichoderma* spp. could compete and sequester ions of iron (the ions are essential for the plant pathogen,) by releasing compounds known as siderophores (Srinivasan et al. 1992). There are several reports demonstrating control of a wide range of plant pathogens including *Fusarium* spp. by *Trichoderma* spp. by elicitation of induced systemic or localized resistance which occur due to the interaction of bioactive molecules such as proteins avr-like proteins and cell wall fragments released by the action of extracellular enzymes during mycoparasitic reaction. Thangavelu and Musataffa, (2010) reported that the application of *T. viride* NRCB1 as rice chaffy grain formulation and challenge inoculation with *Foc* in cv. Rasthali resulted in the induction of defense related enzymes such as Peroxidase and Penylalanine Ammonia lyase (PAL) and also the total phenolic content significantly higher (>50%) as compared to control and *Foc* alone inoculated banana plants and the induction was maximum at 4-6th day after treatment. They suggested that this increased activities of these lytic enzymes and thus increased content of phenols in the *T. viride* applied plants might have induced resistance against *Foc* by either making physical barrier stronger or chemically impervious to the hydrolytic enzymes produced by the pathogen (Thangavelu and Mustaffa, 2010). Morpurgo et al. (1994) reported that the activity of peroxidase was at least five times higher in the roots and corm tissues of *Foc* resistant banana variety than in the susceptible variety. Inoculation of resistant plants with *Foc* resulted in 10-fold increase in PO activity after seven days of inoculation, whereas the susceptible variety exhibited only a slight increase in PO activity.

3. *Pseudomonas* spp.

Pseudomonas spp. are particularly suitable for application as agricultural biocontrol agents since they can use many exudates compounds as a nutrient source (Lugtenberg et al.1999a); abundantly present in natural soils, particularly on plant root systems, (Sands & Rovira, 1971); high growth rate, possess diverse mechanisms of actions towards phytopathogens including the production of a wide range of antagonistic metabolites (Lugtenberg et al. 1991; Dowling & O'Gara, 1994; Dunlap et al.1996; Lugtenberg et al., 1999b), easy to grow *in vitro* and subsequently can be reintroduced into the rhizosphere (Lugtenberg et al. 1994; Rhodes & Powell, 1994) and capable of inducing a systemic resistance to pathogens (van Loon et al . 1998; Pieterse et al. 2001).

Several studies have investigated the ability of *P. fluorescens* to suppress Fusarium wilt disease of banana. Fluorescent pseudomonad species such as *Pseudomonas fluorescens* (Sakthivel and Gnanamanickam 1987), *Pseudomonas putida* (de Freitas and Germida 1991), *Pseudomonas chlororaphis* (Chin-A-Woeng et al. 1998) and *Pseudomonas aeruginosa* (Anjaiah et al. 2003) have been used to suppress pathogens as well as to promote growth and yield in many crop plants. Sivamani and Gnanamanickam (1988) reported that the seedlings of *Musa balbisiana* treated with *P.fluorescens* showed less severe wilting and internal discoloration due to *Foc* infection in green house experiments. The bacterized seedlings also showed better root growth and enhanced plant height.

Thangavelu et al. (2001) demonstrated that *P. fluorescens* strain pf10, which was isolated from the rhizosphere of banana roots, was able to detoxify the fusaric acid produced by *Foc* race-1 and reduced wilt incidence by 50%. Dipping of suckers in the suspension of *P. fluorescens* along with the application of 500 g of wheat bran and saw dust inoculation (1: 3) of the respective bio-control agent effectively reduced Fusarium wilt incidence in banana (Raghuchander et al.1997). Rajappan et al. (2002) reported that the talc based powder formulation of *P. fluorescens* strain pf1 was effective against *Foc* in the field. *Pseudomonas fluorescens* strain WCS 417, known for its ability to suppress other Fusarium wilt diseases, reduced the disease incidence by 87·4% in Cavendish bananas in glasshouse trials (Nel et al. 2006). Saravanan et al. (2003) demonstrated that either basal application of neem cake at 0.5 kg/plant + sucker dipping in spore suspension of *P. fluorescens* for 15 min+ soil application of *P. fluorescens* at 10 g/plant at 3,5 and 7 months after planting or the basal application of neem cake at 0.5 kg/plant + soil application of *P. fluorescens* at 10 g/plant at 3, 5 and 7 months after planting showed the greatest suppression of wilt disease in two field trials conducted in Tamil Nadu, India.

Fishal et al. (2010) assessed the ability of two endophytic bacteria originally isolated from healthy oil palm roots, *Pseudomonas* sp. (UPMP3) and *Burkholderia* sp. (UPMB3) to induce resistance in susceptible Berangan banana against *Fusarium oxysporum* f. sp. *cubense* race 4 (FocR4) under glasshouse conditions. The study showed that pre-inoculation of banana plants with *Pseudomonas* sp UPMP3 recorded 51% reduction of Fusarium wilt disease severity, whereas, the combined application of UPMP3+UPMB3 and single application of UPMB3 alone recorded only 39 and 38% reduction of Fusarium wilt disease severity respectively. Ting et al. (2011) reported that among six endobacteria isolates, only two isolates (*Herbaspirillum* spp and *Pseudomonas* spp.) produced volatile compounds which were capable of inhibiting the growth of *Foc* race 4. The compounds were identified as 2-pentane 3-methyl, methanethiol and 3-undecene. They found that the isolate *Herbaspirillum* spp. recorded 20.3% inhibition of growth of *Foc* race 4 as its volatile compounds contained all the three compounds whereas *Pseudomonas* isolate AVA02 recorded only 1.4% of growth inhibition of race 4 *Foc* as its volatile compounds contained only methanethiol and 3-undecene. They concluded that the presence of all these three compounds especially 2-pentane 3-methyl and also in high quantity is very important for the antifungal activity against *Foc*. Of the 56 fluorescent pseudomonad isolates obtained from banana rhizosphere, *Pseudomonas aeruginosa* strain FP10 displayed the most potent antibiosis towards the *Foc*. This strain was found to produce IAA, siderophores and phosphate-solubilizing enzyme which indicated that this strain is having potential of plant-growth-promoting ability. The presence of DAPG gene (ph1D) in the strain FP10 was confirmed by PCR and the production of DAPG was confirmed by TLC, HPLC and FT-IR analyses. The *in-vivo* bioassay carried out showed that the banana plants received with pathogen and the strain FP10 exhibited increased height (30.69cm) and reduced vascular discolouration (24.49%), whereas, the pathogen *Foc* alone-inoculated plants had an average height of 21.81 cm and 98.76% vascular discolouration (Ayyadurai et al. 2006).

Saravanan and Muthusamy (2006) reported that soil application of talc-based formulation of *P. fluorescens* at 15 g/plant in banana, suppressed Fusarium wilt disease significantly (30.20 VDI) as compared to pathogen *Foc* alone-inoculated plants (88.89 VDI). It was found that the ability of *P. fluorescens* to suppress Fusarium wilt pathogens depends on their ability to produce antibiotic metabolites particularly 2, 4- Diacetylphloroglucinol (DAPG). The metabolite DAPG extracted from the rhizosphere of *P. fluorescens* applied to soil showed significant inhibition of growth and spore germination of *Foc*. They also showed that the quantity of DPAG production was less in the extracts of soil, inoculated with *P. fluorescens* and challenge inoculated with *F. oxysporum* f. sp. *cubense* as compared to *P. fluorescens* alone inoculated soil.

In plants pretreated with *P. fluorescens* and challenged with pathogen *Foc,* there was reduction in the number of *Foc* colonies (14 numbers) as compared to the plants treated with *Foc* alone (41 number). A 72% reduction in the pathogen infection was noticed as a result of *P.fluorescens* treatment. Colonies of *P.fluorescens* in plants challenged with *F. oxysporum* were reduced to 33 in number, perhaps due to competition for infection loci (Sukhada et al. 2004). Electron microscopic studies revealed that in the root samples of bacteria treated and pathogen challenge inoculated plants, there was extensive fungal proliferation in the cortex and had wall appositions made of electron-dense materials lining the host cortical cell wall. The wall appositions formed were highly significant in restricting the further growth of the fungus. They opined that electron-dense materials might have been produced either by the bacteria or the host tissue in response to the attacking pathogen. Massive depositions of unusual structures at sites of fungal entry was also noticed, which clearly indicated that bacterized root cells were signalled to mobilize a number of defence structures for preventing the spread of pathogen in the tissue (Sukhada et al. 2004). Pre-inoculated *P. fluorescens* helped the banana plant to resist pathogen attack to some extent due to the structural modification of the root system and due to the accumulation of newly formed electron-dense molecules, which may be providing the defense mechanism to the host plant. Treatment of 'Maçã' banana (*Musa spp.; group ABB*) with endophytic diazotrophic bacteria *Herbaspirillum* (BA234) and *Burkholderia* (AB202) also resulted in significant reduction of *Foc* unit propagules as well as increase in biomass of the plant in four and two months after plant inoculation with AB202 and BA234 respectively suggesting that these endophytic diazotrophic bacteria may be used as potential bio-fertilizer and bio-control agents for banana (Weber et al. 2007).

4. *Bacillus* spp.

Bacillus subtilis has been identified as a potential biological control agent. These strains could produce a wide range of antifungal compounds, such as subtilin, TasA, subtilosin, bacilysin, mycobacillin and some enzymes, which can degrade fungal cell wall (Berg et al. 2001). It was suggested that these antibiotic production plays a major role in plant disease suppression (Knox et al. 2000; Leelasuphakul et al. 2006). In addition, some antagonistic mechanisms of these *Bacillus* species involves in the competition for nutrients and space, the induction of plant resistance, etc. (Guerra-Cantera et al., 2005; Van loon et al., 1998).

Sun et al. (2011) isolated an antagonistic *Bacillus* strain, KY-21 from the soil of banana's rhizosphere and tested against *Foc* both under *in-vitro* and *in-vivo* conditions. Under lab condition, mycelium growth of the pathogen was seriously inhibited after treatment with the fermentation filtrate of KY-21. The microscopic examination of mycelium revealed that the tips of the hypha were deformed into spherical structures that were remarkably constricted by dual culture. Besides, the inoculation of banana plants with *Bacillus* strain, KY-21 also increased the activities of polyphenol oxidase (PPO) and peroxidase (POD) significantly compared to control. The *in-vivo* biocontrol assays showed that at 60 days after *Foc* inoculation, the plantlets treated with KY-21 exhibited 35% severe wilt symptom and 18.3% severe vascular discoloration as against 68.4% and 48.3% of severe wilt symptom and severe vascular discoloration respectively in control plantlets. Besides, plantlets inoculated with KY-21 showed significantly reduced development of disease as compared to the control.

5. *Actinomycetes*

Actinomycetes particularly *Streptomyces* spp. are important soil dwelling microorganisms, generally saprophytic, spend majority of their life cycle as spores and are best known for their ability to produce antibiotics. They may influence plant growth and protect plant roots against invasion by root pathogenic fungi (Crawford et al. 1993). Streptomyces species have been used extensively in the biological control of several formae speciales of *F. oxysporum*, which caused wilt disease in many plant species (Reddi and Rao 1971; Lahdenpera and Oy, 1987; Smith et al. 1990). *Streptomyces violaceusniger* strain G10 isolated from a coastal mangrove (*Rhizophora apiculata* (Blume)] stand, was shown to exhibit strong *in-vitro* antagonism toward several plant pathogenic fungi including *Foc* race 4. Under *in-vivo* bioassay, treating the planting hole and roots of tissue-culture-derived 'Novaria' banana plantlets with *Streptomyces* sp. strain g10 suspension (10^8 cfu/ml), resulted in 47% reduction of leaf symptom index (LSI) and 53% of rhizome discoloration index (RDI) with reduced wilt severity when the plantlets were inoculated with 10^4 spores/ml *Foc* race 4 compared to untreated plantlets. However, the reduction in disease severity was not significant when plantlets were inoculated with a higher concentration (10^6 spores/ml) of Foc race 4 (Getha et al. 2005). Getha and Vikineswary (2002) studied the interaction between *Streptomyces violaceusniger* strain g10 and *F. oxysporum* f.sp. *cubense* and demonstrated the production of antifungal metabolites especially antibiotics by the antagonists which caused swelling, distortion, excessive branching and lysis of hyphae and inhibition of spore germination of *Foc* pathogen by the antagonist.

Among 242 actinomycete strains, isolated from the interior of leaves and roots of healthy and wilting banana plants, *Streptomyces griseorubiginosus*-like strains were the most frequently encountered strains. The screening of these strains for antagonistic activity against *Fusarium oxysporum* f. sp. *cubense* revealed that 50% of the *Streptomyces* strains isolated from healthy trees especially from the roots had antagonistic activities against *Foc* and only 27% of strains isolated from wilting trees showed the same activity (Cao et al.

2004). Similarly in 2005, out of 131 endophytic actinomycete strains isolated from banana roots, the most frequently isolated and siderophore producing endophytic *Streptomyces* sp. strain S96 was found to be highly antagonistic to *Foc*. The subsequent *in vivo* biocontrol assays carried out showed that the disease severity index of Fusarium wilt was significantly reduced and mean fresh weight of plantlets increased compared to those grown in the absence of the biocontrol strain S96 (Cao et al. 2005).

6. General mode of action of antagonistic bacteria

Generally biocontrol agents can antagonize soil-borne pathogens through the following strategies: (1) Competition for niches and nutrients (niche exclusion), (2) Production of secondary metabolites which are used in direct antagonism (3) Growth promotion by changing the physiology of the plant and (4) Induction of resistance to disease

Antagonistic bacteria are more effective against root pathogens only if they have a strong ability to colonize the root system (Weller, 1988) and also the fungal hyphae. This is widely believed to be essential for biocontrol (Weller et al. 1983; deWeger et al. 1987; Parke, 1990). The scanning and transmission electron microscopy study revealed that colonization on banana roots, on the hyphal surface and macrospores of *Foc* fungus race 4, by the endophyte *Burkholderia cepacia*. The study also showed that *B. cepacia* exists mainly in the intercellular space of the banana root tissues. Benhamou et al. (1996) provided evidence that root colonization by the endophytic bacterium *Pseudomonas fluorescens*, involved in a sequence of events that included bacterial attachment to the plant roots, proliferation along the elongation root, and local penetration of the epidermis. M'Piga et al. (1997) also confirmed the entry of *P. fluorescens* into the root system and their colonization inside. Once inside the host tissue, these bacteria produce an array of antifungal metabolites like siderophores and different antibiotics like phenazine-1 carboxylic acid, and 2, 4-diacetylphloroglucinol preventing the further advancement of the fungus (Beckman et al. 1982; Mueller & Beckmann,1988) by inducing severe cell disturbances in pathogenic fungi (Dowling & O'Gara, 1994). Sukhada et al. (2004) also located the colonies of *P. fluorescens* and *Foc* in banana using respective FITC-conjugated antibodies. They found that the bacterial population was relatively greater towards the cortex region of the root as compared to the stele region. In plants pretreated with *P. fluorescens* and challenged with *Foc*, there was reduction in the number of *Foc* colonies (14 numbers) as compared to the plants treated with *Foc* alone (41 number).

Competition for nutrients such as carbon, nitrogen or iron is one of the mechanisms through which biocontrol strains can reduce the ability of fungal pathogens to propagate in the soil (Alabouvette, 1986; Buyer & Leong, 1986; Leong, 1986; Loper & Buyer, 1991; Fernando et al., 1996; Handelsman & Stabb, 1996). Already established (pre-emptive competitive exclusion) or aggressively colonizing biocontrol bacteria can therefore prevent the establishment and subsequent deleterious effects of a pathogen. Most organisms, including fluorescent *Pseudomonas* species, take up ferric ions through high-affinity iron chelators, designated as siderophores that are released from bacterial cells under Fe^{3+} limiting conditions. The role of siderophores produced by pseudomonads has been well correlated with the biocontrol of disease suppressive soils and on the plant growth by supplying the plant with sequestered iron. Kloepper et al. (1980) reported that inhibition of the wilt pathogen was attributable to iron deprivation caused by pseudomonad siderophores compounds produced in low-iron environments that function in iron transport. It is suggested that the management of Fe availability in the infection court, through Fe competition, can induce suppressiveness to a Fusarium wilt pathogen.

Dowling and O'Gara (1994) reported that bacterial endophytes like *P. fluorescens* produced an array of antifungal metabolites like siderophores and different antibiotics like phenazine-1 carboxylic acid, and 2, 4-diacetylphloroglucinol that could induce severe cell disturbances in a number of pathogenic fungi. These compounds have direct effect on the growth of the pathogens. Biocontrol bacteria producing chitinase (Shapira et al., 1989; Dunne et al., 1996; Ross et al., 2000), protease (Dunlap et al., 1997; Dunne et al., 1998), cellulase (Chatterjee et al., 1995) or β glucanases (RuizDuenas & Martinez, 1996; Jijakli & Lepoivre, 1998) were shown to suppress plant diseases as these enzymes are involved in the breakdown of fungal cell walls by degrading cell wall constituents such as glucans and chitins, resulting in the destruction of pathogen structures or propagules. The bacteria also play a major role in growth promotion by producing phytohormones such as auxins, gibberellins, cytokinins and ethylene (García de Salamone et al., 2001; Remans et al., 2008). Besides promoting growth, they induce resistance in plants against pest and disease. There are two types of induced resistance exist called Systemic acquired resistance (SAR) and Induced systemic resistance (ISR). SAR is dependent on the salicylic acid pathway and is mainly associated with pathogen attack or in response to the exogenous application of chemicals such as salicylic acid and produces pathogenesis-related (PR) proteins such as β -1,3-glucanases, endo-chitinases and thaumatin-like proteins (Ward et al. 1991; Uknes et al. 1992; Rahimi et al. 1996; Van Pelt-Heerschap et al. 1999).Bacteria induced defenses in plants are expressed through structural and biochemical mechanisms. Structural mechanisms include the reinforcement of plant cell walls by deposition of newly formed molecules of callose, lignin and phenolic , occlusion of colonized vessels by gels, gums and tyloses (He et al. 2002; Jeun et al. 2004 Gordon and Martyn 1997; Olivain and Alabouvette, 1999). Whereas, the biochemical mechanism of resistance includes accumulation of secondary metabolites such as phytoalexins and production of PR proteins such as β -1,3-glucanases and chitinases. In the case of induced systemic resistance (ISR), the resistance induced only after the colonization of plant roots by bacteria. After colonization, they produce secondary metabolites and volatiles and defense related enzymes (Stougard, 2000; Han et al., 2006), which give resistance to plants. The level of defense related enzymes are known to play a crucial role in the degree of host resistance. Peroxidase (PO) and Polyphenol oxidase (PPO) are believed to be one of the most important factors of the plant's biochemical defense against pathogens, and are actively involved in the self-regulation of plant metabolism after infection (Kavitha and Umesha, 2008; Dutta et al., 2008). Peroxidase is involved in substrate oxidation and cell wall lignifications; the PPO can oxidize phenolic compounds to quinines. Both of these defense mechanisms are associated with disease resistance. ISR elicited by PGPR has shown promise in managing a wide spectrum of plant pathogens in several plant species under greenhouse and field environments (Radjacommare et al., 2004; Thangavelu et al. 2004; Murphy et al., 2003). Fishal et al. (2010) observed increased accumulation of resistance-related enzymes such as peroxidase (PO), phenylalanine ammonia lyase (PAL), lignithioglycolic acid (LTGA), and pathogenesis-related (PR) proteins (chitinase and β-1, 3-glucanase) in banana plantlets treated with endophytic bacteria UPMP3 and UPMB3 singly or as mixture under glasshouse conditions.

7. Non-pathogenic Fusarium (npFo)

Several endophytic isolates of non-pathogenic *F. oxysporum* (npFo) derived from symptomless banana roots provided some degree of protection against *Foc* race-4 for the Cavendish cultivar Williams in the green house (Gerlach et al.1999). Similarly, pre-treatment of banana plants with endophytic bacterial strain UPM39B3 (*Serratia*) and fungal strain UPM31P1 (*Fusarium oxysporum*), isolated from the roots of wild bananas either singly or in combination resulted in significant increase in plant growth parameters in the FocR4 inoculated plants than the diseased plantlets that were not infected with endophytes (Ting et al. 2009). It was also observed that the diseased plantlets benefited from the improved plant growth were able to survive longer than diseased plantlets without endophytes. Nel et al. (2006) evaluated several npFo and *Trichoderma* isolates obtained from suppressive soils in South Africa for the suppression of Fusarium wilt disease under glass house conditions. The results of the study indicated that two of the nonpathogenic *F. oxysporum* isolates, CAV 255 and CAV 241 recorded 87.4 and 75.0% reduction of Fusarium wilt incidence respectively. Forsyth et al. (2006) isolated three non-pathogenic *F. oxysporum* isolates from the roots of banana grown in Fusarium wilt suppressive soils and evaluated for their capability for suppressing Fusarium wilt of banana in glasshouse trials. The results showed that among the three npFo isolates examined, one isolate BRIP 29089, was associated with a significant reduction in internal disease symptom development, with 25 % of plants showing mild vascular discoloration caused by *Foc* race 1 and race 4 in Lady Finger and Cavendish (cv. Williams) group of banana respectively. Interestingly, Cavendish plants treated with isolate BRIP 45952, and inoculated with *Foc*, displayed a significant increase in internal symptom development, with 50 % of the plants showing severe vascular discolouration. Hence, it is important to understand that npFo can either reduce or increase the disease severity based on the nature of the strains used and hence one should be cautious while selecting strain for disease control (Forsyth et al. 2006). Ting et al. (2008) demonstrated the potential of endophytic microorganisms in promoting the growth parameters (plant height, pseudostem diameter, root mass and total number of leaves) of their host plant by artificially introducing five isolates of bacterial and fungal strains isolated from the roots of wild bananas into both healthy and diseased banana plantlets (Berangan cv. Intan). The results indicated that among the five isolates tested the bacterial isolate UPM39B3 (*Serratia*) and fungal isolate UPM31P1 (*Fusarium oxysporum*) showed tolerance towards Fusarium wilt via improving vegetative growth of the plant. This "tolerance" to disease may also be attributed to direct inhibition of the pathogen through the production of antifungal compounds (White and Cole 1985; Koshino et al. 1989). Thangavelu and Jayanthi (2009) selected two npFo isolates (Ro-3 and Ra-1) out of 33 obtained from banana rhizosphere soil based on mycelial growth and spore germination under *in-vitro* condition. These two npFo isolates were evaluated under both pot culture and field conditions by application: (i) at planting; (ii) at planting + 2 months after planting; and (iii) at planting + 2 months after planting + 4 months after planting; in tissue-cultured as well as in sucker derived plants of cv. Rasthali (Silk-AAB). The result showed that soil application of Ro3 npFo isolate three times in both tissue-cultured and sucker derived plants of banana registered 89% reduction of Fusarium wilt severity and significant increase in plant growth parameters when compared with Foc alone inoculated banana plants.

The modes of actions of non- pathogenic Fusarium isolates suggested commonly are: competition for nutrients (Couteaudier and Alabouvette, 1990), competition for infection sites at the root surface or inside the roots (Fravel et al. 2003) production of secondary metabolites, which cause antibiosis and antixenosis and induced resistance (Clay, 1991; Dubois et al. 2006). Some endophytes with growth promoting properties are also useful in enhancing tolerance to diseases by growth promotion (Ting et al. 2009).

Although the non-pathogenic Fusarium isolates are useful in controlling the Fusarium wilt disease, the main concern are: i) whether the biocontrol agent is truly nonpathogenic, ii) whether it may be pathogenic on a species of plant on which it has not yet been tested and iii) whether the biocontrol agent could become pathogenic in the future.

8. Biocontrol agents for tissue cultured plants

In the case of micro-propagated banana plants, its usage as planting material leads to a reduction in the spread of *Foc*, but at the same time, resulted in enhanced susceptibility to *Foc* under field conditions (Smith et al. 1998) due to the loss of native endophytes during tissue culture, including beneficial plant growth promoting rhizobacteria and fungi (Nowak, 1998; Smith et al.1998). Therefore, biotization of tissue culture plantlets with native effective non-pathogenic endophytic microbes including mycorrhizal fungi during first or second stage hardening but before planting, enhance plant resistance to tissue cultured plants against Fusarium wilt (Nowak, 1998). Lian et al. (2009) reported that re-introduction of naturally occurring endophytes to tissue culture banana plantlets resulted in a substantial reduction in the infection and severity of Fusarium wilt disease (67%) as well as increased plant growth parameters (height, girth, leaf area). Arbuscular mycorrhiza (AM) fungi are the most beneficial symbiotic fungi, increases nutrient uptake ability of the plant roots, by enhancing the water transport in the plant thus increasing the growth and yield. Besides, these fungi have also been shown to provide physical barrier against invading pathogens and thus reduce disease severity in short-term green house studies. The application of *Glomus* spp to micropropagated banana plantlets (Grand Naine) reduced the internal and external symptoms of *Foc* race 4 and enhanced plant development and nutrient uptake of the plants (Jaizme-vega et al. 1998). Jie et al. (2009) re-introduced mixture of naturally-occurring uncultivated endophytes (dominated by γ-Proteobacteria) isolated from native healthy banana plant into tissue culture banana plantlets led to 67% suppression rate of wilt disease at the fifth month after pathogen infection on plantlets in the greenhouse. In addition to disease suppression, growth of host plantlets was also promoted with the inoculation of these endophytes both in pathogen- infected and healthy control plants. They proposed that the suppression of wilt disease was due to increased activities of PPO, POD and SOD enzymes in the plantlets inoculated with endophytic communities.

9. Suppressive soil for the biological control of Fusarium wilt

Suppressive soils are sites where, despite the presence of a virulent pathogen and susceptible host, disease either does not develop, or the severity and spread of disease through the site is restricted (Alabouvette et al. 1993). This type of suppressive soils for

Fusarium wilt has been reported in many regions of the world. Although the suppression has generally been shown to be due to soil physical structure (type of soil, drainage condition, presence of montmorillonoid soils and pH) nutritional status and microbial composition (Fungi, bacteria and Actinomycetes) and biological factors also said to play a major role (Scher and Baker, 1982; Alabouvette et al. 1993). Biological control of Fusarium wilts of numerous crops by application of antagonistic fungi and bacteria isolated from suppressive soils has been accomplished during the last two decades all over the world (Leeman et al., 1996; Lemanceau et al., 1992; Park et al., 1988; Raaijmakers et al., 1995). Most of the studies have found that non-pathogenic strains of *F. oxysporum* are associated with the natural suppressiveness of soil to Fusarium wilt diseases (Smith and Snyder, 1971; Alabouvette, 1990; Postma & Rattink, 1992). These npFo colonize the plant rhizosphere and roots without inducing any symptoms in the plants (Olivain and Alabouvette, 1997). Nel et al. (2006) evaluated the ability of non-pathogenic *F. oxysporum* and *Trichoderma* isolates from suppressive soils in South Africa to suppress Fusarium wilt of banana in the glasshouse. The results revealed that only npFo isolates CAV 255 and CAV 241, reduced Fusarium wilt incidence by 87·4 and 75·0%, respectively. Smith et al. (1999) proposed that application of biocontrol agents isolated from banana roots grown in Fusarium wilt suppressive soil of tissue culture plantlets in the nursery. By application of these biocontrol agents, the banana roots had a better chance of protection against *Foc*. Generally, the microbial activity in suppressive soil is influenced by type of clay minerals present in the soil. In tropical America, a close relationship was found between suppression of Fusarium wilt and presence of clay (montmorillonoid type) soils, where as in the Canary Islands, suppression was associated with host mineral nutrition (Ploetz, 2000).

10. Integrated approach of Fusarium wilt management

In general, most of the available approaches for biocontrol of plant diseases are involved in the use of a single biocontrol agent to a single pathogen (Raupach and Kloepper, 1998). This has led to inconsistent performance of biocontrol agents and poor activity in all soil environments in which they are applied or against all pathogens that attack the host plant. To overcome these problems, applications of mixtures of biocontrol agents having multiple mode of actions are advocated particularly under field conditions, where they are highly influenced by abiotic and biotic conditions (Duffy et al., 1996; Raupach and Kloepper, 1998; Guetsky et al., 2001). Integration of biocontrol with agronomic practices may also improve the efficacy of the biocontrol organisms and the health of the host plants, which may be sensitive to environmental changes. Under this situation, compatible interactions are an important pre-requisite for the successful development of an integrated approach for the control of plant diseases. In the case of banana, integration of multiple control methods was more effective than single method for controlling Fusarium wilt disease in banana. Saravanan et al. (2003) carried out both *in-vitro* and *in-vivo* studies with biocontrol agents along with organic manures to develop integrated disease management practices to control Fusarium wilt disease. They found that basal application of neem cake at 0.5 kg/plant + sucker dipping in spore suspension of *Peudomonas fluorescens* for 15 min+soil application of *P. fluorescens* at 10 g/plant at 3,5 and 7 months after planting showed the greatest suppression of wilt disease and this was on par with basal application of neem cake at 0.5

kg/plant + soil application of *P. fluorescens* at 10 g/plant at 3,5 and 7 months after planting. They also reported that *Trichoderma viride* applied as soil or sucker dipping or their combinations or along with the neem cake also had a significant reduction in disease index, but less than that of *P. fluorescens*. Raghuchander et al. (1997) reported that dipping of suckers in the suspension of *T.viride* along with application of 500 g of wheat bran and saw dust inoculation (1: 3) of the respective bio control agent effectively reduced Fusarium wilt incidence in banana. Kidane and Laing (2010) developed integrated method of controlling Fusarium wilt by integrating biological and agronomic control methods. Single and combined applications of non-pathogenic, endophytic *Fusarium oxysporum* N16 strain by dipping their roots in a spore suspension containing 10^7 cfu ml^{-1}, *Trichoderma harzianum* Eco-T® (Plant Health Products (Pty) Ltd. KwaZulu-Natal, South Africa) @ 4L^{-pt} at a concentration of 10^5 conidia ml^{-1} at the time of planting, monthly application of plants with 4 L of silicon solution per plant containing 900mg silicon L^{-1} and placing coarse macademia husks at the bottom of banana plants as mulching were tested against *F. oxysporum* f. sp. *cubense* on bananas under greenhouse and field conditions. The results showed that treatments involving combinations of nonpathogenic *F. oxysporum*, *T. harzianum* Eco-T®, silicon and mulch had significantly higher number of leaves, stem height and girth size than single applications of the treatments. They found that the mulching increased the growth of feeder roots and created a conducive microenvironment, thereby increased the microbial activity in the soil. The combined application of non-pathogenic Fusarium strain along with silicon also resulted in reduction of corm disease index by more than 50% and shoot yellowing and wilting by 80%. Therefore, integration of biocontrol with agronomic practices improved the efficacy of the biocontrol organisms and the health of the host plants. Recently Zhang et al. (2011) evaluated the effects of novel bio-fertilizers, which combined an amino acid fertilizer and mature pig manure compost with the antagonists *Paenibacillus polymyxa* SQR21, *Trichoderma harzianum* T37 and *Bacillus subtilis* N11 (isolated from the healthy banana roots) in a severely Fusarium wilt diseased field for the suppression of Fusarium wilt of banana as pot experiments. The results showed that the bio-organic fertilizers which contained the bio-agents significantly suppressed the incidence of wilt disease (by 64–82%), compared to the control. The best biocontrol effect was obtained in the treatment with the BIO2 that contains *Bacillus subtilis* N11. The reason for more effect might be due to the application of the antagonists in combination with suitable organic amendments.

Botanical fungicides are also gaining momentum as these are considered as an alternative source for chemicals in the management of soil borne pathogens. The active principles present in both bio-agents and botanicals may either act on the pathogen directly or induce systemic resistance in the host plants resulting in reduction of disease development (Paul and Sharma, 2002). Akila et al. (2011) tested two botanical fungicides from *Datura metel*-Wanis 20 EC and Damet 50 EC along with *Pseudomonas fluorescens*, Pf1 and *Bacillus subtilis*, TRC 54 individually and in combination for the management of Fusarium wilt under greenhouse and field conditions. Combined application of botanical formulation and biocontrol agents (Wanis 20 EC + Pf1 + TRC 54) reduced the wilt incidence significantly under greenhouse (64%) and field conditions (75%). The reduction in disease incidence was positively correlated with the induction of defense-related enzymes peroxidase and polyphenol oxidase.

Sl. no	Name of biocontrol agents	Mode of action	References
1.	*Trichoderma viride*	Induction of defense related enzymes, production of antibiotics	Thangavelu and Mustaffa, 2010
2.	*Pseudomonas* spp.	Production of volatiles (2-Pentane 3-methyl, methanethil and 3-undecene, antibiotics DAPG and Siderophore production.	Ting et al. 2011
3	*Pseudomonas aeruginosa*	Production of antibiotics (2,4-Diacetyl Phloroglucinol	Saravanan and Muthusamy, 2006
4	*P. fluorescens*	Competition for space, cell wall appositions lining the cortical cell wall	Sukhada et al. 2004
5.	*Bacillus* spp.	Antibiotics, induction of defense related enzymes such as Peroxidase and Polyphenol oxidase.	Sukhada et al. 2004
6.	*Streptomyces violaceusniger*	Production of Antibiotics	Getha et al. 2005
7.	*Streptomyces violaceusniger*	Production of Antibiotics	Getha and Vikineswary, 2002
8	Non-pathogenic Fusarium	Plant growth promotion	Ting et al. 2009
9	*Serratia* sp.	Plant growth promotion	Ting et al. 2008
10	*F. oxysporum*	Plant growth promotion	Ting et al. 2008
11	γ-Proteobacteria	Increase in Polyphenol oxidase, Peroxidase, Superoxide dismutase,	Jie et al. 2009
12.	*P. fluorescens*	Induction of defense related enzymes such as Peroxidase &Polyphenol oxidase	Akila et al. 2011
13.	*Bacillus subtilis*	Induction of defense related enzymes such as Peroxidase &Polyphenol oxidase	Akila et al. 2011

Table 1. Summary of Bio-control agents used in the management of Fusarium wilt disease of banana with their mode of action.

A

B

Fig. 1. A) External symptoms (yellowing and buckling of leaves) of Fusarium wilt infected banana plant. B) Brown vascular discoloration in the Pseudostem C) Brown vascular discoloration in the corm of Fusarium wilt infected plant D) Microscopic view of both macro and micro conidia of *Foc*.

11. Conclusion

Although several biocontrol agents including botanicals have been tried against Fusarium wilt disease, still this lethal disease could not be controlled completely. Besides most of the biocontrol experiments were conducted either under lab condition or green house conditions and only in few cases, field experiments were conducted. Therefore, most of the bioagents tested against Fusarium wilt of banana have not yet registered and reached the end users ie. banana growers. This is mainly because of lack of confidence on the efficacy and consistency of the bioagents in controlling the disease. Therefore, for evolving consistent and effective biological control methods for the management of Fusarium wilt disease are i) the *Foc* pathogen present in a particular area or country must be characterized thoroughly up to VCG level and the bio-agents isolated must be screened under both *in vitro* and *in vivo* conditions ii) the bio-agents having multiple mode of actions and functions should be selected rather than selecting bioagents with one or two mode of actions. In addition, mixture of bioagents of different genera or mixture of fungal and bacterial bioagents along with or without fungicides or botanicals have to be tried to improve the level and extent of disease control under different environmental and soil conditions iii) the compatibility between bioagents or tolerance of bioagents to chemicals or botanicals must be tested, iv) suitable method of mass production and delivery system which support more number of propagules and long shelf life, easy to prepare and adopt must be selected, v) mass produced bioagents should be applied at right quantity (the initial inoculum level of bioagents should be more than the inoculum level of the pathogen) at the right place (at the soil around the rhizosphere) at the right time (before planting or at the time of planting and also at 2nd and 4th month after planting as booster application) and at the appropriate physiological state, vi) mass production and delivery system should be compatible with the production system of banana, vii) application of bioagents with other organic amendments which can support the survival and multiplication of bio-agents and vii) integration of biological control with other cultural or agronomic practices so that the Fusarium wilt disease can be controlled effectively.

12. References

Akila, R., Rajendran, L., Harish, S,. Saveetha, K., Raguchander, T., Samiyappan. R., 2011. Combined application of botanical formulations and biocontrol agents for the management of *Fusarium oxysporum* f. sp. *cubense* (Foc) causing Fusarium wilt in banana. Biological Control 57, 175–183.

Alabouvette, C., 1986. Fusarium wilt suppressive soils from the Chateaurenard region: reviews of a 10 year study. Agronomie 6, 273–284.

Alabouvette, C., 1990. Biological control of *Fusarium* wilt pathogens in suppressive soils. In: Horn by, D. (Ed.), Biological Control of Soil-borne Plant Pathogens. CAB International, Wallingford, pp. 27–43.

Alabouvette, C., Lemanceau, P., Steinberg, C., 1993. Recent advances in the biological control of *Fusarium* wilts. Pesticides Science 37, 365–373.

Amsellem, Z., Zidack, N. K., Quimby, P. C., Jr., & Gressel, J. 1999. Long-term dry preservation of viable mycelia of two mycoherbicidal organisms. Crop Protection, 18, 643–649.

Anjaiah, V., Cornelis, P., Koedam, N., 2003. Effect of genotype and root colonization in biological control of *Fusarium* wilts in pigeonpea and chickpea by *Pseudomonas aeruginosa* PNA1. Canadian Journal of Microbiology 49, 85–91.

Anonymous,1977. *Fusarium oxysporum* f. sp. *cubense*, Distribution maps of plant diseases. Map No. 31, 4th ed. Commonwealth Mycological Institute, Kew, England.

Ayyadurai, N., Ravindra Naik., P Sreehari Rao. M., Sunish Kumar, R., Samrat, S.K., Manohar, M., Sakthivel, N., 2006. Isolation and characterization of a novel banana rhizosphere bacterium as fungal antagonist and microbial adjuvant in micropropagation of banana. Journal of Applied Microbiology 100, 926–937

Bancroft, J. 1876. Report of the board appointed to enquire into the cause of disease affecting livestock and plants. In: Votes and Proceedings 1877, Vol 3, Queensland, pp. 1011-1038

Bastasa, G.N., Baliad, A.A., 2005. Biological control of Fusarium wilt of abaca (Fusarium oxysporum) with Trichoderma and yeast. Philippine Journal of Crop Science (PJCS) 30(2), 29-37

Beckman, C. H., Mueller, W.C., Tessier, B.J., Harrisson, N.A., 1982. Recognition and callose deposition in response to vascular infection in Fusarium wilt-resistant or susceptible tomato plants. Physiological Plant Pathology 20, 1-10.

Benhamou,N., Chet, I., 1993. Hyphal interactions between Trichoderma harzianum and Rhizoctonia solani: Ultrastructure and gold cytochemistry of the mycoparasitic process. Phytopathology, 83, 1062-1071

Benhamou, N.,. Belanger, R. R., Paulitz, T., 1996. Ultrastructural and cytochemical aspects of the interaction between Pseudomonas fluorescens and Ri T-DNA transformed pea roots: host response to colonization by Phythium ultimum Trow, Planta 199, 105-117.

Berg. G., Fritze. A., Roskot. N., Smalla. K., 2001. Evaluation of potential biocontrol rhizobacteria from different host plants of Verticillium dahliae Kleb. Journal of Applied Microbiology 91, 963-971.

Buyer, J, S., Leong, J., 1986. Iron transport-mediated antagonism between plant growth-promoting and plant-deleterious Pseudomonas strains. Journal of Biological Chemistry 261, 791-794.

Cao, L., Qiu, Z., Dai, X., Tan, H., Lin, Y., Zhou, S., 2004. Isolation of endophytic actinomycetes from roots and leaves of banana (Musa acuminata) plants and their activities against Fusarium oxysporum f. sp. Cubense. World Journal of Microbiology & Biotechnology 20, 501-504.

Cao, L., Qiu, Z., You, J., Tan, H., Zhou,S., 2005. Isolation and characterization of endophytic streptomycete antagonists of fusarium wilt pathogen from surface-sterilized banana roots. FEMS Microbiology Letters 247, 147-152.

Carefoot. G. L., Sprott, E. R., 1969. 'Famine on the Wind.' (Angus and Robertson: London)

Chatterjee, A., Cui, Y., Liu, Y., Dumenyo, C. K., Chatterjee, A. K. 1995. Inactivation of rsmA leads to overproduction of extracellular pectinases, cellulases, and proteases in Erwinia carotovora subsp. carotovora in the absence of the starvation/cell density-sensing signal, N-(3-oxohexanoyl) - L-homoserine lactone. Applied Environmental Microbiology 61, 1959-1967.

Chin-A-Woeng, T.F.C., Bloemberg, G.V., Vander Bij, A.J., Vander Drift, K.M.G.M., Schripsema, J., Kroon, B., Scheffer, R.J., Keel, C., 1998. Biocontrol by phenazine-1-carboxamide-producing Pseudomonas chlororaphis PCL1391 of tomato root rot caused by Fusarium oxysporum f. sp. radicis lycopersici. Mol Plant Microbe Interact 11, 1069-1077.

Clay, K., 1991. Endophytes as antagonists of plant pests. In: Andrews J. H., Hirano, S. S., (eds. Huang, T.Y., 1991) Soil suppressive of banana Fusarium wilt in Taiwan. Plant Microbial ecology of leaves. Springer, New York, 331-357.

Couteaudier, Y., Alabouvette, C., 1990. Survival and inoculum potential of conidia and chlamydospores of Fusarium oxysporum f.sp. lini in soil. Canadian Journal of Microbiology 36, 551-556.

Crawford, D.L., Lynch, J.M., Whipps, J.M., Ousley, M.A., 1993. Isolation and characterization of actinomycete antagonists of a fungal root pathogen. Applied Environmental Microbiology 59, 3899-3905.

de Freitas, J.R., Germida, J.J., 1991. *Pseudomonas cepacia* and *Pseudomonas putida* as winter wheat inoculants for biocontrol of *Rhizoctonia solani*. Canadian Journal of Microbiology 37, 780–784.

de Weger, L.A., van der Vlught, C.I.M., Wijfjes, A.H.M., Bakker, P.A.H.M., Schippers, B., Lugtenberg, B.J.J., 1987. Flagella of a plant growth-stimulating *Pseudomonas fluorescens* strain are required for colonization of potato roots. Journal of Bacteriology 169, 2769–2773.

Domsch, K. H., Gams, W., Anderson, T. H., 1980. Compendium of Soil Fungi, Vol. 1. Academic Press, New York.

Dowling, D.N., O'Gara, F., 1994. Metabolites of Pseudomonas involved in the biocontrol of plant disease. Trends in Biotechnology 3, 121–141.

Dubois, T., Gold, C. S., Paparu, P., Athman S., Kapindu, S., 2006. Tissue culture and the *in vitro* environment. Enhancing plants with endophytes: potential for ornamentals? In: Teixeira S. J., (ed) Floriculture, ornamental and plant biotechnology: advances and topical issues, 3rd edn. Global Science Books, London, 397–409.

Duffy, B.K., Simon, A., Weller, D.M., 1996. Combination of *Trichoderma koningii* with fluorescent pseudomonads for control of take-all on wheat. Phytopathology 86: 188–194.

Dunlap, C., Crowley, J. J, Moënne-Loccoz, Y., Dowling, D.N, de Bruijn FJ, O'Gara F. 1997. Biological control of *Pythium ultimum* by *Stenotrophomonas maltophilia* W81 is mediated by an extracellular proteolytic activity. Microbiology 143, 3921–3931.

Dunlap, C., Delaney, I., Fenton, A., Lohrke. S., Moënne-Loccoz, Y., O'Gara, F., 1996. The biotechnology and application of *Pseudomonas* inoculants for the biocontrol of phytopathogens, 441– 448. In: Stacey, G., Mullin, B., Gresshoff, P.M., eds. *Biology of plant microbe interactions*. St Paul, MN, USA: International Society for Molecular Plant–Microbe Interactions.

Dunne, C., Delany, I., Fenton, A., O'Gara, F., 1996. Mechanisms involved in biocontrol by microbial inoculants. Agronomie 16, 721–729.

Dunne, C., Moenne, L.Y., McCarthy, J., Higgins, P., Powell, J., Dowling, D., O'Gara, F., 1998. Combining proteolytic and phloroglucinol-producingbacteria for improved biocontrol of *Pythium*-mediated damping-off of sugar beet. Plant Pathology 47, 299–307.

Dutta, S., Mishra, A. K, Dileep Kumar, B.S., 2008. Induction of systemic resistance against fusarial wilt in pigeon pea through interaction of plant growth promoting rhizobacteria and rhizobia. Soil Biol. Biochem., 40: 452-461.

Fernando, W.G.D., Watson, A. K., Paulitz, T.C., 1996. The role of *Pseudomonas* spp. and competition for carbon, nitrogen and iron in the enhancement of appressorium formation by *Colletotrichum coccodes* on velvetleaf. European Journal of Plant Pathology 102, 1–7.

Fishal, E.M.M., Meon, S., Yun, W.M., 2010. Induction of Tolerance to Fusarium Wilt and Defense-Related Mechanisms in the Plantlets of Susceptible Berangan Banana Pre-Inoculated with *Pseudomonas* sp. (UPMP3) and *Burkholderia* sp. (UPMB3). Agricultural Sciences in China 9, 1140-1149.

Forsyth, L. M., Smith, L.J., Aitken, E., A. B., 2006. Identification and Characterization of non-pathogenic *Fusarium oxysporum* capable of increasing and decreasing Fusarium wilt severity. Mycological Research 30, 1-7.

Fravel, D., Olivain, C., Alabouvette, C., 2003. *Fusarium oxysporum* and its biocontrol. New Phytologist 157, 493–502.

García de Salamone, I.E., Hynes, R.K., Nelson, L. M., 2001. Cytokinin production by plant growth promoting rhizobacteria and selected mutants. Canadian Journal of Microbiology 47, 404–411.

Gerlach, K.S., Bentley, S., Moore, N.Y., Aitken, E.A.B., Pegg, K.G., 1999. Investigation of non-pathogenic strains of *Fusarium oxysporum* for suppression of Fusarium wilt of banana in Australia. In: Alabouvette C, ed. Second International Fusarium Workshop. Dijon, France.

Getha K., Vikineswary, S., Wong, W., Seki, T., Ward, A., Goodfellow, M., 2005. Evaluation of *Streptomyces* sp. strain G10 for suppression of Fusarium wilt and rhizosphere colonization in pot grown banana plantlets. Journal of Industrial Microbiology and Biotechnology 32, 24-32.

Getha, K., Vikineswary, S., 2002. Antagonistic effects of *Streptomyces violaceusniger* strain G10 on *Fusarium oxysporum* f.sp. *cubense* race 4: Indirect evidence for the role of antibiosis in the antagonistic process. Journal of Industrial Microbiology & Biotechnology. 28, 303 – 310.

Gordon, T.R., Martyn, R. D. 1997. The evolutionary biology of *Fusarium oxysporum*. Annu Rev Phytopathol 35, 111–28.

Guerra-Cantera MARV, Raymundo, A.K., (2005). Utilization of a polyphasic approach in the taxonomic reassessment of antibiotic and enzyme-producing *Bacillus* spp. isolated from the Philippines. World. J. Microb. Biot., 21: 635-644

Guetsky, R., Shtienberg, D., Elad, Y., Dinoor, A., 2001. Combining biocontrol agents to reduce the variability of biological control. Phytopathology 91, 621–627.

Han, S.H., Lee, S.J., Moon, J.H., Yang, K.Y., Cho, B.H., Kim, K.Y., Kim, Y.W., Lee, M.C., Anderson, A.J., Kim, Y.C., 2006. GacS-dependent production of 2R, 3R butanediol by Pseudomonas chlororaphis O6 is a major determinant for eliciting systemic resistance against Erwinia carotovora but not against *Pseudomonas syringae* pv. *tabaci* in tobacco. Interaction 19, 924–930.

Handelsman, J., Stabb, E.V., 1996. Biocontrol of soilborne plant pathogens. Plant Cell 8, 1855–1869.

He, C.Y., Hsiang, T., Wolyn, D.J., 2002. Induction of systemic disease resistance and pathogen defence responses in *Asparagus officinalis* inoculated with non-pathogenic strains of *Fusarium oxysporum*. Plant Pathology 51, 225–30.

Herbert, J.A., Marx, D., 1990. Short-term control of Panama disease in South Africa. Phytophylactica 22, 339–340.

Hwang, S.C., 1985. Ecology and control of *Fusarium* wilt of banana. Plant Protection Bulletin (Taiwan) 27, 233-245.

Jaizme-Vega M.C., Hernández, B.S., and Hernández, J.M., 1998. Interaction of arbuscular mycorrhizal fungi and the soil pathogen *Fusarium oxysporum* f.sp. *cubense* on the first stages of micropropagated Grande Naine banana. Acta Horticulturae 490, 285–95.

Jeun, Y.C., Park, K.S., Kim, C., Fowler, W.D., Kloepper, J.W., 2004. Cytological observations of cucumber plants during induced resistance elicited by rhizobacteria. Biol Control 29, 34–42.

Jie, L., Zifeng, W., Lixiang, C., Hongming, T., Patrik, I., Zide, J., Shining, Z., 2009. Artificial inoculation of banana tissue culture plantlets with indigenous endophytes originally derived from native banana plants. Biological control. 51, 427-434.

Jijakli, M.H., Lepoivre, P., 1998. Characterization of an exo-beta-1, 3-glucanase produced by *Pichia anomala* strain K, antagonist of *Botrytis cinerea* on apples. Phytopathology 88, 335–343.

Kidane, E.G., Laing, M.D., 2010. Integrated Control of Fusarium Wilt of Banana (*Musa* spp.) In. Proc. IC on Banana & Plantain in Africa Eds: T. Dubois et al. Acta Horticulture, 879, 315-321.

Kavitha, R., Umesha, S., 2008. Regulation of defense-related enzymes associated with bacterial spot resistance in tomato. Phytoparasitica 36, 144-159.

Kloepper, J.L., Leong, J., Teintze, M., Schroth, M.N., 1980. *Pseudomonas* siderophores: a mechanism explaining disease-suppressive soils. Curr. Microbiol 4, 317–320.

Knox O.G.G., Killham, K., Leifert, C., 2000. Effects of increased nitrate availability on the control of plant pathogenic fungi by the soil bacterium Bacillus subtilis. Appl. Soil Ecol 15, 227-231.

Koshino, H., Terada, S., Yoshihara, T., Sakamura, S., Shimanuki, T., Sato, T., Tajimi, A., 1989. A ring B aromatic sterol from stromata of *Ephichloe typhina*. Phytochemistry, 28, 771-772.

Lahdenpera, M. L., Oy, K., 1987. The control of Fusarium wilt on carnation with a Streptomyces preparation. Acta Horticult 216, 85– 92.

Lakshmanan, P., Selvaraj, P., Mohan, S., 1987. Efficiency of different methods for the control of Panama disease. Trop. Pest Manage 33, 373–376.

Larkin, R.. Fravel, D., 1998. Efficacy of various fungal and bacterial biocontrol organisms for the control of Fusarium wilt of tomato. Plant Disease 82, 1022-1028.

Leelasuphakul, W., Sivanunsakul, P., Phongpaichit, S., 2006. Purification, characterization and synergistic activity of β-1,3- glucanase and antibiotic extract from an antagonistic Bacillus subtilis NSRS 89-24 against rice blast and sheath blight. Enzym. Microb. Technol 38, 990-997.

Leeman, M., Vanpelt, J.A., Den Ouden, F.M., Heinsbroek, M., Bakker, P.A.H.M., Schippers, B., 1996. Iron availability affects induction of systematic resistance to fusarium wilt to radish by *Pseudomonas fluorescens*. Phytopathology 86, 149–155.

Lemanceau, P., Alabouvette, C., 1991. Biological control of *Fusarium* diseases by fluorescent *Pseudomonas* and non-pathogenic *Fusarium*. Crop Protection 10, 279-286.

Lemanceau, P., Bakker, P.A.H.M., DeKogel, W.J., Alabouvette, C., Schippers, B., 1992. Effect of pseudobactin 358 production of *Pseudomonas putida* wcs 358 on suppression of fusarium wilt of carnations by non pathogenic Fusarium oxysporum FO47. Applied Environmental Microbiology 58, 2978-2982.

Leong, J., 1986. Siderophores: their biochemistry and possible role in the biocontrol of plant pathogens. Annual Review of Phytopathology 24, 187–209.

Lian, J., Wang, Z., Cao, L., Tan, H., Inderbitzin, P., Jiang, Z., Zhou, S., 2009. Artificial inoculation of banana tissue culture plantlets with indigenous endophytes originally derived from native banana plants. Biological Control 51, 427–434.

Loper, J.E., Buyer, J.S., 1991. Siderophores in microbial interactions on plant surfaces. Molecular Plant–Microbe Interaction 4, 5–13.

Lugtenberg, B.J.J., de Weger, L.A., Bennett, J.W., 1991. Microbial stimulation of plant growth and protection from disease. Current Opinions in Biotechnology 2, 457–464.

Lugtenberg, B.J.J., de Weger, L.A., Schippers, B., 1994. Bacterization to protect seed and rhizosphere against disease. BCPC Monograph 57, 293–302.

Lugtenberg, B.J.J., Dekkers, L.C., Bansraj, M., Bloemberg, G.V., Camacho, M., Chin-A-Woeng, T.F.C., van den Hondel, C., Kravchenko, L., Kuiper, I., Lagopodi, A.L., Mulders, I., Phoelich, C., Ram, A., Tikhonovich, I., Tuinman, S., Wijffelman, C., Wijfjes A., 1999b. *Pseudomonas* genes and traits involved in tomato root colonization. In: De Wit PJGM, Bisseling, T., Stiekema, W.J., eds 1999. IC-MPMI Congress Proceedings: biology of plant–microbe interactions, Vol. 2. St Paul, MN, USA: International Society for Molecular Plant–Microbe Interactions, 324–330.

Lugtenberg, B.J.J., Kravchenko, L.V., Simons, M. 1999a. Tomato seed and root exudate sugars: composition, utilization by *Pseudomonas* bio-control strains and role in rhizosphere colonization. Environmental Microbiology 1, 439–446.

M'Piga, P., Belanger, R.R., Paulitz, T.C., Benhamou, N., 1997. Increased resistance to *Fusarium oxysporum* f. sp. *radicis-lycopersici* in tomato plants treated with the endophytic bacterium *Pseudomonas fluorescens* strain 63–28. Physiology Molecular Plant Pathology 50, 301–320.

Marois, J.J., Mitchel, D.J., Somada, R.M. 1981. Biological control of Fusarium crown and root rot of tomato under field condition. Pytopathology 12, 1257-1260.

Molina, A.B., Valmayor, R. V., 1999. Banana production systems in South East Asia. Bananas and Food security, Pica C., Foure, E., Frison, E.A., (eds.), INIBAP, Montpellier, France, 423-436.

Moore, N.Y., Pegg, K.G., Bentley, S., Smith, L.J., 1999. Fusarium wilt of banana: global problems and perspectives. In: Molina, A.B., Masdek, N.H.N. Liew, K.W, (eds). Banana Fusarium Wilt Management: Towards Sustainable Cultivation. Proceedings of the International Workshop on Banana Fusarium Wilt Disease. Kuala Lumpur, Malaysia: INIBAP, 11–30.

Morpurgo, R., Lopato, S.V., Afza, R., Novak, F.J., 1994. Selection parameters for resistance to *Fusarium oxysporum* f.sp.*cubense* race 1 and race 4 on diploid banana (Musa acuminata Colla). Euphytica 75, 121-129.

Mueller, W.C., Beckman C.H., 1988. Correlated light and EM studies of callose deposits in vascular parenchyma cells of tomato plants inoculated with *Fusarium oxysporum* f.sp. *lycopersici*. Physiological and Molecular Plant Pathology 33, 201–208.

Murphy, J.F., Reddy, M.S., Ryu, C.M., Kloepper, J.W., Li, R., 2003. Rhizobacteria mediated growth promotion of tomato leads to protection against Cucumber mosaic virus. Phytopathology 93, 1301–1307

Nel, B., Steinberg, C., Labuschagne, N., Viljoen, A., 2006. The potential of nonpathogenic *Fusarium oxysporum* and other biological control organisms for suppressing fusarium wilt of banana. Plant Pathology 55, 217–223.

Nelson, P. E., Toussoun, T. A., Marasas, W.F.O., 1983. *Fusarium* species: An illustrated Manual for identification. Pennsylvania State University Press, University Park.

Nowak, J., 1998. Benefits of *in vitro* 'biotization' of plant tissue cultures with microbial inoculants. *In vitro* cell development biology-Plant 34, 122-130

Olivain C, Alabouvette C., 1999. Process of tomato root colonization by a pathogenic strain of *Fusarium oxysporum* f. sp. *lycopersici* in comparison with a non-pathogenic strain. New Phytologist 141, 497–510.

Olivain, C., Alabouvette, C., 1997. Colonization of tomato root by a non-pathogenic strain of *Fusarium oxysporum*. New Phytologist 137, 481-494.

Papavizas, G.C., 1985. Trichoderma and Gliocladium: biology, ecology and potential for biocontrol. Annu Rev Phytopathol 23, 23–54.

Park, C.S., Paulitz, T.C., Baker, R., 1988. Biocontrol of Fusarium wilt of cucumber resulting from interactions between *Pseudomonas putida* and non-pathogenic isolates of *Fusarium oxysporum*. Phytopathology 78, 190–4.

Parke, J. L., 1990. Population dynamics oi *Pseudomona.s cepacia* in the pea spermosphere in relation to biocontroi of *Pythium*. Phytopathology 80, 1307-1311.

Paul, P.K., Sharma, P.D., 2002. Azadirachta indica leaf extract induces resistance in barley against leaf stripe disease. Physiology and Molecular Plant Pathology 61, 3–13.

Paul, P.K., Sharma, P.D., 2002. *Azadirachta indica* leaf extract induces resistance in barley against leaf stripe disease. Physiology and Molecular Plant Pathology 61, 3–13

Pieterse, C.M.J., van Pelt J.A., van Wees S.C.M., Ton, J., Leon-Kloosterziel K.M., Keurentjes J.J.B., Verhagen B.W.M., van Knoester, M,, dSI, Bakker, P.A.H.M., van Loon, L.C., 2001. Rhizobacteria-mediated induced systemic resistance: triggering, signalling and expression. European Journal of Plant Pathology 107,51–61.

Ploetz, R. C., 2005. Panama disease, an old enemy rears its ugly head: Parts 1 and 2. In: Plant Health Progress, APSnet: Online doi:10.1094/PHP-2005-1221-01-RV.

Ploetz, R. C., Pegg, K. G., 1997. *Fusarium* wilt of banana and Wallace's line: Was the disease originally restricted to his Indo-Malayan region? Australasian Plant Pathology 26, 239-249.

Ploetz, R. C., Pegg, K. G., 2000. Fusarium wilt. Pages 143-159 In: Diseases of Banana, Abacá and Enset. D. R. Jones, ed. CABI Publishing, Wallingford, UK.

Ploetz, R.C., 2000. Panama disease: a classic and destructive disease of banana. Plant Health Progress 10, 1–7.

Postma, J., Rattink, H., 1992. Biological control of Fusarium wilt of carnation with a non-pathogenic isolate of *Fusarium oxysporum*. Canadian Journal of Botany 70, 1199–205.

Raaijmakers, J.M., Leeman, M., van Oorschot, M.M.P., er Sluis, I.V., Schippers, b., bakker, P.A.h.m., 1995. Dose-response relationships in biological control of *Fusarium* wilt of radish by *pseudomonas* spp. Phytopathology 85, 1075-1081.

Radjacommare, R., Ramanathan, A., Kandan, A., Harish, S., Thambidurai, G., Sible, G.V., Ragupathy, N., Samiyappan, R., 2004. PGPR mediates induction of pathogenesis – related (PR) proteins against the infection of blast pathogen in resistant and susceptible fingermillet cultivars. Plant and Soil 266, 165–176.

Raguchander, T., Jayashree, K., Samiyappan, R., 1997. Management of *Fusarium* wilt of banana using antagonistic microorganisms. Journal of Biological Control 11, 101–105.

Rahimi, S., Perry, R. N., Wright, D.G. 1996. Identification of pathogenesis related proteins induced in leaves of potato plants infected with potato cyst nematodes, Globodera species. Physiological Molecular Plant Pathology 49, 49–59.

Rajappan, K., Vidhyasekaran, P., Sethuraman, K., Baskaran, T. L., 2002. Development of powder and capsule formulations of *Pseudomonas fluorescens* strain Pf-1 for the control of banana wilt. Zeitschrift für Pflanzenkrankheiten und Pflanzenschutz 109, 80–87.

Raupach, G.S., Kloepper, J.W., 1998. Mixtures of plant growth-promoting rhizobacteria enhance biological control of multiple cucumber pathogens. Phytopathology 88, 1158–1164.

Reddi, G. S., Rao. A. S., 1971. Antagonism of soil actinomycetes to some soil - borne plant pathogenic fungi. Indian Phytopathol 24, 649–657.

Remans, R., Beebe, S., Blair, M., Manrique, G., Tovar, E., Rao, I., Croonenborghs, A., Torres-Gutierrez, R., El-Howeity, M., Michiels, J., Vanderleyden, J., 2008. Physiological and genetic analysis of root responsiveness to auxin-producing plant growth-promoting bacteria in common bean (*Phaseolus vulgaris* L.). Plant and Soil 302, 149–161.

Rhodes, D.J., Powell, K.A., 1994. Biological seed treatments – the development process. BCPC Monograph 57, 303–310.

Ross, I. L., Alami, Y., Harvey, P.R., Achouak, W., Ryder, M.H., 2000. Genetic diversity and biological control activity of novel species of closely related pseudomonads isolated from wheat field soils in South Australia. Applied Environmental Microbiology 66, 1609–1616.

RuizDuenas, F.J., Martinez, M.J., 1996. Enzymatic activities of *Trametes versicolar* and *Pleurotus eryngii* implicated in biocontrol of *Fusarium oxysporum* f. sp. *lycopersici*. Current Microbiology 32, 151–155.

Sakthivel, N., Gnanamanickam, S.S., 1987. Evaluation of *Pseudomonas fluorescens* for suppression of sheath rot disease and for the enhancement of grain yields in rice (Oryza sativa L.). Applied Environmental Microbiology 53, 2056–2059.

Sands, D.C., Rovira, A.D., 1971. *Pseudomonas fluorescens* biotype G, the dominant fluorescent pseudomonad in South Australian soils and wheat rhizospheres. Journal of Applied Bacteriology 34, 261–275.

Saravanan, T., Muthusamy, M., 2006. Influence of *Fusarium oxysporum* f. sp. *cubense* (e.f. smith) Snyder and Hansen on 2, 4- diacetylphloroglucinol production by pseudomonas fluorescens migula in banana rhizosphere. Journal of plant protection research 46, 241-254

Saravanan, T., Muthusamy, M., Marimuthu, T., 2003. Development of integrated approach to manage the Fusarial wilt of banana. Crop Protection 22, 1117–1123.

Scher, F. M., Baker, R., 1982. Effect of *Pseudomonas putida* and a synthetic iron chelator on induction of soil suppressiveness to Fusarium wilt pathogen. Phytopathology 72, 1567-1573.

Shapira, R., Ordentlich, A., Chet, I., Oppenheim, A.B., 1989. Control of plant diseases by chitinase expressed from cloned DNA in *Escherichia coli*. Phytopathology 79, 1246–1249.

Sivamani, E., Gnanamanickam, S. S., 1988. Biological control of *Fusarium oxysporum* f.sp. *cubense* in banana by inoculation with *Pseudomonas fluorescens*. Plant Soil 107, 3 9.

Sivan, A., Chet, I., 1986. Biological control of *Fusarium* spp. in cotton, wheat and muskmelon by *Trichoderma harzianum*. J.Phytopathol. 116, 39–47.

Smith, J., Putnam, A., Nair, M., 1990. *In vitro* control of Fusarium diseases of *Asparagus officinalis* L. with a *Streptomyces* or its polyene antibiotic, faeriefungin. J Agric Food Chem 38, 1729–1733.

Smith, M. R., Hamil, S. D., Doogan, V. J., Daniells, J. W., 1999. Chracterization and early detection of an off type from micropropagated Lady Finger bananas. Australian Journal of Experimental Agriculture 39,1017-023.

Smith, M., Wiley, A., Searle, C., Langdon, P., Schaffer, B., Pegg, K., 1998. Micropropagated bananas are more susceptible to *Fusarium* wilt than plants grown from conventional material. Australian Journal of Agricultural Research 49, 1133–1139.

Smith, S. N., Snyder, W. C., 1971. Relationship of inoculum density and soil types to severity of *Fusarium* wilt of sweet potato. Phytopathology 61, 1049-1051.

Srinivasan, U., Staines, H. J., Bruce, A.,1992. Influence of media type on antagonistic modes of *Trichoderma* spp. against wood decay basidiomycetes, Mater. Org. 27, 301–321.

Stougard, J., 2000. Regulators and regulation of legume root nodule development. Plant Physiology 124, 531–540.

Stover, R. H., 1962. Fusarial Wilt (Panama Disease) of Bananas and Other *Musa* Species. Commonwealth Mycological Institute, Kew, England.

Su, H. J., Hwang, S. C., Ko, W. H., 1986. Fusarial wilt of Cavendish bananas in Taiwan. Plant Disease 70, 814–818.

Sukhada, M., Manamohan, M., Rawal, R.D., Chakraborty, S., Sreekantappa, H., Manjula, R., Lakshmikantha, H.C., 2004. Interaction of *Fusarium oxysporum* f.sp. *cubense* with *Pseudomonas fluorescens* precolonized to banana roots. World Journal of Microbiology & Biotechnology 20, 651–655.

Sun, J.B., Peng, M., Wang, Y.G., Zhao P.J., Xia Q.Y., 2011. Isolation and characterization of antagonistic bacteria against *Fusarium* wilt and induction of defense related enzymes in banana. African Journal of Microbiology Research 5, 509-515.

Suslow, T. V., 1982. Role of root-colonizing bacteria in plant growth. In: Mount, M. S., and G.H. Lacy (eds), Phytopathogenic prokaryotes. Vol. I, pp. 187-223. Academic Press, Inc., New York,

Thangavelu, R. 2002. Characterization of *Fsarium oxysporum* schlecht. f.sp. *cubense* (e.f. smith) snyd. & hans. and Molecular Approaches for the Management of Fusarium Wilt of Banana. Ph.D. thesis. Tamil Nadu Agricultural University, Coimbatore,Tamil Nadu, India,. 254 pp.

Thangavelu, R., and Mustaffa M.M., 2010. A Potential isolate of *Trichoderma viride* NRCB1and its mass production for the effective management of *Fusarium* wilt disease in banana. Tree and Forestry Science and Biotechnology 4 (Special issue 2), 76-84.

Thangavelu, R., Palaniswami, A., Ramakrishnan, G., Sabitha, D., Muthukrishnan, S., Velazhahan, R., 2001. Involvement of Fusaric acid detoxification by *Pseudomonas fluorescens* strain Pf10 in the biological control of Fusarium wilt of banana caused by *Fusarium oxysporum* f.sp. *cubense*. Journal of Plant Disease and Protection 108, 433-445.

Thangavelu, R., Palaniswami, A., Velazhahan, R., 2004. Mass production of *Trichoderma harzianum* for managing *Fusarium* wilt of banana. Agriculture, Ecosytems and Environment 103, 259–263.

Thangavelu,R., Jayanthi, A., 2009. RFLP analysis of rDNA-ITS regions of native non-pathogenic *Fusarium oxysporum* isolates and their field evaluation for the suppression of Fusarium wilt disease of banana. Australasian Plant Pathology 38, 13–21.

Ting, A.S.Y., Mah, S.W., Tee, C.S., 2011. Detection of potential volatile inhibitory compounds produced by endobacteria with biocontrol properties towards *Fusarium oxysporum* f. sp. *cubense* race 4. World J Microbiol Biotechnol. 27, 229–235.

Ting, A.S.Y., Meon,S., Kadir, J., Son Radu,S., Singh, G., 2008. Endophytic microorganisms as potential growth promoters of banana. BioControl 53, 541–553

Ting, A.S.Y., Sariah, M., Kadir, J., Gurmit, S., 2009. Field evaluation of Non- pathogenic *Fusarium oxysporum* isolates UPM31P1 and UPM39B3 for the control of *Fusarium* wilt in 'Pisang Berangan' (*Musa*, AAA). In. Proceedings on Banana crop protection for Sustainable Production and Improved Livelihoods (Eds. D. Jones and I. Van den Berg. Acta Horticulture 828, 139-143.

Uknes, S., Mauch-Mani, B., Moyer, M., Potter, S., Williams, S., Dincher, S., 1992. Acquired resistance in Arabidopsis. Plant Cell 4, 645–56.

University of Sydney. 2003. Disease management: Biological control. http://bugs.bio.usyd.edu.au/plantpathology

Van loon, L.C., Bakker, P.A., Pieterse, C.M., 1998. Systemic resistance induced by rhizospere bacteria. Annu. Rev. Phytopathol 36, 453-483.

Van Pelt-Heerschap, H., Smit-Bakker, O., 1999. Analysis of defense-related proteins in stem tissue of carnation inoculated with a virulent and avirulent race of Fusarium oxysporum f. sp. dianthi. Eur J Plant Pathol 105,681–91.

Ward, E. R, Uknes, S.J, Williams, S.C, Dincher, S.S, Wiederhold, D.L, Alexander, D.C, 1991. Coordinate gene activity in response to agents that induce systemic acquired resistance. Plant Cell 3, 1085–94.

Wardlaw, C.W., 1961. Banana diseases, including Plantains and Abaca. Longmans, Green and Co. Ltd, London, 648.

Weber, O.B., Celli R. Muniz, C.R., Aline O. Vitor, A.O., Freire, F.C.O., Valéria M. Oliveira, V.M., 2007. Interaction of endophytic diazotrophic bacteria and *Fusarium oxysporum* f. sp. *cubense* on plantlets of banana 'Maça' Plant Soil 298, 47–56

Weindling, R. 1941. Experimental consideration of the mold toxin of *Gliocladium* and *Trichoderma*. Phytopathology 31, 991-1003

Weller, D.M., 1983. Colonization of wheat roots by a fluorescent pseudomonad suppressive to take-all. Phytopathology 73, 1548- 553.

Weller, D. M., 1988. Biological control of soilbome plant pathogens in the rhizosphere with bacteria. Ann. Rev, Phytopathol. 26, 379- 407.

Weller, D.M., Raaijmakers, J.M., McSpadden Gardener,B.B., Thomashow, L.S., 2002. Microbial populations responsible for specific soil suppressiveness to plant pathogens. Annual Review of Phytopathology 40, 309–48.

White, J. F. Jr., Cole, G, T., 1985. Endophyte-host association in forage grasses. III. *In-vitro* inhibition of fungi by *Acremonium coenophialum*. Mycologia, 77, 487-489.

Viljoen, A., 2002. The status of Fusarium wilt (Panama disease) of banana in South Africa. South African Journal of Science 98, 341–344.

Yedidia, I., Benhamou, N., Kapulnik, Y., Chet, I., 2000. Induction and accumulation of PR proteins activity during early stages of root colonization by the mycoparasite *Trichoderma harzianum* strain T-203. Plant Physiol. Biochem. 38, 863-873.

Zhang, N., Wu, K., He, X., Li, S., Zhang, Z., Shen, B., Yang, X., Zhang, R., Huang, Q., Shen, Q., 2011. A new bioorganic fertilizer can effectively control banana wilt by strong colonization with *Bacillus subtilis* N11. Plant Soil. 344, 87–97

11

Effect of Nutrition and Soil Function on Pathogens of Tropical Tree Crops

Peter McMahon
Department of Botany, La Trobe University, Bundoora Vic
Australia

1. Introduction

Crops grown in the tropics are subject to different kinds of disease pressure from those produced in temperate regions. The greater biodiversity found in the tropics, including diversity of fungi, is reflected by the larger number of pathogen species in tropical regions (see Ploetz, 2007; Wellman, 1968, 1972). Perennial crops, and tropical perennials in particular, have features in common that may predispose them to pathogen infections. Pathogen inocula, such as microsclerotia, may build up from year to year in perennial crops (Pennypacker, 1989). Also, tropical conditions are usually suitable for the year-round survival and propagation of pathogen species, unlike temperate climates which have a cooler season when pathogen populations die off or are reduced. Tropical perennial crops often include susceptible genotypes on the farm and the presence of susceptible host material encourages the production of inoculum and the initiation of new infections (Ploetz, 2007). Ploetz (2007) remarks that the presence of susceptible hosts is a particularly important barrier to disease control in tropical perennials.

Diseases in the tropics may be complicated by interactions between different pathogens, or between pathogens and insect pests (Holliday, 1980; Ploetz, 2006; Vandermeer et al., 2010; Anonymous, 2010). Disease complexes involving a number of fungal pathogens or fungi and nematodes are common in tropical situations. Interactions between pathogens and environmental stress may also occur. Crops can become more susceptible to pathogen infections when weakened by environmental stress such as drought, temperature extremes, and exposure to sunlight or wind (Agrios, 2005). Stressed plants, or plants sustaining damage caused by insects or other pathogens, may also be susceptible to attack by secondary pathogens or pathogens that infect through wounds (Palti, 1981). Nutrient deficiencies may increase the susceptibility of crops to disease. In tropical perennial crops, poor plant nutrition is likely to be a particularly important contributing factor to production losses (Schroth et al., 2000). In addition to lower production due to nutrient deficiency, low nutrition may predispose plants to diseases, increasing losses further. Nutrient deficiency causes the plant to become weakened and generally more susceptible to infection. Under such conditions, infection by weakly pathogenic species that would normally cause few problems may become more serious. The incidence and severity of particular diseases may also be linked to deficiencies of particular nutrients. However, much more research has been conducted on the relationships between particular nutrients and diseases in annual crops, than in perennial species, particularly tropical perennials.

Many tropical perennials are grown in an agroforestry situation with other crop species. Unlike the situation with annual cropping, there are fewer opportunities to include fallow periods or rotations in perennial systems during which inoculum loses its viability and the system can in general 'recover'. In tropical perennial systems, soil-borne pathogens, such as nematodes, may build up over time. An important aim in the management of these systems is to achieve a position of equilibrium between pests/diseases and the predators and parasites that keep them in check (Schroth et al., 2000). Furthermore, as perennial crops are present more or less permanently in the system, they remove nutrients on a continuous basis, without a fallow period during which soil fertility can be restored. In this light, the role of the nutrient status of tropical perennials in mitigating disease is an important topic and deserves attention from researchers.

Many diseases of tropical perennial crops are "new encounter" diseases which develop following production in new areas outside of the region of the crop's origin (Ploetz, 2007). At first, such plantings may enjoy a mainly disease-free period with high productivity. They are removed from the pressures of co-evolved pathogens and pests in their region of origin. However, such a 'honeymoon' period ends when new fungal pathogens (as well as other pathogens and pests) transfer from hosts indigenous to the region in which the crop is being produced (Keane and Putter, 1992). The indigenous hosts often remain unidentified. This is largely because the fungus causing the new disease often resides asymptomatically on its original host plant (for example, as an endophyte) (Ploetz, 2007). New encounter diseases may cause devastating losses. Unlike co-evolved pathogens in the region of the crop's origin, new encounter pathogens have few antagonists that could reduce disease incidence or severity. Poor growing conditions and poor farm management may further exacerbate the situation.

2. Role of plant nutrition in mitigating disease

Most studies on the role played by individual nutrients in preventing or reducing disease have been conducted on temperate crop species, or on tropical annuals such as rice. Little attention has been paid to the role of nutrition in alleviating diseases of tropical tree crops. However, for some diseases of tropical perennial species, a link is often observed between a deficient nutrient status caused by low soil fertility or poor plant nutrition and disease severity (Desaeger et al., 2004). Generally, plants stressed by various environmental limitations may be weakened and more vulnerable to disease and these include nutrient-deficient plants. Nutrient-deficient plants may be particularly susceptible to infection by facultative pathogens (Palti, 1981). Pathogens that are mild in normal conditions of plant growth and exist mainly as saprophytes or endophytes, such as some *Fusarium* spp. and *Alternaria* spp., may cause severe disease under conditions of nutrient stress or aluminium toxicity (Desaeger et al., 2004).

Adequate nutrition helps to mitigate pest/disease damage by replacement of root and shoot tissues (Marschner, 1995). However, studies conducted on other crop species, especially temperate crops, have elucidated how particular nutrients, including micronutrients, may enhance disease prevention. Further studies are needed to ascertain whether nutrient elements have similar roles in tropical perennial crops.

Macronutrient elements

In tropical perennials crops, as in annuals, disease is often a consequence of inadequate nutrition, particularly of *nitrogen* (N). Low supplies of N may predispose plants to infections by facultative parasites such as *Fusarium* spp. However, most research on the effect of N supply on disease has been conducted on temperate annuals (Agrios, 2005; Jones et al., 1989; Palti, 1981). For example, diseases in species of the Solanaceae family including Fusarium wilt, Alternaria early blight, *Pseudomonos solanacearum* wilt, *Sclerotium rolfsii* and Pythium damping off are increased under low N conditions (see Agrios, 2005). Vascular wilts caused by *F. oxysporum* in the annual crops tomato, cotton and pea, as well as Alternaria blights of a number of crops, may also be increased under low N conditions (Palti, 1981).

In contrast to cases where a low N supply predisposes crops to disease, research on annual crops has demonstrated that an excessive N supply increases disease or damage caused by some pests and pathogens (Jones et al., 1989; Palti, 1981 p.136). Pest and pathogen attack of above ground parts of the plant may be encouraged by high N in the presence of low K and P (Desaeger et al., 2004); a high N/K ratio encourages insect herbivory by increasing the content of free amino acids in plant tissues (Marschner, 1995). A number of studies with cereals and other crops have shown that obligate pathogens in particular, such as *Puccinia* spp. causing rust and other biotrophs, can be encouraged by a high N supply (see Palti, 1981). For example, an increase in rice blast disease (*Magnaporthe grisea*) was observed in upland rice, which had been treated with the green manure of an alley crop with a high N content (Maclean et al., 1992). Possibly excess N might also favour the development of infections by obligate fungal parasites in tropical perennials although there is little evidence for this. For example, a poor nutrient status in coffee plants has been reported to predispose them to rust infection (Waller et al., 2007 p. 302).

The form of N supplied can be a significant factor in plant disease. A supply of ammonium-N may predispose plants to certain diseases, while nitrate-N is favourable for the development of others (Palti, 1981). For example, Fusarium wilt severity in some crops is greater when N is supplied as ammonium, while Verticillium wilts are enhanced by a nitrate-N supply (see Section 4.1). Possibly this is connected to a pH effect in the rhizosphere. Fusarium wilts are favoured by acidic soils and Verticillium wilts by a higher soil pH (see Palti, 1981). Uptake of ammonium-N occurs in exchange for protons (H^+ ions), causing a decrease in pH in the rhizosphere while nitrate uptake has the opposite effect on pH as OH^- ions are pumped out by roots in exchange for NO_3^- ions (Rice, 2007). Pathogenic fungi may be particularly sensitive to localised changes in the rhizosphere, such as pH fluctuations.

Some tropical perennial crops, notably banana, coffee, coconut and cocoa, have a high demand for *potassium* (K), suggesting that K deficiency may occur in areas that produce these crops over a long-term (see Section 3, below). This is particularly the case in areas planted with perennials that receive insufficient levels of fertiliser. In agroforestry farming systems, mulch from woody biomass (e.g. pruned branches) can be a good source of K and, conversely, as K is sequestered by woody species, this may create K-deficiency in sites which already have low levels of K in the soil (Beer et al., 1998). Potassium is a mobile element with multiple functions in the plant. It acts as a counter-ion for anion transport, regulates stomatal aperture and the water potential of plant cells, affects cell wall plasticity, as well as other roles (Rice, 2007). It promotes wound healing and decreases frost injury (Palti, 1981). Potassium deficiency has been found to be linked to diseases in a number of temperate crops (see Palti, 1981) and a high K supply can improve resistance of plants to fungal and bacterial pathogens (Marscher, 1995; Perrenoud, 1977; 1990). The mechanism of resistance in some disease-resistant genotypes might be related to a greater efficiency in K uptake (Prabhu et al., 2007). The N/K ratio can affect resistance: if it is too high cells have thinner cell walls and weaker membranes and are more prone to pathogen attack (Perrenoud, 1990; Potash Institute, www.ipipotash.org). For similar reasons, cereals may become more prone to lodging. A low potassium/chlorine (K/Cl) ratio in plant tissues, which might result from the application of chloride-containing compounds such as ammonium fertilisers, may predispose plants to disease (e.g. wheat rust caused by *Puccinia* spp. or other diseases, see Prabhu et al., 2007; Jones et al., 1989). K-deficiency increases the concentration of soluble sugars in leaf tissues providing a substrate for many pathogens (Potash Institute, www.ipipotash.org). It is likely that the susceptibility of tropical perennial crops to some pathogens is also increased under conditions of K deficiency. In a study on tea plants, for example, a high K supply reduced nematode and borer damage (Muraleedharan and Chen, 1997). Another study showed that supplying K reduced Fusarium wilt in oil palm (Turner et al., 1970). However, few research studies have been conducted that could confirm a link between K nutrition and disease incidence or severity in tropical perennial crops.

Phosphorus (P)-deficiency is especially limiting to production of perennial crop in many tropical soils. Most P in the soil is in a fixed form (unavailable to plants) and the proportion of fixed P is increased at low soil pH levels (see Rice, 2007). Very low levels of available P are found in acid tropical soils (Mengel and Kirkby, 1982 p. 471). Woody biomass is very low in P, unlike K, and, therefore, external sources of P are often necessary in farm management (Beer et al., 1998). Phosphorus nutrition improves crop vigour and may decrease severity of diseases through new growth (Smyth and Cassell, 1995; Buresh, 1997). Improved root growth by P nutrition may allow the plant to 'escape' attack by soil-borne fungal pathogens or nematodes (Prabhu et al., 2007). Foliar application of phosphates may decrease diseases such as powdery mildew (Reuveni and Reuveni, 1998). Incidence of anthracnose (caused by *Colletotrichum lindemuthianum*) in susceptible cowpea cultivars was found to be higher in plants grown without applied P than in plants grown with P supplied at rates up to 80 kg of P fertiliser/ha (Adebiton, 1996). In the same study, disease severity in all of the cowpea cultivars tested was also decreased by P amendment and chickpea genotypes with resistance to Ascochyta blight had higher tissue concentrations of P and K than susceptible genotypes, which had a higher N content.

Mycorrhizal fungi, which form symbiotic associations with the roots of tropical perennials, such as coffee and banana, play a crucial role in accessing sources of P for their host plants. These fungi can access sources of P in the soil that are unavailable to non-mycorrhizal plant roots. As well as decreasing the impacts of plant pathogens (Azco'n-Aguilar and Barea, 1996), mycorrhizal plants have higher contents of certain nutrients, such as P. An example of this was demonstrated by greenhouse experiments conducted on coffee by Vaast et al. (1997). Coffee plants inoculated at an early stage with AM fungi had higher tissue P contents than non-mycorrhizal plants. High P tissue contents were maintained following inoculation with the nematode pathogen *Pratylenchus coffeae* and these plants also had fewer root lesions than non-mycorrhizal plants inoculated with the nematode or mycorrhizal plants that had been inoculated with the AM fungi at a later stage. Care needs to be taken with supplying inorganic P to crops as an excessive external P supply can inhibit mycorrhizal development. This may lead to a shortage of other nutrients, such as zinc, that mycorrhizal roots are efficient at accessing for the plant (Andrade et al., 2009).

An adequate supply of *calcium* (Ca) has been demonstrated to enhance resistance to a number of diseases in annual crop species caused by pathogens such as *Rhizoctonia solani, Sclerotium* spp., *Botrytis* spp., *Fusarium oxysporum* and the nematodes *Meloidogyne* spp. and *Pratylenchus* sp. (Agrios 2005; Jones et al., 1989). Resistance of lucerne to nematodes was shown to increase with supplied Ca (see Palti, 1981 p. 142). A large proportion of Ca in plants is present in the apoplast and, influences cell structural properties, especially of the cell wall (Rice, 2007). Increased levels of Ca-pectate complexes in the cell wall are likely to increase resistance to vascular wilt pathogens because this form of pectate is resistant to breakdown by endopolygalacturonase enzymes produced by fungi to degrade pectin in the xylem vessel walls. Ca-pectin complexes might also impede the progress of wilt pathogens growing within the xylem (Corden, 1965; Pennypacker, 1989; Waggoner and Dimond, 1955). However, Ca also has metabolic functions within the symplast as a secondary messenger in signalling pathways (Rice, 2007). Calcium possibly plays a significant role in mechanism(s) of disease resistance in fruits. A relatively large proportion of Ca taken up by plants is distributed to fruits and low Ca has been linked to increased incidence of fruit diseases such as brown-eye spot in coffee berries (see Section 4.2). Groundnut pods have a high Ca demand and pod rot caused by *Pythium* and *Rhizoctonia* spp. has been linked to a low Ca content. High rates of magnesium and K application can reduce the Ca content of pods, increasing disease severity (Prabhu et al., 2007)

Sulphur (S) is a component of defense-related peptides and proteins such as glutathione and phytoalexins. Application of S to deficient soil reduced leaf spot, caused by *Pyrenopeziza* in oil seed rape and stem canker caused by *Rhizoctonia solani* in potato (Haneklaus, 2007). The effect of S nutrition on diseases of tropical perennials is largely unknown. However, deposits of elemental sulphur (S) were observed in the xylem of cocoa plants in response to infection with *Verticillium dahliae* (Resende et al, 1996; Cooper and Williams, 2004). Similar findings were made in tomato (Williams et al., 2002 – see Haneklaus, 2007). Elemental S is toxic to some fungal pathogens and may be considered to be a phytoalexin in its own right (Resende et al., 1996). *Magnesium* (Mg) is an essential component of chlorophyll and, therefore, the photosynthetic systems of plants. However, a direct relation between Mg and plant disease has been less commonly demonstrated than with the other macronutrient elements. Magnesium, with K, plays a role in phloem-loading of sugars (Cakmak et al., 1994). Magnesium also activates enzymes such as glutathione synthetase.

Micronutrient elements

Micronutrients have a diverse range of functions in plants: for example, as enzyme co-factors with redox roles and, in the case of elements such as boron and silicon, in tissue strengthening or structural functions. The numerous biochemical functions of micronutrients are reflected by their roles in a diverse range of mechanisms of disease resistance. *Zinc* (Zn) nutrition appears to be involved in resistance to many diseases. The mechanisms involving Zn in disease resistance are unclear but Zn acts as a co-factor for numerous enzymes (Rice, 2007). Stimulation of root growth by Zn may account for some observed cases of disease resistance (Duffy, 2007). Zinc application to soils reduces attack by root pathogens of tomato, including *Fusarium solani, Rhizoctonia solani* and *Macrophoma phaseoli*, and also Rhizoctonia root rots of wheat, chickpea, cowpea and medicago (Duffy, 2007; Gaur and Vaidge, 1983; Kalim et al., 2003; Streeter et al., 2001). In tropical perennials, the role of Zn in disease resistance remains to be investigated. However, Zn-deficiency in rubber (*Hevea brasiliensis*) predisposes the tree to infection with *Oidium heveae* (Duffy, 2007). Zinc has been reported to alleviate Phytophthora diseases. Low Zn levels in soils and leaf tissues were associated with a high incidence of Phytophthora pod rot (or black pod) of cocoa in Papua New Guinea (Nelson et al., 2011). Supplying *Manganese* (Mn) has been shown to alleviate various diseases in a number of crop plant species (Palti, 1981; Thompson and Huber, 2007). Manganese occurs in different redox states and while it is present in healthy tissues as the Mn^{2+} ion, it accumulates at sites of pathogen attack in the Mn^{4+} form, for example in rice affected by blast (Thompson and Huber, 2007). *Iron* (Fe) has an essential role in plant cells as a co-factor in redox reactions and other functions. Fe is mainly available to plants as its reduced ion, Fe^{2+}. Verticillium wilt in mango caused by *V. albo-atrum*, and in groundnut caused by *V. dahliae* was mitigated in both cases by the application of Fe in chelated form (see Palti, 1981 p. 142). On the other hand, control of Fusarium wilt in tomato was favoured by low Fe (Woltz and Jones, 1981). Similarly, Fusarium wilt has been shown to be lower at low levels of Mn. *F. oxysporum* has a particularly high demand for some micronutrient elements, especially Mn, Fe and Zn (Jones et al., 1989; Woltz and Jones, 1981). The supply of Mn, Fe and possibly other nutrients, to *F. oxysporum* strains causing vascular wilt may therefore increase disease incidence and/or severity (see section 4.1, below).

The availability of other micronutrients to plants has been linked to disease alleviation in particular instances. *Copper* (Cu) deficiency decreases lignification in the xylem and has been linked to lodging in cereals (Evans et al., 2007). Copper has direct toxic effects on pathogens as well. A Cu supply protects grapes and hops from Downy mildew, caused by *Plasmopara viticola* and *Pseudoperonospora humuli*, respectively (see Evans et al., 2007). *Nickel* (Ni), like Fe and Zn, is a co-factor of some enzymes, such as ureases, which break down urea into less toxic forms (Rice, 2007). Nickel application has been shown to reduce brown spot in rice (caused by *Cochliobolus miyabeanus* syn. *Helminthosporium oryzae*). Supplying *molybdenum* (Mo) reduced late blight in potato and Ascochyta blight in beans and peas (Palti, 1989 p. 143). As a co-factor of nitrate reductase, this element plays a particularly important role in the reduction of nitrate to ammonium (Rice, 2007).

Silicon (Si), now regarded as an essential micronutrient, has been shown to enhance disease resistance in many instances (Datnoff et al., 2007). In sugar cane, ring spot was alleviated by Si amendments (see Datnoff et al., 2007). Low Si in rice has been linked to susceptibility to a number of pathogens including *Pyricularia, Sclerotium oryzae, Cochliobolus* and *Xanthomonas oryzae* (Palti, 1981 and references within). A supply of Si enhances resistance to rice blast (Datnoff et al., 2007). Supplying Si to coffee reduced leaf disease and nematode infections in roots (see Sections 4.2 and 4.5). Possibly Si, with other nutrients such as Ca and *boron* (B), influences cell wall properties and enhances mechanical strengthening of tissues (Rice, 2007). Shen et al. (2010) tested the effect of potassium silicate on *in vitro* growth of some plant pathogens, including *Fusarium oxysporum, Rhizoctonia solani* and *Pestalotiopsis clavispora*, finding no influence if the media pH was maintained at the same level as the control. They suggested that the mechanism by which Si confers resistance may be related to provision of a physical barrier to pathogen infection or to the induction of a defense response in the host, rather than to a chemical effect. Silicon and other elements, including Ca and B, may be of particular importance in resistance to facultative pathogens, wound invading pathogens and nematode infections.

3. Management of soils supporting tropical perennial crops

Since many tropical soils are nutrient-poor and the replacement of nutrients by mineral or organic fertilisers is often inadequate, tropic perennial crops are particularly prone to nutrient stress. Growing perennial crops can lead to nutrient deficiencies if the soils in which they are grown are not adequately amended with mineral or organic fertilisers. Long-term cropping of one or a few species in tropical soils can also have other impacts on soil properties, such as the soil pH, that in turn may cause nutrient deficiencies. In addition, poor sanitation (e.g. removal of infected plant material), flooding (which may spread inoculum as well as causing plant stress) and inappropriate canopy management are common exacerbating factors leading to increases in disease incidence and severity on a farm (Kohler et al., 1997). Importantly, disease may also have the effect of reducing the plants' nutrient status or impairing water uptake. For example, coffee tree roots and the roots of other tree crops infected with nematodes or fungi may have impaired water and chemical uptake mechanisms causing wilting and nutrient deficiency (Nelson et al., 2002; Waller et al., 2007 p. 279).

Table 1 presents data on nutrient uptake by some tropical perennial crops based on previous studies. The data refer only to the nutrient content of marketable products; the nutrients contained in waste (e.g. discarded tea leaves and coffee or cocoa beans) are not included. In the case of nutrients removed by cocoa in Nigeria (reported by Wessel, 1985), it can be seen that the pod husks, which are normally discarded, remove high amounts of K (77 kg ha^{-1}y^{-1} in the husks of pods producing one ton of dry beans). Substantial amounts of Ca are also removed with cocoa pods and other fruit products.

		Nutrients removed (kg ha^{-1}y^{-1})		
	Yield	N	P	K
Coffee	1000 kg beans	40	2	42
Rubber	Latex	6-36	1-7	5-31
Tea	1000 kg leaves (dry)	41	3	21
Banana	40 – 70 tonnes	225-450	20-40	800-1200
Coconut	6920 nuts	96	19	115
Cocoa	1000 kg beans (dry)	23	4	8
	Cocoa pod husks	17	2	77

Table 1. Estimated removal of N, P and K from the soil by the harvested products of some tropical tree crops. Yields indicated for each crop are estimates of the quantity of produce obtained per hectare each year (sources: Krauss, 2003; Wessel, 1985). Note that where the quantities removed were given for the oxides of P and K in the original data, these figures have been converted to indicate the respective quantities of the elements removed.

Soil acidity or low pH, common in the tropics, may be increased under particular crops, especially long-term perennial crops, or by the application of some kinds of mineral fertiliser (Jones et al., 1989). Soil pH decreases when forest soils are turned over to perennial crops, such as coffee and cocoa (Beer, 1988; Beer et al., 1998; Hartemink, 2005). Lowering soil pH decreases the availability of basic cations, particularly Ca and Mg. However, increased soil acidity increases the availability of other cations to plant roots. These include the cations of Mn and aluminium (Al) which can reach toxic levels as their uptake by plants increases. The proportion of soil P that is fixed and unavailable for plant uptake increases in acid conditions (Mengel and Kirkby, 1982). Liming can reduce the severity of a number of diseases perhaps by increasing the availability of a number of nutrients to crops, as well as providing a source of Ca, reducing Al toxicity and improving soil structure (Palti, 1981 p. 142). The increase in pH in limed soils also favours the growth of bacteria, including actinomycetes, which include species that are antagonistic to fungal pathogens (Palti, 1981 p. 29; Jones et al., 1989).

Shade and nutrition

Tropical perennials produced in agroforestry systems are affected by other species on the farm, including shade trees in the case of shade-requiring tree crops such as cocoa and coffee (Schroth et al., 2001). Importantly for such shade-requiring species, managed shade can reduce incidence and severity of some pests and diseases. Shade may also reduce stress to tree crops by preventing extremes in temperature, water loss etc. that may result from exposure (Staver et al., 2001). This, in turn, mitigates diseases that become more severe in stressed plants. Removal of shade can increase photosynthesis and, therefore, raise the productivity of tree crops such as coffee and cocoa. The removal of shade from coffee farms, for example, can provide double the yields of shaded coffee in the short-term (Waller et al., 2007). However, this may be followed by impacts from other problems, including increased susceptibility to diseases such as brown-eye spot (see Section 4.2), wind and storm damage, frost damage at higher altitudes, increased evapotranspiration (and water loss) and lower levels of soil organic matter (Waller et al., 2007 p. 313). Conditions such as overbearing dieback and sunscorch of coffee (see Section 4.4) may result from shade removal. Other twig and leaf blights, such as anthracnose caused by *Colletotrichum gloeosporioides* on cocoa grown in Indonesia become more severe following shade removal (Agus Purwantara, pers. comm.). Conversely, excessive shade and inadequate pruning can provide suitable conditions for other coffee and cocoa pathogens, such as *Corticium* spp. (causing web blight and pink disease), *Phytophthora palmivora* (causing pod rot and other diseases in cocoa) and *Mycena citricolor* (causing South American leaf spot in coffee). Shade trees may be sources of other pathogens with wide host ranges such as the root pathogens *Armillaria* and *Ganoderma* spp.

Shade trees have a mixed effect on the plant nutrition of other crops in the agroforestry system (Schroth et al., 2001). They may compete with crops for water and nutrients in the soil and sequester nutrients in their biomass (Palm, 1995). However, they also provide inputs of nutrients to the system through leaf litter or by nitrogen-fixation. In Central America, *Cordia alliodora* shade trees on each hectare of coffee produce 5.7 tons of leaf litter per year, containing 114 kg N, 7 kg P and 54 kg K (Beer, 1988). Forest trees providing shade for cocoa in West Africa produced 5 tons of leaf litter per hectare each year, containing 79 kg N and 4.5 kg P (Murray, 1975). Legume shade trees in cocoa and coffee agroforestry systems provide approximately 60 kg N ha^{-1}y^{-1} by biological fixation of N_2 (Beer, 1988). However, some legumes may cause decreases in soil pH. Somarriba and Beer (2011) reported that timber species grown with cocoa did not impact cocoa production. Shade trees with relatively deep roots can remobilise nutrients in the system (Schroth et al., 2001). Beer et al. (1998) cite reports of lower leaching rates of N under shaded coffee (9 kg ha^{-1}y^{-1}) than under unshaded coffee (24 kg ha^{-1}y^{-1}).

4. Diseases of tropical perennial crops in relation to nutrient and other growing conditions

The main types of disease that impact production and performance of major tropical perennials and the conditions that influence disease incidence and/or severity, particularly nutrition and soil function, are outlined below. Most of the diseases described are caused by new encounter pathogens, while some such as witches' broom disease of cocoa in South America and coffee wilt disease, which has caused severe losses and tree death in East Africa, are caused by co-evolved pathogens. While each group of diseases (e.g. dieback diseases) has common features that relate to their management and control, it should be noted that pathogens from widely separate taxonomic groups may cause similar symptoms common to a particular type of disease. Therefore, control measures to reduce plant diseases need to take into account the taxon of the pathogen as well as the type of disease it causes. Diseases of tropical crops are often complex being associated with more than one pathogen or pest, or transmitted by vectors. Conversely, a particular pathogen species may cause more than one disease. Therefore, the disease groupings outlined below may overlap with each other considerably.

4.1 Vascular wilts

Vascular wilts can cause serious losses for a number of tropical perennial crops. Vascular wilts may be soil-borne with infections initiated in the roots, as in many wilts caused by *Fusarium oxysporum*, or else they may be initiated in the phyllosphere, especially by infections via wounds as occurs in many Ceratocystis wilts, caused by *Ceratocystis fimbriata* and closely related species on a wide range of crop hosts. Coffee wilt (tracheomycosis) attributed to *Gibberella xylarioides (anamorph: Fusarium xylarioides)* has become a serious disease problem in Robusta coffee-growing areas of East Africa (Rutherford, 2006; Flood, 2010). An adequate nutrient supply to coffee trees is recommended as part of an integrated strategy to manage this disease (e.g. see www.nyrussell.com/all-about-coffee). Wounding caused by nematodes in the roots or by insects in aerial parts may increase the possibility of infection by fungi causing soil-borne diseases, including wilts. In coffee, for example, infection with nematodes can predispose plants to infection with *F. oxysporum* (Kohler et al., 1997; Nelson, 2002).

Insects may also transmit disease by carrying spores to wounded tissues or by producing insect frass containing spores that can be dispersed by wind. For example, species of Ambrosia beetles are attracted to cocoa tissues infected with *Ceratocystis* spp. and insect frass containing their spores may be disseminated to other plants by wind. Vascular wilts have in some cases been linked to environmental factors, such as drought stress. Drought-stressed plants are more susceptible to Ceratocystis wilt (Harrington, 2004). Vascular wilt diseases are complicated by the fact that plants may be more susceptible to the causal pathogen in conditions of water or nutrient deficiency but that, additionally, the pathogen may impair transport in the vascular tissues thus causing water or nutrient deficiency in the plant tissues.

Wilts caused by host-specific races of *F. oxysporum* occur in a number of crop species. Panama disease of banana caused by *F. oxysporum* f. sp. cubense is particularly devastating (Ploetz, 2006). Following infection of the root, the fungus enters the vascular system and impedes water and nutrient transport to the upper plant. The leaves of infected banana plants become yellow and dry and the plant eventually wilts. The pathogen can be spread through flooding and, therefore, drainage and prevention of over-irrigation are particularly necessary. As the pathogen can be harboured by the bunch stalks, sanitation measures should be applied (Kohler et al., 1997). Some studies on *F.oxysporum* wilts suggest that nutrient availability to both the plant and the pathogen may affect the level of disease severity. The factor(s) causing the wilt are unclear: in Fusarium wilt of tomato it has been attributed to physical blockage or by the secretion of toxins (Walker, 1972 p. 300). Reducing soil acidity by liming is used to reduce *F. oxysporum* infection of coffee plants (Kohler et al., 1997). Possibly, liming reduces the availability of some micronutrients to the fungus. As mentioned previously, *F. oxysporum* has a particularly high demand for some micronutrients, including Mn, Fe and Zn. Mn may be in particular demand by the pathogen. Jones and Woltz (1972) showed that the control of Fusarium wilt of tomato by liming could be reversed by supplying Mn in chelated form to the limed soils. Liming may also help to control Fusarium wilts by changing the soil microbial populations. Under conditions of high pH, bacteria, including actinomycete, populations increase, including those of antagonistic species (Jones et al., 1989). Increased calcium supply to the plant has been shown to reduce the severity of Fusarium wilt of tomato (Walker, 1972 p. 303). The mechanism involved may be related to the increase in resistance conferred by Ca to enzymatic breakdown of pectate compounds by the fungus, which might also account for the effect of Ca supply on reducing Verticillium wilt (see below).

As mentioned previously, the form of N that is present in the soil, whether as nitrate or ammonium ions can influence the incidence and severity of vascular wilts (see Section 2). Application of N as ammonium fertiliser can increase Fusarium wilt: this has been suggested to be an effect of increased acidity or related to a reduction in the K/Cl ratio (Jones et al., 1989). A study on Fusarium wilt of banana by Nasir et al. (2003) suggested that the effect of the form of N supplied on disease severity is not related to the activity of the pathogen. When banana plantlets were transplanted into soils infested with *F. oxysporum*, an increase of wilt disease severity and the invasion of roots by the pathogen were found to be independent of *F. oxysporum* activity in the soil (Nasir et al, 2003). Amendment of soil with chicken manure increased disease severity, but not *F. oxysporum* activity; it appeared that the increase in disease and pathogen invasion was a consequence of supplying N as the ammonium-form but that this was not connected to pathogen activity in the soil. Another study demonstrated that a lower rate of germination of *F. oxysporum* chlamydospores occurred in soils that had an adequate supply of Ca, particularly in relation to other basic cations, Mg and K (Chuang, 1988; 1991). The same study indicated that higher soil pH values and populations of actinomycetes also decreased spore germination rates. Domingues et al. (2001) compared banana field plots which differed in their capacity to suppress Fusarium wilt. Soils that suppressed disease had a lower proportion by weight of water-stable aggregates, than conducive soils. They hypothesised that the higher proportion of water-stable aggregates in the conducive soils favoured anaerobiosis, which increased availability of reduced Fe ions for the pathogen, which as mentioned previously, has a high demand for micronutrients, such as Fe.

Verticillium wilt caused by *Verticillium dahliae* also occurs in some perennial crops. While Fusarium wilt is encouraged by acidic soils, Verticillium wilt is favoured by a higher soil pH. A build-up of inoculum can occur in perennial crop species affected by *V. dahliae*; in the case of Verticillium wilt of pistachio, new infections are initiated by microsclerotia, which can survive for long periods on pistachio roots. A higher incidence of Verticillium wilt has been found in pistachio trees under conditions of K deficiency (Pennypacker, 1989). This may be a consequence of impaired root growth and, therefore, increased exposure to microslerotia (Pennypacker, 1989). K deficiency in cotton is enhanced by infection with *V. dahliae*, perhaps due to a decrease in the ability of the roots to take up K after infection with the fungus (Bell, 1989).

An increase in severity of Verticillium wilt of cotton in response to P supplied as superphosphate has been reported – possibly P promotes pathogen activity within the plant (Bell, 1989). A study on cocoa affected by Verticillium wilt in Nigeria indicated that P supply had no effect on infections (Emechebe, 1980). Calcium nutrition may enhance resistance to Verticillium wilt in some cases (Bell, 1989). This could be related to an increase in the levels of Ca-pectin complexes, which are resistant to breakdown by a pectin-degrading enzyme produced by the fungus (see Section 2). Possibly, Ca-pectin complexes are involved in impeding growth of the fungus through the xylem (Pennypacker, 1989).

4.2 Diseases causing lesions on leaves and fruits

Leaf and fruit spots

Some leaf/fruit diseases, e.g. infections by the Mycosphaerellaceae pathogen, *Cercospora mangiferae* (syn. *Stigmina mangiferae*), causing angular leaf spot in mango and other pathogens of mango are influenced by the growing conditions of the host. Similarly, brown-eye leaf spot in coffee caused by *Cercospora coffeicola* , becomes more serious under conditions of environmental stress. Brown-eye spot is encouraged if coffee is grown in unshaded conditions, particularly if the soil is nutrient-poor (see Kohler et al., 1997; Nelson, 2008a). Deficiency in N and K in particular, may accentuate brown-eye spot (Wrigley 1988). In an experiment with coffee seedlings grown in pots, Pozza et al. (2001) showed that supplying bovine manure to coffee trees, as well as other types of fertiliser reduced the severity of brown-eye leaf spot. Santos et al. (2008) compared the incidence of brown-eye spot on ten-year old coffee in neighbouring plots that were managed either conventionally using inorganic fertiliser applications, or by organic methods in which only organic amendments were applied to the soil. In a period of two consecutive years, disease incidence was higher in the plot under conventional management (28% and 29%) than in the plot managed using organic methods (9% and 12%), although higher berry yields were obtained in the conventionally managed plots. Possibly, the higher rates of disease in the conventionally-treated trees were a consequence of nutrient-deficiency, particularly of Ca and Mg. The leaf concentrations of Ca and Mg were lower in the conventionally-treated plants. Possibly this was due to the higher yields from the trees receiving conventional treatment (so that these trees had a higher nutrient requirement than the lower yielding organic trees).

Infections in citrus-growing areas in Cameroon and other tropical countries in Africa by the Mycosphaerellaceae pathogen, *Phaeoramularia angolensis*, causing leaf and fruit lesions creates serious losses for farmers. In a survey of disease-affected areas, Ndo et al. (2010) found that disease severity was lower in citrus species grown on volcanic soils in Cameroon than in other soil-types. Altitude was also a key factor affecting disease severity. Since volcanic soils are generally nutrient-rich, this suggests that plant nutrition plays a role in mitigating this disease. In the case of *Dothiorella gregaria*, a Botryosphaeriaceae pathogen that causes fruit spot on mandarin, Ca and Zn contents of plant tissues may influence disease severity. Isolates of this pathogen were inoculated into mandarin by da Silva Moraes et al. (2007) who found that the highest lesion rates were obtained on plants with the lowest Ca and Zn contents.

Particularly serious leaf diseases of banana include black sikatoga (or black leaf streak) caused by *Mycosphaerella fijiensis* (syn. *Paracercospora fijiensis*), yellow sikatoga (*Mycosphaerella musicola* syn. *Pseudocercospora musae*), freckle caused by *Guignardia musae* (syn. *Phyllosticta musarum*) and Black Cross (*Phyllachora musicola*) which may be associated with previous infections with *Cordana musae*. Black sikatoga disease is reduced under shade (Ploetz, 2003; Stover, 1972). The disease may increase under conditions of poor nutrition and, therefore, improved host nutrition is one recommended control measure (Nelson et al., 2006; Mobambo et al., 1994); adequate phosphorus nutrition may be particularly necessary for disease alleviation.

Algal leaf spot disease occurs on a number of tropical perennials including breadfruit, citrus, guava, cocoa, mango, soursop and black pepper. The causal agent is *Cephaleuros virescens* or other species of the same genus. Orange and green spots can be seen on leaves and young stems, particularly in trees weakened by stress or in periods of high rainfall (Nelson, 2008b). In avocado, poor plant nutrition, lack of soil drainage and still conditions (e.g. under dense canopies) are predisposing factors for algal infection (Nelson, 2008b).

A study in Brazil indicated a link between N nutrition and infection of citrus trees with the bacterial pathogen, *Xylella fastidiosa* causing variegated chlorosis disease (see Huber and Thompson, 2007). Nitrification increased disease severity but where a groundcover grass species was planted between rows of citrus trees, a decrease in the disease was demonstrated; this was explained by the inhibiting effect of the ground cover crop on nitrification. Nitrification decreases concentrations of ammonium ions in the soil and this, in turn, decreases Mn availability to the plant. Inhibition of nitrification caused an increase in Mn uptake by 50%. Thus in this case it appears that an increase in availability of Mn confers resistance to the host rather than favouring the pathogen, as in some Fusarium wilts.

Rusts

Rust in banana is caused by *Uredo musae* and in coffee by *Hemileia vastatrix*. In some annual crop species, rusts may be encouraged by high levels of N fertilisation (see Section 2). However, coffee rust has been reported to be more severe on plantations grown under nutrient poor conditions (Waller et. al., 2007, p. 302). Applications of silicon (Si) have been shown to decrease the level of coffee rust on coffee seedlings. Martinati et al. (2008) found that the number of rust lesions on leaves of Si-treated coffee (*C. arabica*) seedlings was decreased in proportion to the dosage of Si (as potassium silicate) supplied to the soil by up to 66% compared to the control, which received no Si.

In an epidemiological study on coffee rust in Honduras, Avelino et al. (2006) reported that the intensity of coffee rust infection was dependent on the production situation, rather than regional differences in environmental parameters, such as rainfall levels. They showed that coffee rust was associated with acid soils, soils treated with mineral fertilisers and increased yields. Possibly, an acidifying effect of the mineral fertilisers was the main factor accounting for higher levels of disease.

Anthracnose

Anthracnose diseases affect leaves, shoots and fruit of a variety of tropical perennial crops on farms and can also create severe post-harvest problems. The Ascomycotina species, *Glomerella cingulata* (anamorph: *Colletotrichum gloeosporioides*), an endophyte on plant species such as cocoa, *Theobroma cacao* (Rubini et al., 2005), is the most common causal pathogen, infecting a number of tropical tree crops such as avocado, mango, coffee and kauri (*Agathis* sp). Symptoms include blackened and sunken lesions on fruit, and marginal necrosis of young or flush leaves. On avocado, fruit and leaves are infected by *C. gloeosporioidies*, the leaves developing large light brown lesions and, in wet weather, pink spore masses. Anthracnose is transmitted mainly by rainsplash (Kader, 2002) and heavy rainfall increases the severity of anthracnose in Robusta coffee. Rust infections can predispose Arabica coffee to anthracnose infection (Kohler et al., 1997). As an endophyte and facultative parasite, this pathogen may be asymptomatic in healthy plants of some crop species, only becoming pathogenic under conditions of stress, such as over-exposure to sun in the case of shade-requiring crops. As with other latent species (see Prakash, 2000; Kohler et al., 1997), the removal of dead twigs and branches before flowering is a crucial measure to control infection by the fungus. Possibly, nutrition plays a role in the control of anthracnose. Anthracnose in the orchid, *Cymbidium* sp., caused by *Colletototrichum orchidacearum*, for example, was reduced by applying macronutrients, especially K and P (Yi et al., 2003). Acosta-Ramos et al. (2003) took an integrated management approach, including soil and foliar application of nutrients, to mango trees in an orchard in Mexico and recorded decreases in the incidence of both anthracnose, caused by *C. gloeosporioides*, and stem end rot, caused by *Lasiodiplodia theobromae*.

In papaya and mango, a post-harvest problem caused by anthracnose infection of the fruit is particularly serious and can be controlled by dipping fruit in hot water or a fungicide (such as benomyl). Anthracnose infects all parts of the mango plant: new leaf flushes are particularly susceptible and, in wet weather, flowers are susceptible to blossom blight. In guava, which incurs serious losses from anthracnose disease, infections may be associated with fruit fly damage and with scab damage caused by *Sphaceloma perseae* (Ploetz, 2007). Passion fruit is also a host of anthracnose. Passion fruit produced under poor growing conditions may also be infected with *Alternaria* sp. causing brown spot.

4.3 Fruit rots

Conditions predisposing perennial crops to fruit rots include high rainfall, poor drainage and poor sanitation practices, whereby infected fruits (which provided sources of new inoculum) are not disposed of correctly. This is particularly relevant in the case of *Phytophthora palmivora*, the straminopile (formerly oomycete) pathogen, which is responsible for fruit rots in a wide range of host species in the tropics including breadfruit, cocoa, black pepper, papaya and vanilla. On coconut and betel nut trees it causes bud rot, inflicting serious losses to these crops. The pathogen is dispersed mainly by motile zoospores, which require the presence of external water and infection (Guest, 2007). Hence, incidence of fruit rots caused by *P. palmivora* increases dramatically under conditions of high moisture during heavy rainfall periods in the wet season, poor drainage or slightly cooler conditions at higher altitudes that reduce evapotranspiration rates. In many host species, it also infects other parts of the plant including leaves and stems. Inoculum from one infected part of the plant can initiate infections in other tissues. On cocoa farms, insects, including ants and beetles, also transmit disease, carrying spores from infected sites to initiate new infections (Konam and Guest, 2004). Therefore, sanitation is a crucial control measure for Phytophthora diseases. A link between low levels of zinc nutrition and pod rot incidence was reported from Papua New Guinea (Nelson et al., 2011). Other *Phytophthora* species also cause fruit rots; particularly devastating losses in cocoa in West Africa are caused by *P. megakarya*, for example (Guest, 2007). Similarly devastating losses are caused in South America by the basidiomycete, *Moniliophthora roreri* that causes frosty pod. A fruit rot in cocoa is also caused by *Lasiodiplodia theobromae* in southern and Southeast Asia. On avocado, the same pathogen species causes browning of the fruit from the stem end (Kohler et al., 1997) and fruit rot in papaya (Kader, 2002). *L. theobromae* generally requires wounding to initiate infection (Kader, 2002). A possible relation to plant nutrition has been raised by studies of some Phytophthora fruit rots. For example, brown rot of citrus caused by *Phytophthora citrophthora*, is enhanced under when N fertiliser is provided as ammonium-N (Menge and Nemec, 1997).

Verticillium theobromae causes cigar end rot of banana. Sanitation (including removal of dead flowers), canopy aeration and exposure to light are recommended management methods for this disease (Nelson et al. 2006). A serious fruit rot of guava is caused by *Pestalotiopsis disseminate* that infects the fruit from the stalk end producing white fruiting bodies that later become brown. This disease is linked with poor nutrition; hence proper soil amendment to improve fertility is a recommended control measure. Similarly, *Pestalotiopsis* sp. infection of coconut can be reduced by improved growing conditions through the application of fertiliser (Kohler et al., 1997). *Pestaliotopsis* (Fig. 1) has been isolated from cocoa leaves that have symptoms of vascular-streak dieback – possibly it is a secondary pathogen.

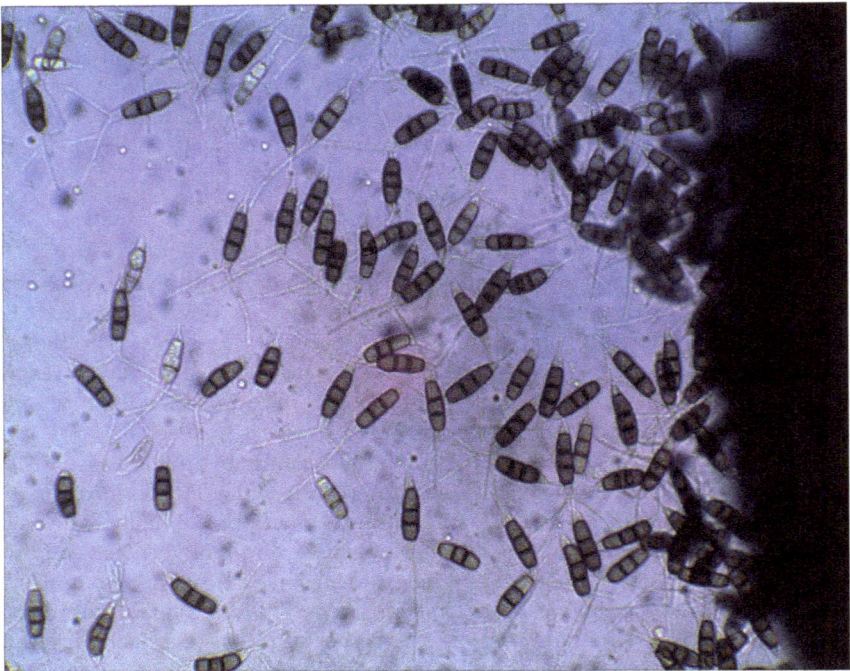

Fig. 1. Conidia of *Pestalotiopsis* sp. isolated from cocoa leaves in Sulawesi, Indonesia; magnified x400 (*author's photo*).

4.4 Dieback diseases

Dieback diseases are caused by a range of pathogens. Some diseases may involve the interaction of more than one pathogen species and may be prevalent in ageing trees or trees weakened by environmental stress (Ploetz, 2006). This is particularly the case for infections by facultative parasites, such as some Botryosphaeriaceae species which can survive in a latent form in dead wood or are wound invaders. Similarly, the weakly aggressive pathogen, *Fusarium decemcellulare* has been linked to dieback and cushion gall of cocoa. Other tree crops are also infected with this pathogen (see Ploetz, 2006). Infection is possibly facilitated by wounds caused by insects (Holliday 1980). The pathogen can interact with other pathogens such as *Lasiodiplodia theobromae* (syn. *Botryodiplodia theobromae*) and *Phytophthora palmivora* (Holliday, 1980).

In coffee, overbearing dieback is linked to poor nutrition, nitrogen-deficiency in particular, which has the effect of lowering soluble sugar reserves in the stem. Exposure to the sun, resulting in plant stress and excessive cropping, poor root function (e.g. resulting from pathogen attack) and weed competition can all be factors predisposing trees to this condition. Infection by opportunistic fungi such as *Colletotrichum gloeosporioides* and *F. oxysporum* follows, causing dieback of the stem (Flood, 2010; Waller et al., 2007). The disease can be managed by supplying shade, applying N and decreasing crop density (Waller et al., 2007 p 284). Sunscorch damage is also associated with infection of berries by pathogens such as *F. stilboides* and *Cercospora coffeicola* (Waller et al., 2007 p. 285).

In the case of pink disease caused by the basidiomycete, *Corticium salmonicolor* (syn. *Erythricium salmonicolor*), a lack of light caused by heavy shade and insufficient pruning, encourages infections in a number of species including citrus, cocoa, coffee, rubber, tea and black pepper. On cocoa, symptoms of the disease include a pink to creamy white crust on the bark and cracking of the bark with gum exudates. Sudden death of the whole branch may occur with the leaves remaining attached. A number of timber (forest) trees are also hosts to this pathogen. Wet conditions promote infections of tree crops.

The most severe dieback disease affecting cocoa in the Southeast Asian region is vascular-streak dieback (VSD) caused by another basidiomycete species, *Ceratobasidium theobromae* (syn. *Oncobasidium theobromae*). Infection is initiated in particularly wet conditions on young leaves by wind-borne spores, which are very short-lived (Keane et al., 1972; Keane, 1981; Prior, 1985; Guest and Keane, 2007). Following germination of the spores, the fungal hyphae penetrate the xylem in the leaf by a mechanism that remains unknown. The pathogen can grow via the xylem to lower parts of the branch causing leaf chlorosis and fall, and eventual dieback of the branch (Fig. 2). In susceptible varieties tree death may result, but most cocoa genotypes exhibit partial resistance to VSD. The identification of a second species, *C. ramicola*, in VSD-infected tissue (Samuels et al., 2011) raises the possibility that more than one pathogen is causing the disease. A significant and widespread change in VSD symptoms has occurred since 2004. After 2004, necrosis of infected leaves was observed (rather than only chlorosis as previously seen), with the infected leaves remaining attached to the branch longer; cracks in the midrib of the infected leaves were also observed, which allowed emergence and sporulation of the fungus from the leaves themselves, in addition to the petiolar scars on the stems as observed prior to 2004 (Purwantara et al., in process). Disease severity has also increased so that many farmers in Indonesia now identify VSD as their primary problem and the main reason given for changing their cocoa farms over to other crops, including maize, neelam and oil palm (pers. comm. Ade Rosmana, Hasanuddin University). Possibly the change in symptoms and increased severity of the disease is the result of an environmental factor interacting with *C. theobromae* infections or VSD-infected trees have become susceptible to infection by a secondary pathogen(s); trees weakened by stress, such as poor nutrition, might be predisposed to such a secondary infection (Mossu, 1990 cited by Schroth et al., 2000). An alternative explanation for the change in symptoms and severity of the disease is that a new strain of the pathogen, *C. theobromae*, has emerged. Further work is underway to elucidate the pathogen-environmental relationship that might lie behind the changed VSD symptoms.

A devastating disease of cocoa is witches' broom, caused by a co-evolved pathogen of cocoa, a basidiomycete species, *Moniliophthora perniciosa*. The disease causes distortion of growth in the shoots, creating a broom-like appearance. Although the pathogen co-evolved with cocoa, it has caused most damage in plantations in Bahia, Brazil away from its centre of origin in the Amazon rainforest. Improved management, including the appropriate use of fertilisers, combined with the introduction of resistant cocoa genotypes has been effective in mitigating the impact of this disease (Keane and Putter, 1992). Dieback in cocoa caused by *L. theobromae* has been observed in the Cameroons (Mbenoun et al., 2008) and in India (Kannan et al. 2010). Vascular streaking has been observed in both cases. Kannan et al. (2010) isolated the pathogen and reinoculated seedlings, which showed disease symptoms after 20 days. Infection by *L. theobromae* is facilitated by

Fig. 2. Left: Cocoa tree in Sulawesi, Indonesia infected with VSD. Right: LS of a stem of infected cocoa showing hyphae of the causal organism, *C. theobromae*, in a xylem vessel, magnified x400 (*author's photos*).

wounding or insect damage. In passion fruit, for example, infection is facilitated by tunnelling by a species of beetle. Generally, preventing insect damage or other forms of wounding may reduce infection. Removal of dead twigs or branches is also a recommended control measure since *L. theobromae* is a facultative saprophyte (Kohler et al., 1997). Thread blight or black rot in coffee is caused by a basidiomycete, *Ceratobasidium noxium* (formerly *Pellicularia koleroga* syn. *Corticium koleroga*), which infects branches of coffee trees causing blackening and drying of leaves, leading to branch dieback. The pathogen also infects cocoa, citrus and woody species. *Marasmiellus scandens* causes white thread blight of cocoa. Infection by *M. scandens* is may be associated with stem borers, especially Scolytidae beetles (observation by Asman, Hasanuddin University).

Various causes of mango decline have been reported. Mango decline is associated with opportunistic fungi such as *Botryosphaeria ribis*, *Physalospora* sp. and others (see Zheng et al., 2002). *L. theobromae* has also been linked to mango decline (Shahbaz et al., 2009), while a form of mango decline in Brazil is caused by *Ceratocystis fimbriata* (Ploetz, 2007). Mango decline symptoms are reported to include interveinal chlorosis in leaves, stunting and terminal and marginal necrosis with dieback of young stems, internal softening of the fruit and even tree death; these symptoms may be linked to Mn and Fe deficiency (Crane and Campbell, 1994). Both Mn and Fe may be deficient in plants growing in high pH soil. Another mango disease that has recently become a serious problems in some regions (e.g northern Australia) is mango malformation caused by *Fusarium* spp., which cause distortion in the growth of shoots and buds.

Many dieback diseases are caused by a combination of factors, including nutrient deficiency, drought and wounding or transmission of inoculum by insects. A fungus surviving in a host plant as an endophyte or as a saprophyte on dead tissue may switch to being a pathogen in stressed plants (Shulz and Boyle, 2005). Nutrient deficiency could be a key predisposing factor for this switch to occur. Generally, adequate host nutrition, as well as shade management and sanitation (such as adequate pruning and the removal of dead twigs), are crucial preventative measure for some forms of dieback.

4.5 Root and collar rots, cankers and nematode infections

Root and collar rots

Root and collar rots are particularly severe in soils with low organic matter content, poor soil structure with a high level of compaction and poor drainage (Desaeger et al., 2004). If roots are restricted to the upper soil layers due to claypans or hardpans, this results in a greater exposure of roots to soil-borne pathogens. Similarly, old roots may leave pathogen inocula in cracks and fissures in the soil, increasing contact with new roots; an example of this is found in cases of *Eucalyptus marginate* and other native Australian species infected by the soil-borne pathogen *Phytophthora cinnamomi* (see Desaeger et al., 2004).

The basidiomycete, *Marasmiellus inoderma* infects all parts of the banana plant causing stem rot in particular. The pathogen has a number of hosts and is encouraged in marginal, nutrient-poor, soils with a high clay content and by poor drainage; soil improvement by organic amendments and better plant nutrition are recommended for managing the disease (Nelson et al., 2006). In breadfruit, *Lasiodiplodia theobromae* (syn. *Botryodiplodia theobromae*) causes a collar rot characterised by external white strands. As mentioned above, this is a wound invading fungus and drought-stressed trees may be particularly predisposed to infection. *Phellinus noxius* (Basidiomycota) also infects bread fruit trees at the base of the trunk creating a brown encrustation (that includes soil particles) sometimes with a white margin and gum exudates. Other hosts of this pathogen include cocoa, coffee, *Leucaena* sp., mango, oil palm and forest trees, such as *Tectonia*, *Swietenia* and laurel (Kohler et al., 1997). Diseases caused by the fungus include brown root and collar rot in cocoa: the fungus may encircle the whole trunk, causing sudden death of the tree with leaves remaining attached. As the pathogen is dispersed between roots in the soil, adjacent trees may be infected and killed. Therefore, removal of tree stumps and all large roots before planting is a necessary preventative control measure. Rigidoporus root rot (or white root rot) caused by the basidiomycete, *Rigidoporus microporus* causes serious crop losses to rubber, mango, durian and other tree crops. It is a serious pathogen of rubber trees in Malaysia. Red root disease caused by *Ganoderma philippii* (syn. *G. pseudoferreum*) another basidiomycete, is also an important rubber pathogen in Malaysia and India. The use of arsenic-containing sprays as a control method for root rots creates environmental concerns. As for other root pathogens, sanitation of areas prior to planting is a necessary control measure.

Ganoderma orbiforme (syn. *G. boninense*) is the most severe pathogen on oil palm in southeast Asia (Susanto et al., 2005; Flood et al., 2005; www.dfid.gov.uk). Sanitation is particularly important and diseased trees, including the roots, are dug out mechanically, and shredded for composting. In a recently established trial in Sumatra, Indonesia, shredded plant material and empty fruit bunches are being used to prepare compost using microbial promoters, particularly *Trichoderma* spp., which the trial aims to test by soil application in order to assess their effect on root rot disease (Agus Purwantara, pers. comm.). Srinivasulu (2003) reported a higher incidence of *Ganoderma* spp. infection of coconut growing on sandy and red soils (which had a low organic matter content) than on black soils (with a higher organic matter content). Amendment of soil with calcium nitrate has been used to reduce Ganoderma basal stem rot in coconut palms (Kandan et al., 2010). In a plot of coconut trees affected by this disease, Kharthikaya et al. (2006) demonstrated that a combined treatment of frequent irrigation, soil applications of neem cake, *Trichoderma viride, Pseudomonas fluorescens* and a fungicide prevented the spread of the pathogen, *G. lucidum*, and led to the recovery of 42% of diseased palm trees.

Phytophthora spp. cause root rots in crops such as avocado and citrus. Root rot in avocado caused by *P. cinnamomi* becomes particularly severe under conditions of flooding (Ploetz, 2007). Phytophthora root rot in citrus is associated with citrus leaf miner damage and Diaprepes root weevil (Ploetz, 2007). The form of nitrogen available to citrus trees appears to affect the severity of this disease. Root rot of citrus was shown to increase in the presence of ammonium-N but decreased by supplying nitrate-N (Menge and Nemec, 1997). Root rots are also caused by *Rosellinia* spp. on crops such as avocado, citrus and banana – they are favoured by acidic soil conditions (Ploetz, 2007).

Nematodes

Nematodes are generally favoured by coarse-textured soils that are low in organic matter and biological activity (Desaeger et al., 2004). For example, bananas became more susceptible to nematodes when grown in degraded soil that had lost much of its original organic matter (Page and Bridges, 1993) and nematode attack on maize was more damaging in unfertilised, than in fertilised plots (Desaeger et al., 2004). The intensity of crop production can also influence nematode populations. In Costa Rica, Avelino et al. (2009) examined the conditions that influence populations of two nematode species, *Meloidogyne exigua* (root-knot nematode) and *Pratylenchus coffeae* colonising roots of coffee. The two species had specific preferences of altitude and soils, with low *M. exigua* populations being associated with non-sandy soils with a high K and Zn content, but high populations of both species occurred on farms which had inter-row planting distances of less than 0.9 m, irrespective of environmental conditions. This, the authors suggest, indicates that intensification of coffee production provides conditions favourable for nematode reproduction and transmission.

Nutrient supply and organic amendments can have direct impacts on nematode populations and infection. In a guava growing area of Brazil, the numbers of juveniles of the root-knot nematode of guava, *M. enterolobii* (syn. *M. mayaguensis*) in naturally infested areas, were decreased by manure application to the soil (Souza et al., 2006). As mentioned above (Section 4.2), silicon (Si) supplied to coffee plants was shown to decrease the number of rust lesions on leaves; Si also has an effect on resistance to nematodes. Silva et al. (2010) found that a lower number of *M. exigua* galls and eggs occurred in the roots of coffee (*C. arabica*) plants that had been inoculated with the nematode and provided with Si, compared to control plants which were not supplied with Si.

Nematode infections may predispose tree crops to infections by other pathogens creating disease complexes (Desaeger et al., 2004). Generally, nematodes may have a number of roles in facilitating disease development acting as vectors and wounding agents. They may affect the susceptibility of the host to other pathogens, or influence rhizosphere ecology (Desaeger et al., 2004). Melendez and Powell (1969) reported that nematode infection of roots of tobacco plants caused a soil-inhabiting *Trichoderma* sp. to become pathogenic. Vascular wilt pathogens are particularly encouraged by endoparasitic nematodes, while cortical root pathogens are encouraged by ectoparasitic nematodes (Hillocks and Waller, 1997). The nematode species, *Radopholus similis* causes root rot in, among other crops, banana, avocado, coconut, coffee and sugar cane. In banana, *Fusarium oxysporum* infections of the root were associated with roots infested with *R. similis* (Blake, 1966). Gomes et al. (2011) reported that the causes of root rot in guava decline in Brazil included colonisation by nematodes (*M. enterolobii* syn. *M. mayaguensis*) and a *Fusarium* sp., identified upon isolation as *F. solani*. Infestation of roots by the nematode appeared to predispose roots to infection by *F. solani*, as the latter species was only isolated from nematode-infected roots. The authors found that *F. solani* isolates inoculated into the roots of guava plants initiated infections in trees that had been pre-inoculated with the nematode, but not those that had been physically damaged using a knife. Therefore, it appears that physical damage alone by the *M. enterolobii* did not account for the predisposition of guava colonised by the nematodes to *F. solani* infection. Khan et al. (1995) had earlier shown that inoculation of papaya with both the nematode *M. incognita* and *F. solani* caused a greater decrease in plant growth than inoculation with either species alone. They also showed that the level of root rot (caused by *F. solani*) could be decreased by the application of NPK fertiliser.

Phytophthora *stem canker*

Canker infections are initiated on species such as cocoa and durian by the same causal pathogen of fruit rots and leaf blights, *Phytophthora palmivora*. Other *Phytophthora* species may be associated with cankers, but *P. palmivora* is the most prevalent species of the genus on tropical tree crops. On cocoa, cankers are moist, wine-red lesions under the bark that expand in diameter during the wet season (Guest, 2007; McMahon et al., 2010). If cankers girdle the whole stem, sudden death results with the leaves still attached to the tree. Particularly wet conditions encourage Phytophthora stem canker, which can be a serious problem in areas prone to water logging or flooding. The presence of susceptible hosts, lack of pruning or management to enhance air circulation and lack of sanitation of sources of inoculum, such as infected pods can all lead to increased incidence of the disease. Since old and apparently weaker trees are more susceptible (author's observation) it is possible that poor nutrition might encourage cankers to develop. Low zinc (Zn) nutrition has been suggested to predispose plants to Phytophthora infection (Nelson et al., 2011).

5. Soil amendments with microbial species to control disease

Application of composts to soils has been shown to have a suppressive effect on soil-borne diseases such as damping off, root rots and wilts, both in controlled glasshouse experiments and in the field (Noble and Coventry, 2005). Loss of disease suppression occurs if the composted materials are sterilised indicating that their suppressive effect is mainly biological (Bonanomi et al., 2010; Noble and Coventry, 2005). In Papua New Guinea, the time in which *P. palmivora* inoculum remained viable was found to be shorter in soils under cocoa leaf litter mulch than under grass litter: possibly conditions in the soils under leaf litter were more favourable to microorganisms antagonistic to the pathogen (Konam and Guest, 2002). McDonald et al. (2007) showed that the factor(s) conferring disease suppression of *P. cinnamomi* in avocado orchards can be transferred from suppressive to conducive soils. Suppressive soils may be effective in reducing disease due to either a high total microbial activity or to the presence of particular antagonistic species (Weller et al., 2002). Reeleder et al. (2003) identified biological parameters as being the best predictors of the capacity of soils for disease suppression. These include microbial biomass, substrate respiration, fluorescein diacetate (FDA) activity and populations of bacteria, including fluorescent pseudomonads, and of antagonist fungi, particularly *Trichoderma* spp. But the mechanisms leading to pathogen suppression and the antagonistic organisms involved remain little understood. The employment of molecular techniques to track changes in microbial communities in amended soils and to select potential biocontrol agents has been proposed by some researchers (Noble and Coventry, 2005; Reeleder et al, 2003).

Various studies have demonstrated disease suppression following soil amendments with organic materials or by treatments with antagonist microorganisms isolated from soils (Table 2). Root rot caused by *P. nicotianae* in Florida citrus orchards was reduced by treatments with composted municipal waste (Widmer et al., 1998). However, the suppressive activity of the compost was lost after storage for three months or more. Also, some sources of waste contained toxins that impaired plant growth. Vawdrey et al (2002) tested different soil amendments for their effect on root rot of papaya, caused by *Phytophthora palmivora*. They found in both pot and field experiments that a sawdust/urea preparation was more effective in reducing disease and *P. palmivora* populations, than the other organic materials tested, such as molasses. A lower soil moisture level in sawdust/urea treated soils might partly explain the suppressive effect of this treatment (Vawdrey et al., 2002). Organic amendments based on preparations from the neem tree (*Azadirachta indica*) have been shown to be effective in disease suppression, and to have nemiticidal properties (Agbenin, 2004). Applications of neem cake and other organic materials in combination with *Trichoderma harzianum*, reduced both the populations of *P. meadii* and their infection of cardamon plants (Bhai and Sarma, 2009). Peng et al. (1999) compared the effect of conducive and suppressive soils on disease severity of Fusarium wilt in Cavendish banana plantlets. The conducive soil had higher populations of filamentous fungi than the suppressive soil, which had greater numbers of bacteria and actinomyctes. They reported that, compared to the conducive soil, the suppressive soil reduced *Fusarium oxysporum* chlamydospore germination by 41% and decreased disease severity by over 50%. However, supplying Ca compounds or Fe in a chelated form to both suppressive and conducive soils decreased Fusarium wilt in the plantlets by 33-50%.

A number of studies have been conducted reporting the isolation of potential microbial biocontrol agents from the soils in which tropical perennials are grown. Examples are the identification of five *Trichoderma* species, selected from 25 isolates from mango orchards in Mexico, which had an inhibitory effect on the growth of *Fusarium* spp. *in vitro* (Michel-Acev et al., 2001), the isolation of *Trichoderma* spp. demonstrated to have *in vitro* inhibitory effects on the cocoa pathogen, *Moniliophthora perniciosa* (Rivas-Cordero, 2010), on *Ganoderma orbiforme* (syn. *G. boninense*) (Siddiquee et al., 2009) and on *Lasiodiplodia theobromae* and *Colletotrichum musae,* isolated from infected banana (Samuels, 2006; Sangeetha 2009). Lower levels of disease caused by *G. orbiforme* in infected oil palm seedlings were recorded

Crop	Disease (causal pathogen)	Type of treatment	Impact on disease	Notes	Source
Oil palm	Basal stem rot (*G. orbiforme*)	*T. harzianum* conidia spray	Disease severity index 5% cf. 95% (control)	Oil palm seedlings treated	Izzati et al. , 2008
Banana	Nematode infection (*R. similis*)	Non-pathogenic *F. oxysporum* isolate	No. of root lesions reduced	Defence-related genes upregulated	Paparu, 2008
Coffee	Nematode infection (*P. coffea*)	Pre-inoculation of coffee with AM fungi	Reduce number of lesions	High P tissue level in mycorrhizal plants	Vaast and Craswell, 1997
Tea	Rot (*Armillaria* sp.)	*T. harzianum* applied to detached tea stems	Reduce incidence by 52% up to 100%	Four *T. harzanium* isolates effective	Otieno et al., 2003
Papaya	Root rot (*P. palmivora*)	Sawdust/urea mix	Root rot severity index 1.2 compared to 5 (control)	Treatment also lowered soil moisture content	Vawdrey et al., 2002
Cardamon	Root rot (*P. meadii*)	Neem cake plus *T. harzianum*	Reduced infection	*T. harzianum* applied with manure also effective	Bhai and Sarma, 2009
Citrus	Root rot (*P. nicotianae*)	Composted municipal waste	Reduce incidence from 95% to 5%	But some wastes toxic to plants	Widmer et al., 1998
Mango	Leaf spot (*P. mangiferae, L. theobromae, M. mangiferae*)	Apply *B. subtilis* isolate *in vivo*	Reduce leaf spot diameter by over 70%	*In vitro* inhibition by *B. subtilis* also verified	Okigbo and Osuinde, 2003

Table 2. Examples of impacts on diseases by soil amendments and other treatments applied to tropical trees crops.

following treatment of the seedlings with *T. harzianum* conidia (Izzati et al. 2008). Tea stems inoculated with *T. harzianum* demonstrated resistance to infection by *Armillaria* sp. (Otieno et al., 2003). Muleta et al. (2007) found that among isolates of rhizobacteria isolated from soils under coffee, some *Pseudomonas* and *Bacillus* species strongly inhibited the *in vitro* growth of *Fusarium* spp., including *F. stilboides* and *F. oxysporum*. Such rhizobacteria have potential as biocontrol agents of coffee wilt diseases. Using an integrated approach to management of *Fusarium* spp. and the citrus nematode, *Tylenchulus semipenetrans*, in Egypt, Abd-Elgawad et al. (2010) showed that the application of bacterial isolates contributed to a reduction of populations of these pathogens.

In a study of leaf spot disease of mango in Nigeria, Okigbo and Osuinde (2003) demonstrated the pathogenicity of three fungi species isolated from the leaf lesions (*Pestalotiopsis mangiferae*, *Lasiodiplodia theobromae* and *Macrophoma mangiferae*) by inoculating them individually onto healthy mango leaves. In addition, the authors isolated a bacterium identified as *Bacillus subtilis* from soil under mango trees and showed that it inhibited growth *in vitro* of the three causal pathogens and also reduced disease severity *in vivo* when applied to soil in the field. In tea plants infected with *L. theobromae*, *in vivo* control of the disease was demonstrated by pre-treatment of the plants with bacteria that had been isolated from the tea rhizosphere and shown to have *in vitro* antagonistic activity (Purkayastha et al., 2010). Stirling et al. (1992) isolated fluorescent pseudomonads from avocado soils suppressive to *P. cinnamomi* (causing root rot) and demonstrated their *in vitro* antagonism to the pathogen. *In vivo* control of the root-knot nematode on coffee roots, *Meloidogyne incognita*, by the application of an obligate bacterial parasite of the nematode, a strain of *Pasteuria penetrans*, was demonstrated by Carneiro et al. (2007). The colonisation of banana roots by the nematode *Radopholus similis* could be decreased by inoculation of the banana plants with a non-pathogenic isolate of *F. oxysporum* (Paparu et al., 2008). The authors demonstrated that inoculation of the fungus caused the up-regulation of a number of defence-related genes in the host plant (the expression of some of these genes was also increased by inoculation with the nematode). However, the means by which biocontrol agents exert antagonistic effects towards pathogens may include a variety of other mechanisms, including direct antibiosis and competition.

Role of mycorrhizae

Most plants form symbiotic associations with fungi, forming mycorrhizae. In tropical perennial crop species such associations mainly occur with arbuscular mycorrhizal (AM) fungi, but associations with other taxonomic groups of fungi, forming ectomycorrhizae, are also found. In fact, it could be said that, under natural conditions, plants have mycorrhizae rather than roots (Azco'n-Aguilar and Barea, 1996). Mycorrhizal fungi have an irreplaceable role in supplying nutrients to the plants, particularly of phosphorus (P), which is often unavailable for direct uptake by plant roots. They can also reduce disease severity, particularly of diseases caused by soil-borne pathogens, such as nematodes, in addition to conferring tolerance to drought and salinity (Andrade et al., 2009). Increased tolerance to nematodes has been reported in perennial crop species inoculated with mycorrhizal fungi. Vaast et al. (1997) reported enhanced resistance to *Pratylenchus coffeae* in coffee plants inoculated with AM fungi, with fewer lesions occurring in the AM fungi-inoculated roots (see Section 2). Similarly, an increase in resistance to nematodes was reported in banana plants that had been inoculated with mycorrhizal fungi (Elsen et al., 2003). In a study in citrus orchards in Thailand, increased growth, P uptake and resistance to root rot caused by *Phytophthora nicotianae* was shown to result from the inoculation of citrus trees with a species of AM fungus, *Glomus etunicatum*, isolated from local citrus orchard (Watanarojanaporn et al., 2011). A second AM fungus isolate, identified as *Acaulospora tuberculata*, also conferred resistance to *P. nicotianae* disease. Mycorrhizal fungi may compete for infection sites with the pathogen and/or they may impede access to nutrients by the pathogen (Azco'n-Aquila and Barea, 1996).

6. Conclusion

Soil function, plant nutritional status and cultural management practices have a strong influence on the incidence and severity of many diseases of tropical perennials. Diseases are influenced, not only by the general nutritional status of the plant (i.e. by an adequate supply of macronutrients), but also by individual nutrients. Soil pH is particularly important for a number of reasons, including its effect on availability of cations to the plant, pathogen or antagonists. It also influences microbial ecology, with a number of potential antagonists among bacterial and actinomycete species being favoured by a higher soil pH. Since tropical perennial crops and the use of inorganic fertiliser can lower soil pH, liming and/or compost treatments are strategies that can be adopted for disease mitigation in these crops. Composting farm waste, not only returns nutrients to the farm, but also improves farm sanitation as it kills larvae of pests and pathogen spores and other forms of inoculum. Nutrient elements supplied in mineral fertilisers can interact and care needs to be taken in their application. As mentioned earlier, some fertilisers can increase concentrations of Cl in plant tissues and decrease the K/Cl ratio. Also mentioned previously (Section 2), applications of Mg or K in excess can reduce the uptake of other basic cations, particularly Ca. To complement the use of soil amendments as a way to increase production and decrease pest and pathogen damage, it is important that tropical perennial crops are managed properly with cultural methods such as pruning, shade regulation, soil drainage and sanitation.

7. Acknowledgement

Preparation of this review paper was supported by funding provided by the Australian Centre for International Agricultural Research (ACIAR).

8. References

Abd-Elgawad, M. M., N. S. El-Mougy, N. G. El-Gamal, M. M. Abdel-Kader, and M. M. Mohamed. "Protective Treatments against Soilborne Pathogens in Citrus Orchards." *Journal of Plant Protection Research* 50, no. 4 (2010): 477-84.

Acosta-Ramos, M., D. H. Noriega-Cantu, D. Nieto-Angel, and D. Teliz-Ortiz. "Effect of Integrated Mango (*Mangifera Indica* L.) Management on the Incidence of Diseases and Fruit Quality." *Revista Mexicana de Fitopatologia* 21, no. 1 (2003): 46-55.

Adebitan, S. A. "Effects of Phosphorus and Weed Interference on Anthracnose of Cowpea in Nigeria." *Fitopatologia Brasileira* 21, no. 2 (1996): 173-79.

Agbenin, O. N. "Potentials of Organic Amendments in the Control of Plant Parasitic Nematodes." *Plant Protection Science* 39, no. 1 (2004): 21-25.

Agrios, G. N. *Plant Pathology*. Amsterdam: Elsevier Academic Press, 2005.

Andrade, S. A. L., P. Mazzafera, M. A. Schiavinato, and A. P. D. Silveira. "Arbuscular Mycorrhizal Association in Coffee." *The Journal of Agricultural Science* 147, no. 2 (2009): 105-15.

Anonymous. "On Organic Coffee Farm, Complex Interactions Keep Pests under Control." *U.S. News & World Report*, no. Journal Article (2010): 1.

Avelino, J. , M. Bouvret, L. Salazar, and C. Cilas. "Relationships between Agro-Ecological Factors and Population Densities of *Meloidogyne Exigua* and *Pratylenchus Coffeae* Sensu Lato in Coffee Roots, in Costa Rica." *Applied Soil Ecology* 43, no. 1 (2009): 95-105.

Avelino, J. , H. Zelaya, A. Merlo, A. Pineda, M. Ordonez, and S. Savary. "The Intensity of a Coffee Rust Epidemic Is Dependent on Production Situations." *Ecological Modelling* 197, no. 3-4 (2006): 431-47.

Azco'n-Aguilar, C., and J .M. Barea. "Arbuscular Mycorrhizas and Biological Control of Soil-Borne Plant Pathogens - an Overview of the Mechanisms Involved." *Mycorrhiza* 6 (1996): 457-64.

Beer, J. "Litter Production and Nutrient Cycling in Coffee(*Coffea Arabica*) and Cacao (*Theobroma Cacao*) Plantations with Shade Trees." *Agroforestry Systems* 7 (1988): 103-14.

Beer, J., R. Muschler, D. Kass, and E. Somarriba. "Shade Management in Coffee and Cacao Plantations." *Agroforestry Systems* 38, no. 1 (1998): 139-64.

Bell, A. A. "The Role of Nutrition in Diseases of Cotton." In *Soilborne Plant Pathogens: Management of Diseases with Macro- and Microelements* edited by Arthur W. Engelhard, 167-204. St. Paul, Minn: APS press, 1989.

Bhai, R. S., and Y. R. Sarma. "Effect of Organic Amendments on the Proliferation Stability of *Trichoderma Harzianum* and Suppression of *Phytophthora Meadii* in Cardamom Soils in Relation to Soil Microflora." *Journal of Biological Control* 23, no. 2 (2009): 163-67.

Blake, C.D. "The Histological Changes in Banana Roots Caused by *Radopholus Similis* and *Helicotylenchus Multicinctus*." *Nematologica* 12, no. 1 (1966): 129-37.

Bonanomi, G., V. Antignani, M. Capodilupo, and F. Scala. "Identifying the Characteristics of Organic Soil Amendments That Suppress Soilborne Plant Diseases." *Soil Biology & Biochemistry* 42, no. 2 (2010): 136-44.

Buresh, R. J., P.C. Smithson, and D.T. Hellums. "Building Soil P Capital in Africa." In *Replenishing Soil Fertility in Africa*, edited by R.J. Buresh, P.A. Sanchez and F.G. Calhoun, 111-49. Madison, Wisconsin: Soil Science Society of America, 1997.

Cakmak, I. , C. Hengeler, and H. Marschner. "Changes in Phloem Export of Sucrose in Leaves in Response to Phosphorus, Potassium and Magnesium Deficiency in Bean Plants." *Journal of Experimental Botany* 45 (1994): 1251-57.

Carneiro, R. M. D. G., L. F. G. de Mesquita, P.A. S. Cirotto, Fabiane C Mota, Maria Ritta A Almeida, and Maria Celia Cordeiro. "The Effect of Sandy Soil, Bacterium Dose and Time on the Efficacy of *Pasteuria Penetrans* to Control *Meloidogyne Incognita* Race 1 on Coffee." *Nematology* 9, no. Part 6 (2007): 845-51.

Chuang, T-Y. "Studies on the Soils Suppressive to Banana Fusarium Wilt Ii. Nature of Suppression to Race 4 of Fusarium-Oxysporum-F-Sp-Cubense in Taiwan Soils." *Plant Protection Bulletin* 30, no. 2 (1988): 125-34.

Chuang, T Y. "Suppressive Soil of Banana Fusarium Wilt in Taiwan." *Plant Protection Bulletin* 33, no. 1 (1991): 133-41.

Cooper, Richard M, and Jane S Williams. "Elemental Sulphur as an Induced Antifungal Substance in Plant Defence." *Journal of experimental botany* 55, no. 404 (2004): 1947-53.

Corden, M. E. "Influence of Calcium Nutrition on Fusarium Wilt of Tomato and Polygalacturonase Activity." *Phytopathology* 55 (1965): 222-24.

Crane, J. H., and C. W. Campbell. "The Mango." *Fact Sheets, Horticultural Sciences Department, Florida Cooperative Extension Service* (1994).

da Silva Moraes, Wilson, Hilario Antonio de Castro, Juliana Domingues Lima, Eloisa Aparecida das Gracas Leite, and Mauricio de Souza. "Susceptibility of Three Citrus Species to *Dothiorella Gregaria* Sacc. In Function of the Nutriconal State." *Ciencia Rural* 37, no. 1 (2007): 7-12.

Datnoff, L. E., F. A. Rodrigues, and K.W. Seebold. "Silicon and Plant Disease." In *Mineral Nutrition and Plant Disease*, edited by L. E. Datnoff, Wade H. Elmer and D. M. Huber, 233-46. St. Paul, Minn: American Phytopathological Society, 2007.

Desaeger, J. , Meka R. Rao, and J. Bridge. "Nematodes and Other Soilborne Pathogens in Agroforestry." In *Below-Ground Interactions in Tropical Agroecosystems: Concepts and Models with Multiple Plant Components*, edited by Meine van Noordwijk, Georg Cadisch and C.K. Ong: CABI, 2004.

Dominguez, J., M. A. Negrin, and C. M. Rodriguez. "Aggregate Water-Stability, Particle-Size and Soil Solution Properties in Conducive and Suppressive Soils to Fusarium Wilt of Banana from Canary Islands (Spain)." *Soil Biology & Biochemistry* 33, no. 4-5 (2001): 449-55.

Duffy, B. "Zinc and Plant Disease." In *Mineral Nutrition and Plant Disease*, edited by L. E. Datnoff, Wade H. Elmer and D. M. Huber, 155-78. St. Paul, Minn: American Phytopathological Society, 2007.

Edington, L.V. , and J.C. Walker. "Influence of Calcium and Boron Nutrition on Development of Fusarium Wilt of Tomato." *Phytopathology* 48 (1958): 324-26.

Elsen, A., H. Baimey, R. Swennen, and D. De Waele. "Relative Mycorrhizal Dependency and Mycorrhiza-Nematode Interaction in Banana Cultivars (*Musa* Spp.) Differing in Nematode Susceptibility." *Plant and Soil* 256, no. 2 (2003): 303-13.

Emechebe, A.M. "The Effect of Soil Moisture and of N, P and K on Incidence of Infection with Cacao Seedlings Inoculated with *Verticillium Dahliae*." *Plant & Soil* 54, no. 1 (2006): 143-47.

Engelhard, Arthur W. *Soilborne Plant Pathogens: Management of Diseases with Macro- and Microelements*. St. Paul, Minn: APS press, 1989.

Evans, I., E. Solberg, and D. M. Huber. "Copper and Plant Disease." In *Mineral Nutrition and Plant Disease*, edited by L. E. Datnoff, Wade H. Elmer and D. M. Huber, 177-88. St. Paul, Minn: American Phytopathological Society, 2007.

Flood, J. *Coffee Wilt Disease*. Wallingford: CAB International, 2010.

Flood, J., L. Keenan, S. Wayne, and Y. Hasan. "Studies on Oil Palm Trunks as Sources of Infection in the Field." *Mycopathologia* 159, no. 1 (2005): 101-07.

Gaur, R. B., and P. K. Vaidya. "Reduction of Root Rot of Chickpea by Soil Application of Phosphorus and Zinc." *International Chickpea Newsletter* 9 (1983): 17-18.

Gomes, Vicente Martins, Ricardo Moreira Souza, Vicente Mussi-Dias, Silvaldo Felipe da Silveira, and Claudia Dolinski. "Guava Decline: A Complex Disease Involving Meloidogyne Mayaguensis and Fusarium Solani." *Journal of Phytopathology* 159, no. 1 (2011): 45-50.

Guest, D. I., and P.J. Keane. "Vascular-Streak Dieback: A New Encounter Disease of Cacao in Papua New Guinea and Southeast Asia Caused by the Obligate Basidiomyceteoncobasidium Theobromae." *Phytopathology* 97, no. 12 (2007): 1654-57.

Guest, David. "Black Pod: Diverse Pathogens with a Global Impact on Cocoa Yield." *Phytopathology* 97, no. 12 (2007): 1650-53.

Haneklaus, S., E. Bloem, and E. Schnug. "Sulfur and Plant Disease." In *Mineral Nutrition and Plant Disease*, edited by L. E. Datnoff, Wade H. Elmer and D. M. Huber, 101-18. St. Paul, Minn: American Phytopathological Society, 2007.

Harrington, T. "Ceratocystis Wilt: Taxonomy and Nomenclature." CABI Publishing, http://www.public.iastate.edu/~tcharrin/CABIinfo.html (2004).

Hartemink, A.E. "Nutrient Stocks, Nutrient Cycling and Soil Changes in Cocoa Ecosystems: A Review." *Advances in Agronomy* 86 (2005): 227-52.

Hillocks, R. J. , and J.M. Walller, eds. *Soilborne Diseases of Tropical Crops*: CABI Publishing, 1997.

Holliday, Paul. *Fungus Diseases of Tropical Crops*. New York: Cambridge University Press, 1980.

Huber, D. M., and I. A. Thompson. "Nitrogen and Plant Disease." In *Mineral Nutrition and Plant Disease*, edited by L. E. Datnoff, Wade H. Elmer and D. M. Huber, 31-44. St. Paul, Minn: American Phytopathological Society, 2007.

Izzati, Mohd Zainudin Nur Ain, and Faridah Abdullah. "Disease Suppression in Ganoderma-Infected Oil Palm Seedlings Treated with Trichoderma Harzianum." *Plant Protection Science* 44, no. 3 (2008): 101-07.

Jones, P. J., A. W. Engelhard, and S. S. Woltz. "Management of Fusarium Wilt of Vegetables and Ornamentals by Macro- and Microelement Nutrition." In *Soilborne Plant Pathogens: Management of Diseases with Macro- and Microelements*, edited by Arthur W. Engelhard, 18-32. St. Paul, Minn: APS press, 1989.

Jones, P. J., and S. S. Woltz. "Effect of Soil Ph and Micronutrient Amendments on *Verticillium* and *Fusarium* Wilt of Tomato." *Plant Disease Report* 56 (1972): 151-53.

Kader, Adel A., ed. *Postharvest Technology of Horticultural Crops*. 3rd ed: University of California, 2002.

Kalim, S., Y. P. Luthra, and S. K. Gandhi. "Role of Zinc and Manganese in Resistance of Cowpea Rot." *Journal of Plant Diseases & Protection* 110 (2003): 235-43.

Kandan, A., R. Bhaskaran, and R. Samiyappan. "Ganoderma - a Basal Stem Rot Disease of Coconut Palm in South Asia and Asia Pacific Regions." *Archives of Phytopathology & Plant Protection* 43, no. 15 (2010): 1445-49.

Kannan, C., M. Karthik, and K. Priya. "*Lasiodiplodia Theobromae* Causes a Damaging Dieback of Cocoa in India." *Plant Pathology* 59, no. 2 (2010): 410-10.

Karthikeyan, G., T. Raguchander, and R. Rabindran. "Integrated Management of Basal Stem Rot/Ganoderma Disease of Coconut in India." *Crop Research* 32, no. 1 (2006): 121-23.

Keane, P. J., and C. A. J. Putter. *Cocoa Pest and Disease Management in Southeast Asia and Australasia*. Vol. 112. Rome: Food and Agriculture Organization of the United Nations, 1992.

Keane, P.J. "Epidemiology of Vascular-Streak Dieback of Cocoa." *Annals of Applied Biology* 98 (1981): 227-41.

Keane, P.J., M.T. Flentje, and K.P. Lamb. "Investigation of Vascular-Streak Dieback of Cocoa in Papua New Guinea." *Australian Journal of Biological Sciences* 25 (1972): 553-64.

Keane, Philip, and David Guest. "Vascular-Streak Dieback: A New Encounter Disease of Cacao in Papua New Guinea and Southeast Asia Caused by the Obligate Basidiomyceteoncobasidium Theobromae." *Phytopathology* 97, no. 12 (2007): 1654-57.

Khan, Tabreiz A., and Shabana T. Khan. "Effect of Npk on Disease Complex of Papaya Caused by *Meloidogyne Incognita* and *Fusarium Solani*." *Pakistan Journal of Nematology* 13, no. 1 (1995): 29-34.

Kohler, F. , F. Pellegrin, G. Jackson, and E. McKenzie. *Diseases of Cultivated Crops in Pacific Island Countries*: South Pacific Commission, 1997.

Konam, J.K., and D. I. Guest. "Role of Flying Beetles (Coleoptera: Scolytidae and Nitidulae) in the Spread of Phytophthora Pod Rot of Cocoa in Papua New Guinea." *Australasian Plant Pathology* 33 (2004): 55-59.

Konam, John K., and David I. Guest. "Leaf Litter Mulch Reduces the Survival of *Phytophthora Palmivora* under Cocoa Trees in Papua New Guinea." *Australasian Plant Pathology* 31, no. 4 (2002): 381-83.

Krauss, A. "Importance of Balanced Fertilization to Meet the Nutrient Demand of Industrial and Plantation Crops." In *International Workshop Importance of potash fertilizers for sustainable production of plantation and food crops* Colombo, Sri Lanka,1-2 December 2003, : International Potash Institute, Basel, Switzerland, 2003.

Maclean, R. H., J. A. Litsinger, K. Moody, and A. K. Watson. "The Impact of Alley Cropping *Gliricidia Sepium* and *Cassia Spectabilis* on Upland Rice and Maize Production." *Agroforestry Systems* 20, no. 3 (1992): 213-28.

Marschner, Horst. *The Mineral Nutrition of Higher Plants*. London: Academic Press, 1995.

Martinati, J. C., R. Harakava, S. D. Guzzo, and S. M. Tsai. "The Potential Use of a Silicon Source as a Component of an Ecological Management of Coffee Plants." *Journal of Phytopathology* 156, no. 7-8 (2008): 458-63.

Mbenoun, M., E. H. Momo Zeutsa, G. Samuels, F. Nsouga Amougou, and S. Nyasse. "Dieback Due to Lasiodiplodia Theobromae, a New Constraint to Cocoa Production in Cameroon." *Plant Pathology* 57, no. 2 (2008): 381-81.

McDonald, V., E. Pond, M. Crowley, B. McKee, and J. Menge. "Selection for and Evaluation of an Avocado Orchard Soil Microbially Suppressive to *Phytophthora Cinnamomi*." *Plant & Soil* 299, no. 1-2 (2007): 17-28.

McMahon, P. J., A. Purwantara, A. Wahab, M. Imron, S. Lambert, P. J. Keane, and D. I. Guest. "Phosphonate Applied by Trunk Injection Controls Stem Canker and Decreases Phytophthora Pod Rot (Black Pod) Incidence in Cocoa in Sulawesi." *Australasian Plant Pathology* 39, no. 2 (2010): 170-75.

Melendez, K., and N.T. Powell. "The Influence of *Meloidogyne* on Root Decay in Tobacco Caused by *Pythium* and *Trichoderma*." *Biological Abstracts* 59 (1969).

Menge, J. A., and S. Nemec. "Citrus." In *Soilborne Diseases of Tropical Crops*. Wallingford, Oxon, UK: CAB International, 1997.

Mengel, Konrad, and Ernest A. Kirkby. *Principles of Plant Nutrition*. Bern: International Potash Institute, 1982.

Michel-Aceves, A. C., O. Rebolledo-Dominguez, R. Lezama-Gutierrez, M. E. Ochoa-Moreno, J. C. Mesina-Escamilla, and G. J. Samuels. "*Trichoderma* Species in Soils Cultivated with Mango and Affected by Mango Malformation, and Its Inhibitory Potential on *Fusarium Oxysporum* and *F. Subglutinans*." *Revista Mexicana de Fitopatologia* 19, no. 2 (2001): 154-60.

Mobambo, K. N., K. Zuofa, F. Gauhl, M. O. Adeniji, and C. Pasberggauhl. "Effect of Soil Fertility on Host Response to Black Leaf Streak of Plantain (Musa Spp, Aab Group) under Traditional Farming Systems in Southeastern Nigeria." *International Journal of Pest Management* 40 no. 1 (1994): 75-80.

Mossu, G. *Le Cacaoyer*, Le Technicien D'agriculture Tropicale Paris Maisonneuve et Larose 1990.

Muleta, D, F Assefa, and U Granhall. "In Vitro Antagonism of Rhizobacteria Isolated from *Coffea Arabica* L. Against Emerging Fungal Coffee Pathogens." *Engineering in Life Sciences* 7, no. 6 (2007): 577-86.

Muraleedharam, N., and Z.M. Chen. "Pests and Diseases of Tea and Their Management." *Journal of Plantation Crops* 25, no. 1 (1997): 15-43.

Murray, D.B. "Shade and Nutrition." In *Cocoa*, edited by G.A.R. Wood and R.A. Lass, 105-24. London: Longman, 1975

Nasir, N., P. A. Pittaway, and K. G. Pegg. "Effect of Organic Amendments and Solarisation on Fusarium Wilt in Susceptible Banana Plantlets, Transplanted into Naturally Infested Soil." *Australian Journal of Agricultural Research* 54, no. 3 (2003): 251-57.

Ndo, Eunice Golda Daniele, Faustin Bella-Manga, Sali Atanga Ndindeng, Michel Ndoumbe-Nkeng, Ajong Dominic Fontem, and Christian Cilas. "Altitude, Tree Species and Soil Type Are the Main Factors Influencing the Severity of *Phaeoramularia* Leaf and Fruit Spot Disease of Citrus in the Humid Zones of Cameroon." *European Journal of Plant Pathology* 128, no. 3 (2010): 385-97.

Nelson, P.N. , M.J. Webb, S. Berthelsen, G. Curry, Yinil, D., and Fidelis C. *Nutritional Status of Cocoa in Papua New Guinea*. Edited by Australian Centre for International Agricultural Research. Vol. 76, Aciar Technical Reports Canberra: Australian Centre for International Agricultural Research (2011)

Nelson, S. "Cercospora Leaf Spot and Berry Blotch of Coffee." *Plant Disease* (2008a), www.ctahr.hawaii.edu/oc/freepubs/pdf/PD41.pdf.

Nelson, S. , D. Scmitt, and V.E. Smith. "Managing Coffee Nematode Decline." *Plant Disease* (2002), www.ctahr.hawaii.edu/oc/freepubs/pdf/PD.23.pdf.

Nelson, S.C. "Cephaleuros Species, the Plant-Parasitic Green Algae." *Plant Disease* (2008b).

Nelson, S.C., R. C. Ploetz, and A. K. Kepler. "Musa Species (Banana and Plantain)." *Species Profiles for Pacific Island Agroforestry* (2006).

Noble, R., and E. Coventry. "Suppression of Soil-Borne Plant Diseases with Composts: A Review." *Biocontrol Science & Technology* 15, no. 1 (2005): 3-20.

Okigbo, Ralph N., and Maria I. Osuinde. "Fungal Leaf Spot Diseases of Mango (*Mangifera Indica* L.) in Southeastern Nigeria and Biological Control with *Bacillus Subtilis*." *Plant Protection Science* 39, no. 2 (2003): 70-77.

Otieno, Washington, Michael Jeger, and Aad Termorshuizen. "Effect of Infesting Soil with *Trichoderma Harzianum* and Amendment with Coffee Pulp on Survival of *Armillaria.*" *BIOLOGICAL CONTROL* 26, no. 3 (2003): 293-301.

Page, S.L.J., and J. Bridge. "Plant Nematodes and Sustainability in Tropical Agriculture." *Experimental Agriculture* 29, no. 2 (1993): 139-54.

Palm, C.A. "Contribution of Agroforestry Trees to Nutrient Requirements of Intercropped Plants." In *Agroforestry: Science, Policy and Practice*, edited by Fergus L. Sinclair, pp105-24, 1995.

Palti, J. *Cultural Practices and Infectious Crop Diseases*. Vol. 9. New York: Springer-Verlag, 1981.

Paparu, Pamela, Thomas Dubois, Danny Coyne, and Altus Viljoen. "Defense-Related Gene Expression in Susceptible and Tolerant Bananas (*Musa* Spp.) Following Inoculation with Non-Pathogenic *Fusarium Oxysporum* Endophytes and Challenge with *Radopholus Similis.*" *PHYSIOLOGICAL AND MOLECULAR PLANT PATHOLOGY* 71, no. 4-6 (2008): 149-57.

Peng, H X, K Sivasithamparam, and D W Turner. "Chlamydospore Germination and Fusarium Wilt of Banana Plantlets in Suppressive and Conducive Soils Are Affected by Physical and Chemical Factors." *Soil Biology & Biochemistry* 31, no. 10 (1999): 1363-74.

Pennypacker, B. W. "The Role of Mineral Nutrition in the Control of Verticillium Wilt." In *Soilborne Plant Pathogens: Management of Diseases with Macro- and Microelements*, edited by Arthur W. Engelhard, 33-45. St. Paul, Minn: APS press, 1989.

Perrenoud, S. "Potassium and Plant Health." *Soil Science* 127, no. 1 (1977): 63.

Perrenoud, S. "Potassium and Plant Health." *IPI - Research Topics* (1990).

Ploetz, R. C., ed. *Diseases of Tropical Fruit Crops*. Wallingford, Oxon, UK: CAB International 2003.

Ploetz, R. C. "Diseases of Tropical Perennial Crops: Challenging Problems in Diverse Environments." *PLANT DISEASE* 91, no. 6 (2007): 644-63.

Ploetz, Randy C. "Fusarium-Induced Diseases of Tropical, Perennial Crops." *Phytopathology* 96, no. 6 (2006): 648-52.

Pozza, Adelia A A, Paulo T G Guimaraes, Edson A Pozza, Marcelo M Romaniello, and Janice G de Carvalho. "Effect of Substrata and Fertilizations of Coffee Seedlings in Tubes in Production and Brown Eye Spot Intensity." *Summa Phytopathologica* 27, no. 4 (2001): 370-74.

Prabhu, A. S., N. K. Fageria, D. M. Huber, and F. A. Rodrigues. "Potassium and Plant Disease." In *Mineral Nutrition and Plant Disease*, edited by L. E. Datnoff, Wade H. Elmer and D. M. Huber, 57-78. St. Paul, Minn: American Phytopathological Society, 2007.

Prakash, O M. "Integrated Management of Mango Seedling (Planting Material) Diseases: Present Status." *Biological Memoirs* 26, no. 1 (2000): 37-43.

Prior, C. "Cocoa Quarantine: Measures to Prevent the Spread of Vascular-Streak Dieback in Planting Material." *Plant Pathology* 34 (1985): 603-08.

Purkayastha, G D, A Saha, and D Saha. "Characterization of Antagonistic Bacteria Isolated from Tea Rhizosphere in Sub-Himalayan West Bengal as Potential Biocontrol Agents in Tea." *Journal of Mycology & Plant Pathology* 40, no. 1 (2010): 27-37.

Reeleder, R D. "Fungal Plant Pathogens and Soil Biodiversity." *Canadian Journal of Soil Science* 83, no. 3 (2003): 331-36.

Resende, M L V, J Flood, J D Ramsden, M G Rowan, M H Beale, and R M Cooper. "Novel Phytoalexins Including Elemental Sulphur in the Resistance of Cocoa (Theobroma Cacao L.) to Verticillium Wilt (Verticillium Dahliae Kleb.)." *Physiological & Molecular Plant Pathology* 48, no. 5 (1996): 347-59.

Reuveni, R, and M Reuveni. "Foliar-Fertilizer Therapy: A Concept in Integrated Pest Management." *Crop Protection* 17, no. 2 (1998): 111-18.

Rice, R. W. "The Physiological Role of Minerals in Plants." In *Mineral Nutrition and Plant Disease*, edited by L. E. Datnoff, Wade H. Elmer and D. M. Huber, 9-30. St. Paul, Minn: American Phytopathological Society, 2007.

Rivas Cordero, M., and D. Pavone Maniscalco. "Diversity of *Trichoderma* Spp. On *Theobroma Cacao* L. Fields in Caraboro State, Venezuela, and Its Biocontrol Capacity on *Crinipellis Perniciosa* (Stahel) Singer." *Interciencia* 35, no. 10 (2010): 777-83.

Rubini, M.R. , R.T. Silva-Ribeiro, A.W.V. Pomella, C.S. Maki, W.L. Araujo, D.R. dos Santos, and J.L. Azevedo. "Diversity of Endphytic Fungal Community of Cacao (*Theobroma Cacao* L.) and Biological Control of *Crinipellis Perniciosa*, Causal Agent of Witches' Broom Disease." *International Journal of Biological Science* 1 (2005): 24-33.

Rutherford, Mike A. "Current Knowledge of Coffee Wilt Disease, a Major Constraint to Coffee Production in Africa." *Phytopathology* 96, no. 6 (2006): 663-66.

Samuels, Gary J., Adnan Ismaiel, Ade Rosmana, Muhammad Junaid, David Guest, Peter McMahon, Philip Keane, Agus Purwantara, Smilja Lambert, Marianela Rodriguez-Carres, and Marc A. Cubeta. "Vascular Streak Dieback of Cacao in Southeast Asia and Melanesia: In Planta Detection of the Pathogen and a New Taxonomy." *Fungal biology*, no. Journal Article (2011).

Samuels, Gary J., Carmen Suarez, Karina Solis, Keith A. Holmes, Sarah E. Thomas, Adnan Ismaiel, and Harry C. Evans. "Trichoderma Theobromicola and T. Paucisporum: Two New Species Isolated from Cacao in South America." *Mycological Research* 110, no. Pt 4 (2006): 381-92.

Sangeetha, Ganesan, Swaminathan Usharani, and Arjunan Muthukumar. "Biocontrol with *Trichoderma* Species for the Management of Postharvest Crown Rot of Banana." *Phytopathologia Mediterranea* 48, no. 2 (2009): 214-25.

Santos, Florisvalda da Silva, Paulo Estevao de Souza, Edson Ampelio Pozza, Julio Cesar Miranda, Sarah Silva Barreto, and Vanessa Cristina Theodoro. "Progress of Brown Eye Spot (*Cercospora Coffeicola* Berkeley & Cooke) in Coffee Trees in Organic and Conventional Systems." *Summa Phytopathologica* 34, no. 1 (2008): 48-54.

Schroth, G. , J. Lehmann, M. R. L. Rodrigues, E. Barros, and J. L. V. Macêdo. "Plant-Soil Interactions in Multistrata Agroforestry in the Humid Tropics." *Agroforestry Systems* 53 (2001): 85-102.

Schroth, G., U. Krauss, L. Gasparotto, J. A. Duarte Aguilar, and K. Vohland. "Pests and Diseases in Agroforestry Systems of the Humid Tropics." *Agroforestry Systems* 50, no. 3 (2000): 199-241.

Schulz, Barbara, and Christine Boyle. "The Endophytic Continuum." *Mycological Research* 109, no. 6 (2005): 661-86.

Shahbaz, M., Z. Iqbal, A. Saleem, and M.A. Anjum. "Association of *Lasiodiplodia Theobromae* with Different Decline Disorders in Mango (*Mangifera Indica* L.)." *Pakistan Journal of Botany* 41, no. 1 (2009): 359-68.

Shen, G H, Q H Xue, M Tang, Q Chen, L N Wang, C M Duan, L Xue, and J Zhao. "Inhibitory Effects of Potassium Silicate on Five Soil-Borne Phytopathogenic Fungi in Vitro." *Journal of Plant Diseases & Protection* 117, no. 4 (2010): 180-84.

Siddiqui, I A, S S Shaukat, and M Hamid. "Role of Zinc in Rhizobacteria-Mediated Suppression of Root-Infecting Fungi and Root-Knot Nematode." *Journal of Phytopathology* 150, no. 10 (2002): 569-75.

Silva, R V, R D L Oliveira, K J T Nascimento, and F A Rodrigues. "Biochemical Responses of Coffee Resistance against Meloidogyne Exigua Mediated by Silicon." *Plant Pathology* 59, no. 3 (2010): 586-93.

Smyth, T.J., and D.K. Cassell. "Synthesis of Long-Term Soil Management Research on Ultisols and Oxisols in the Amazon." In *Soil Management: Experimental Basis for Sustainability and Environmental Quality*, edited by R. Lal and B. A. Stewart. Boca Raton: Lewis Publishers, 1995.

Somarriba, Eduardo, and John Beer. "Productivity of *Theobroma Cacao* Agroforestry Systems with Timber or Legume Service Shade Trees." *AGROFORESTRY SYSTEMS* 81, no. 2 (2011): 109-21.

Souza, Ricardo M, Maciel S Nogueira, Inorbert M Lima, Marcelo Melarato, and Claudia M Dolinski. "Management of the Guava Root-Knot Nematode in Sao Joao Da Barra, Brazil, and Report of New Hosts." *Nematologia Brasileira* 30, no. 2 (2006): 165-69.

Srinivasulu, B, K Aruna, D V R Rao, and H Hameedkhan. "Epidemiology of Basal Stem Rot (Ganoderma Wilt) Disease of Coconut in Andhra Pradesh." *Indian Journal of Plant Protection* 31, no. 1 (2003): 48-50.

Staver, C., F. Guharay, D. Monterroso, and R. G. Muschler. "Designing Pest-Suppressive Multistrata Perennial Crop Systems: Shade-Grown Coffee in Central America." *Agroforestry Systems* 53, no. 2 (2001): 151-70.

Stirling, A. M., A. C. Hayward, and K. G. Pegg. "Evaluation of the Biological Control Potential of Bacteria Isolated from a Soil Suppressive to *Phytophthora Cinnamomi*." *AUSTRALASIAN PLANT PATHOLOGY* 21, no. 4 (1992): 133-42.

Stover, R. H. . *Banana, Plantain and Abaca Diseases*. Kew, Surrey, UK: Commonwealth Mycological Institute 1972

Streeter, T. C., Rengel. Z., S. M. Neate, and R. D. Graham. "Zinc Fertilisation Increases Tolerance to *Rhizoctonia Solani* (Ag 8) in *Medicago Trunculata*." *Plant & Soil* 228 (2001): 233-42.

Susanto, A., P. S. Sudharto, and R. Y. Purba. "Enhancing Biological Control of Basal Stem Rot Disease (*Ganoderma Boninense*) in Oil Palm Plantations." *Mycopathologia* 159, no. 1 (2005): 153-57.

Thompson, I. A., and D. M. Huber. "Manganese and Plant Disease." In *Mineral Nutrition and Plant Disease*, edited by L. E. Datnoff, Wade H. Elmer and D. M. Huber, 139-54. St. Paul, Minn: American Phytopathological Society, 2007.

Turner, P. D. "Some Factors in the Control of Root Diseases of Oil Palm." In *Root Diseases and Soil-Borne Pathogens*, edited by T. A. Tousson, R. V. Bega and P. E. Nelson, 194-200. Berkely, LA: Univ of California, 1970.

Vaast, P., E. P. Caswell-Chen, and R. J. Zasoski. "Influences of a Root-Lesion Nematode, *Pratylenchus Coffeae*, and Two Arbuscular Mycorrhizal Fungi, *Acaulospora Mellea* and *Glomus Clarum* on Coffee (*Coffea Arabica* L.)." *Biology and Fertility of Soils* 26, no. 2 (1997): 130-35.

Vandermeer, John, Ivette Perfecto, and Stacy Philpott. "Ecological Complexity and Pest Control in Organic Coffee Production: Uncovering an Autonomous Ecosystem Service." *BioScience* 60, no. 7 (2010): 527-37.

Vawdrey, L. L., T. M. Martin, and J. De Faveri. "The Potential of Organic and Inorganic Soil Amendments, and a Biological Control Agent (*Trichoderma* Sp.) for the Management of Phytophthora Root Rot of Papaw in Far Northern Queensland." *Australasian Plant Pathology* 31, no. 4 (2002): 391-99.

Waggoner, P.E., and A.E. Dimond. "Production and Role of Extracellular Pectic Enzymes of *Fusarium Oxysporum* F. Lycopersici." *Phytopathology* 45 (1955): 79-87.

Walker, J. C. *Plant Pathology*. New York: McGraw-Hill, 1972.

Waller, J. M., M. Bigger, and R. J. Hillocks. *Coffee Pests, Diseases and Their Management*. Cambridge, MA: CABI Pub, 2007.

Watanarojanaporn, Nantida, Nantakorn Boonkerd, Sopone Wongkaew, Phrarop Prommanop, and Neung Teaumroong. "Selection of Arbuscular Mycorrhizal Fungi for Citrus Growth Promotion and *Phytophthora* Suppression." *Scientia Horticulturae* 128, no. 4 (2011): 423-33.

Weller, David M., Jos M. Raaijmakers, Brian B. McSpadden Gardener, and Linda S. Thomashow. "Microbial Populations Responsible for Specific Soil Suppressiveness to Plant Pathogens." *Annual review of phytopathology* 40, no. Journal Article (2002): 309-09.

Wellman, F.L. "More Diseases on Crops in the Tropics Than in Temperate Zones." *Ceiba* 14 (1968): 17-28.

Wellman, F.L. *Tropical American Plant Disease*. Metuchen, NJ: The Scarecrow Press, 1972.

Wessel, M. "Shade and Nutrition of Cocoa." In *Cocoa*, edited by G.A.R. and Lass Wood, R.A. Essex: Longman Scientific and Technical, 1985.

Widmer, T L, J H Graham, and D J Mitchell. "Composted Municipal Waste Reduces Infection of Citrus Seedlings by *Phytophthora Nicotianae*." *PLANT DISEASE* 82, no. 6 (1998): 683-88.

Williams, Jane S, Sharon A Hall, Malcolm J Hawkesford, Michael H Beale, and Richard M Cooper. "Elemental Sulfur and Thiol Accumulation in Tomato and Defense against a Fungal Vascular Pathogen." *Plant Physiology* 128, no. 1 (2002): 150-59.

Woltz, S. S., and J. P. Jones. "Nutritional Requirements of *Fusarium Oxysporum*: Basis for a Disease Control System." In *Diseases, Biology and Taxonomy*, edited by P.E. Nelson, T.A. Tousson and R.J. Cook, 340-49. University Park, PA: Penn State Univ Press, 1981.

Wrigley, G. *Coffee*: Harlow: Longman Scientific and Technical., 1988.

Yi, Qi-fe, Fu-wu Xin, and Xiu-lin Ye. "Effects of Increasing Phosphate and Potassium Fertilizers on the Control of Cymbidium Anthracnose." *Journal of Tropical & Subtropical Botany* 11, no. 2 (2003): 157-60.

Zheng, Q., and R. Ploetz. "Genetic Diversity in the Mango Malformation Pathogen and Development of a Pcr Assay." *Plant Pathology* 51, no. 2 (2002): 208-16.

12

Horizontal or Generalized Resistance to Pathogens in Plants

P.J. Keane
Department of Botany, La Trobe University, Victoria
Australia

1. Introduction

The threat to world wheat production and the panic among the agricultural science community caused by the emergence of the 'super virulent' wheat stem rust (*Puccinia graminis tritici*) race Ug99 in East Africa (Singh et al., 2011) is a reminder that the name and ideas of the South African J.E. van der Plank should not be forgotten. Based on his long experience with resistance to *Phytophthora infestans* in potatoes, he developed in his seminal books, *Plant Diseases: Epidemics and Control* (1963) and *Disease Resistance in Plants* (1968), the quantitative study of disease epidemics and the associated concepts of 'vertical' and 'horizontal' resistance to emphasize the two contrasting types of resistance to disease in crops. He contended that our preoccupation through the 20th Century with the more scientifically fascinating and precise vertical resistance, controlled by identifiable genes with a major effect, had resulted in the unfortunate neglect of the more mundane and nebulous horizontal resistance, mostly inherited quantitatively, even though it is evident that the former is unstable in the field while the latter is more stable and consistently useful. The saga of the scientific study of disease resistance shows our human tendency to dig where the light shines brightest, not where we know the potatoes are buried. This tendency continues unabated with the preoccupation of molecular biologists with vertical resistance, often discussed currently as if it is the only form of resistance.

Before their domestication, plants co-evolved with their parasites and underwent natural selection for resistance to them. Since the dawn of agriculture, plants with a degree of resistance to pathogens and insect pests have been selected by farmers, either consciously or unconsciously. In the genetically diverse crops of early agriculture, when plants of particular species began to be crowded together and so made more vulnerable to pest and disease attack, plants with great susceptibility would have been selected against in competition with more resistant types. They would have contributed fewer offspring to the next generation. Traditional farmers would have learned very early that it was better to select seed or vegetative propagating material from the healthiest plants and they still do so today. It is highly likely that they were selecting for partial or quantitative (van der Plank's 'horizontal') resistance. They weren't selecting for resistance to particular pests or diseases but rather for general plant health. They selected for pest and disease resistance as they selected for higher yield and other quantitatively inherited traits such as size and quality of the harvested product and adaptation to the environment. This process has been replicated by Simmonds (1964, 1966) who re-created the domesticated potato common in Europe (*Solanum tuberosum* ssp. *tuberosum*) by repeated mass selection of true seed of the best types from the highly variable wild Andian potato (*S. tuberosum* ssp. *andigena*). The potatoes were exposed to late blight epidemics during the selection process, and it was shown that the selected materials were more resistant to the disease than the standard domesticated potato (Thurston, 1971).

Through the 1800s there were reports of wheat farmers noticing in their fields occasional 'off-types' with complete resistance or immunity (probably 'vertical resistance' in van der Plank's terms) to a prevalent disease, although little practical use was made of this until it was understood that it was inherited (Biffen, 1905, p.40). It would have been possible to spot these completely resistant types and to distinguish them from 'escapes' only when disease levels were very high and practically all plants in a crop were heavily diseased. In the late 1800s, plants that appeared to be resistant in the field were selected, multiplied and promoted for use on a wider scale. A farmer in South Australia, James Ward, planted a South African variety called De Toit and noticed that it was generally "as rusty as a horse nail" except for a few plants that were rust-free (Callaghan & Millington, 1956) and were later thought to have arisen from a contaminant (Farrer, 1898). Ward saved the seed of these plants and increased it to produce a commercial variety named "Ward's Prolific". Because of its rust resistance and other characters, this became the most widely grown wheat in South Australia at the time. Rees et al. (1979) showed that in the 1970s this variety had little horizontal resistance and so it is likely that Ward had selected a type of extreme resistance that has since been overcome by the pathogen. Another South Australian farmer, Daniel Leak, noticed an occasional rust-free plant in a crop of heavily rusted Tuscany wheat, and selected and multiplied this as a commercial variety called 'Leak's Rust-proof' (Williams, 1991). There were many similar attempts at selecting rust resistant wheat, resulting in other varieties like 'Anderson's Rust-proof' and 'Kalm's Rust-proof' for which rust resistance was claimed (Cobb 1890). Farmers were developing a general appreciation of inherited variation in disease resistance. The Australian wheat breeder, William Farrer (1898), was one of the first to declare that resistance to wheat stem rust was inherited (Biffen, 1905). With the discovery of Mendel's work on the genetics of particular traits of the garden pea, his methods were soon applied to the trait of extreme disease resistance that had been observed in certain crops. R.H. Biffen, working at Cambridge University with resistance to stripe rust (*Puccinia striiformis*) in wheat, was the first to take up these studies, beginning in 1902 and summarizing his findings in papers entitled 'Mendel's laws of inheritance and wheat breeding' (Biffen, 1905) and 'Studies in the inheritance of disease resistance' (Biffen, 1907). He showed that a high level of disease resistance or immunity was inherited as a simple Mendelian character, and so could be crossed readily into well adapted varieties. Thus began our enchantment with the use of 'resistance genes' to protect our crops from disease and the search for them in crop varieties and in the centres of evolution and diversification of the crops (Vavilov, 1951).

Potato late blight caused by *Ph. infestans* has been central to modern plant pathology since the catastrophic epidemics in 1845-47 in Western Europe and Ireland that triggered the terrible Irish famine of the period and led to our understanding that a fungus could invade and cause disease in a healthy plant. While no completely resistant 'off-types' were noticed amongst the heavily blighted potato crops of the 1800s, a wild *Solanum* species collected from Mexico was shown to be immune to late blight (Salaman, 1910). In breeding experiments, individuals of this species stood out as completely healthy amongst genotypes of the cultivated potato that were severely diseased. This stimulated collecting expeditions to Central America to discover sources of resistance in wild relatives of the cultivated potato. *Solanum demissum*, a wild potato species from Mexico, was shown to have several resistance (R) genes that conferred immunity to late blight. This immunity was evident as a hypersensitive necrotic response in the leaf tissues invaded by the pathogen. At the time, it was considered to be the solution to the late blight problem in Europe and North America (Reddick, 1934). These R genes were cross-bred into the domesticated potato (*S. tuberosum* ssp. *tuberosum*) that formerly had no identifiable genes for immunity to late blight. However, with their widespread use in the field these varieties soon succumbed completely to the disease (Thurston, 1971). They were immune to some isolates (races) of the fungus but were completely susceptible to other races that increased in the pathogen population in response to the selective pressure resulting from the widespread use of a particular resistant variety. It was said that their resistance "broke down" with the selection of 'virulent' races of the pathogen that could invade the varieties with R genes. Van der Plank (1963) said that these varieties showed 'vertical resistance', named after the extreme vertical differences evident in the graphic plot of the degree of resistance (or, conversely, the amount of disease) on the Y-axis against a series of races of the pathogen along the X-axis (Figure 1). The sharp contrast between varieties with high levels of disease and those with very low levels and often no disease at all (immunity), shown by the abrupt vertical jumps in van der Plank's bar graphs, allowed identification of Mendelian genes with strong effects on resistance; hence the genes are often referred to as 'major resistance genes' or just 'resistance genes'. Varieties with different resistance genes gave completely different plots of resistance (or amount of disease) against races (Figure 1).

Varieties lacking R genes, or whose R genes were matched by the virulence of the prevailing pathogen races, gave more-or-less similar amounts of disease when inoculated with different races. The amount of disease could be high or low. Van der Plank said that varieties showing low levels of disease had 'horizontal resistance', the plot of degree of resistance or amount of disease against a series of races being more-or-less horizontal, or at least not showing extreme variation from complete resistance to great susceptibility evident with vertical resistance (Figure 1; van der Plank, 1963, Figs. 14.1, 14.2). The plot for two varieties may be displaced up or down, but the more resistant variety is more resistant to all races. The graph may not be completely horizontal; it may show some up and down displacements depending on the relative 'aggressiveness' of the races, but the displacements are the same for different varieties. The fundamental difference between the two types of resistance is that vertical resistance in the host varieties shows a sharp differential interaction (a strong statistical interaction) with the pathogen races; i.e. the amount of vertical resistance is specific for a particular race (very high for one race in Figure 1, very low for the other). It is 'race-specific'. Horizontal resistance is not race-specific to the extreme degree evident in vertical resistance (Figure 1). Because it is race-specific, the effect of vertical resistance is prone to being lost rapidly due to selection of virulent races in the pathogen population. Lacking this sharp interaction, horizontal resistance tends to be more stable, more 'durable'. That is its big advantage. Researchers working with late blight resistance in potato concluded that "R-gene hypersensitivity cannot be relied upon as a permanent protection against *Phytophthora infestans* and so the necessity of providing a degree of field protection in new cultivars is generally recognised by potato breeders."(Malcolmson, 1976). The 'field protection' referred to here is horizontal resistance.

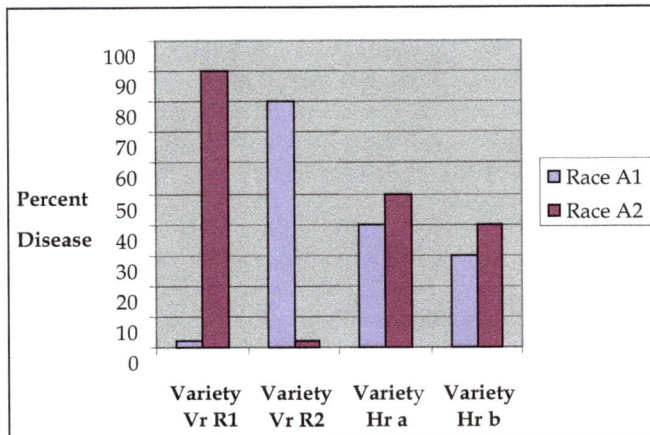

Fig. 1. Plot of percent disease in two varieties with vertical resistance (Vr R1 with resistance gene R1; Vr R2 with resistance gene R2) and two varieties lacking vertical resistance (R genes) but expressing some horizontal resistance (Hr a, Hr b) infected with two pathogen races (A1 with avirulence against R1; A2 with avirulence against R2). Vertical resistance shows a strong interaction with the races (i.e. is 'race-specific'). Horizontal resistance does not (i.e. is 'non-race-specific') although it shows significant main effects of Variety (variety Hr a is more susceptible to both races than Hr b) and Race (race A2 is more aggressive than A1 on both varieties, and this is also evident in the vertically resistant varieties where the R gene is ineffective).

Horizontal resistance has had a much longer history in human knowledge than vertical resistance and has had a greater profusion of names. Once, probably all observed resistance was of this type. It is probably the resistance that keeps 'minor pathogens' consistently 'minor' and consequently is not much studied for these pathogens because they are of minor importance. For the 'major' pathogens, farmers would often recognize that certain varieties 'got less disease' than other varieties. For particular diseases, 'slow-rusting', 'slow-mildewing' (for powdery mildews), 'slow-blighting' (for potato late blight) or 'slow-blasting' (for rice blast) are older terms that accurately describe horizontal resistance, an essential feature of which is the slowing down of epidemics. In natural plant communities or crops with a diversity of resistance genes (e.g. traditional mixed crops), vertical resistance will also slow down epidemics, but in current monocultures vertical resistance tends to prevent epidemics until such time as it is matched by virulence in a large proportion of the pathogen population. Several terms have been used to distinguish this general resistance from the race-specific resistance discovered in the early 1900s. These have included 'partial resistance', 'quantitative resistance', 'generalized resistance', 'field resistance', 'adult-plant resistance', 'durable resistance', and 'tolerance'. All have their problems. 'Race-non-specific resistance', usefully abbreviated to 'non-specific resistance', best captures the essential nature of this resistance as highlighted by van der Plank, although at a fine level a degree of race-specificity has been shown to apply to it (Parlevliet, 1995). Field resistance was commonly used for potato varieties lacking R genes, but is not a good general term as the resistance of a crop variety in the field could be due to a combination of vertical and horizontal resistance. 'General resistance' can refer to the resistance of a variety to several pests and diseases and again could be the result of both vertical and horizontal resistance. Sometimes 'tolerance' is used for 'horizontal resistance', but this is certainly incorrect. Tolerance has a special meaning: it refers to plant varieties that suffer less damage for a given degree of infection compared with a disease sensitive variety (Caldwell et al., 1958; Schafer, 1971). This has the same meaning as 'rust-enduring' referred to by N.A. Cobb (1894), one of the earliest students of rust resistance. Certainly, horizontal resistance is usually 'partial' and 'quantitative' and can be expressed in a gradient from very little disease to quite a lot, depending on the host genotype, the aggressiveness of the pathogen and the environment. But, for some diseases it can be complete. Thus, the non-prescriptive term, 'horizontal resistance', is perhaps the best. 'Generalized resistance' as applied to potatoes is also apt. Robinson (1976) favoured the terms 'vertical' and 'horizontal' resistance because they were somewhat abstract, and in fact had application beyond disease resistance; e.g. to fungicide use whereby copper-based fungicides could be regarded as 'horizontal' in effect because they knocked out several enzyme systems and the fungi could not adapt to them, while the new, highly specific fungicides like benomyl are 'vertical' because they knock out only one narrow function and the fungi can adapt to them (they 'break down' in the same way that vertical resistance breaks down).

Until it is matched by a virulent race, vertical resistance tends to be complete or to reduce reproduction of the pathogen to a tiny amount. That is its great attraction and that is how it was first noticed (Callaghan & Millington, 1956; Biffen, 1907; Salaman, 1910). In the 1800s it was observed by farmers as stark 'off-types' in crops of rusted wheat. The domesticated wheat species probably had vertical resistance genes against stem rust, while the domesticated potato in Europe did not have vertical resistance genes against late blight until they were bred into it from a wild relative. Horizontal resistance is harder to detect and measure, although it is likely to be selected unconsciously by observant farmers who collect their seed from their healthiest looking plants. It tends to be partial or quantitative in its expression (i.e. there is some disease and some sporulation of the pathogen) but the rate of development of the disease epidemic is reduced compared with that on a susceptible variety under similar environmental conditions. Van der Plank's interest in disease epidemiology (the quantitative study of populations of pathogens and crops) led to his understanding of the importance of partial resistance. There has been a recent shift in thinking about pests and diseases – we no longer talk about their 'control', which implies their elimination from a crop; rather we now talk about pest and disease 'management' which implies acceptance of some level of their presence as long as they cause little economic loss. Under the modern concept of Integrated Pest Management (IPM), moderate levels of pest/disease resistance in a crop may be sufficient if applied synergistically with cultural control methods and minimal, targeted use of pesticides. Under the IPM approach, resistance does not have to be complete; partial resistance may be all that is required, and horizontal resistance becomes important. Modern plant breeders and pathologists talk about "avoiding high degrees of susceptibility". The idea of IPM was promoted initially by entomologists in response to the phenomenon of 'breakdown of insecticide efficacy' that is analogous to the 'breakdown of disease resistance'. Entomologists have realized the futility of the 'scorched earth' approach to the use of insecticides in an attempt to eliminate pests from crops. They now accept that there is a threshold level of pest infestation below which little economic damage is caused. Entomologists are now emphasizing crop resistance (mostly of a horizontal nature) to insect pests after a long period of total reliance on insecticides (and consequent neglect and decline of resistance to insect pests in crops), while the repeated problem of the breakdown of disease resistance in many crops, exemplified by the emergence of wheat stem rust race Ug99, has led to a re-awakening of plant pathologists to the merits of horizontal resistance after a long period of pre-occupation with vertical resistance. It appears that both entomologists and plant pathologists are emerging from a period of bedazzlement by scientific and technical 'revolutions' (the 'insecticide revolution' and the 'resistance gene revolution') that showed great promise in the laboratory but lost their effectiveness following their widespread use as 'stand-alone' control measures in the farmers' fields. It is now realized that these brilliant technical developments, that offered an illusion of a simple and complete 'revolution' in control of pests and diseases, have to be incorporated into the ecological complexity of crop growth and production in the field. Rather than 'revolutions', a steady evolution of stable IPM methods, based on a foundation of steady evolution of horizontal resistance, is required. Of course, in addition to cultural methods, IPM methods may include targeted use of pesticides and vertical resistance, which, in IPM, are supported by the other methods that prolong and enhance their value. This approach is far from new. In the 1890s, Australia and North America produced vast quantities of very cheap wheat that flooded the European markets. It is obvious that the production was highly successful in most years. The potentially devastating wheat stem rust was managed effectively for long periods by a combination of use of early maturing varieties (that avoided the worst of the epidemics late in the growing season and allowed disease escape), drought adapted varieties (that allowed wheat to be grown in drier climates not conducive to the rust), the use of varieties that had probably been selected unconsciously over the years for a degree of horizontal resistance, and later, the addition of vertical resistance (e.g. in a variety like Thatcher in North America) and the deliberate incorporation of horizontal resistance by cross-breeding with tetraploid wheats.

2. Vertical or specific resistance

Researchers working to develop and deploy stem rust resistance genes (Sr genes) in wheat varieties in North America (Stakman & Levine, 1922) and Australia (Waterhouse, 1936, 1952; Watson, 1958; Watson and Luig, 1963) had exactly the same experience of 'breakdown of resistance' controlled by single resistance genes (Sr genes) as the researchers working with R genes in potato. This is well documented for several countries by Person (1967). The situation was starkest in the spring wheat crops in Australia (Table 1). Eureka, the first commercial wheat variety bred with a vertical resistance gene (Sr6) against stem rust in Australia, was released in 1938 and because of its resistance and other qualities was very popular, increasing to constitute nearly 20% of the wheat area in northern New South Wales by 1945 (Watson, 1958). However, races of stem rust able to attack the variety (i.e. virulent on Sr6) increased from practically nil in 1938 to make up 72% of the rust isolates collected in the area in 1945. The resistance of Eureka was seen to have 'broken down', and the variety rapidly lost popularity. The wheat variety Gabo and some others with resistance gene Sr11 were released in 1942 and by 1950 made up 62% of the same wheat area. But races of stem rust with virulence on Sr11 increased from nil to 91% by 1950 and the resistance of Gabo was broken. The period of the 'boom and bust' cycle of resistance breeding had begun and the fascinating phenomenon of pathogen-resistance specificity was revealed.

Year	Variety (Resistance gene)	Dominant stem rust race	Disease/Resistance situation in the field
1938	Eureka (Sr6) released	126-Avirulent on Sr6	No disease/Resistance effective
1942		126-Virulent on Sr6	Disease widespread/Resistance broken down
1942	Gabo types (Sr11) released	222-Avriulent on Sr11	No disease/Resistance effective
1948		222-Virulent on Sr11	Disease widespread/Resistance broken down
1950	Festival (Sr9b) released	21-Avirulent on Sr9b	No disease/Resistance effective
1959		21-Virulent on Sr9b	Disease widespread/Resistance broken down
1958	Mengavi (Sr36) released	34-Avirulent on Sr34	No disease/Resistance effective
1960		34-Virulent on Sr34	Disease widespread/Resistance broken down

Table 1. The interaction between the wheat cultivars Eureka, Gabo, Festival and Mengavi and wheat stem rust (*Puccinia graminis tritici*) races in Australia, showing the repeated breakdown of vertical resistance as new races evolved (Watson and Luig, 1963; after Knott, 1989).

Many researchers prefer the term 'race-specific resistance', which accurately describes the statistical interaction which is the essential feature van der Plank sought to highlight in his definition of 'vertical resistance' and is the basis of resistance breakdown. Races that can specifically invade (i.e. are virulent on) varieties with certain resistance genes cause the breakdown of that resistance when they increase to a high proportion of the total rust population in the field. 'Specific resistance' is a neat abbreviation as long as it is understood as 'race-specificity' not 'species-specificity'. The use of the term 'virulence' in plant pathology, where it is used to describe the ability of a pathogen race to invade a plant with a particular resistance gene, is different from its use in medicine, where it is used to describe what in plant pathology would be called the aggressiveness of a pathogen. In plant pathology, the pathogenicity of an organism consists of its virulence (ability to infect varieties with particular vertical resistance genes) and its aggressiveness (the amount of disease it causes on varieties it is able to infect). 'Virulence' is the ability of a pathogen to overcome vertical resistance, while 'aggressiveness' is the ability to overcome horizontal resistance. In much current plant pathology writing, there is a tendency to revert to the medical meaning of 'virulence', which requires the development of another term for the ability of a pathogen to match vertical resistance.

Another set of terms is used to describe the interaction of a specialized pathogen and its host. If a pathogen can infect and sporulate more-or-less normally on a plant, the interaction is said to be 'compatible'; if the interaction results in hypersensitive necrosis that largely excludes the pathogen, it is said to be 'incompatible', a term used very early in describing the effect of vertical resistance (see Hayes et al., 1925)(Table 2). These terms are particularly apt for biotrophic pathogens such as rusts and powdery mildews, where there is an intimate parasitic symbiosis between the pathogen and its host. In the same sense, the normal interactions of a plant and its mycorrhiza and endophyte symbionts would be said to be 'compatible'. The importance of the 'basic compatibility' required for a symbiont to live in its host has been well explored by Heath (1981). In fact, the copious molecular investigations of incompatibility would be better directed to trying to understand the mechanisms of compatibility – how does a biotrophic symbiont like a rust or a powdery mildew obtain nutrients from its host plant and why can't this feeding relationship be replicated in a Petri dish?

Pathogen	Plant host	
	Resistance gene RR or Rr	Double mutant rr (Susceptible regardless of A/a)
Avirulence gene AA or Aa	**Incompatible interaction** Hypersensitive necrosis; pathogen does not complete its life-cycle 1	**Compatible interaction** Pathogen infects and completes its life cycle
	Horizontal resistance is masked	**Horizontal resistance evident**
Double mutant aa (Gives virulence against R)	**Compatible interaction** Pathogen infects and completes 2 its life cycle	**Compatible interaction** Pathogen infects and completes its life cycle
	Horizontal resistance evident	**Horizontal resistance evident**

Table 2. Summary of the gene-for-gene interaction involved in vertical resistance. Resistance is usually dominant to susceptibility in the plant. Avirulence is usually dominant to virulence in the pathogen. The only starkly unique interaction occurs when an A gene matches an R gene. Arrow 1 indicates the change induced by breeding an R gene into a plant variety; arrow 2 indicates the change associated with the breakdown of resistance. Note that for wheat stem rust, the R gene is given the symbol Sr, for wheat stripe (yellow) rust it is given the symbol Yr etc.

Our knowledge of vertical resistance has a short history, beginning with the discovery of Mendel's genetics and the subsequent work of R.H. Biffen at Cambridge University in 1902. It has been shown to be expressed almost universally as a hypersensitive necrosis of host cells contacted by the pathogen during the early stages of infection, in some cases occurring rapidly in the first few cells contacted and so evident only under a microscope (infection type 0 in the scheme of Stakman and Levine, 1922, for cereal rusts), and in others occurring only after the pathogen has invaded a large patch of cells which dies and so is evident to the naked eye as a necrotic fleck (infection types 0; and 1 in cereal rusts). In some cases, the pathogen may develop to the extent of a small amount of sporulation before the lesion becomes necrotic (infection type 2). It has been shown to be controlled by a 'gene-for-gene' interaction between the pathogen and the host (Flor, 1956). Flor worked with flax rust (*Melampsora lini*), which completes its sexual reproductive cycle on flax (*Linum usitatissimum*), rather than with wheat stem rust which requires two hosts (wheat and barberry) to complete its sexual cycle. In a brilliant study, he showed that for every rust resistance gene in flax there is a corresponding gene for virulence (actually, avirulence if genes are named after their dominant allele) in flax rust (Table 2). Resistance is usually dominant to susceptibility and resistance genes occur as multiple alleles at a restricted number of loci in the plant while avirulence is usually dominant to virulence and avirulence genes occur at separate loci in the rust. That is, the number of resistance genes that can be expressed in the plant is restricted while there is no such restriction on the number of possible virulences that can be expressed in the rust. This ensures that the pathogen will always be able to overcome the resistance expressed in the host. While the meticulous and exhaustive work of Flor with flax rust has been repeated for very few other diseases, there is evidence that the relationship occurs in many highly specialized parasitic relationships (Sidhu, 1975), including several insect-plant relationships (Broekgaarden et al., 2011). It is usually expressed after the formation of haustoria in the highly specialized biotrophic parasites such as the rusts and powdery mildews and so appears to require intimate molecular contact between the pathogen and host. Rust resistance genes have been cloned from four of the five resistance loci in flax and they all appear to code for similar proteins, in the Nucleotide Binding Site-Leucine Rich Repeat (NBS-LRR) class (Ellis et al., 2007). Many resistance genes for a wide range of pathogens (including *Phytophthora,* rusts, powdery mildew, downy mildew, viruses, nematodes and bacteria) in a wide range of hosts (including potato, lettuce, tomato, barley, maize and *Arabidopsis*) have been found to fit into the same or similar class (Martin et al., 2003; Nimchuk et al., 2003). The resistance genes against the highly specialized phloem-feeding insect parasites of plants also appear to fall into this class (Broekgaarden et al., 2011). Thus, the recent molecular studies appear to confirm that vertical resistance involves a particular molecular system for recognition and rapid response to an invading parasite. On the other hand, the avirulence genes cloned from flax rust code for small secreted proteins ('effector proteins') that show no similarity between loci, providing evidence that their main function is probably something to do with the normal metabolism of the fungus and not to make them 'avirulent' on their host. Their 'avirulence' arises from the fact that they happen to be recognized by the resistance-coded proteins, leading to inhibition of the fungus. If avirulence genes code for proteins with a function in the normal life of the pathogen, this would explain the commonly observed phenomenon of stabilizing selection (Flor, 1956; van der Plank, 1968), whereby races expressing virulence (recessive mutants of A genes, often assumed to be non-functional) at one or more loci tend to be less fit than avirulent races on hosts that have no resistance genes. If avirulence genes code for a variety of normal functions in a pathogen, they could vary in the likelihood of their virulent (double recessive) mutant rising to prominence in the pathogen population, via mutation and selection for fitness, and so overcoming the particular resistance in the plant population (Luig, 1983). For example, if an avirulence gene (AvrX) coded for an essential function in the pathogen, then loss of this function in the homozygous virulent mutant (avrX avrX) may mitigate against the selection and buildup of this mutant, even if it matches the resistance RX in the host population. This could help explain the phenomenon of 'weak' and 'strong' vertical resistance genes as proposed by van der Plank (1968) and observed commonly in the field – i.e. some vertical resistance genes break down much more rapidly than others. The genes that are overcome rapidly are matched by races in which the mutation to virulence has little cost in fitness of the pathogen. If the mutation to virulence against a particular gene imposes a high cost in general fitness of the pathogen, the virulent race will not build up rapidly and the resistance gene will not be rapidly overcome; it will be seen to be 'strong'.

The ecological significance of the gene-for-gene relationship is that the rust is always able to match the resistance in the host – it can have unrestricted expression of virulence genes able to match any resistance genes that may occur in the plant population. The evolutionary significance is that the rust and the flax can co-exist and co-evolve. The evidence for this is that the host and the pathogen still exist: the plant has not driven the rust to extinction and the rust has not driven the plant to extinction. Mathematical studies have shown that the gene-for-gene system as described by Flor and elucidated further by Person et al. (1962) can be the basis of co-evolution when both the plant and the parasite are genetically variable and adaptive over time (i.e. are outbreeding) and genetically diverse in space (i.e. occur in populations of the species consisting of several different genotypes)(Mode, 1958). Geneticists call this 'balanced polymorphism' (Person, 1967). The gene-for-gene relationship (vertical resistance) probably evolved as a system that protected natural, genetically diverse, outbreeding and adaptive plant populations from excessive disease on the basis of the well documented 'mixture' or 'multiline' effect; pathogen races sporulating on a particular plant would not have been able to attack the immediate neighbors which had other resistance genes. This controls the pathogen population so that it doesn't overly reduce the fitness of the host population (otherwise the host could be outcompeted by other species and become extinct) and the parasite is also able to survive. This system functions as long as the pathogen does not build up a 'super race' with virulence against all the genes in the plant population. The stabilizing selection first observed by Flor (1956) and referred to above would tend to reduce the chance of such a race developing.

Thus, we can hypothesize that vertical resistance evolved in the outbreeding, genetically diverse, wild ancestors of crop species before their domestication, and that, as Mode (1958) said, the systems of vertical resistance we see in crop species today "are the relics of ancient systems of polymorphism, stemming from the time when wheat, barley and flax reproduced by outbreeding." (i.e. before their complete domestication). Evidence for this is the fact that the regions of evolution and diversification of crops (the Vavilov Regions) are the repositories of the vertical resistance genes in those crops (Leppik, 1970); that is where plant breeders, inspired by Vavilov (1951), have gone to find new resistance genes. It is possible that vertical resistance continued to play a role in stabilization of disease in traditional agriculture, where crop diversity was maintained. The rapid breakdown of vertical resistance in modern agriculture is due to the fact that we are now using the genes in planting systems that lack the genetic diversity in space and time of the ancestral wild and the early domesticated plant communities (Browning, 1974; Simmonds, 1979). It is a fascinating fact of agricultural botany that domestication has transformed many of our most important crop species from outbreeders in the wild to inbreeders in agriculture, and we are increasingly transforming our agricultural systems from polycultures to monocultures. The constant trend in modern industrial agriculture, driven, against all ecological wisdom, by global economics and centralized, powerful agricultural institutions, has been the steady elimination of this diversity in crop populations, including the critically important repositories of diversity in the regions of crop evolution (the Vavilov Centres). Thus, in modern agriculture the use of vertical resistance has coincided with a tendency to remove the genetic diversity that probably underpinned it in the wild pathosystems and in traditional mixed agricultural systems. In the deployment of resistance genes, crops are in fact becoming global monocultures, hence the problem and the panic created by Ug99.

3. Horizontal or non-specific resistance

Many words have been written and much heated argument generated in trying to define horizontal resistance precisely. The diversity of terminology applied to disease resistance has been summarized by Robinson (1969, 1976). Much has been said above about vertical resistance. This is necessary in a chapter about horizontal resistance if we define horizontal resistance as any resistance that is not vertical, as originally proposed by Black and Gallegly (1957) who defined field resistance (i.e. horizontal resistance) in the potato as "all forms of inherited resistance that plants possess with the exception of hypersensitivity as controlled by R-genes". Black restated this view in 1970 – "Field resistance to blight may be defined as the degree of resistance exhibited by a plant to all races of the fungus to which it is not hypersensitive." Such a definition is clear when 'specific resistance' and 'non-specific resistance' are substituted for 'vertical resistance' and 'horizontal resistance', respectively, as many like to do. Race-non-specific (horizontal) resistance is any resistance that is not race-specific, i.e. that does not operate on the gene-for-gene recognition system of Flor (1956) involving a hypersensitive necrosis response in the plants, and, on initial evidence, a particular molecular interaction as described by Ellis et al. (2007). It is the resistance expressed when there are no genes for vertical resistance in the plant or when the resistance has been overcome. Hayes et al. (1925) and Stakman and Levine (1922), in describing the infection types in wheat stem rust, considered that the presence or absence of hypersensitive necrosis marked the divide between resistance and susceptibility; it was then recognized that there are "different levels of susceptibility" (Parlevliet, 1995). Such a definition opens up a pandora's box of possible phenotypes, genotypes and mechanisms of horizontal resistance, which is why its definition has been so difficult and contentious. Just about any attempt at precision in defining it raises exceptions that defy the particular definition (Robinson, 1976).

Vertical resistance determines the basic compatibility or incompatibility of an interaction between a plant and its parasite (Table 2). Disease develops normally or it does not. However, once a parasite establishes basic compatibility with its host (i.e. is able to invade and reproduce normally) it is logical to suppose that there are very many points in the subsequent compatible symbiotic infection process that may determine whether the invasion and reproduction is fast and prolific or slow and limited. This is especially so in the highly specialized biotrophic pathogens that depend for their nutrition on an intimate physical and physiological association with live host cells, usually occurring through highly specialized haustoria formed within the cells. Its spores have to germinate on the leaf surface, germ tubes have to locate stomata and form appressoria and penetration pegs through stomatal pores (in rusts) or grow over the surface and directly penetrate the cuticle (in powdery mildews), then the infection hyphae have to penetrate cell walls, form haustoria and establish the active metabolic process of deriving nutrients from the host and becoming a sink for nutrients within the plant. It then has to invade further, forming many more haustoria, and eventually sporulate on the surface of the plant (for powdery mildews) or break through the epidermis to form a pustule of spores (for rusts). During all of these interactions there are opportunities for the physiological processes or morphological structures of the plant to hinder or slow down the interaction, and this could depend on very many genes that play a part in normal plant metabolism and structure. The pathogen will invade fast and sporulate prolifically, allowing it to create a destructive epidemic in the host population, if it encounters no great physical or physiological obstructions during the process of obtaining nutrients and colonizing the plant tissue and sporulating on the surface of the plant. It will invade more slowly if it encounters any physical or physiological obstructions during the parasitism. These obstructions are likely to be fortuitous, related to normal functions in the plant that, primarily, have nothing to do with resistance; they will exist whether or not the pathogen is present. For example, the proportion of peduncle tissue occupied by sclerenchyma in a wheat variety may restrict invasion and sporulation by stem rust (Hursh, 1924; Hart, 1931). A plant may just have tougher structures that are not damaged by the invading pathogen. This is especially important in stem pathogens, where invasion of a weak stem may result in the collapse of the whole plant. This is clearly evident in all damping-off diseases caused by *Pythium* species. Pythiums can only invade soft, immature hypocotyls, causing collapse (damping-off) of the plants. Once the hypocotyls become lignified they are resistant. This is horizontal resistance. A variety in which lignification is delayed could have less resistance than a variety that is lignified early. It is now possible to alter or reduce the lignification of pasture grasses in order to improve their digestibility to livestock; it has been observed that plants altered in this way become highly susceptible to rusts and insect attack, indicating that lignification of cell walls may be linked to the horizontal or generalized resistance of plants to parasites and herbivores (P. Dracatos, pers. comm.). It is important to note that impediments apart from the basic determination of compatibility may occur also in the pre-penetration and penetration phases of infection, as noted below in potato varieties expressing horizontal resistance prior to the incorporation of R genes. For example, the waxiness of the leaf surface may determine the proportion of spores landing on a leaf that are able to locate stomata and penetrate the leaf. Partial resistance to *Puccinia hordei* in barley has been shown to act before haustoria are formed (Niks, 1988). The cuticle thickness and the rate of vacuolization of epidermal cells may determine the proportion of powdery mildew spores that can establish infections (Schlosser, 1980). These are all expressions of horizontal resistance. It is to this multitude of processes that molecular biologists might profitably look for ways to enhance the resistance in plants, rather than perpetuating the preoccupation with vertical resistance genes whose effects are always likely to be overcome by mutations and selection in the pathogen.

The inheritance of horizontal resistance is best discussed in contrast to the inheritance of vertical resistance. Vertical resistance is invariably controlled by easily identifiable Mendelian genes with strong effects and is very well understood, now even down to the molecular expression of some of the genes involved (Ellis et al., 2007). In most cases, the inheritance of horizontal resistance is complex, different in different diseases and poorly understood, except to say that it is mostly additive and quantitative; it is about as well understood as the genetics of any other quantitative character such as 'yield'. This is not surprising given that horizontal resistance is likely to consist of any aspect of a plant's biology that slows down the growth and sporulation of a pathogen invading the plant in a basically compatible interaction. The obstructions the parasite may encounter are numerous and varied, and so their modes of inheritance will be numerous and varied. A single mechanism may be of great importance in the inhibition, and so the resistance may be dominated by the single gene that controls that mechanism (which may be called a 'resistance gene', e.g. the gene Rpg1 for durable resistance to stem rust in barley; Steffenson, 1992). Or it may be due to many aspects of the interaction, in which case it would be recognized as having 'quantitative' or 'additive' or 'complex' inheritance, and said to be controlled by 'polygenes' or inherited 'polygenically'. It is entirely possible that factors in the host that fortuitously inhibit the pathogen may be overcome by adaptation in the pathogen population (Parlevliet, 1995). That is, there may be variants of the pathogen that are not inhibited as much as other variants, and these variants may have a selective advantage because of their slightly greater fitness. They may sporulate more than other isolates and so contribute more offspring to the next generation. However, this adaptive process is not expected to be as rapid as that involved in the breakdown of vertical resistance. When it occurs, it is more likely to be expressed as a slow 'erosion' rather than a rapid 'breakdown' of resistance, as noted by Toxopeus (1956) and Niederhauser (1962) for late blight in potatoes. The difference between the more inhibited and less inhibited pathogen phenotypes is likely to be a matter of degree, not the extreme differences evident in vertical resistance, and so the selective pressures changing the pathogen population are likely to be far less. There are commonly several points of inhibition; being part of the normal functioning of the plant, they are likely to function independently in resistance, and so the adaptation of a pathogen variant at one point of inhibition is not likely to affect other points of inhibition.

Horizontal resistance can be accumulated by continually crossing and selecting varieties with resistance with little detailed understanding of the genes involved, in the same way that yield and environmental adaptation of a crop have been built up steadily from generation to generation with little understanding of the many genes involved. This is shown in some of the examples given below. Resistance to vascular streak dieback in cocoa was selected in Papua New Guinea even before the cause of the disease was known. Because the genes controlling horizontal resistance are not primarily 'resistance genes' but just the genes involved in the normal processes of the plant, van der Plank (1968) has suggested that 'there may be large untapped reserves of horizontal resistance in many crops'. Parlevliet (1995) concludes that the search for quantitative (horizontal) resistance in alien species is unlikely to be fruitful and advises that "Fortunately, there is in most crop-pathogen systems no need for these procedures, as quantitative resistance appears to be present sufficiently within the crop species whenever scientists look for it." There are many examples where crossing of susceptible or resistant varieties results in transgressive segregation of resistance, whereby some progeny are more resistant than either parent (and some are more susceptible)(Skovmand et al., 1978). Robinson (1979) suggests that 'good sources' of resistance are not necessary for breeding for horizontal resistance, which, because it is not based on R genes but rather on the normal processes of a plant, can be built up from the normal range of genetic resources in a species. In fact, the preoccupation with genes and gene-transfer (by crossing and back-crossing) in conventional breeding (and now in genetic engineering using recombinant DNA methods) is inimical to the development of horizontal resistance, which usually requires the accumulation of many unknown genes, better served by recurrent selection methods (Robinson, 1979).

The early students of vertical resistance were well aware of horizontal resistance and greatly valued it, probably because they were the first witnesses of the catastrophic breakdown of vertical resistance. Hayes et al. (1925), based on observations of wheat stem rust, described the difference between vertical and horizontal resistance very early in the development of our understanding of disease resistance – "It is known definitely that there are two types of resistance: (1) a true protoplasmic resistance which varies very little, and (2) a morphological resistance which varies with the age of the host and the conditions under which it is grown." They considered that the former was due to a "real physiological incompatibility between the resistant plants and the invading fungus" and that "the struggle between host and parasite was short and decisive and involved only a few cells in the most resistant plants ---- In susceptible varieties, however, the fungus apparently does not injure the host cells immediately but actually seems to stimulate them to increase physiological activity." Stakman and Harrar (1957) in their important textbook *Principles of Plant Pathology* recognized that "There are various types of resistance in plants. The more kinds a variety has, the more likely it is to be generally resistant. The high degree of specificity between certain physiological races of pathogens and certain varieties of plants has been emphasized repeatedly. A variety may be immune from one race but completely susceptible to another. If more can be learned about the kinds of resistance that are effective against all physiological races, however, it might be possible to breed varieties that have at least some resistance to all races. --- For example, some varieties of wheat have physiological resistance to many races of the stem rust fungus. If it is possible to add resistance to entrance because of stomatal characters, to extension in the tissues because of tough cells and barriers of sclerenchyma, and to the rupture of the epidermis by the sporulating mycelium, the variety should be much more resistant than those which have only one or a few of the many characters that can contribute to resistance. Even though the specific contribution of each character might be relatively slight, the combination of all of them might be effective under a wide range of conditions." The insights of these early students of disease resistance have often been forgotten.

It is worth discussing resistance to late blight in potato in more detail as this clearly shows the contrast between vertical and horizontal resistance, as thoroughly reviewed by Thurston (1971). The potato now widely grown throughout the world, *S. tuberosum* ssp. *tuberosum*, but thought to have originated in the Andian region of South America, had no vertical resistance to *Ph. infestans* until resistance genes were bred into it in Europe by crossing with a wild relative, *S. demissum*, which is native to Central America and clearly had a co-evolved vertical resistance pathosystem with *Ph. infestans* (controlled by R genes, eleven of which have been transferred into the potato in the 20th Century; Malcolmson & Black, 1966). It appears that *Ph. infestans* evolved as a parasite on *S. demissum* and other wild species in Central America and not on *S. tuberosum* ssp. *tuberosum* and *S. tuberosum* ssp. *andigena* in the Andes, and is in fact a 'new-encounter' pathogen on *S. tuberosum* ssp. *tuberosum* (Leppik, 1970). Varieties like Maritta and Kennebec that have the R1 resistance gene transferred from *S. demissum* show complete resistance expressed as hypersensitive necrosis to pathogen races with avirulence on R1, but are susceptible to races that have virulence against R1 (Table 2); however, Kennebec is more susceptible than Maritta to these virulent races (van der Plank 1963). Maritta has more background horizontal resistance than Kennebec to races that can infect both varieties. Many potato varieties lack R genes, and these vary in horizontal resistance to *Ph. infestans*. The variety Capella has a very high degree of horizontal resistance: it can become infected (i.e. it can be said to be 'compatible' with *Ph. infestans* – the fungus can grow and reproduce in the variety), but the fungus takes longer to produce lesions, the lesions are smaller, the fungus sporulates sparsely on the lesions, the plants remain green overall, the epidemic develops slowly, and the variety still yields well (van der Plank, 1963). This is in sharp contrast to the very susceptible varieties being grown in Ireland at the time of the great Irish Potato Famine in 1845-47, where the disease developed very fast, spread throughout western Europe in a matter of months, killed plants rapidly, decimated yield and rotted even the few tubers that were formed. Beginning in 1912, many observers documented the occurrence of degrees of resistance to late blight in the field before the time when R genes were bred into the potato (Thurston, 1971). This was often referred to as 'general resistance'. It was noted that inhibition of disease in these early resistant varieties could be due to inhibition at several stages of the process of pathogenesis, most notably resulting in (i) a reduced number of infections for a given inoculum dose (i.e. inhibition acting prior to penetration of the leaf by the fungus), (ii) a reduced rate of growth of mycelium in the plant tissues, (iii) a delay in sporulation, and (iv) a reduced number of sporangia produced per unit area of lesion. While vertical resistance is invariably associated with the sudden collapse and death of host cells during the initial establishment of parasitism (especially establishment of haustoria), horizontal resistance can be associated with death of invaded host tissue much later in the parasitic process. Van der Plank (1968, p.185) described how necrosis often occurs in the centres of developing lesions on potato varieties with horizontal resistance to *Ph. infestans*, resulting in a narrower zone of sporulation on the lesion than in a susceptible variety; he considered that it was possible to judge the degree of horizontal resistance of a variety by the amount of necrosis evident in sporulating lesions. Necrosis that appears to reduce the amount of sporulation on lesions is also evident in wheat varieties with horizontal (adult-plant) resistance to stripe rust (*Puccinia striiformis*) and Robusta coffee with horizontal resistance to leaf rust (*Hemileia vastatrix*).

Recent molecular studies have found that factors controlling general (horizontal) resistance to potato late blight occur on almost every potato chromosome and have confirmed that this resistance is, indeed, polygenic (Gebhardt & Valkonen, 2001). Several Quantitative Trait Loci (QTL) for use in marker assisted breeding for horizontal resistance have been located. The degree of general resistance observed in the early varieties was often seen to be affected by environmental factors and the developmental stage of the plant. The well documented history of general resistance in potatoes allows the conclusion that this resistance has, indeed, been durable. For example, Thurston (1971) documents the history of the variety Champion, which was first widely grown in Ireland in 1877 and was clearly popular because of its resistance to late blight, constituting 70% of the potato plantings in 1898. In 1953 Muller & Haigh reported that Champion still had a very high level of resistance. However, following the discovery of the R genes in *S. demissum* and their transfer into *S. tuberosum* ssp. *tuberosum*, as Thurston (1971) commented, "For several decades, almost all potato breeders dropped their work on general resistance and concentrated on obtaining commercial potato varieties with R-genes." Following the failure of R genes to provide long-term resistance, potato breeders turned back to horizontal resistance (Toxopeus, 1964). Van der Plank (1971) documented the fact that six potato varieties released without R genes in the 1920s and 1930s maintained their resistance rating of 6-9 (on a scale of 3=very susceptible to 10=very resistant) over a 30-year period from 1938 to 1968. Black (1970) showed that such resistance could be accumulated rapidly through crossing and selecting appropriate resistant material. In fact, Black turned back to *S. demissum* as a source of horizontal resistance, maintaining that it was mainly horizontal resistance that protected the wild potato species from *Ph. infestans* in Mexico. He showed that by crossing and selecting agronomically useful potato varieties in the presence of late blight, high levels of horizontal resistance could be accumulated. He established that the crossing of two moderately resistant varieties could result in some highly resistant progeny (due to transgressive segregation). As he said, "it is possible for two resisters --- to possess different resistance factors, and thereby to produce on hybridization a proportion of seedlings of greater resistance than either parent." In fact, Black's 1970 paper is a compact manual for breeding for horizontal resistance. It shows how easy breeding for a quantitative character can be, involving steady accumulation of resistance rather than the game of snakes and ladders associated with breeding for vertical resistance. It avoids the bewildering work of collecting, identifying and naming new pathogen races, and involves working on races only to the extent that inoculation of test plots must be done with pathogen races that have virulence on all the vertical resistance genes that occur in the parent plants.

4. Examples of disease management with horizontal resistance

4.1 Rusts of maize

In the Americas, the two co-evolved rust pathogens of maize, common rust (*Puccinia sorghi*) and tropical rust (*P. polysora*), are regarded as minor diseases. At least one of them is found infecting nearly every maize plant throughout its natural range in Central America, but there is no report of serious rust epidemics on maize in the region (Borlaug, 1972). Certainly, the maize rusts "have been much less important in limiting corn production in the tropics and subtropics than has *Puccinia graminis tritici* on wheat under similar conditions." (Borlaug, 1965). In the extensive and productive corn belt of the United States, the common species, *P. sorghi*, has caused little damage even though the conditions of vast areas of intensive cultivation, continuous presence of the pathogen and environmental conditions conducive to the rust are ideal for epidemic development (Hooker, 1967). There are vertical resistance genes against the maize rusts but these have been unimportant, and the minor status of the rusts has been maintained by horizontal resistance. The fact that maize is outbreeding has facilitated the continuous bulk selection in maize for horizontal resistance, whereas the inbreeding small grain cereals have not allowed this process except following conscious cross-breeding to create the genetic diversity required for selection of improved types. Experience with the maize rusts is evidence that the long list of 'minor diseases' observed for each crop species and listed in the various compendia are kept to their 'minor' status by horizontal resistance (Hooker, 1967; Simmonds, 1991).

Van der Plank (1968, p. 155) has described the local, on-farm and highly effective selection for horizontal resistance to tropical rust (*P. polysora*) in maize in Africa. Maize was probably first introduced to Africa soon after Columbus crossed the Atlantic in 1492 and began the introduction to the Old World of American crops. Given the amount of shipping contact between Africa and America over the centuries, there were undoubtedly numerous introductions, resulting in great genetic diversity of the crop in Africa. The fact that maize is outbreeding would also have ensured its genetic diversity. *Puccinia sorghi* was introduced very early and remained of no importance, as in its centre of evolution (Harlan, 1976). Maize became a staple crop and thrived in Africa for at least four centuries in the absence of the tropical rust with which it had co-evolved in America. When *P. polysora* eventually found its way to Africa in 1949 (Schieber, 1971), it caused devastating epidemics, killed plants and massively reduced maize yields, and swept across the continent in a way that suggested a grand epidemic and great susceptibility in the maize populations, the like of which has never been reported in America (Borlaug, 1972). Van der Plank (1968) presented evidence that horizontal resistance to tropical rust had declined greatly in maize in Africa during its

400-year separation from the rust, resulting in a destructive epidemic when the pathogen was eventually introduced. In fact, *P. polysora* was barely mentioned in the plant pathology literature before it became destructive on maize in Africa. The pathogen spread eastward across Africa and into Southeast Asia and Melanesia. However, the destructiveness of the disease declined after the initial epidemic (Cammack, 1960), and now throughout this extended range it is regarded as of little importance although it can be found on most maize plants. It is evident that selection by farmers of resistant types from genetically variable populations, as has been done since time immemorial in America, resulted in rapid accumulation of horizontal resistance to *P. polysora* in African and Asian maize populations. Farmers would have selected seed preferentially from the resistant survivors of the epidemic; often they would have had no choice since highly susceptible genotypes were killed (Harlan, 1976) or would not have produced much seed. Van der Plank (1968) and Robinson (1976) argued that this experience showed how rapidly and effectively adequate levels of horizontal resistance could be accumulated by bulk selection from genetically diverse crop populations. In contrast to what happened on the farms, researchers conducting seedling tests concluded that there was no resistance in African maize. They were looking for vertical resistance. They were looking for 'genes for resistance' (Stanton & Cammak, 1953).

4.2 Leaf rust (*Hemileia vastatrix*) of coffee

The contrast between vertical and horizontal resistance has been evident in the quest to control leaf rust (*Hemileia vastatrix*) on Arabica coffee (*Coffea arabica*), the species grown in the highlands of many tropical countries to produce high-flavor coffee. Leaf rust has long been a devastating disease on Arabica coffee. It destroyed the plantations in Ceylon (Sri Lanka) in the period 1870 - 1890, reducing the industry from the world's major supplier of coffee to nil (Large 1962), and has caused serious problems since its spread to all coffee-producing countries, including the Americas following its introduction from Africa in 1970. Much of the damage results from the premature defoliation of leaves with even moderate amounts of infection. Severe defoliation eventually kills the coffee bushes. Beginning in 1911 with Kent's selection in India, a succession of resistance genes (S_H genes 1 to 6) was used in an attempt to control the disease, but with their widespread use in the field these all succumbed rapidly to selection of virulent rust races (having virulence genes 1 to 6)(Rodrigues, 1984). This was the typical expression of vertical resistance.

Quantitative resistance to leaf rust has been found in Arabica coffee, for example in Ethiopia and in particular material from eastern Sudan and Kenya (van der Graaff, 1986). This was expressed as differences in latent period, number of lesions, and period of leaf retention after infection. Transgressive segregation for resistance was observed in some crosses and there was no doubt that the resistance was inherited quantitatively. However, most interest has centred on the resistance of a less important commercial species of coffee, *Coffea canephora* (especially the varieties known as 'Kouillou' and 'Robusta'). This species is adapted to the tropical lowlands, where it has become commercially important (e.g. in Brazil and Indonesia) although it is regarded as having inferior flavor to Arabica coffee. Leaf rust commonly infects Kouillou coffee in Brazil but it is not regarded as a serious problem (Eskes 1983), despite the fact that the warm, humid lowland environment appears ideal for the activity of the rust, which is more damaging on Arabica coffee at lower than at higher altitudes. Following the introduction of *C. canephora* to Java in 1900 after the devastation of Arabica coffee by rust, the Robusta variety showed high levels of resistance, which has been maintained and even increased by selection and breeding to the present time (Kushalappa & Eskes, 1989). On trees in Indonesia, older leaves commonly have some lesions, but these never cover the leaves and they never appear to cause premature defoliation. No one seems concerned about the disease. The resistance is partial, is quantitatively inherited, and has been stable over a very long time; there are no reports of a sudden destructive upsurge of rust on the lowland species. It is horizontal resistance. Like the horizontal resistance of potato to *P. infestans*, it is associated with necrosis of large areas of tissue which appears to limit sporulation. This resistance has played a big part in the management of rust in Arabica coffee in recent decades (Kushalappa & Eskes, 1989). The horizontal resistance of Robusta coffee was incorporated into Arabica coffee in a rare hybridization between the tetraploid, self-compatible *C. arabica* and the diploid, cross-pollinated *C. canephora*, discovered in 1927 in East Timor (Rodrigues, 1984) where plantings of highlands Arabica coffee overlapped with plantings of the lowlands Robusta coffee. Tetraploid progeny of this hybridization were selected as 'Hibrido de Timor' and planted widely as a rust resistant Arabica flavor type in East Timor. Later in Brazil, the compact high quality variety Caturra was crossed with the Timor hybrid to produce the agronomically acceptable Catimor lines of Arabica coffee with the flavor of Arabica and the rust resistance of Robusta (Rodrigues, 1984). Catimor lines are now grown widely around the world to manage rust, apparently without any catastrophic loss of resistance. Similar types of full flavoured, rust-resistant coffee, known as 'Arabusta coffee', presumably of a similar origin, are now grown commercially in Indonesia (e.g. in the Toraja region of South Sulawesi). Moreno-Ruiz & Castillo-Zapata (1990) have described in detail the development of the rust resistant, compact variety 'Colombia' from 'Hibrido de Timor' in Colombia.

Arabica coffee can still be found growing wild and semi-domesticated in the highlands of Ethiopia, where it and the rust co-evolved. Here we can see the ecology of a crop species and its co-evolved pathogen in wild, ancestral communities of the species. It is evident that the wild coffee forests consist of a mixture of genotypes with different resistance genes, and probably types with a moderate degree of horizontal resistance, such that coffee rust is not seen as being epidemic there, certainly not to the extent seen in commercial Arabica coffee plantations in various countries since the 1870s (van der Graaff, 1986).

4.3 Blackleg disease of canola caused by *Leptosphaeria maculans*

The value of horizontal resistance was shown by its straightforward selection in the oil-seed crop canola (*Brassica napus*, bred for seed with low levels of toxic erucic acid and glucosinolates) to control blackleg disease (caused by the ascomycete *Leptosphaeria maculans*) that had practically destroyed the crop in southern Australia in the 1970s (Salisbury et al., 1995). This brilliant work enabled the establishment of a highly productive canola industry and added a crucial crop to the wheat–legume rotation that has sustained dryland cropping in Australia. The fungus invades the laminae of cotyledons and initial leaves as a biotroph but tissues behind the hyphal front die and the fungus eventually sporulates on the dead tissue. The fungus grows down the petiole and into the stem where it invades and eventually kills tissues of the stem cortex. Stem cankering ('blackleg') is the main cause of damage to the plants; in the most susceptible types stems may be completely girdled and the plant tops may collapse and die before maturity. There is variation in the pathogenicity of isolates of the pathogen; some weakly pathogenic types can form lesions on cotyledons and leaves but not stem cankers while highly pathogenic types progress to form damaging stem cankers. The latter predominate in Australia and selection of disease resistance was essential for the survival of the crop.

Salisbury and co-workers selected horizontal resistance to blackleg by exposing a wide range of canola genotypes to the pathogen in nursery plots heavily contaminated with infested crop residues. The more resistant types survived the blackleg epidemics that developed and were selected for further breeding work. These had mature plant resistance which was evidently inherited polygenically. The resistance was partial: the pathogen invaded and colonized the cotyledons and leaves of resistant types but stem cankering was reduced or eliminated, although under heavy disease pressure, the resistant types could still suffer significant amounts of disease. Continued improvement in resistance was achieved by crossing of partially resistant types and further selection using the same method of field testing (Salisbury, 1988), with the result that the canola varieties produced had "the highest levels of blackleg resistance of any spring canola varieties in the world" and when these were grown with appropriate cultural control measures, losses were negligible. In this initial work little attention appears to have been paid to pathogen races even though it was known that the fungus reproduced sexually and was highly variable. It was not necessary to do so. In a disease such as blackleg, it is possible to imagine that any plant characters associated with strengthening of the stem base may well contribute to resistance or tolerance to the disease; a stronger stem may be less liable to invasion by the fungus, and, if invaded, may be less liable to collapse leading to the death of the plant. This is the sort of resistance seen in many plant species to weak pathogens such as *Pythium*: as tissues of the hypocotyl and lower stem mature, they become completely resistant simply by dint of their increased mechanical strength.

There are many lessons to be learned from this work. Firstly, inoculations of seedlings in a glasshouse gave different results than field tests. Later Salisbury & Ballinger (1996) showed that the resistance of seedlings tested in a glasshouse and of developing plants tested in the nursery plots were under different genetic control. The resistance evident in the blackleg nurseries was effective in the field. The basis for the success of this work was the field testing of resistance against the prevailing races of the highly variable fungus. There is evidence that this resistance can be eroded over time under severe disease pressure (Salisbury et al., 1995), but it has been relatively stable and subject to steady improvement in breeding programs. There has been no spectacular breakdown. In stark contrast, when in a separate program several varieties with vertical resistance (immunity expressed as hypersensitive necrosis) to the disease, controlled by a single dominant resistance gene bred into canola from a related species, *Brassica rapa* ssp. *sylvestris*, were released, the resistance broke down within three years (Sprague et al., 2006). Races of the pathogen virulent on these resistant varieties were selected from among the highly variable *L. maculans* population. Researchers then had to worry about the races of the pathogen and its high variability.

It appears that these vertically resistant varieties had been developed without a background of the horizontal resistance selected by Salisbury. When the resistance broke down, disease severity (measured as percent of the stem cross section blackened) was very high in the varieties with vertical resistance compared with nearby older varieties with only horizontal resistance. The availability of varieties with immunity gave the farmers a false sense of security and encouraged them to plant the crop more intensively than previously, placing immense selective pressure on the pathogen. It is clear that the management of this disease in the future should rely on the horizontal resistance selected in the 1970s and since built up by regular crossing and selection among resistant varieties, combined with cultural control measures such as crop rotation and separation of new plantings from the previous crops (Marcroft et al., 2004). If vertical resistance is used, it must be added to a background of horizontal resistance. If there is evidence of erosion of horizontal resistance, this can be addressed by a program of steady improvement in resistance as practiced during the 1980s.

4.4 Vascular streak dieback of cocoa caused by *Oncobasidium theobromae* in Southeast Asia and Melanesia

It is now rare to see complete susceptibility to a pathogen in the field. Historical records sometimes give an indication of it (as in the Irish Potato Famine of 1845-47, or the coffee leaf rust epidemics in Southeast Asia in the 1890s), or we can glimpse it when a very susceptible variety is inoculated in a glasshouse, but in general we grow up seeing only crops that have been selected for a relatively high degree of horizontal resistance. These are the survivors of epidemics past. In 1969 in Papua New Guinea the author witnessed the extreme susceptibility of cocoa (*Theobroma cacao*) to a dieback disease later shown to be caused by the indigenous basidiomycete, *Oncobasidium* (*Ceratobasidium*) *theobromae*, which invades only the xylem and causes vascular streaking after which the disease was named (Keane & Prior, 1991). This new-encounter pathogen killed a large proportion of the genetically diverse cocoa plantings established in Papua New Guinea in the 1950s and 60s, leaving only the types with a degree of resistance that enabled them to survive the destructive epidemic. Farmers in the field and agronomists at the Lowlands Agricultural Experiment Station, Keravat, East New Britain Province, had no choice but to propagate from the survivors and, in so doing, selected types with disease resistance that has ever since sustained the industry in Papua New Guinea and throughout Southeast Asia where cocoa has become a major crop despite the presence of the disease (Indonesia is now the third largest producer of cocoa in the world). In fact, this resistance was selected by farmers and agronomists even before the cause of the disease was known, following the fundamental process of natural selection that has undoubtedly sustained wild plant species through evolutionary time and domesticated species since the dawn of agriculture. Some cocoa clones being tested on the Experiment Station were highly susceptible and became extinct - the fungus grew through their xylem so rapidly that it penetrated into the lower stems and roots, completely blocked the xylem, and killed the trees. Others were only slightly affected. Resistant genotypes become infected but the disease progresses more slowly in the xylem, doing less damage to the trees, and the fungus sporulates less. Resistance is quantitatively inherited and has high heritability (Tan & Tan, 1988). It has been relatively easy to select for in breeding programs. The epidemics are much reduced compared with those seen in the 1960s and the resistance is adequate to control the disease as part of an IPM program that includes heavy pruning of cocoa and shade trees to remove infected branches and maintain an open, drier canopy. It has been durable for over 50 years and is still important wherever cocoa is grown throughout the region. Vertical resistance has not been found for this disease, and it is postulated that it is unlikely to occur in this new-encounter pathogen that has had, at most, a history of 300 years of contact with cocoa since the first introduction of the crop to Southeast Asia from Central America.

4.5 Foliar disease of *Eucalyptus* in Australia

Eucalyptus globulus (blue gum) is undergoing a rapid process of domestication. It is fast becoming one of the few indigenous Australian species to be added to the pantheon of the world's domesticated plants and is now one of the most widely planted tree species in the temperate zones. While foliar diseases are of little concern in native forests, they can be destructive in plantations of a single species such as blue gum (Park et al., 2000). One of the most serious diseases has been Mycosphaerella leaf blight caused by species of the ascomycete *Mycosphaerella*, which have co-evolved with *Eucalyptus*. The fungi initially invade the leaf tissues biotrophically and then cause sudden death of the invaded area to produce a necrotic blight on which the ascocarps are formed. Young, soft, expanding leaves are much more susceptible to infection than older, fully expanded, harder leaves. When collections of blue gum provenances from throughout its natural range were compared at one location favourable for Mycosphaerella leaf blight, significantly different degrees of disease incidence and severity occurred on the different provenances (Carnegie et al., 1994). Provenances from cold locations where the disease was likely to be less active tended to be very susceptible, while those from warmer areas with more summer rainfall where the disease was likely to be active were much more resistant. There had apparently been greater selection for disease resistance in locations where it was of more benefit to the host. This is horizontal resistance. It is partial, being assessed on a continuous scale from low to high percent leaf area affected, and is quantitatively inherited (Dungey et al., 1997). It has not been seen to be associated with hypersensitive necrosis of leaf tissue. It can be readily selected for in breeding programs and will be important in the development of improved varieties of blue gum for places where the disease is serious.

5. Resistance to stem rust and stripe rust in wheat

Some history of the early selection of 'off-types' of wheat with apparent high levels of resistance to stem rust during severe epidemics in Australia is referred to above. In 1894, a farmer, H.J. Gluyas, from the northern wheat belt of South Australia selected from Ward's Prolific an 'off-type' which he called "Early Gluyas"; from 1910 to about 1940 this was an important variety in the drier areas of Australia and became an important parent in the wheat breeding programs that developed from the turn of the 20th Century (Callaghan & Millington, 1956). An early contribution to rust control in Australia was the selection of early maturing varieties of wheat by William Farrer (Callaghan & Millington, 1956). These tended to escape stem rust, which built up and did most of its damage late in the growing season, and could be grown in drier areas where the disease was less of a problem. As well as aiming to produce early maturing varieties, Farrer (1898) also aimed to produce rust resistant varieties and this was one of his selection criteria. It has since been shown that some of his varieties did indeed have resistance to some of the rust races common up until 1926 (Waterhouse, 1936). Farrer's most famous variety was Federation, derived from crosses between Indian varieties, Canadian Fife wheats, and a high yielding commercial variety of the time, Purple Straw. This had stiff, short straw, was a good yielder, and matured early. It was first released in 1901, and from 1910 to 1925 was the most widely grown wheat in Australia. Dundee, a variety derived from Federation with similar characteristics, was in the top two or three most popular varieties in New South Wales and Victoria in 1938, on the dawn of the era of breeding and deployment of varieties with identified vertical resistance genes. The relatively long-lived popularity of the Federation-type wheats may indicate that, although they were regarded as being susceptible to stem rust, they may have had a degree of horizontal resistance that enabled them to continue to yield well and remain popular with farmers. The fact that they had short, stiff straw could have contributed to this. Farrer had been aware that varieties with erect, stiffer leaves tended to suffer less rust infection. He attempted to combine the (partial) rust resistance of late maturing varieties with earliness (Guthrie, 1922). His methods showed an awareness of quantitative genetics and what is now called 'transgressive segregation'. While it is often stated that he did not develop rust-resistant varieties, this assessment is usually made through the lens of vertical resistance that came to dominate breeding for stem rust resistance in Australia after his death. One of his varieties, Bomen, was still regarded as a valuable variety in the rust-prone northern districts of New South Wales in the early 1920s, and in fact won a Royal Agricultural Society prize for the best crop in 1921 when stem rust was a serious problem (Guthrie, 1922). It is possible that, given his breeding intentions and his methods of selecting for quantitative characters, Farrer did select a degree of horizontal resistance which underpinned the evident longevity of some of his varieties. After his death, his methods which favored the selection of horizontal resistance were replaced by the selection of the Mendelian genes for vertical resistance which has continued to dominate wheat breeding to the present time.

As discussed above for Australia (Watson and Luig, 1963; Table 1) and summarized for North America and Kenya by Person (1967), the use of vertical resistance to control wheat stem rust up until about 1960 resulted in the rapid breakdown of resistance as new races in the rust population adapted to the successive deployment of one or two resistance genes in particular varieties of wheat. This resistance broke down rapidly in Australia and Kenya (within about 5 years), and was also lost in North America, although there it was longer-lived (being effective for more than 10 years in some of the most important varieties). The breeding and deployment of vertical resistance involved a massive effort in surveying the races in the rust populations as the researchers attempted to keep track of the adaptation of the rust to the new varieties, producing bewildering lists of races. In fact, trying to review the race changes in the rust populations is truly confusing, as noted by Waterhouse (1952). The race names bear no relationship to the resistance genes they are matching, but rather are named after the pattern of virulence shown on sets of 'differential' varieties with different resistance genes (McIntosh et al., 1995). Only the fully initiated can easily keep track of the virulence genes that are being expressed in the rust populations. Van der Plank (1983) criticized the current concept of a pathogen race as stretching the bounds of taxonomic practice. If a new resistance gene is found in a host, the number of possible pathogen races increases exponentially (following Flor's gene-for-gene hypothesis, the potential number of races is 2 to the power of the number of resistance genes) and the previously described races have to be re-described to include their virulence on the new gene. Each race has a unique set of characteristics which consists of its virulence, its aggressiveness, and its overall fitness to survive in the environment (Luig, 1983); together these characteristics contribute to its ability to cause epidemics and so each race has to have a taxonomic identifier of some sort. For example, the most common races in Canada during the 1920s had a longer uredinial period on standard varieties than the less common races (Newton et al., 1932), and this trait would have contributed to their survival in the rust population. The two major races of stem rust in North America from the 1930s to the 1960s, race 56 and race 15B, differed greatly in their aggressiveness as well as their virulence (Katsuya & Green, 1967). On varieties on which both races were virulent, race 56 gave a higher number of infections per unit amount of inoculum, especially at higher temperatures (20-25°C), and had a 2-day shorter latent period than race 15B. Uredinia of race 56 expanded faster although race 15B ultimately had larger uredinia, and race 56 produced more spores per uredinium than race 15B. These differences are important characteristics of the races. They indicate variation in the ability of the races to invade a plant after basic compatibility has been established, equivalent to variation in horizontal resistance in the plants.

In Australia, relatively stable control of the rust in the most rust-prone areas of northern New South Wales and southern Queensland was achieved after about 1960 by assembling combinations of several resistance genes (up to five) in particular varieties so that mutants virulent for one or two genes were still blocked by other unmatched resistance genes (Watson, 1970; McIntosh, 1976; Park, 2007). It was also considered that races with multiple virulences were likely to be less fit than races with simple virulence (Flor, 1956; van der Plank, 1968; Leonard, 1969), and so were unlikely to build up rapidly in the rust population (Watson, 1970). The release of varieties with just one or two resistance genes was avoided so that the rust was denied possible stepping stones for developing full virulence on the varieties with several resistance genes. However, even some of the multiple resistances broke down - e.g. Sr7a, Sr11, Sr17, Sr36 in the variety Mendos (Luig, 1983; Park, 2007), and Sr5, Sr6, Sr8, Sr12 in the variety Oxley (Luig, 1983). However, the strategy was largely successful and McIntosh (1976) was able to conclude that "The sacrifice for almost 35 years of rust resistance has been a regular turnover of cultivars and the loss of effectiveness of a number of resistance genes." Park (2007) considered that particular combinations of genes such as Sr2, Sr24 and Sr26 had been particularly effective. The success was built on a constant effort in surveying the occurrence of rust races and breeding of new varieties with appropriate resistance genes.

There was a general view among wheat pathologists and breeders that horizontal resistance to stem rust was "uncommon in bread wheats" (Watson, 1974) or "yet to be clearly demonstrated" (Knott, 1971), and that it was less likely to have been accumulated over time in the inbreeding crops like wheat than in an outbreeding crop like maize (van der Plank, 1968; Knott 1971). However, Knott (1968) acknowledged the importance of the resistance of Hope and H-44 which can be considered to show horizontal resistance. Scraps of evidence of horizontal resistance can be seen in the early preoccupation with vertical resistance. For example, of the group of varieties with Sr11 involved in the second breakdown of resistance noted in Australia (Table 1), Waterhouse (1952) ranked Gabo as being more resistant than Yalta, both of which had Sr11. After the breakdown of the Sr11 resistance, Yalta was rapidly eliminated from the rust liable areas of New South Wales while Gabo, "being somewhat less susceptible" was still grown successfully (Watson, 1958). It is likely that Gabo had a greater degree of horizontal resistance than Yalta. It was generally known that there were potentially useful forms of resistance other than the vertical resistance expressed as hypersensitivity. Watson (1958) recognized the potential of what he called the "morphological resistance" in Webster that had been transferred into the variety Fedweb and had remained effective against all local races of stem rust. In fact, the resistance of Fedweb lasted from 1938 to 1964 (Park, 2007). Watson (1974) recognized that there were two known types of non-specific resistance that had been transferred into bread wheat (*Triticum aestivum*) from other wheat species. Resistance from *T. turgidum* var. *dicoccum* (Yaroslav

Emmer) had been transferred into the varieties Hope and Renown, and from *T. turgidum* var. *durum* (Iumillo) into the famous variety Thatcher. These resistances appeared to be controlled mainly by single genes, but Watson thought there were other unidentified genes involved. Watson (1974) thought non-specific resistance (presumably from the above sources) had performed well in the cultivars Warigo and Selkirk in the 1973-74 wheat stem rust epidemic in Australia. Warigo had been an exceptional variety during the period of repeated release and breakdown of varieties with single resistance genes (1938 – 1960; Table 1). Released in 1943 and known to have Sr17 (a recessive resistance; McIntosh et al., 1995), its resistance lasted an exceptional 16 years until 1959 (Park, 2007). It had Yaroslav Emmer in its parentage and Watson (1974) assumed that this was part of its success. Now it is known also to contain Sr2, another recessive gene from Yaroslav Emmer that has conferred durable resistance on many varieties (McIntosh et al., 1995). Rees et al. (1979) documented a wide range of horizontal resistance in wheat varieties including some of the older ones used in Australia.

In North America there were two spectacular resistance breakdowns leading to major epidemics in 1935 (associated with the coming to dominance in the rust population of race 56) and 1954 (associated with the dominance of race 15B) (Person, 1967), but it appears that stem rust was generally controlled, except for these two major epidemics, through the use of multiple resistance genes (e.g. Thatcher had Sr5, Sr9g, Sr12 and Sr16; Kolmer et al., 2011; Luig, 1983), and incorporation into the background of many varieties of the horizontal resistance (referred to as 'adult plant resistance') derived from tetraploid wheats (Hare & McIntosh, 1979). The resistance of Hope (with resistance from Yaroslav Emmer) and Thatcher (with resistance from Iumillo) may have helped to partly protect the spring wheat crops in North America during the severe epidemic of 1954. Selkirk, which became the leading spring wheat variety after the 1954 epidemic, had six identified resistance genes (Sr2, Sr6, Sr7b, Sr9d, Sr17, Sr23; Luig, 1983), including Sr2 and Sr17, the two identified recessive genes from Yaroslav Emmer. The evidence that the main North American varieties may have had a degree of horizontal resistance could also account for the longevity of the vertical resistance of these varieties from 1935 through to the 1960s. Certainly, Stakman & Christensen (1960) recognized the occurrence of important levels of resistance in wheat varieties that were susceptible at the seedling stage. It was noted during the stem rust epidemic of 1954 that the amount of damage on the durum wheat variety Stewart was three times that on the bread wheat variety Lee, and this was attributed to the earliness and non-specific resistance of Lee (Loegering et al., 1967).

The ability of cereals to slow down the development of rusts, even though the plants were considered basically susceptible, was recognized by pre-eminent breeders and pathologists many years ago (Farrer, 1898; Stakman & Harrar, 1957). A famous early example was the oat species *Avena byzantina* known as Red Rustproof, which was introduced to the southern United States and recognized as being resistant to crown rust (*Puccinia coronata*) in the 1860s (Luke et al., 1972). It was partially resistant, not immune, and has remained so for over 100 years. It is 'late-rusting'. It is also 'slow-rusting', expressed as a low percent of leaf area infected during the growing season. The degree of resistance varies between varieties of the species. Luke et al. (1972) had no hesitation in recognizing this as 'horizontal resistance'. They were also inclined to call it 'generalized resistance'. Wilcoxson (1981) comprehensively reviewed the biology of slow rusting in cereals and discussed the evidence for long-lived, slow-rusting against stem rust in some well known wheat varieties such as Lee,

McMurachy, Kenya 58, Thatcher and Idaed 59. In crosses between fast- and slow-rusting varieties, Skovmand et al. (1978) showed that transgressive segregation occurred in all crosses, and that slow-rusting was quantitatively inherited with a narrow-sense heritability of 80%. In Europe, farmers and plant breeders over many years used a satisfactory level of horizontal (partial) resistance to protect spring barley against leaf rust (*Puccinia hordei*)(Parlevliet, 1981). Slow rusting in cereals generally involved decreased frequency of penetration, slower invasion of host tissue, longer latent period, smaller pustules, lower sporulation rate, and shorter period of sporulation, singly or, more commonly, in combination (Kuhn et al., 1978; Parlevliet, 1979). The detailed physiology of these effects is not understood, but early workers determined that some physical features of cereals could affect rust development. For example, Hursh (1924) found evidence that the proportion of sclerenchyma to collenchyma in the upper peduncle was correlated with the resistance to stem rust of Sonem Emmer and Kota wheat compared with Little Club. Horizontal resistance, being partial, is strongly affected by environmental factors. It has long been known that excessive nitrogenous fertilization makes cereals more susceptible to rusts (Hursh, 1924). Hart (1931) showed that Webster had several morphological features that increased its resistance to stem rust compared with very susceptible varieties. These included a higher proportion of sclerenchyma in the peduncle and the degree of lignification and relative toughness of the epidermis which often prevented uredinia breaking through to the surface. The resistance in Webster has since been attributed to the gene Sr30, which has been overcome in the Australian variety Festiguay (Knott & McIntosh, 1978). However, it is unlikely that the set of morphological features that inhibit rust infection in Webster (Hart, 1931) is controlled by only one gene.

The known sources of horizontal resistance (often referred to as 'adult plant resistance' or 'durable resistance') against stem rust in wheat are very narrow, consisting mainly of the two tetraploid wheats, *T. turgidum* var. *dicoccum* (cv. Yaroslav Emmer) and *T. turgidum* var. *durum* (cv. Iumillo) and the bread wheat variety Webster (McIntosh et al., 1995). Leppik (1970) lists a range of wild wheat species discovered through the activities of the Russian collecting expeditions that are possible sources of resistance. An unfortunate downside of the spread of dwarf wheat varieties around the world from the 1960s has been the loss of the genetic diversity in the local land races that had probably undergone selection for horizontal resistance. But building up horizontal resistance in a crop does not necessarily involve just searching for sources of resistance, but rather crossing and selection of existing varieties in such as way that genes involved in the normal functioning of the plant that happen to interfere with pathogen growth and development are accumulated and resistance is built up.

Stripe (yellow) rust (*Puccinia striiformis*) was first detected in Australia in 1979 and its history of control by resistance has been very different from that of stem rust. The first response to the incursion, which caused heavy losses in some of the most widely grown varieties such as Zenith, was to deploy vertical resistance genes (YrA and Yr6). However, the effectiveness of these genes was lost very rapidly (Wellings & McIntosh, 1990). It was observed that some wheat varieties such as Condor, Egret and Olympic, although clearly susceptible at the seedling stage, showed varying degrees of adult-plant resistance and suffered much lower losses than the very susceptible varieties such as Zenith (McIntosh & Wellings, 1986). These had similar temperature-sensitive, partial, adult-plant resistance to that observed in some prominent varieties such as Cappelle-Desprez in Europe (Johnson, 1978), and Gaines,

Nugaines and Luke in the Pacific Northwest of the United States (Milus & Line, 1986). The degree of resistance increased as the plants matured, and was greater at higher than lower temperatures (Qayoum & Line, 1985). The resistance was quantitatively inherited, and some crosses showed transgressive segregation (Milus & Line, 1986). This resistance has been long-lived. In Australia, it has been incorporated into many varieties (e.g. Meering and its successors, developed from Condor) and has proved to give long-lasting resistance to stripe rust (Park & Rees, 1989). While it has been identified with Yr18 (McIntosh et al., 1995), additive genes have also been found (Park, 2008) and it has been attributed to a 'Y18 complex' (Ma & Singh, 1996). It is therefore evident that control of stripe rust has relied largely on horizontal resistance.

With the emergence of race Ug99 and its derivatives, virulent on several important Sr genes (Sr24, Sr36, Sr21, Sr31, Sr38) that have been widely distributed around the world from the CIMMYT program in Mexico, wheat breeders are considering turning back to horizontal resistance to control stem rust (Schumann & Leonard, 2011). The emergence of this race has shown the dangers of relying on vertical resistance while steadily eroding the diversity of the global genetic resources of a crop. Before central agencies distributed and promoted particular wheat genotypes around the world, there would have been much greater genetic diversity in the crops. From country to country, from valley to valley and from farmer to farmer there would have been variation in the planting material, including variation in the deployment of resistance genes. An epidemic in one area would not necessarily have threatened another area. With the centralization of breeding for rust resistance in crops such as wheat and the influence of central agencies like CIMMYT in distributing resistance genes, there has been a global narrowing in the base of vertical resistance, such that new races originating in East Africa can threaten the wheat crops of many countries. In fact, it is likely that the global varieties have steadily replaced the local landraces that, in the absence of vertical resistance, probably relied on horizontal resistance to survive and yield in the face of rust infection. Further, the destructive nature of race Ug99 in East Africa suggests that the varieties it is attacking have little horizontal resistance. From the emphasis placed on vertical resistance in wheat breeding (the perpetual search for resistance genes), it is highly likely that horizontal resistance has not been maintained at a high level in the major wheat varieties, in stark contrast to the situation in maize.

6. Resistance of plants to insects

Resistance of plants to insect attack has many of the same characteristics as resistance to microbial pathogens, except that the plant-insect interaction is complicated by the behavioral biology of insects that is lacking in pathogens. Thus, in many reviews, resistance to insects is divided into three components: (i) non-preference, (ii) antibiosis, and (iii) tolerance (Painter, 1958). Tolerance is used here in precisely the same sense as in plant pathology – it is the ability of a plant to survive a certain level of insect attack without suffering significant loss of yield. Entomologists have developed the concept of a threshold level of infestation, below which the insect does not cause significant loss of yield and is not worth worrying about. Non-preference involves the avoidance by insects of particular plants and attraction to others for oviposition or feeding. This involves inherited traits of the plants (e.g. chemical stimuli, colours, morphologies) that can be just as important in breeding for resistance as mechanisms of antiobiosis. However, it is based on the behavior of the insects and has no equivalent in pathogens. More recently, an indirect mechanism of protection of plants from insects has been recognized (Broekgaarden et al., 2011). This involves the attraction of predator and parasitoid insects to plants releasing volatile chemicals as a result of attack by herbivorous insects. The predation and parasitism then reduces the populations of the pest, effectively protecting the plant. Again, there is no equivalent in plant pathology. Some plants like *Acacia* species produce extra-floral nectaries that attract ants that in turn protect the plant from insect herbivores. These indirect mechanisms are amenable to selection in order to improve the protection of plants from insect pests.

Antiobiosis refers to properties of plants that directly reduce the amount of insect infestation on the plant (equivalent to reducing the amount of pathogen infection). This is similar to resistance against plant pathogens, and can include both vertical (race-specific) and horizontal (race-non-specific) resistance as defined for pathogens by van der Plank (1963, 1968). There is strong evidence that plants have vertical resistance to some of the highly specialized insect pests such as phloem feeding aphids. This resistance has the same attributes as vertical resistance to pathogens. It is often controlled by single, identifiable genes and shows a gene-for-gene interaction as in pathogens, in fact involving the same family of resistance proteins in the plants (NBS-LRRs)(reviewed in Broekgaarden et al., 2011). Long ago, hessian fly was considered to have a gene-for-gene interaction with wheat (Sidhu, 1975; Gallun et al., 1975). Wheat-hessian fly and medicago-bluegreen aphid interactions involve a hypersensitive necrosis reaction like that in plant diseases. And, as with plant diseases, there is ample evidence that this vertical resistance to insects breaks down with its widespread deployment in the field.

Most of the resistance of plants to insects discussed in the literature fits the category of horizontal resistance (reviewed in Yencho et al., 2000). This can be explained by the fact that most insect pests have a far less intimate association with their host than the microbial pathogens: most insects just eat the host tissue. Most resistance in plants against insect attack is partial, inherited quantitatively, and is relatively stable. In fact, horizontal resistance to insects is easier to understand than horizontal resistance to pathogens because the mechanisms are more obvious and easily observed. If an insect is attracted to feed on a plant, and there is no immediate hypersensitive response that prevents it, then there are likely to be a multitude of constitutive or induced factors involved in the insect-plant interaction that either allow the insect to feed unimpeded and rapidly build up its population to damaging levels or restrict its feeding and so control its population on the plant. In reviewing the topic, Beck (1965) concluded that "It is doubtful that any example of resistance can be explained on the basis of a single simple biological characteristic of the plant. The multiplicity of factors exerting influences on the insect-plant relationship precludes the formulation of meaningful all-inclusive generalizations." Plants produce a wide range of secondary plant compounds such as alkaloids, tannins, essential oils, flavones and phenolics that can inhibit the build-up of insect populations, while not necessarily making the plants immune to attack (Beck, 1965; Levin, 1976). Many morphological and physical properties of plants such as density of sticky secretory glandular trichomes, density of hooked trichomes, and tissue toughness due to silica or lignin content may reduce herbivory and/or digestibility and consequently the build-up of insect populations on the plants (e.g. Tingey, 1979). In response to insect attack, solanaceous plants produce proteinase inhibitors that enhance their resistance to insects (Heath et al., 1997). Most of these traits are likely to be inherited quantitatively (Yencho et al., 2000). Also, like horizontal resistance to plant pathogens, they are liable to erosion if the insect population is able to adapt to particular mechanisms. Thus, many secondary plant compounds that probably initially conditioned resistance to particular insects have become specific attractants for insects that have adapted to their presence. This process has been especially evident in relation to the glucosinolates in the Brassicaceae (Hopkins et al., 2009).

7. The contribution of molecular biology

Several studies of the molecular basis of vertical resistance against several types of parasites (bacteria, fungi and insects) have provided evidence that the gene-for-gene interaction involves a specific molecular system, and that the different R genes, both within and across species, fall within similar gene families (Ellis et al., 2007; Broekgaarden et al., 2011). In other words, the basic incompatibility process involves the expression of variants of the same interactive molecular system. This is a special molecular recognition system that could only have developed through a long period of co-evolution between the host and parasite in a situation where the two types of dominant genes (R-resistance, A-avirulence) had a selective advantage when interacting in genetically diverse populations of host and parasite in the region of evolution of the crop. There is evidence that this specific recognition system does not occur in new–encounter diseases in which there has been insufficient time for such a system to develop. This accounts for the fact that in such diseases (e.g. vascular streak dieback of cocoa, possibly late blight of *S. tuberosum* ssp. *tuberosum*) only horizontal and not vertical resistance has been found naturally.

Another exciting development from the use of recombinant DNA technology is the use of DNA molecular markers for important plant traits such as yield and resistance (Young 1996; Yencho et al., 2000). Mapping and development of DNA markers for Quantitative Trait Loci linked to horizontal resistance could enable the Vertifolia Effect of van der Plank (1963, 1968) and referred to by Black (1970) to be avoided; that is, it could enable the selection of vertical resistance while ensuring that the background horizontal resistance is not lost. While R genes can be added to a variety, and combined as in the effective strategy for wheat stem rust in Australia (McIntosh, 1976), the varieties can be monitored using DNA analysis to ensure that the underlying horizontal resistance is not lost. This is an important development as the presence of a high level of horizontal resistance in vertically resistant varieties reduces the chances of rapid breakdown of vertical resistance, and it reduces the damage done if the resistance does break down. If the vertical resistance is underpinned by good horizontal resistance, breakdown of vertical resistance, as seen with the emergence of a race like Ug99, is not likely to be catastrophic.

It is hypothesized that horizontal resistance is due to any aspect of plant biology that happens to slow down the invasion and sporulation of a pathogen in a basically compatible interaction. This understanding opens up a vast array of functions that could be altered by recombinant DNA techniques in a subtle way that may confer partial resistance on the host, rather than continuing the preoccupation with vertical resistance which we know the pathogen can overcome.

8. Conclusion

Our personal experience conditions how we see the world, including the world of science with which we work. van der Plank (1963) developed his ideas from life-long experience of breeding for resistance to late blight in potato, in which R genes were not effective. They broke down rapidly. Breeding for resistance to late blight relied more on horizontal resistance, and van der Plank was impressed by horizontal resistance. Workers with wheat stem rust grew up with direct personal experience of one of the great biological phenomena discovered in our time, vertical resistance involving the gene-for-gene recognition of a plant species and its co-evolved parasite. This resistance was genetically simple and made a spectacular difference; addition of a single Mendelian gene could convert a very susceptible variety into an immune one and completely protect a crop that had suffered regular devastating epidemics down through history. It is no wonder that researchers were excited by it and worked so hard to exploit it. Even the breakdown of resistance with which they had to contend was such a striking phenomenon, with incredible practical importance on the farms, that this only added to the excitement of the endeavor; researchers not only had to track the resistance genes in the host but also the virulence genes in the pathogen. All the emphasis was placed on making vertical resistance work, with considerable success in the case of wheat stem rust through breeding several resistance genes into each variety, keeping track of virulence changes in the rust populations, and continuously breeding varieties with new resistance genes. Horizontal resistance to wheat rusts was paid little attention during the grand quest to make vertical resistance work in practice. The inbreeding nature of wheat and the other small grain cereals made it harder to accumulate horizontal resistance as occurred in some outbreeding crops such as maize and cocoa. As a consequence, the researchers involved were not enthusiastic about van der Plank's synthesis and many continue to ignore his insights. The present author, introduced to plant pathology through study of vascular streak dieback of cocoa in Papua New Guinea and Southeast Asia, saw the functioning and value of horizontal resistance at first hand. It was easily selected for in an outbreeding, genetically diverse crop exposed to severe epidemics. Indeed, resistance was selected by farmers and agronomists before the cause of the disease was known. Although the pathogen reproduces only sexually and is therefore likely to be highly variable, it has not been necessary, and indeed the biology of the fungus has made it impossible, to be concerned about 'races'. It has proved durable, protecting cocoa from a potentially devastating pathogen and allowing the region to develop over 50 years into the second most important cocoa producing region after West Africa. As a result, this author has been impressed by van der Plank's concept of horizontal resistance.

Long ago, the father of the early work on vertical resistance in the cereals, E.C. Stakman (1957, 1958, 1964), called for greater use of horizontal resistance against cereal rusts. Hooker (1967) suggested that, in developing disease resistance in crops, "perhaps man did not properly assess the resources at his disposal or employed the wrong tactics in their usage." In the light of his experience with the maize rusts he concluded that "If the system prevailing in maize and maize rust is applicable to other host-pathogen systems, then genes for specific hypersensitive-based resistance should be avoided or used only as a minor supplement to a high level of generalized resistance. As many modes of generalized resistance as possible should be combined to produce multimodal resistant varieties." In van der Plank's (1968) and Robinson's (1979) terms, horizontal resistance should be built up in crops as a primary objective and as the foundation of disease management, with vertical resistance being added as necessary, along with cultural control measures and targeted use of pesticides, as part of an IPM strategy.

9. References

Beck, S.D. (1965). Resistance of plants to insects. *Annual Review of Entomology* 10, 207-232.

Biffen, R.H. (1905). Mendel's laws of inheritance and wheat breeding. *Journal of Agricultural Science* 1, 4-48.

Biffen, R.H. (1907). Studies in the inheritance of disease-resistance. *Journal of Agricultural Science* 2, 109-127.

Black, W. (1970). The nature and inheritance of field resistance to late blight (*Phytophthora infestans*) in potatoes. *American Potato Journal* 47, 279-288.

Black, W. & Gallegly, M.E. (1957). Screening of *Solanum* species for resistance to physiological races of *Phytophthora infestans*. *American Potato Journal* 34, 273-281.

Borlaug, N.E. (1965). Wheat, rust, and people. *Phytopathology* 55, 1088-1098.

Borlaug, N.E. (1972). A cereal breeder and ex-forester's evaluation of the progress and problems involved in breeding rust resistant forest trees: moderator's summary. *US Department of Agriculture Miscellaneous Publication* 1221, pp. 615-642, Washington DC.

Broekgaarden, C., Snoeren, T.A.L., Dicke, M. & Vosman, B. (2011). Exploiting natural variation to identify insect-resistance genes. *Plant Biotechnology Journal* 2011, 1-7.

Browning, J.A. (1974). Relevance of knowledge about natural ecosystems to development of pest management programs for agroecosystems. *Proceedings of the American Phytopathological Society* 1, 191-199.

Caldwell, R.M.; Schafer, J.F.; Compton LE. & Patterson, F.L. (1958). Tolerance to cereal leaf rusts. *Science* 128, 714-715.

Callaghan, A.R. & Millington, A.J. (1956). *The Wheat Industry in Australia*. Angus and Robertson, Sydney, Australia.

Cammack, R.H. (1960). *Puccinia polysora*: a review of some factors affecting the epiphytotic in West Africa. *Report of 5th. Commonwealth Mycological Conference, 196*, pp. 134-138.

Carnegie, A.J, Keane, P.J., Ades, P.K. & Smith, I.W. (1994). Variation in susceptibility of *Eucalyptus globulus* provenances to Mycosphaerella leaf disease. *Canadian Journal of Forest Research* 24, 1751-1757.

Cobb, N.A. (1890). Contributions to an economic knowledge of the Australian rusts (*Uredineae*). *Agricultural Gazette, New South Wales* 1(3), 185-214.

Dungey, H.S., Potts, B.M., Carnegie, A.J. & Ades, P.K. (1997). *Mycosphaerella* leaf disease: genetic variation in damage to *Eucalyptus nitens, E. globulus* and their F_1 hybrids. *Canadian Journal of Forest Research* 27, 750-759.

Ellis, J.G.; Dodds, P.N. & Lawrence, G.J. (2007). Flax rust resistance gene specificity is based on direct resistance-avirulence protein interactions. *Annual Review of Phytopathology* 45, 289-306.

Eskes, A.B. (1983). Incomplete resistance to coffee leaf rust. In: *Durable Resistance in Crops*, F. Lamberti, J.M. Waller and N.A. van der Graaff (eds.), pp. 291-315, Plenum Press, New York.

Farrer, W. (1898). The making and improvement of wheats for Australian conditions. *Agricultural Gazette, New South Wales* 9, 131-168, 241-260.

Flor, H.H. (1956). The complementary genic systems in flax and flax rust. Advances in Genetics 8, 29-54.

Gallun, R.L., Starks, K.J. & Guthrie, W.D. (1975). Plant resistance to insects attacking cereals. *Annual Review of Entomology* 20, 337-357.

Gebhardt, C. & Valkonen, J.P.T. (2001). Organization of genes controlling disease resistance in the potato genome. *Annual Review of Phytopathology* 39, 79-102.

Guthrie, F.B. (1922). *William J. Farrer and the Results of his Work. Science Bulletin No. 22*. Department of Agriculture, New South Wales, Sydney.

Hare, R.A. & McIntosh, R.A. (1979). Genetic and cytogenetic studies of durable adult-plant resistances in 'Hope' and related varieties to wheat rusts. *Zeitschrift Pflanzenzuchtung* 83, 350-367.

Harlan, J.R. (1976). Diseases as a factor in plant evolution. *Annual Review of Phytopathology* 14, 31-51.

Hart, H. (1931). *Morphologic and Physiologic Studies on Stem-Rust Resistance in Cereals*. Technical Bulletin No. 266, United States Department of Agriculture, Washington D.C.

Hayes, H.K., Stakman, E.C. & Aamodt, O.S. (1925). Inheritance in wheat of resistance to black stem rust. *Phytopathology* 15, 371-386.

Heath, M.C. (1981). A generalized concept of host-parasite specificity. *Phytopathology* 71, 1121-1123.

Heath, R.L., McDonald, G., Christeller, J.T., Lee, M., Bateman, K., West, J., van Heeswijck, R. & Anderson, M.A. (1997). Proteinase inhibitors from *Nicotiana alata* enhance plant resistance to insect pests. *Journal of Insect Physiology* 43, 833-842.

Hooker, A.L. (1967). The genetics and expression of resistance in plants to rusts of the genus *Puccinia*. *Annual Review of Phytopathology* 5, 163-182.

Hopkins, R.J., van Dam, N.M. & van Loon, J.J.A. (2009). Role of glucosinolates in insect-plant relationships and multitrophic interactions. *Annual Review of Entomology* 54, 57-83.

Hursh, C.R. (1924). Morphological and physiological studies on the resistance of wheat to *Puccinia gaminis tritici* Erikss. and Henn. *Journal of Agricultural Research* 27, 381-411.

Johnson, R. (1978). Practical breeding for durable resistance to rust diseases in self-pollinating cereals. *Euphytica* 27, 529-540.

Katsuya, K. & Green, G.J. (1967). Reproductive potentials of races 15B and 56 of wheat stem rust. *Canadian Journal of Botany* 45, 1077-1091.

Keane, P.J. & Prior, C. (1991). *Vascular-Streak Dieback of Cocoa*. Phytopathological papers No. 33, International Mycological Institute, Wallingford, Oxon. ISBN 0-85198-733-8

Knott, D.R. (1968). The inheritance of resistance to stem rust races 56 and 15B-1L (Can.) in the wheat varieties Hope and H-44. *Canadian Journal of Genetics and Cytology* 10, 311-320.

Knott, D.R. (1971). Can losses from wheat stem rust be eliminated in North America? *Crop Science* 11, 97-99.

Knott, D.R. (1989). *The Wheat Rusts-Breeding for Resistance*. Springer-Verlag, Berlin.

Knott, D.R. & McIntosh, R.A. (1978). The inheritance of stem rust resistance in the common wheat cultivar Webster. *Crop Science* 17, 365-369.

Kolmer, J.A., Garvin, D.F. & Jin, Y. (2011). Expression of a Thatcher wheat adult plant stem rust resistance QTL on chromosome arm 2BL is enhanced by *Lr34*. *Crop Science* 51, 526-533.

Kuhn, R.C., Ohm, H.W. & Shaner, G.E. (1978). Slow leaf-rusting resistance in wheat against twenty-two isolates of *Puccinia recondita*. *Phytopathology* 68, 651-656.

Kushalappa, A.C. & Eskes, A.B. (1989). *Coffee Rust: Epidemiology, Resistance, and Management*. CRC Press, Florida.

Large, E.C. (1940). *The Advance of the Fungi*. Dover Publications, New York.

Leonard, K.J. (1969). Selection in heterogeneous populations of *Puccinia graminis* f.sp. *avenae*. *Phytopathology* 59, 1851-57.

Leppik, E.E. (1970). Gene centers of plants as sources of disease resistance. *Annual Review of Phytopathology* 8, 323-344.

Levin, D.A. (1976). The chemical defenses of plants to pathogens and herbivores. *Annual Review of Ecology and Systematics* 7, 121-159.

Loegering, W.Q., Hendrix, J.W. & Browder, L.E. (1967). *The Rust Diseases of Wheat*. Agriculture Handbook No. 334, United States Department of Agriculture, Washington, D.C.

Luig, N.H. (1983). *A Survey of Virulence Genes in Wheat Stem Rust,* Puccinia graminis *f.sp.* tritici. Verlag Paul Parey, Berlin and Hamburg.

Luke, H.H., Chapman, W.H. & Barnett, R.D. (1972). Horizontal resistance of Red Rustproof oats to crown rust. *Phytopathology* 62, 414-417.

Ma, H. & Singh, R.P. (1996). Contribution of adult plant resistance gene *Yr18* in protecting wheat from yellow rust. *Plant Disease* 80, 66-69.

Malcolmson, J.F. (1976). Assessment of field resistance to blight (*Phytophthora infestans*) in potatoes. *Transactions of the British Mycological Society* 67, 321-325.

Malcolmson, J.F. & Black, W. (1966). New R genes in *Solanum demissum* Lindl. And their complementary races of *Phytophthora infestans* (Mont.) de Bary. *Euphytica* 15, 199-203.

Marcroft, S.J., Sprague, S.J., Pymer, S.J., Salisbury, P.A. & Howlett, B.J. (2004). Crop isolation, not extended rotation length, reduces blackleg (*Leptosphaeria maculans*) severity of canola (*Brassica napus*) in south-eastern Australia. *Australian Journal of Experimental Agriculture* 44, 601-606.

Martin, G.B., Bogdanove, A.J. & Sessa, G. (2003). Understanding the functions of plant disease resistance proteins. *Annual Review of Plant Biology* 54, 23-61.

McIntosh, R.A. (1976). Genetics of wheat and wheat rusts since Farrer. *The Journal of the Australian Institute of Agricultural Science* 42, 203-216.

McIntosh, R.A. & Wellings, C.R. (1986). Wheat rust resistance – the continuing challenge. *Australasian Plant Pathology* 15, 1-8.

McIntosh, R.A., Wellings, C.R. and Park, R.F. (1995). *Wheat Rusts: an Atlas of Resistance Genes*. CSIRO Publications, Melbourne. ISBN 0 643 05428 6

Milus, E.A. & Line, R.F. (1986). Number of genes controlling high-temperature, adult-plant resistance to stripe rust in wheat. *Phytopathology* 76, 93-96.

Mode, C.J. (1958). A mathematical model for the co-evolution of obligate parasites and their hosts. *Evolution* 12, 158-165.

Moreno-Ruiz, G. & Castillo-Zapata, J. (1990). *The Variety Colombia: a Variety of Coffee with Resistance to Rust* (Hemileia vastatrix *(Berk. & Br.).* Technical Bulletin No. 9, Centro Nacional de Investigacioes de Café, Chinchina, Caldes, Colombia.

Muller, K.O. & Haigh, J.C. (1953). Nature of "field resistance" of the potato to *Phytophthora infestans* de Bary. *Nature* 171, 781-783.

Newton, M., Johnson, T. & Gussow, H.T. (1932). *Studies in Cereal Diseases VIII. Specialization and Hybridization of Wheat Stem Rust,* Puccinia graminis tritici, *in Canada.* Bulletin No. 160, New Series, Department of Agriculture, Dominion of Canada.

Niederhauser, J.S. (1962). Evaluation of multigenic "field resistance" of the potato to *Phytophthora infestans* in 10 years of trials at Toluca, Mexico. *Phytopathology* 52, 746 (Abstr.).

Niks, R.E. (1988). Failure of haustorial development as a factor in slow growth and development of *Puccinia hordei* in partially resistant barley seedlings. *Physiological and Molecular Plant Pathology* 28, 309-322.

Nimchuk, Z., Eulgem, T., Holt, B.F.III & Dangl, J.L. (2003). Recognition and response in the plant immune system. *Annual Review of Genetics* 37, 579-609.

Painter, R.H. (1958). Resistance of plants to insects. *Annual Review of Entomology* 3, 267-290.

Park, R.F. (2007). Stem rust of wheat in Australia. *Australian Journal of Agricultural Research* 58, 558-566.

Park, R.F. (2008). Breeding cereals for rust resistance in Australia. *Plant Pathology* 57, 591-602.

Park, R.F. & Rees, R.G. (1989). Expression of adult plant resistance and its effect on the development of *Puccinia striiformis* f.sp. *tritici* in some Australian wheat cultivars. *Plant Pathology* 38, 200-208.

Park, R.F., Keane, P.J., Wingfield, M.J & Crous, P.W. (2000). Fungal diseases of eucalypt foliage. In: *Diseases and Pathogens of Eucalypts,* Keane, P.J., Kile, G.A., Podger, F.D. & Brown, B.N. (eds.), pp. 153-239, CSIRO Publishing, Collingwood. ISBN 0 643 06523 7

Parlevliet, J.E. (1979). Components of resistance that reduce the rate of epidemic development. *Annual Review of Phytopathology* 17, 203-222.

Parlevliet, J.E. (1981). Race-non-specific disease resistance. In: *Strategies for the Control of Cereal Diseas,* Jenkyn, J.F. & Plumb, R.T. (eds.), Blackwell, London.

Parlevliet, J.E. (1995). Present problems in and aspects of breeding for disease resistance. In: *Molecular methods in Plant Pathology,* Singh, R.P. & Singh, U.S. (eds.), CRC, Lewis Publishers, Boca Raton, London, Tokyo.

Person, C. (1967). Genetic aspects of parasitism. *Canadian Journal of Botany* 45, 1193-1204.

Person, C., Samborski, D.J. & Rohringer, R. (1962). The gene-for-gene concept. *Nature* 194, 561-562.

Qayoum, A. & Line, R.F. (1985). High-temperature, adult-plant resistance to stripe rust of wheat. *Phytopathology* 75, 1121-1125.

Reddick, D. (1934). Elimination of potato late blight from North America. *Phytopathology* 24, 555-557.

Rees, R.G., Thompson, J.P. & Mayer, R.J. (1979). Slow rusting and tolerance to rusts in wheat. I The progress and effects of epidemics of *Puccinia graminis tritici* in selected wheat cultivars. *Australian Journal of Agricultural Research* 30, 403-419.

Robinson, R.A. (1969). Disease resistance terminology. *Review of Applied Mycology* 48, 593-606.

Robinson, R.A. (1976). *Plant Pathosystems.* Springer Verlag, Berlin.

Robinson, R.A. (1979). Permanent and impermanent resistance to crop parasites; a re-examination of the pathosystem concept with special reference to rice blast. *Zeitschrift Pflanzenzuechtung* 83, 1-39.

Rodrigues, C.J. Jr. (1984).Coffee rust races and resistance. In: *Coffee Rust in the Americas,* R.H Fulton (ed.), pp. 41-58, The American Phytopathological Society, St. Paul, Minnesota.

Salaman, R.N. (1910). The inheritance of colour and other characters in the potato. *Journal of Genetics* 1, 7-46.

Salisbury, P.A. (1988). Blackleg resistance in rapeseed. *Plant Protection Quarterly* 3, 47.

Salisbury, P.A. & Ballinger, D.J. (1996). Seedling and adult plant evaluation of race variability in *Leptosphaeria maculans* on *Brassica* species in Australia. *Australian Journal of Experimental Agriculture* 36, 485-488.

Salisbury, P.A., Ballinger, D.J, Wratten, N., Plummer, K.M. & Howlett, B.J. (1995). Blackleg disease on oilseed *Brassica* in Australia: a review. *Australian Journal of Experimental Agriculture* 35, 665-672.

Schafer, J.F. (1971). Tolerance to plant disease. *Annual Review of Phytopathology* 9, 235-252.

Schieber, E. (1971). Distribution of *Puccinia polysora* and *P. sorghi* in Africa and their pathogenicity on Latin American maize germ plasm. *FAO Plant Protection Bulletin* 19, 25-31.

Schlosser, E.W. (1980). Preformed internal chemical defenses. . In Horsfall, J.G. & Cowling, E.B., *Plant Disease. An Advanced Treatise* Vol.V, 161-177, Academic Press, N.Y.

Schumann, G.L. & Leonard, K.J. (2000). Stem rust of wheat (black rust). *The Plant health Instructor.* DOI:10.1094/PHI-I-2000-0721-01. Updated 2011.

Sidhu, G.S. (1975). Gene-for-gene relationships in plant parasitic systems. *Scientific Progress, Oxford* 62, 467-485.

Simmonds, N.W. (1964). Studies of the tetraploid potatoes. II. Factors in the evolution of the Tuberosum group. *Journal of the Linnaean Society of London (Botany)* 59, 43-56.

Simmonds, N.W. (1966). Studies of the tetraploid potatoes. III. Progress in the experimental re-creation of the Tuberosum group. *Journal of the Linnaean Society of London (Botany)* 59, 279-288.

Simmonds, N.W. (1979). *Principles of Crop Improvement.* Longman, ISBN 0-582-44630-9, New York.

Simmonds, N.W. (1991). Genetics of horizontal resistance to diseases of crops. *Biological Reviews* 66, 189-241.

Singh, R.P.; Hodson, D.P.; Huerta-Espino, J.; Jin, Y.; Bhavani, S.; Njau, P.; Herrera-Foessel, S.; Singh, P.K.; Singh, S. & Govindan, V. (2011). The emergence of Ug99 races of the stem rust fungus is a threat to world wheat production. *Annual Review of Phytopathology* 49, 1-17.

Skovmand, B., Wilcoxson, R.D., Shearer, B.L. & Stucker, R.E. (1978). Inheritance of slow rusting to stem rust in wheat. *Euphytica* 27, 95-107.

Sprague, S.J., Marcroft, S.J., Hayden, H.L. & Howlett, B.J. (2006). Major gene resistance to blackleg in *Brassica napus* overcome within three years of commercial production in southeastern Australia. *Plant Disease* 90, 190-198.

Stakman, E.C. (1964).Will the fight against wheat rust ever end? *Zeitschrift Pflanzenkrankheiten Plantzenschutz* 71, 67-73.

Stakman, E.C. & Levine, M.N. (1922). The determination of biologic forms of *Puccinia graminis* on *Triticum* spp. *Minnesota University Agricultural Experiment Station Technical Bulletin* 8.

Stakman, E.C. & Harrar, J.G. (1957). *Principles of Plant Pathology.* The Ronald Press Co., New York.

Stakman, E.C. & Rodenhiser, H.A. (1958) Race 15B of wheat stem rust – what it is and what it means. *Advances in Agronomy* 10, 143-165.

Stakman, E.C. Christensen, J.J. (1960). The problem of breeding resistant varieties. In Horsfall, J.G. & Dimond, A.E., *Plant Pathology* Vol.3, 567-624, Academic Press, N.Y.

Steffenson, B.J. (1992). Analysis of durable resistance to stem rust in barley. *Euphytica* 63, 153-167.

Tan, G.Y. & Tan, W.K. (1988). Genetic variation in resistance to vascular-streak dieback in cocoa (*Theobroma cacao*). *Theoretical and Applied Genetics* 75, 761-766.

Thurston, H.D. (1971). Relationship of general resistance: late blight of potato. *Phytopathology* 61, 620-626.

Tingey, W.M. (1979). Breeding for arthropod resistance in vegetables. pp. 495-522 in *Biology and Breeding for Resistance to Arthropods and Pathogens in Agricultural Plants.* M.K. Harris (ed.), Texas A & M University, College Station, Texas.

Toxopeus, H.J. (1956). Reflections on the origin of new physiological races of *Phytophthora infestans* and the breeding of resistance in potatoes. *Euphytica* 5, 221-237.

Toxopeus, H.J. (1964). Treasure-digging for blight resistance in potatoes. *Euphytica* 13, 206-222.

van der Graaff, N.A. (1986). Coffees, *Coffea* spp. Chapter 6 in *Breeding for Durable Resistance in Perennial Crops.* FAO Plant Production and Protection Paper 70, Fao, Rome.

van der Plank, J.E. (1963). *Plant Disease: Epidemics and Control.* Academic Press.

van der Plank, J.E. (1968). *Disease Resistance in Plants.* Academic Press.

van der Plank, J.E. (1971). Stability of resistance to *Phytophthora infestans* in cultivars without R genes. *Potato Research* 14, 263-270.

van der Plank, J.E. (1983). Durable resistance in crops: should the concept of physiological races die? In: *Durable Resistance in Crops,* F. Lamberti, J.M. Waller and N.A. van der Graaff (eds.), pp. 41-44, Plenum Press, New York.

Vavilov, N.I. (1951). *The Origin, Variation, Immunity and Breeding of Cultivated Plants.* Translatedfrom Russian by K. S. Chester, The Ronald Press Co., New York.

Waterhouse, W.L. (1936). Some observations on cereal rust problems in Australia. *Proceedings of the Linnean Society of New South Wales* 77, 209-258.

Waterhouse, W.L. (1952). Australian rust studies. IX. Physiologic race determinations and surveys of cereal rusts. *Proceedings of the Linnean Society of New South Wales* 61, v-xxxviii.

Watson, I.A. (1958). The present status of breeding disease resistant wheats in Australia. *Agriculture Gazette, New South Wales Department of Agriculture* 69 (12), 1-31.

Watson, I.A. (1970). Changes in virulence and population shifts in plant pathogens. *Annual Review of Phytopathology* 8, 209-230.

Watson, I.A. (1974). Losses from wheat stem rust in Australia – are they inevitable? *Australian Plant Pathology Society Newsletter* 3(3), 64-65.

Watson, I.A. & Luig, N.H. (1963). The classification of *Puccinia graminis* var. *tritici* in relation to breeding resistant varieties. *Proceedings of the Linnean Society of New South Wales* 88, 235-258.

Williams, R.F. (1991). *To Find the Way. History of the Western Fleurieu Peninsula.* The Yankalilla and District Historical Society Inc. ISBN 0646062565

Wellings, C.R. & McIntosh, R.A. (1990). *Puccinia striiformis* f.sp. *tritici* in Australasia: pathogenic changes during the first 10 years. *Plant Pathology* 39, 316-325.

Wilcoxson, R.D. (1981). Genetics of slow rusting in cereals. *Phytopathology* 71, 989-993.

Yencho, G.C., Cohen, M.B. & Byrne, P.F. (2000). Applications of tagging and mapping insect resistance loci in plants. *Annual Review of Entomology* 45, 393-422.

Young, N.D. (1996). QTL mapping and quantitative disease resistance in plants. *Annual Review of Phytopathology* 34, 479-501.

13

Epidemiology and Control of Plant Diseases Caused by Phytopathogenic Bacteria: The Case of Olive Knot Disease Caused by *Pseudomonas savastanoi* pv. *savastanoi*

José M. Quesada[1], Ramón Penyalver[2] and María M. López[2,*]
[1]*Departamento de Protección Ambiental, Estación Experimental del Zaidín, Consejo Superior de Investigaciones Científicas (CSIC), Granada,*
[2]*Centro de Protección Vegetal y Biotecnología, Instituto Valenciano de Investigaciones Agrarias (IVIA), Valencia Spain*

1. Introduction

Pseudomonas savastanoi pv. *savastanoi* (Gardan et al., 1992) (hereafter Psv, according to Vivian & Mansfield (1993)) is the causal agent of olive knot disease. It is considered one of the most serious diseases affecting olive trees (*Olea europaea* L.) in most olive growing regions worldwide and mainly in Mediterranean countries, where this crop has been growing for centuries. The disease can lead to severe damage in olive groves, causing serious losses in terms of production. This is probably the first disease clearly described in antiquity by Theophrastus (370-286 BC) (Iacobellis, 2001) and its bacterial etiology was known through the work of Savastano since 1887 (Smith & Rorer, 1904). However, there are currently many unknown facts about the epidemiology of this disease or its chemical control. Here we describe the most relevant studies performed on the epidemiology and chemical control of olive knot.

2. Biology of infection

Psv causes the formation of hyperplastic growth in olive trees, producing spherical knots on the trunk and branches, and less frequently on leaves and fruits (Sisto & Iacobellis, 1999; Smith, 1920). See details in figure 1. Psv infections in fresh wounds of olive trees start with a small cavity caused by the collapse of adjoining plant cells and are more frequent on trunks and branches, and rare on leaves and fruits. Subsequently, a proliferation of tissue follows the periphery of the cavity resulting in knot development (Smith, 1920; Surico, 1977). Tumor development is dependent on bacterial production of phytohormones indoleacetic acid and cytokinins (Comai & Kosuge, 1980; Iacobellis et al., 1994; Rodríguez-Moreno et al., 2008; Smidt & Kosuge, 1978; Surico et al., 1985). Besides, recent results have revealed that Psv strains contain two copies of all the genes involved in indoleacetic acid synthesis (Matas et al., 2009; Pérez-Martínez et al., 2008). It has been reported that olive knots are also dependent on the *hrp/hrc* genes (Sisto et al., 2004), which encode the biosynthesis of a functional Type III Secretion System (TTSS). Recently, remarkable progress has been made in research into several aspects of the host-pathogen interaction of the causal agent of olive knot (Pérez-Martínez et al., 2008; Matas, 2010; Pérez-Martínez et al., 2010). Several putative virulence factors in Psv have been identified, including TTSS protein effectors and a variety of genes encoding known *P. syringae* virulence determinants (Pérez-Martínez et al., 2008). Analyses of TTSS protein effectors of Psv have recently shed light on the role of TTSS in pathogenicity and host range (Matas, 2010; Pérez-Martínez et al., 2010).

Fig. 1. Typical olive knot symptoms caused by *Pseudomonas savastanoi* pv. *savastanoi* on twigs (upper left), leaf (upper right), branches (lower left) and fruits (lower right).

Anatomical studies of knots have been performed in olive (Smith, 1920; Surico, 1977), oleander (Wilson & Magie, 1964; Wilson, 1965) and more recently, in buckthorn (Temsah et al., 2007a) and myrtle (Temsah et al., 2007b) by light microscopy. Rodríguez-Moreno et al. (2009) performed the first real-time monitoring of Psv disease development and the first illustrated description of the ultrastructure of Psv induced knots. They examined knot sections using a green fluorescent protein tagging a Psv strain, coupled with epifluorescence microscopy and scanning confocal electron microscopy. Additionally, scanning and TEM (transmission electron microscopy) were used for a detailed ultrastructural analysis within knot tissues (Rodríguez-Moreno et al., 2009).

Infection by Psv and subsequent knot formation in young twigs of oleander (*Nerium oleander*) requires vascular cambium activity (Wilson, 1965). The host invasion by the bacterium begins with the colonization of the infection site, followed by the disintegration and breakdown of adjacent plant cells that results in the formation of a large cavity around the area colonized by the bacteria. Curiously, this bacterium produces cell wall degrading enzymes *in vitro* such as cellulase, cellobiase, xylanase and peptinase (Magie, 1963). In a second phase, intact cells surrounding the pathogen suffer the effect of the hormones that Psv produces and increase in size (hypertrophy) followed by an abnormal cell division (hyperplasia). Finally, there is a differentiation of certain cells of the hyperplastic area, elements of xylem and phloem.

During infection of young olive stems after inoculation, the bacteria multiply by a succession of phases which include a population increase, a stationary phase and a population decline. There is a clear parabolic trend whose maximum value depends on cultivar susceptibility (Varvaro & Surico, 1978). Pathogen multiplication inside tissues of micropropagated olive plants can reach densities of 10^7 to 10^8 cfu/ knot (Rodríguez-Moreno et al., 2008), values very similar to those previously described with 1-2 years old seedlings by Penyalver et al. (2006). The first reaction of tissue from the inoculated slit of a young stem is to renew or quickly increase cambium activity, although this depends on whether the inoculation takes place in winter, summer or spring (Surico, 1977). The increased activity of the cambium promotes the formation of two new tissue masses on both sides of the wound, which grow until their junction and form a knot. Differentiation of phloem and xylem elements, which are organized or not in vascular bundles, occurs within the new parenchyma tissue. Light microscopy shows the presence of vascular bundles of new formation in olive knots, connected with the stem vascular cylinder (Rodríguez-Moreno et al., 2009). Psv has been located in cavities formed after the collapse of intercellular plant cells, as well as in peripheral areas close to the epidermis or invading the newly formed xylem bundles (Rodríguez-Moreno et al., 2009). This could be related to the spread of the pathogen and its external output through plant exudates. Formation of bacterial aggregates, microcolonies and multilayer biofilms has been observed in knot sections by scanning electron microscopy. Besides, TEM analysis of knot sections shows the release of outer membrane vesicles from the pathogen surface (Rodríguez-Moreno et al., 2009).

Subsequently, in old knots, plant cells collapse and form cavities containing large numbers of bacteria. Fisures reaching the knot surface develop inside these cavities, allowing bacteria to escape to the external surface of intact knots (Surico, 1977). However, it remains unclear how knot formation in the host benefits Psv. Knots may represent a favorable environment for bacteria to multiply and also protect them against extreme environmental conditions, such as the usually dry and hot summers of the olive-growing areas (Comai & Kosuge, 1980).

The causative agent of olive knot disease is not the only organism living in knots, because there are also white or yellow saprophytic bacteria characterized as Pantoea agglomerans (García de los Ríos, 1989), other species of enterobacteria and even putative human pathogenic bacteria (Ouzari et al., 2008). The etiologic agent of olive knot disease has only been isolated from 5 to 10% of olive knots, similarly to that observed in other Psv hosts species, such as oleander and ash (García de los Ríos, 1989). Four new bacterial species belonging to the genus Pantoea were proposed in a study of endophytic bacteria from olive knots associated to Psv (Rojas, 1999) and one of them known as Erwinia toletana has been accepted as new species (Rojas et al., 2004). These bacteria would be incorporated to the knot subsequently to the infection caused by the etiologic agent, according to García de los Ríos (1989). Indeed, a symbiotic relationship may exist between Psv and Pantoea (or may be other bacteria) because they are found together not only in knots but also as epiphytic bacteria in infected plants (Ercolani, 1978, 1991; Quesada et al., 2007). Furthermore, preliminary tests have shown that strains of an uncharacterized Erwinia, isolated from Psv-related olive knots, could have a synergistic effect with Psv in the development of typical symptoms of olive knot disease (Fernandes & Marcelo, 2002). Similar results have been reported in olive trees coinoculated with strains of Psv and *P. agglomerans*, which produced larger knots than inoculations with Psv strains alone and this effect could be due to auxin production by *P. agglomerans* (Marchi et al., 2006).

3. General characteristics of phyllosphere habitats

The surfaces of the aerial parts of plants, including stems, buds, flowers and mainly leaves, can be denominated phyllosphere (Lindow & Brandl, 2003). The leaf surface is a large microbial habitat and Morris & Kinkel (2002) estimated that in total terrestrial surface area there would be about 4×10^8 km^2 of leaf surface area, which could be colonized by 10^{26} bacteria.

Leaf-borne microbial communities are diverse and include many different genera of bacteria, filamentous fungi, yeasts, algae and, less frequently, protozoa and nematodes (Lindow & Brandl, 2003). The phyllosphere of field plants is a harsh environment for bacteria. This habitat is severe because they are exposed to extreme microclimatic changes in temperature, relative humidity, wind speed, radiation, etc. in time periods as short as hours. Plant nutrient resources are scarce and accordingly plants like olive trees have a thicker cuticle (Lindow & Andersen, 1996). The phyllosphere may also be exposed to long dry periods or, conversely, heavy rains could dramatically alter the microbiota by the "washing" effect (Hirano & Upper, 2000).

The phyllosphere colonizing microorganisms are called epiphytes and Psv is one of them. A bacterium is considered as epiphytic when it lives and multiplies in the phyllosphere and constitutes the main colonizing group of the leaf surface, with average numbers from 10^6 to 10^7 cfu/cm^2 of leaf (Hirano & Upper, 1983, 2000). Interestingly, not all epiphytic bacteria colonizing the phyllosphere have a strictly commensal relationship with their host plant. This has also been demonstrated in many plant pathogenic bacteria including Psv, which have also an epiphytic resident phase (Ercolani, 1971).

The size and composition of epiphytic bacterial populations vary according to plant characteristics (species, age) and factors related to nutritional and climatic conditions (Hirano & Upper, 2000; Lindow & Brandl, 2003). The size estimation of epiphytic bacterial populations in the laboratory depends on the sampling, bacteriological and statistical procedures used (Hirano & Upper, 2000; Jacques & Morris, 1995). Most studies about epiphytic bacteria have estimated population sizes by washing or sonication to release bacteria from the leaf, followed by plating of serial dilution of the washings (Hirano & Upper, 2000). The estimated population sizes with this technique correspond to a proportion of total microorganisms living in the phyllosphere (Wilson & Lindow, 1992). The error involved in the recovery of epiphytic bacteria on solid medium is not important compared to the high variability of populations in field samples, which could range from 5 to 6 orders of magnitude (Hirano & Upper, 2000).

4. Epiphytic populations of Psv

There are few studies on the epidemiology of olive knot disease and most data of epiphytic Psv populations on the aerial surface of the tree come from studies done in southern Italy

(Ercolani, 1971, 1978, 1979, 1983, 1985, 1991, 1993) and southeastern Spain (Quesada et al., 2007; Quesada et al., 2010a; 2010b). In the aforementioned studies, epiphytic Psv populations in the phyllosphere of olive trees were estimated by washing, followed by plating serial dilution of the washings.

Microbial communities of the olive tree phylloplane can grow embedded in a matrix of exopolysaccharides and form biofilms adhered to the leaf surface (Morris et al., 1997). Furthermore, a great diversity of pigmented bacteria colonizing the surface of olive tree was observed by washing olive leaves and plating the washings. In Italy, Ercolani collected bacteria from the leaf surfaces of olive trees for two sampling periods over several years in the 70s and 80s (Ercolani, 1978, 1991). Phenotypic characterization of these isolates allowed Ercolani to record over 20 bacterial species colonizing the leaf surface. The three highest frequency values of occurrence corresponded to Psv, *Xanthomonas campestris* and *Pantoea agglomerans* with 51, 6.7 and 6%, respectively (Ercolani, 1991).

Spanish studies found that averages of the total bacterial population from leaves and stems were generally significantly higher in Psv-inoculated than in non-inoculated olive trees, suggesting that Psv might have a positive effect on the growth of other epiphytic bacteria or on their ability to colonize olive organs (Quesada et al., 2010a). Populations of *P. agglomerans* could accompany Psv and contribute to the significant differences in total bacterial populations between inoculated and non-inoculated olive trees. Besides, there was a positive correlation between Psv and yellow *P. agglomerans* either on stems or leaf surfaces of naturally infected olive trees (Quesada et al., 2007), and a similar fluctuation of both bacterial populations on the same host. This is of interest because both bacteria produce indoleacetic acid and this can contribute to the epiphytic fitness of Psv in olive trees (Varvaro & Martella, 1993).

The bacterial community composition on the surface of olive leaves is more strongly influenced by the sampling season than by leaf age (Ercolani, 1991). The diversity and size of the total bacterial populations within the olive phyllosphere were lower during the hot dry months and higher during the cold rainy months (Ercolani, 1991). Our observations on one olive orchard showed that seasonal fluctuations of Psv populations felt into the pattern of seasonal shifts described above (Quesada et al., 2007). Interestingly, Psv population sizes in stems and leaf surfaces were correlated (with r^2 values of 0.7 and 0.43, respectively) with rainfall, temperature and relative humidity (Quesada et al., 2007). Therefore, these climatic parameters may exert a more or less strong influence on the Psv population values. Epiphytic bacterial communities on olive leaves were more uniform in mature leaves than in young leaves (Ercolani, 1991). In addition, the olive phyllosphere apparently selects specific genotypes of the bacterial community (Lindow & Brandl, 2003). This gives us an idea of the great variability, in terms of epiphytic populations, existing among leaves of the same tree.

Over 50% of bacterial isolates collected from olive leaves by Ercolani (1978, 1991) were identified as Psv and this bacterium survived and multiplied on the leaf surfaces of olive trees (Varvaro & Ferrulli, 1983). Abu-Ghorrah (1988) observed maximum Psv population levels of about 10^7 cfu/cm^2 in olive trees and the Psv generation time in this host was 24 to 36 hours. In studies of leaves inoculated by spraying a suspension of Psv, the bacteria colonized the lower leaf surface better than the upper surface (Surico, 1993). Basically, they sticked to the vein depressions and to specific structures such as the shields of pectate hairs (Surico, 1993).

The seasonal fluctuation of Psv populations on olive leaves in Italy, recorded over three consecutive years by Ercolani (1971, 1978), showed that Psv populations were higher in spring and fall (about 10^4 cfu/cm^2 of leaf) than in winter and summer (about 10^2-10^3 cfu/cm^2) (Ercolani, 1971, 1978; Varvaro & Surico, 1978). Psv populations on olive leaves in Spain, also recorded over three consecutive years, reached the highest (ca. 10^3-10^4 cfu/cm^2 of leaf) densities mainly in warm and rainy months (mainly spring season) and the lowest (ca. 0-10 cfu/cm^2 of leaf) in hot and dry months (summer season) (Quesada et al., 2007). Significant differences were observed between Psv populations in summer and in the other seasons over the three-year study (Quesada et al., 2007).

Lavermicocca & Surico (1987) simultaneously analyzed Psv populations on olive tree leaves and stems for the first time and during one year, reporting higher frequencies of Psv isolation in stems than in leaves with relatively high epiphytic Psv populations in July (about 10^5 cfu/cm^2) and only 10 cfu/cm^2 in September and March. However, in Spain no significant differences were found between either leaves or stems with respect to the number of analyzed samples where Psv was isolated, detected by PCR, or regarding the average Psv populations over several years (Bertolini et al., 2003a; Quesada et al., 2007; Quesada et al., 2010a, 2010b). Given that in such studies the Psv number were evaluated on stems after they were cut into pieces, some endophytic Psv could be also counted (Penyalver et al., 2006). Furthermore, both types of plant material (stems and leaves) should be analyzed from symptomless shoots to make the evaluation of Psv populations in the phyllosphere more accurate (Bertolini et al., 2003a; Quesada et al., 2007). Psv was also isolated from the surface of olive fruits, but at lower frequency than from leaves, reaching a high Psv population size in September (10^6 bacteria / g fresh weight) (Lavermicocca & Surico, 1987).

Between 70 and 95% of the maximum variance of some microbiological parameters, such as Psv density in the olive tree phyllosphere, was explained by the influence of seven factors, four of which were related with the weather: summer, summer rainfall, winter rainfall and warm fronts (Ercolani, 1985) and the three remaining factors were cambium activity, leaf age and time of flowering. As described for other epiphytic bacteria and hosts (Kinkel, 1997), Psv population densities varied over several orders of magnitude among leaves sampled concurrently from the same shoot, as assessed by the comparison of leaf printing and isolation experiments (Quesada et al., 2007). Due to the low detection level associated with leaf printing (Jacques & Morris, 1995), such results also suggested that Psv probably colonizes low numbers of leaves with high populations, in bulked samples.

The size of Psv populations on each leaf correlated with leaf age, the time when it formed and the time of the year when the sample was taken (Ercolani, 1991). In addition, phenotypically distinct Psv isolates from the phyllosphere, succeed each other in time in the olive tree phyllosphere (Ercolani, 1983). This was discovered because Psv isolates obtained by washing leaves of a certain age at a particular time of the year (over eight years) showed more phenotypic similarity with each other, than with isolates obtained by washing leaves of different ages at different times of the year. Most of the Psv isolates obtained by washing leaves of different ages taken at random in April, were less similar to each other in 60 phenotypic characters (they formed a single group at 65% similarity) than Psv isolates obtained from six-month-old knots in October (one group formed 85% similarity). A similar result was obtained when Psv isolates obtained by washing leaves in October were compared with Psv isolates obtained from six-month-old knots in April (Ercolani, 1993). Psv isolation from these knots was performed six months after washing leaves in April and October as above indicated. Almost all isolates from knots reflected the dominant phenotype of the isolates obtained from the phyllosphere six months earlier. Most Psv isolated by washing leaves in April and October were phenotypically classified near to isolates obtained by washing 13-month-old leaves. According to these authors, senescent leaves (13 months old) could be the main source of bacteria for knot formation in April and October (Ercolani, 1993).

5. Endophytic populations of Psv

Endophytic bacteria are defined as bacteria living in plant tissues without doing substantive harm or gaining benefit other than residency (Kado, 1992). The endophytic colonization of plants is probably fundamental for plant-associated bacteria to develop sustainable epiphytic populations (Manceau & Kasempour, 2002). Although the information about endophytic populations of plant pathogenic bacteria is still scarce, it is very likely that most of the pathogens can undergo an endophytic step during the disease cycle. However, as their numbers are low inside the asymptomatic plants and their distribution is not homogeneous, laborious studies are required to detect them. Consequently, little is known about the real importance of the endophytic phase for most bacterial pathogens.

Psv could also present an endophyte phase spanning a considerable part of its life cycle due to its multiplication in the intercellular spaces, substomatal cavities or in vascular tissues of the plant, without any visible symptoms (Schiff-Giorgini, 1906; Smith 1908, 1920; Wilson & Magie, 1964). Regarding Psv, the bacteria that survive inside and outside the knots could have a greater impact than the bacteria colonizing the olive as a symptomless endophyte. In fact, some studies described the endophytic phase of Psv in olive plants as rare (Wilson & Magie, 1964). According to other authors, Psv could also present an endophytic phase, moving through the intercellular spaces and even in the xylem vessels and infecting areas close to the first infected zone (Penyalver et al., 2006; Schiff-Giorgini, 1906; Smith 1908, 1920; Wilson & Magie, 1964; Wilson & Ogawa, 1979). Further studies are needed to reliably assess the importance of this phase of Psv in olive knot epidemiology. Nowadays, this is relatively easier to address, thanks to an established model system for the study of olive knot disease covering a wide range of aspects. It is formed by a micropropagated olive plant, coming from an *in vitro* germinated seed and a Psv strain (NCPPB 3335) producing the characteristic symptoms of olive knot disease in both woody and micropropagated olive plants (Pérez-Martínez et al., 2007; Rodríguez-Moreno et al., 2008). Besides, Psv strain NCPPB 3335 has been studied in depth and many genetic resources are available because its genome has been sequenced and analyzed using appropriate bioinformatic tools (Rodríguez-Palenzuela et al., 2010; Matas, 2010).

Endophytic Psv populations could contribute to a dramatic increase in Psv numbers and in olive knots in infected olive groves, when the copper-based control treatments are not applied (Quesada et al., 2010b).

6. Epidemiology and disease cycle

Disease caused by Psv populations has an epiphytic-pathogen type cycle. The bacteria have an epiphytic phase in which they multiply on the surface of olive tree stems and leaves without developing symptoms (Ercolani, 1978, 1991; Varvaro & Ferrulli, 1983). Interestingly, Psv populations were recovered from symptomless shoots from non-inoculated control trees prior to the appearance of symptoms, suggesting that these epiphytic bacteria were the potential source of inoculum for infection of healthy plants (Quesada et al., 2010a).

The temperature range in which Psv can initiate infection is between 5 and 37 °C and this would allow the bacteria to cause infections throughout the year. However, optimal conditions for disease development are about 22-25 °C and the subsequent time periods with high infection probability are fall and spring (Protta, 1995). Psv can infect olive trees at any time of the year and trigger knot formation only when conditions are favorable. Thus, when the bacteria infect an olive tree in the fall, knots will begin to develop several months later, but if the infection occurs during the spring, the time required for knot formation may be only two weeks (Wilson, 1935). Field trials performed in California showed that Psv inoculations of olive trees carried out in April caused higher levels of olive knot disease than Psv inoculations carried out in December (Teviotdale & Krueger, 2004).

Dissemination of Psv bacteria from infected (or inoculated) to non-infected (or non-inoculated) trees was suggested (Quesada et al., 2010a). Bacteria could spread over long distances due to the introduction and planting of infected material, or over short distances transported by splashing rain, windblown aerosols, insects and cultural practices (Horne et al., 1912; Wilson, 1935). Currently, bacterial dissemination is facilitated by cultural practices, new plantations with high tree density, frequent severe pruning and with small distance between plants (Tous et al., 2007). Wounds caused by harvesting and pruning, as well as by hail, frost and leaf scars, create niches where infection occurs (Wilson, 1935; Janse, 1982) and olive tree infection by Psv is directly related to the degree of wounding of the trees (Smith et al., 1991). In an assay to evaluate natural dissemination of the bacteria on young plants, knots were not observed in inoculated and non-inoculated trees until 3 and 10 months after inoculations, respectively (Quesada et al., 2010a). The compatible Psv-olive tree interaction facilitates the invasion, infection and multiplication, triggering hypertrophy and hyperplasia of the plant tissues with subsequent knot formation. Bacteria can survive inside the knots from one season to another and when humidity is high enough, exudates containing large amounts of bacteria are emitted in which they can survive as epiphytes (Wilson, 1935). However, the bacteria can only survive in soil for a few days (Wilson & Ogawa, 1979). The disease cycle is summarized in Figure 2.

In 1909, Petri isolated Psv from the intestinal tract and eggs of the olive fly (*Bactrocera oleae*), but there is no other scientific evidence that this, or other insects, can be efficient vectors of olive knot disease. Additionally, there is no conclusive evidence about the role that birds may play as vectors of this disease (Wilson, 1935), although they can transport living bacteria from plant to plant.

Epidemiology and Control of Plant Diseases Caused by Phytopathogenic Bacteria: The Case of Olive Knot
Disease Caused by Pseudomonas savastanoi pv. savastanoi

343

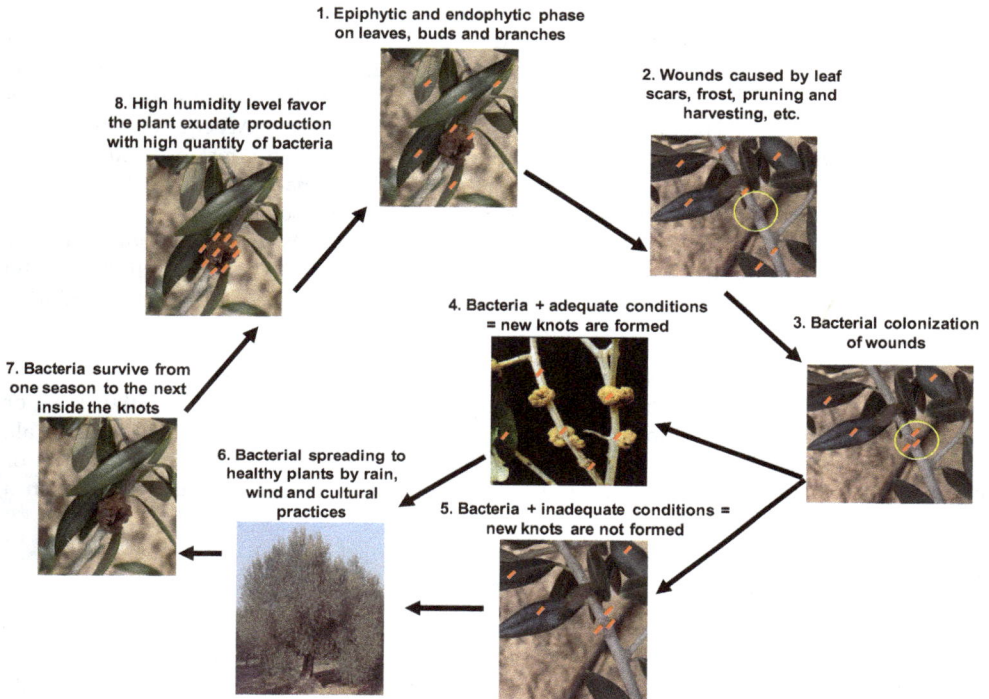

Fig. 2. Disease cycle of olive knot caused by *Pseudomonas savastanoi* pv. *savastanoi* simulated as red bacilli (kindly provided by E. Bertolini, 2003).

7. Effect of olive knot disease on the vigor and yield of olive trees

Although the olive knot disease is widespread throughout most olive-growing areas, there is no accurate estimation of the losses it causes. This is very difficult to measure because many factors can influence the severity of the symptoms. Severe infections can cause death of branches and a progressive weakening, resulting in a loss of tree vigor (Tjamos et al., 1993) and thus of harvest. De Andrés (1991) estimated that Psv-related losses were around 1.3% of national olive production in Spain. Occasionally, this disease has caused the loss of almost the olive local harvest due to the combination of optimal weather conditions for bacterial entry and multiplication, as observed in two Spanish localities in 1987 and 2001 (B. Celada, personal communication) after severe hail storms. Furthermore, this disease is present with variable incidence in many nursery plants, as it limits their commercialization due to the visible symptoms. This is especially important in plants for export because several countries that import plants from the European Union (EU), like Chile, consider Psv as a quarantine organism.

The quantitative effects of olive knot disease on vigor and olive fruit yields are not yet well established because there is only information available from one study in California and another in Spain. In a commercial orchard in California (USA) significant differences were not found in vigor between 40-year-old olive trees lightly (0.10-0.30 knots in 0.3 m of fruit wood) and mildly (0.31-0.50 knots) infected with the disease (Schroth et al., 1973). In contrast, in Spain in a study on non-inoculated and inoculated trees 7-year-old of cv. Arbequina in a high-density grove, vigor was significantly higher in non-inoculated trees (Quesada et al., 2010a). Therefore, vigor was higher in trees of cv. Arbequina where olive knot disease was lower during the study, suggesting a negative effect over time of the disease on plant development. Schroth et al. (1973) in California showed that there was a clear relationship between crop losses and the number of tumors caused by Psv in branches. They observed significant differences in the weight of olive fruits per tree in only one year between lightly and mildly infected olive trees with 121.3 and 94.6 kg, respectively. In the Spanish assays, the different levels of the disease did not significantly affect cumulative olive yield (Quesada et al., 2010a) in young trees.

Furthermore, low oil quality was reported when olive fruits were harvested from olive trees moderately affected with olive knot presenting odors and flavors such as bitter, stale or salty, but the data were lacking of statistical support (Schroth et al., 1968; Tjamos et al., 1993). In another study, olive knot disease incidence did not modify either the chemical or organoleptic characteristics of virgin olive oil extracted from young olive tree fruits in a high-density grove (Quesada et al., unpublished data).

8. Control methods

The methods used to control plant pathogenic bacteria are based on preventive and curative measures and the combination of the two should be used in the context of an integrated control. The five main goals of an integrated plant disease control program are to eliminate or reduce the initial inoculum, reduce the effectiveness of the initial inoculum, increase host resistance, delay disease onset and slow the secondary cycles (Agrios, 2005). The key of any integrated control program is a question of sustainability at different levels (Caballero & Murillo, 2003). In economic terms, it must ensure farmers' profits and at the environmental level, select control methods that minimize environmental impact. And finally, control methods by themselves should ensure sustainability and they must remain effective over time. The monitoring of a strategy of this type is essential to ensure safe and sustainable agriculture.

Disease management of bacterial pathogens in the field is mainly based on preventive procedures, because it is difficult to eradicate pathogens once established. Due to the economic impact of the olive knot disease, growers require adequate control methods to overcome its negative repercussions on yield and even on olive fruit quality (Quesada et al., 2010a; Schroth et al., 1973). Olive knot control should be based on an integrated control strategy, giving priority to the most effective measures that are of preventive type. These are very diverse and can be grouped into regulatory measures, preventing introduction of the pathogen in protected areas and prophylactic measures to reduce or eliminate the pathogen or hinder its establishment in nurseries or orchards (Montesinos & López, 1996).

8.1 Regulatory measures

The production, maintenance and use of certified plant material which is pathogen free, is one of the main preventive measures used to control plant pathogens. In this regard, it is very important that Psv appears in the EU list of pests and diseases that significantly affect plant quality standards, drawn up by the European Commission (Directive N° 92/34/EU). As advised by the European and Mediterranean Plant Protection Organization (EPPO-OEPP), new olive groves should be established using Psv-free certified plant material (EPPO, 2006).

As an example, fifteen years ago, there was scarce reliable information available on the sanitary status of Spanish olive plants with respect to pathogenic bacteria (Bertolini et al., 1998; Padilla, 1997). Although there is now more information and analyses have been performed, there is still a lack of scientific published data available on the status of olive plants in the field or in nurseries, either in Spain or in other olive-growing countries. Government agencies have shown an interest to control the planting material given the increase number of plantations, the notable changes in production technologies and the frequent commercial exchange of olive plants. All these facts emphasize the convenience of providing plant material with certain quality standards and the implementation of certification programs (Cambra et al., 1998).

With respect to this issue the EU has, so far, required the minimum conditions for the nursery plants of type Agricultural Conformitas Comunitatis (or CAC). The implementation of certification systems is the responsibility of each member state, but the European and Mediterranean Plant Protection Organization (EPPO) has developed specifications for certification of olive plants. These referred specifically to health although they were based on studies conducted in Italy and Portugal and thus should be contrasted with the situation in Spain (Chomé, 1998).

Italy was the first country to publish standards for certification of plant material from olive trees in 1993. Certified plants should be free from Psv, *Verticillium dahliae* and six virus (*Olive latent virus 1* (OLV-1), *Olive latent virus 2* (OLV-2), *Cucumber mosaic virus* (CMV), *Arabis mosaic virus* (ArMV), *Cherry leaf roll virus* (CLRV) and Strawberry latent ring spot virus (SLRSV)) (Martelli, 1998). Besides Italy, other countries like Portugal, Israel, Argentina and Spain, have also established certification programs for olive plant material. In the Argentinan Certification program, the mother plants are annually tested and must be free of *P. syringae*, Psv, *Agrobacterium tumefaciens*, *V. dahliae* and *Phytophthora cinnamomi* but viruses are not considered.

There are more than 260 nurseries registered in Spain, mainly located in Andalusia and Valencia, which produced about 5.5 million olive plants in 1999-2000 for new plantations and also for international trade (Chomé, 1998). Given this significant production, the *Real Decreto 1678/1999* (Anonymous, 1999) established quality control and certification requirements for olive seedlings in the certification program of olive plant material in Spain. Currently, the qualification of certified plant material is the responsibility of the competent institutions in each region. To qualify for certification, plant material must meet certain conditions such as having a known origin and having been submitted to cultivar analyses and sanitary tests. Mother plants of the starting material and base material should be officially inspected to verify that they are free of *V. dahliae*, Psv and viruses OLV-1, OLV-2, CMV, ArMV, CLRV and SLRSV. Each year the plants for certification should be sampled and tested for Psv by the responsible official body using approved techniques, which include isolation, serology and nested multiplex RT-PCR. With this technique, Psv and four olive viruses (CMV, CLRV, SLRSV and ArMV) can be detected simultaneously in a sensitive single reaction (Bertolini et al., 2003b). Finally, parent plants of certified nursery stock must be at least free of symptoms of diseases caused by fungi, bacteria and the viruses previously cited.

8.2 Prophylactic measures

Prophylactic measures are designed to reduce or eliminate the pathogen levels or impede its establishment in a crop and these measures can also be of eradicative nature, based on cultivar susceptibility or by direct protection (Montesinos & López, 1996). In the case of Psv, they include all those performed for disinfecting plants, agricultural machinery, or anything in contact with plants.

8.2.1 Eradication

The presence of knots in a tree is related to a high level of disease after several years, and this highlights the need of using preventive control methods or eradication methods to maintain olive trees without knots (Quesada et al., 2010a). In affected plantations the main olive knot disease eradication method would be the uprooting of the affected trees or the use of cultural practices to reduce the inoculum source, performing copper treatments, pruning of infected branches and reduction of number of wounds during the growing season and especially at harvest (Beltrá, 1956; Penyalver et al., 1998; Trapero & Blanco, 1998; Wilson, 1935). This is especially relevant in new plantations with high tree density and frequent severe pruning, where control measures should be accurately monitored (Tous et al., 2007).

The removal of knots is very laborious and may not be entirely effective because new wounds are usually done when knots are removed and new knots can develop in these wounds in the following years, even when treated with preventive chemicals (Wilson, 1935). Pruning of infected branches is more effective than knot removal as fewer wounds are caused to the olive tree and the bacterial inoculum load is minimised (Quesada et al., 2010a; Teviotdale & Krueger, 2004; Wilson, 1935). All cut branches should be burned in the same field to prevent the spread of the disease (Trapero & Blanco, 1998).

In the case of partially contaminated olive groves, healthy trees should be harvested and pruned first (Wilson, 1935). Besides, growers should harvest olives in dry weather only and avoid the use of techniques like knocking the olive tree branches with wooden poles (Krueger et al., 1999). Manual harvesting methods, like the "milking" method or the use of mechanical vibration are more suitable. It is important to assess the index of tree damage in terms of broken branches and compare this to the olive fruit harvested. It has been reported that knocking the olive tree branches with wood poles can break from 13 to 18% of branches while for mechanical vibration this is only 6 to 9%, on complete harvesting (Civantos et al., 2008).

8.2.2 Cultivar susceptibility

The use of resistant cultivars, or low susceptibility cultivars to bacterial plant diseases would be one of the most appropriate disease control methods (Montesinos & López, 1996).

However, in woody crops, such as olive trees, breeders and plant pathologists are hindered by the slow improvement in breeding processes as a result of delayed entry into fruition.

Another drawback is that the information available about cultivar susceptibility to olive knot disease is scarce and mainly comes from field observations, such as those reported in the USA and Spain (Barranco, 1998; Trapero & Blanco, 1998; Wilson, 1935). Very few data are available from comparative inoculation experiments and is limited to several cultivars (five to eight) from Italy, Greece, Morocco, and Portugal (Benjama, 1994; Catara et al., 2005; Hassani et al., 2003; Marcelo et al., 1999; Panagopoulos, 1993; Varvaro & Surico 1978), with the exception of Spain where 29 cultivars were evaluated (Penyalver et al., 2006). Field observations do not always give universally valid information on the intrinsic susceptibility of each cultivar because the initial quantity of bacterial inoculum differs between plants and factors favoring infection can vary in different areas.

Varvaro & Surico (1978) compared the behavior of six Italian olive cultivars inoculated with Psv and found no difference, because more than 95% of the inoculated wounds developed tumors. This was probably due to the high inoculum dose applied (more than 10^6 bacteria per wound) and because the inoculated plants were only one year old. Different doses of eight Psv isolates were inoculated in six olive cultivars in comparative inoculation experiments from Morocco (Benjama, 1994). The results showed that the cultivar Frantoio was the most susceptible among those tested, followed by Ascolana dura, Manzanilla, Picholine marocaine, Dahbia and Gordal Sevillana, which was the least susceptible, although a statistical analysis of the data was not performed. Marcelo et al. (1999) evaluated six Portuguese cultivars and found they differed in the percentage of knots formed at inoculation points, ranging from 36 to 66%. These authors considered that the cultivars Blanqueta, Cobrancosa, Cordovil de Serpa, Galega Vulgar, Redondil and Santulhana were moderately susceptible to olive knot disease, but their data were not statistically analysed. Hassani et al. (2003) evaluated the Italian cultivars Frantoio, Leccino, Moraiolo and Nostrale di Rigali by inoculation of five Psv strains with an inoculum dose of 5×10^7 bacteria per wound and subsequently knot weights were compared. Although they did not indicate the percentage of inoculation sites that developed knots, it is likely that this parameter exceeded 90% in the four cultivars because an excessively high inoculum dose was used.

Penyalver et al. (2006) developed a methodology for evaluation of cultivar susceptibility to Psv and reported that most of that 21 Psv strains evaluated in virulence tests showed a high degree of aggressiveness but also, in some combinations, cultivar-strain interactions were observed. Consequently, strain selection for inoculation is a pre-requisite to obtaining useful data, and at least two strains should be used for accurate evaluations. The methodology was optimized for the first time with 29 olive cultivars. It was concluded that plant material should be genetically homogeneous, at least two or three years old, inoculated in spring or early summer by wounds made with a sterile scalpel. The use of at least two Psv strains with high degree of virulence was also recommended. They should be inoculated at low inoculum doses (10^2 bacteria per wound) to differentiate among different cultivars, as well as at a high dose (10^6 bacteria per wound) to identify the less susceptible cultivars. Ten olive plants should be used per bacterial strain and dose. Five wounds should be performed thus per plant and several measurements of symptoms taken for each combination, although the measurement taken at 90 days was the data included in the analysis of 29 cultivars from the World Olive Germplasm Bank of Spain, located in Cordoba.

Disease severity of a particular cultivar was found to be highly dependent on the pathogen dose applied at the inoculation point. In addition, secondary knot formation in non-inoculated wounds in previously inoculated plants, would suggest pathogen migration in the plant tissues. They also observed a correlation between the number of inoculation sites in which knots developed and the number of secondary knots formed when the initial wounds were inoculated with low bacterial doses. All cultivars developed knots at inoculation points, at least when high inoculum doses were applied. According to the results, six cultivars were classified as highly susceptible (Arbequina, Arróniz, Nevadillo Blanco de Jaén, Pajarero, Picudo and Vallesa). Some cultivars were classified as slightly susceptible (Azapa, Cerezuela, Chemlali, Dulzal de Carmona, Frantoio, FS-17, Gordal de Archidona, Gordal de Hellín, Lechín de Granada, Manzanilla Cacereña, Manzanilla de Sevilla, Nevadillo negro and Villalonga). The remaining cultivars (Changlot Real, Morisca, Gordal sevillana, Lechín de Sevilla, Oblonga, Picual, Ascolana tenera, Royal de Cazorla, Mollar de Cieza and Koroneiki) were classified as moderately susceptible.

So far studies on cultivar susceptibility to olive knot disease suggest that true resistance to this disease is uncommon among cultivated olive cultivars. In contrast, significant differences were observed in the degree of susceptibility among the cultivars tested. In addition, *in vitro* studies of Psv interaction with cell cultures of the cultivar Galega vulgar showed the typical events of a hypersensitive response in inoculated plant cells, such as an increase in reactive oxygen species, the activation of programmed cell death and decreased cell viability (Cruz & Tavares, 2005). High resolution liquid chromatography and mass spectrometry analysis of Psv-related knot extracts from outbreaks in olive trees of cultivar Koroneiki revealed high amounts of phenolic compounds, o-diphenols (oleopurina) and polyamines (spermidine, spermine, putrescine), in addition to auxins (Roussos et al., 2002). These authors postulated that the production of indole-3-acetonitrile and phenolic compounds could be related to the olive tree's defense mechanisms in knots. Cayuela et al. (2006) identified verbascoside as the main phenolic compound produced at significant levels in Psv-related knot extracts in olive trees of the cultivar Picual.

Balanced soil fertilization, avoiding excess nitrogen, may increase plant resistance to infection (Paoletti, 1993). However, in modern olivicultural practices such a balance is hard to maintain because the rapid development of young plants is valued, with early production onset and increased yields from one year to the next. A common mistake made to meet the demands of modern oliviculture is to apply an excess of nitrogen fertilizer, as this increases susceptibility to olive knot disease (Balestra & Varvaro, 1997). It is advisable to perform main fertilization of olive trees in January-February (Baratta & Di Marco, 1981) with low winter temperatures.

8.2.3 Direct control

Direct protection measures are mainly based on chemical or biological principles and are used by the growers when prophylactic measures have failed to stop disease progression in one zone (Montesinos & López, 1996), or are combined with all the other measures in an integrated control strategy.

8.2.3.1 Chemical control

Chemical control of bacterial plant diseases is only effective when they are used in preventive strategies before the onset of infection or very early in the bacterial infection process (Montesinos & López, 1996). Specifically, chemical control of olive knot disease has given inconsistent results in field experiments and may also have low efficacy and even show phytotoxicity to some tissues. This variability is due to several factors, such as the amount of inoculum, timing of treatments, climatic conditions, cultivar susceptibility, treatment application method, or physiological state of the host plant.

Copper compounds are the main preventive chemical treatment recommended against olive knot disease and their use is recommended every year when there is a risk of infection, in spring and fall before the rains, after the leaf fall and especially after hail and frost or other events causing olive injures (Penyalver et al., 1998; Protta, 1995; Smith et al., 1991; Wilson, 1935). A positive correlation has been found between disease incidence and spring rains (Teviotdale & Krueger, 2004) and it was observed that moist winds in coastal areas promote infection (Smith et al., 1991).

The copper-based compounds used in olive groves in Spain include various salts and formulations (hydroxides, oxychlorides, oxides or sulfates) as well as their mixture with organic compounds obtained by chemical synthesis. An interesting example is the combination of cuprocalcic sulfate plus mancozeb because it has a synergistic effect against several bacterial diseases (Hausbeck et al., 2000; Jones et al., 1991; Marco & Stall, 1983). Currently copper oxychloride is the copper compound most commonly recommended against olive knot disease by the Spanish extension services. The active ingredient in these products is the divalent copper ion solubilized and both, bacteria and plant exudates, contain compounds which are capable of solubilizing copper. Generally, these products have a toxic or bacteriostatic effect, only preventing the multiplication of bacteria and most bacteria may die due to the toxic effects of Cu^{++} ions, or enter in the VBNC (Viable But Non Culturable) state in which they are unable to grow on solid medium. This state could be induced by copper ions, as previously reported for several plant pathogenic bacteria (Alexander et al., 1999; Grey & Steck, 2001; Ordax et al., 2005). These preventive chemical treatments are recommended for both to reduce epiphytic Psv populations and prevent their penetration through the plant wounds.

The effectiveness of chemical control of olive knot has been poorly evaluated, both in the field and in experimental assays under controlled conditions. The effect of the treatments against epiphytic inoculum of Psv or the optimal time of application are not well known. Several studies suggest that the management of epiphytic Psv populations probably reduces the incidence of olive knot disease (Ercolani, 1978, 1991; Lavermicocca & Surico, 1987; Quesada et al., 2007, 2010a). In this context, a chemical control program using copper compounds was proposed, based on field observations in California (Horne et al., 1912; Wilson, 1935). The first field experiments described in the literature were conducted in California where several Bordeaux mixture formulations controlled olive knot disease with minimal phytotoxicity symptoms in commercial olive groves with prevalent Psv infections (Krueger et al., 1999; Teviotdale & Krueger, 2004; Wilson, 1935). Assays performed with copper hydroxide showed that a single post-harvest copper application provided only minimal protection against the disease and subsequently, additional sprays in spring were needed to substantially improve its control (Teviotdale & Krueger, 2004). The efficacy of copper hydroxide to control the incidence of knots was higher after three sprays than after two or one single spray. New information has been gathered about the effect of copper compounds on the population dynamics of epiphytic Psv, the possible appearance of copper resistance, or its role in decreasing olive knot incidence under Mediterranean conditions in high-density groves (Quesada et al., 2010b).

The effect of copper oxychloride or cuprocalcic sulfate plus mancozeb treatments on Psv populations and subsequent disease development were evaluated in an olive grove planted with two susceptible cultivars, Arbequina and Picudo, over a four-year period. Unlike the previous studies, to homogenize the knot number per tree before beginning treatments, olive trees were inoculated. The effect of copper on Psv populations was observed after the first application, but the greatest differences between copper-treated and untreated plants were observed in the third year, after five copper applications. Two applications of copper compounds per year, reduced Psv populations effectively. We also found that treatment with copper compounds had a drastic effect on reducing disease incidence (Quesada et al., 2010b). These results for both cultivars, in this high-density grove, supported previous observations by Teviotdale and Krueger (2004) in California, in standard groves. Unlike other plant pathogenic bacteria that develop copper resistance after extensive exposure to copper compounds (Cazorla et al., 2002; Cooksey, 1990; Garret & Schwartz, 1998; Marco & Stall, 1983; Scheck et al., 1996; Sundin et al., 1989, 1994), copper resistance was not detected in the remaining Psv bacteria in copper-treated olives trees.

Chemical treatments based on antibiotics and oil-water emulsion containing hydrocarbons have also been recommended but without encouraging results (Scrivani & Bugiani, 1955; Schroth & Hildebrand, 1968). The use of antibiotics such as streptomycin and terramycin has been successful under experimental conditions (Trapero & Blanco, 1998) but their application against plant pathogenic bacteria is currently forbidden by the EU legislation, although it is permitted in some other countries.

Systemic acquired response, or SAR, is plants' ability to generate defense reactions against external aggression at sites far away from the point of attack. In these distant sites, the genes involved in defense processes are activated (e.g. PR proteins), thereby increasing the resistance of these tissues against possible further attacks (Durrant & Dong, 2004; Kessmann et al., 1994). Some products were recently evaluated for their induction of plant resistance against different pathogens such as acibenzolar-S-methyl (Bion ®), fosetyl-aluminium, calcium prohexadione or harpins. They were assayed to control some bacterial plant diseases like fire blight, citrus canker, apical necrosis of mango, etc. Most of these products do not produce phytotoxicity and their efficacy is sometimes comparable to that of antibiotics or copper-based compounds while others could not control these diseases (Brisset et al., 2000; Cazorla et al., 2006; Graham & Leite, 2004; Scortichini, 2002,). There are reports of acibenzolar-S-methyl-related activation of certain genes involved in defense responses in olive leaves of the cultivar Lechín de Sevilla (Muñoz et al., 2005). However, in one experiment acibenzolar-S-methyl did not reduce either Psv populations or the incidence of olive knot disease after two treatments per year over a four-year period (Quesada et al., 2010b). It is possible that different doses and more product applications could be required for achieve better efficacy. Curiously, vigor of cv. Picudo was significantly higher in olive trees treated with acibenzolar-S-methyl than in untreated trees, although disease incidence was similar in both treated and untreated olive trees (Quesada et al., 2010b).

8.2.3.2 Biological control

Biological control is another alternative to control olive knot disease, but is seldom tested against Psv. To date, biological control agents have been evaluated using isolates of *P. fluorescens* (Blightban) and Psv mutants producing bacteriocins, but without satisfactory results (Krueger et al., 1999; Varvaro & Martella, 1993). Besides, non pathogenic *Pseudomonas* sp. isolated from olive tree rhizosphere proved antagonistic against Psv (Rokni-Zadeh et al., 2008). Recently, *P. fluorescens* and *Bacillus subtilis* isolates from knots and leaves of olive trees showed antagonistic *in vitro* activity against Psv (Krid et al., 2010).

Bacteriocins are excellent candidates for using in agriculture to control plant pathogenic bacteria due to their high specificity. A bacteriocin produced by *P. syringae* pv. *ciccaronei* was shown to inhibit the proliferation and survival of epiphytic Psv form (Lavermicocca et al., 2002, 2003). Effectiveness of assays with two-year-old olive plants in a culture chamber was equivalent to that of copper hydroxide, although it would be interesting to evaluate its efficacy in nursery or field plants and determine their toxicity and persistence before advising their commercial registration.

9. General conclusions

Remarkable progress has been made in several aspects of the host-pathogen interaction of the causal agent of olive knot disease, recently. Additionally, studies on the epidemiology of olive knot disease, as well as on its chemical control, have also been reported to add to the scarce information available. However, further studies are needed to assess reliably the importance of the endophytic phase of Psv in the epidemiology of olive knot as well as the effect of different chemical on disease incidence. Furthermore, studies on the qualitative and quantitative effects of the disease on olive production are insufficient.

The production, maintenance and use of certified and potentially pathogen-free plant material, is one of the main preventive measures used to control plant pathogens and certification schemes based in analytical tests performed on olive plants before leaving the nurseries should be implemented. Although, so far, true resistance to this disease is uncommon among olive cultivars tested, significant differences were observed in the degree of susceptibility to the disease among them. Cultivars tolerant to olive knot and resistant/tolerant to climatic conditions, or avoiding cultural practices which favor olive knot development should be considered for planting in the new commercial fields. This is especially important for high density new plantations.

The olive knot disease integrated control should combine healthy plant material with appropriate cultural practices and the use, like preventive treatments, of chemical compounds. In such a context, copper treatments should be used regularly to achieve effective control.

10. Acknowledgments

Authors wish to thank M. Cambra for critical reading and suggestions. We also thank E. Bertolini for generously give us permission to include Figure 2. The research of J.M. Quesada was supported by a predoctoral fellowship from IFAPA, Andalucía, Spain. This work was supported in part by grant CAO00-007 from INIA, Spain.

11. References

Alexander, E., Pham, D., & Steck, T.R. (1999). The viable-but-nonculturable condition is induced by copper in *Agrobacterium tumefaciens* and *Rhizobium leguminosarum*. *Applied and Environmental Microbiology*, Vol. 65, No. 8, (August 1999), pp. 3754-3756, ISSN 0099-2240

Anonymous. (1999). Ministerio de Agricultura, Pesca y Alimentación. Real Decreto 1678/1999 de 29 de octubre de 1999. BOE 276, 18 de noviembre de 1999, pp. 40077-40079

Abu-Ghorrah, M. (1988). Taxonomie et pouvoir pathogène de *Pseudomonas syringae* pv. *savastanoi*. Thesis (Ph. D.). University of Angers, France

Agrios, G.N. (ed.). (2005). *Plant Pathology*, (5th edition), Elsevier Academic Press, ISBN 0-12-044565-4, San Diego, CA

Balestra, G.M., & Varvaro, L. (1997). Influence of nitrogen fertilization on the colonization of olive phylloplane by *Pseudomonas syringae* subsp. *savastanoi*, *Developments in Plant Pathology. Vol. 9:* Pseudomonas syringae *pathovars and related pathogens*, pp. 88-92, Kluwer Academic Publishers, ISBN 0-7923-4601-7, Dordrecht, The Netherlands

Baratta, B., & Di Marco, L. (1981). Controllo degli attacchi di rogna nella cultivar Nocellara del Belice. *Informatore Fitopatologico*. Vol. 31, No. 1-2, (January/February 1981), pp. 115-116, ISSN 0020-0735

Barranco, D. (1998). Variedades y patrones, In: *El cultivo del olivo*, Barranco, D., Fernández-Escobar, D., & Rallo, L., pp. 56-79, Junta de Andalucía- Mundi-Prensa, ISBN 978-84-8474-234-0, Madrid, Spain

Beltrá, R. (1956). New technique for the identification of *Pseudomonas savastanoi*. *Microbiología España*, Vol. 9, pp. 433-504

Benjama, A. (1994). Étude de la sensibilité variétale de l'olivier au Maroc vis-à-vis de *Pseudomonas syringae* pv. *savastanoi*, agent de la tuberculose. *Cahiers Agricultures*, Vol. 3, No. 6, (November/December 1994), pp. 405-408, ISSN 1166-7699

Bertolini, E., Fadda, Z., García, F., Celada, B., Olmos, A., Gorris, M.T., Del Río, C., Caballero, J., Durán-Vila, N., & Cambra, M. (1998). Virosis del olivo detectadas en España. Nuevos métodos de diagnóstico. *Phytoma-España*, No. 102, (October 1998), pp. 191-193, ISSN 1131-8988

Bertolini, E. (2003). Virosis y bacteriosis del olivo: detección serológica y molecular. Thesis (Ph. D.). Universidad Politécnica de Valencia, Spain

Bertolini, E., Penyalver, R., García, A., Olmos, A., Quesada, J.M., Cambra, M., & López, M.M. (2003a). Highly sensitive detection of *Pseudomonas savastanoi* pv. *savastanoi* in asymptomatic olive plants by nested-PCR in a single closed tube. *Journal of Microbiological Methods*, Vol. 52, No. 2, (February 2003), pp. 261-266, ISSN 0167-7012

Bertolini, E., Olmos, A., Lopez, M.M., & Cambra, M. (2003b). Multiplex nested reverse transcription-polymerase chain reaction in a single tube for sensitive and simultaneous detection of four RNA viruses and *Pseudomonas savastanoi* pv. *savastanoi* in olive trees. *Phytopathology*, Vol. 93, No. 3, (March 2003), pp. 286-292, ISSN 0031-949X

Brisset, M.N., Cesbron, S., Thomson, S.V., & Paulin, J.P. (2000). Acibenzolar-S-methyl induces the accumulation of defense-related enzymes in apple and protects from fire blight. *European Journal of Plant Pathology*, Vol. 106, No. 6, (July 2000), pp. 529-536, ISSN 0929-1873

Caballero, P., & Murillo, J. (2003). *Protección de cultivos. Conceptos actuales y fuentes de información* (First edition), Universidad Pública de Navarra, ISBN 84-9769-014-1, Spain

Cambra, M., López, M.M., Durán-Vila, N., Bertolini, E., Penyalver, R., & Gorris, M.T. (1998). Programa de certificación vegetal en olivo. Incidencia en la producción. *Agricultura y Cooperación*. Vol. 173, pp. 30-31, ISSN 0301-438X

Catara, V., Colina, P., Bella, P., Tessitori, M., & Tirrò, A. (2005). Variabilità di *Pseudomonas savastanoi* pv. *savastanoi* in un'area olivicola della Sicilia e comportamento di alcune varietà di olivo alle inoculazioni [*Olea europaea* L.]. *Tecnica Agricola*, Vol. 57, No. 1/2, (January-June 2005), pp. 41-52, ISSN 0371-5124

Cayuela, J.A., Rada, M., Ríos, J.J., Albi, T., & Guinda, A. (2006). Changes in phenolic composition induced by *Pseudomonas savastanoi* pv. *savastanoi* infection in olive tree: presence of large amounts of verbascoside in nodules of tuberculosis disease. *Journal of Agricultural and Food Chemistry*, Vol. 54, No. 15, (July 2006), pp. 5363-5368, ISSN 0021-8561

Cazorla, F.M., Arrebola, E., Sesma, A., Pérez-García, A., Codina, J.C., Murillo, J., & de Vicente, A. (2002). Copper resistance in *Pseudomonas syringae* strains isolated from mango is encoded mainly by plasmids. *Phytopathology*, Vol. 92, No. 8, (August 2002), pp. 909-916, ISSN 0031-949X

Cazorla, F.M., Arrebola, E., Olea, F., Velasco, L., Hermoso, J.M., Pérez-García, A., Torés, J.A., Farré, J.M., & de Vicente, A. (2006). Field evaluation of treatments for the control of the bacterial apical necrosis of mango (*Mangifera indica*) caused by *Pseudomonas syringae* pv. *syringae*. *European Journal of Plant Pathology*, Vol. 116, No. 4, (December 2006), pp. 279-288, ISSN 0929-1873

Chomé, P.M. (1998). Programa de certificación de plantas de vivero de olivo en España y registro de variedades comerciales. *Phytoma-España*, No. 102, (October 1998), pp. 41-44, ISSN 1131-8988

Civantos, L. (2008). *Obtención del aceite de oliva virgen*, (Third edition), Editorial Agrícola Española, ISBN 978-84-85441-93-8, Madrid, Spain

Comai, L., & Kosuge, T. (1980). Involvement of plasmid deoxyribonucleic acid in indoleacetic acid synthesis in *Pseudomonas savastanoi*. *Journal of Bacteriology*, Vol. 143, No. 2, (August 1980), pp. 950-957, ISSN 0021-9193

Cooksey, D.A. (1990). Genetics of bactericide resistance in plant pathogenic bacteria. *Annual Review of Phytopathology*, Vol. 28, pp. 201-219, ISSN 0066-4286

Cruz, A. Braga da, & Tavares, R.M. (2005). Evaluation of programmed cell death in *Olea europaea* var. Galega vulgar suspension cell cultures elicited with *Pseudomonas savastanoi*, *IX Congresso Luso-Espanhol de Fisiologia Vegetal*, (September 2005), Available from http://hdl.handle.net/1822/3437

De Andrés, F. (1991). *Enfermedades y plagas del olivo* (second edition), Riquelme y Vargas Ediciones, ISBN 8486216192, Jaén, Spain

Durrant, W.E., & Dong, X. (2004). Systemic Acquired Resistance. *Annual Review of Phytopathology*, Vol. 42, pp. 185-209, ISSN 0066-4286

EPPO (2006). Pathogen-tested olive trees and rootstocks. *Bulletin OEPP / EPPO Bulletin*, Vol. 36, No. 1, (April 2006), pp. 77–83, ISSN 1365-2338

Ercolani, G.L. (1971). Presenza epifitica di *Pseudomonas savastanoi* (E. f. Smith) Stevens sull'Olivo, in Pluglia. *Phytopathologia Mediterranea*, Vol. 10, No. 1, pp. 130-132, ISSN 0031-9465

Ercolani, G.L. (1978). *Pseudomonas savastanoi* and other bacteria colonizing the surface of olive leaves in the field. *Journal of General Microbiology*, Vol. 109, (December 1978), pp. 245-257, ISSN 0022-1287

Ercolani, G.L. (1979). Distribuzione di *Pseudomonas savastanoi* sulle foglie dell'olivo. *Phytopathologia Mediterranea*, Vol. 18, pp. 85-88, ISSN 0031-9465

Ercolani, G.L. (1983). Variability among isolates of *Pseudomonas syringae* pv. *savastanoi* from the pylloplane of the olive. *Journal of General Microbiology*, Vol. 129, (April 1983), pp. 901-916, ISSN 0022-1287

Ercolani, G.L. (1985). Factor analysis of fluctuation in populations of *Pseudomonas syringae* pv. *savastanoi* on the phylloplane of the olive. *Microbial Ecology*, Vol. 11, No. 1, (March 1985), pp. 41-49, ISSN 0095-3628

Ercolani, G.L. (1991). Distribution of epiphytic bacteria on olive leaves and the influence of leaf age and sampling time. *Microbial Ecology*, Vol. 21, No. 1, pp. 35-48, ISSN 0095-3628

Ercolani, G.L. (1993). Comparison of strains of *Pseudomonas syringae* pv. *savastanoi* from olive leaves and knots. *Letters in Applied Microbiology*, Vol. 16, No. 4, (April 1993), pp. 199-202, ISSN 0266-8254

Fernandes, A., & Marcelo, M. (2002). A possible synergistic effect of *Erwinia* sp. on the development of olive knot symptoms caused by *Pseudomonas syringae* pv. *savastanoi* in *Olea europaea*, In: *Proceedings of the fourth International Symposium on Olive Growing*, Vitagliano, C. & Martelli, G.P., No. 586, pp. 729-731, Acta Horticulturae (ISHS), ISSN 0567-7572, Valenzano, Italy

García de los Ríos, J.E. (1989). Estudio acerca de la tuberculosis del olivo. Thesis (Ph. D.). Universidad Complutense de Madrid, Spain

García, A., Penyalver, R., & López, M.M. (2001). Sensibilidad de variedades de olivo a *Pseudomonas savastanoi* pv. *savastanoi*, causante de la tuberculosis. *Fruticultura Profesional*, No. 120, pp. 57-58, ISSN 1131-5660

Gardan, L., Bollet, C., Abu-Ghorrah, M.A., Grimont, F., & Grimont, P.A.D. (1992). DNA relatedness among the pathovar strains of *Pseudomonas syringae* subsp. *savastanoi* Janse (1982) and proposal of *Pseudomonas savastanoi* sp. nov. *International Journal of Systematic and Evolutionary Microbiology*, Vol. 42, No. 4, (October 1992), pp. 606-612, ISSN 1466 5026

Garret, K.A., & Schwartz, H.F. (1998). Epiphytic *Pseudomonas syringae* on dry beans treated with copper-based bactericides. *Plant Disease*, Vol. 82, No. 1, (January 1998), pp. 30-35, ISSN 0191-2917

Graham, J.H., & Leite, R.P.Jr. (2004). Lack of control of citrus canker by induced systemic resistance compounds. *Plant Disease*, Vol. 88, No. 7, (July 2004), pp. 745-750, ISSN 0191-2917

Grey, B.E., & Steck, T.R. (2001). The viable but nonculturable state of *Ralstonia solanacearum* may be involved in long-term survival and plant infection. *Applied and Environmental Microbiology*, Vol. 67, No. 9, (September 2001), pp. 3866-3872, ISSN 0099-2240

Hassani, D., Buonaurio, R., & Tombesi, A. (2003). Response of some olive cultivars, hybrid and open pollinated seedling to *Pseudomonas savastanoi* pv. *savastanoi*. *International Conference on Pseudomonas syringae Pathovars and Related Pathogens*, In: Pseudomonas syringae *and related pathogens: biology and genetic*, N.S. Iacobellis et al. (Eds.), pp. 489-494, Kluwer Academic Publishers, ISBN: 1402012276, Dordrecht, Boston

Hausbeck, M.K., Bell, J., Medina-Mora, C., Podolsky, R., & Fulbright, D.W. (2000). Effect of bactericides on population sizes and spread of *Clavibacter michiganensis* subsp. *michiganensis* on tomatoes in the greenhouse and on disease development and crop yield in the field. *Phytopathology*, Vol. 90, No. 1, (January 2000), pp. 38-44, ISSN 0031-949X

Hirano, S.S., & Upper, C.D. (1983). Ecology and epidemiology of foliar bacterial plant pathogens. *Annual Review of Phytopathology*, Vol. 21, pp. 243-269, ISSN 0066-4286

Hirano, S.S., & Upper, C.D. (2000). Bacteria in the leaf ecosystem with emphasis on *Pseudomonas syringae*- a pathogen, ice nucleus, and epiphyte. *Microbiology and Molecular Biology Reviews*, Vol. 64, No. 3, (September 2000), pp. 624-653, ISSN 1092-2172

Horne, T., Parker, B., & Daines, L.L. (1912). The method of spreading of the olive knot disease. *Phytopathology*. Vol. 2, pp. 101-105, ISSN: 0031-949X

Iacobellis, N.S. (2001). Olive knot. In: *Encyclopedia of Plant Pathology*, Vol. 2, O.C. Maloy & T.D Murray (Eds.), pp. 713-715, John Wiley and sons, ISBN 0471298174, New York, USA

Iacobellis, N.S., Sisto, A., Surico, G., Evidente, A., & Di Maio, E. (1994). Pathogenicity of *Pseudomonas syringae* subsp. *savastanoi* mutants defective in phytohormone production. *Journal of Phytopathology*, Vol. 140, No. 3, (March 1994), pp. 238-248, ISSN 1439-0434

Jacques, M.E., & Morris, C.E. (1995). A review of issues to the quantification of bacteria from the phyllosphere. *FEMS Microbiology Ecology*, Vol. 18, No. 1, (September 1995), pp. 1-14, ISSN 0168-6496

Janse, J.D. (1982). *Pseudomonas syringae* subsp. *savastanoi* ex Smith subsp. nov., nom. rev., the bacterium causing excrescences on *Oleaceae* and *Nerium oleander* L. *International Journal of Systematic Bacteriology*, Vol. 32, No. 2, (April 1982), pp. 166-169, ISSN: 1466-5026

Jones, J.B., Woltz, S.S., Jones, J.P., & Portier, K.L. (1991). Population dynamics of *Xanthomonas campestris* pv. *vesicatoria* on tomato leaflets treated with copper bactericides. *Phytopathology*, Vol. 81, No. 7, (July 1991), pp. 714-719, ISSN 0031-949X

Kado, C.I. (1992). Plant pathogenic bacteria, In: *The Prokaryotes: a handbook on the biology of bacteria: ecophysiology, isolation, identification, applications*, Balows, A., Truper, H. G., Dworkin, M., Harder, W. & Schleifer, K.H., Vol. I, pp. 659-674, Springer-Verlag, ISBN 0387972587, New York, USA

Kessmann, H., Staub, T., Hofmann, C., Maetzke, T., Herzog, J., Ward, E., Uknes, S., & Ryals, J. (1994). Induction of systemic acquired disease resistance in plants by chemicals. *Annual Review of Phytopathology*, Vol. 32, pp. 439-459, ISSN 0066-4286

Kinkel, L.L. (1997). Microbial population dynamics on leaves. *Annual Review of Phytopathology*, Vol. 35, pp. 327-347, ISSN 0066-4286

Krid, S., Rhouma, A., Mogou, I., Quesada, J.M., Nesme, X., & Gargouri, A. (2010). *Pseudomonas savastanoi* endophytic bacteria in olive tree knots and antagonistic potential of strains of *Pseudomonas fluorescens* and *Bacillus subtilis*. *Journal of Plant Pathology*, Vol. 92, No. 2, (July 2010), pp. 335-341, ISSN: 1125-4653

Krueger, W.H., Tevitodale, B.L., Scroth, M.N., Metzidakis, I.T., & Voyiaztzis, D.G. (1999). Improvements in the control of olive knot disease, In: *Proceedings of the third International Symposium on Olive Growing*, Metzidakis, I.T. & Voyiaztzis, D.G., No. 474, pp. 567-571, Acta Horticulturae (ISHS), ISSN 0567-7572, Chania, Greece

Lavermicocca, P., & Surico, G. (1987). Presenza epifitica di *Pseudomonas syringae* pv. *savastanoi* e di altri batteri sull'olivo e sull'oleandro. *Phytopathologia Mediterranea*, Vol. 26, pp. 136-141, ISSN 0031-9465

Lavermicocca, P., Lonigro, S.L., Valerio, F., Evidente, A., & Visconti, A. (2002). Reduction of olive knot disease by a bacteriocin from *Pseudomonas syringae* pv. *ciccaronei*. *Applied and Environmental Microbiology*, Vol. 68, No. 3, (March 2002), pp. 1403-1407, ISSN 0099-2240

Lavermicocca, P., Valerio, F., Lonigro, S.L, Lazzaroni, S., Evidente, A., & Visconti, A. (2003). Control of olive knot disease with a bacteriocin, In: Pseudomonas syringae *and related pathogens: Biology and Genetics*, Iacobellis, N.S., Collmer, A., Hutcheson, S.W., Mansfield, J.W., Morris, C.E., Murillo, J., Schaad, N.W., Stead, D.E., Surico, G. & Ullrich, M.S., pp. 451-457, Kluwer Academic Publishers, ISBN 1-4020-1227-6 Dordrecht, The Netherlands

Lindow, E.L., & Andersen, G.L. (1996). Influence of immigration on epiphytic bacterial populations on navel orange leaves. *Applied and Environmental Microbiology*, Vol. 62, No. 8, (August 1996), pp. 2978-2987, ISSN 0099-2240

Lindow, E.L., & Brandl, M.T. (2003). Microbiology of the phyllosphere. *Applied and Environmental Microbiology*, Vol. 69, No. 4, (April 2003), pp. 1875-1883, ISSN 0099-2240

Magie, A.R. (1963). Physiological factors involved in tumor production by the oleander knot pathogen, *Pseudomonas savastanoi*. Thesis (Ph. D.). University of California, Davis, USA

Manceau, C., & Kasempour, M.N. (2002). Endophytic versus epiphytic colonization of plants: what comes first?, In: *Phyllosphere Microbiology*, Lindow, S.E., Hecht-Poinar, E.I. & Elliott, V.J., pp. 115-123, APS Press, ISBN 978-0-89054-286-6, USA

Marcelo, A., Fernández M., Fatima Potes, M., & Serrano, J.F. (1999). Reactions of some cultivars of *Olea europaea* L. to experimental inoculation with *Pseudomonas syringae* pv. *savastanoi*, In: *Proceedings of the third International Symposium on Olive Growing*, Metzidakis, I.T. & Voyiaztzis, D.G., No. 474, pp. 581-584, Acta Horticulturae (ISHS), ISSN 0567-7572, Chania, Greece

Marchi, G., Sisto, A., Cimmino, A., Andolfi, A., Cipriani, M.G., Evidente, A., & Surico, G. (2006). Interaction between *Pseudomonas savastanoi* pv. *savastanoi* and *Pantoea agglomerans* in olive knots. *Plant Pathology*, Vol. 55, No. 5, (October 2006), pp. 614-624, ISSN 0032-0862

Marchi, G., Mori, B., Pollacci, P., Mencuccini, M., & Surico, G. (2009). Systemic spread of *Pseudomonas savastanoi* pv. *savastanoi* in olive explants. *Plant Pathology*, Vol. 58, No. 1, (February 2009), pp. 152-158, ISSN 0032-0862

Marco, G.M., & Stall, R.E. (1983). Control of bacterial spot of pepper initiated by strains of *Xanthomonas campestris* pv. *vesicatoria* that differ in sensitivity to copper. *Plant Disease*, Vol. 67, pp. 779-781, ISSN 0191-2917

Martelli, G.P. 1998. Enfermedades infecciosas y certificación del olivo: Panorama general. *Phytoma-España*, No. 102, (October 1998), pp. 180-186, ISSN 1131-8988

Matas, I.M., Pérez-Martínez, I., Quesada, J.M., Rodríguez-Herva, J.J., Penyalver, R., & Ramos, C. (2009). *Pseudomonas savastanoi* pv. *savastanoi* contains two *iaaL* paralogs, one of which exhibits a variable number of a trinucleotide (TAC) tandem repeat. *Applied and Environmental Microbiology*, Vol. 75, No. 4, (February 2009), pp. 1030-1035, ISSN 0099-2240

Matas, I.M. (2010). Genómica funcional de la interacción *Pseudomonas savastanoi* pv. *savastanoi*- olivo. Thesis (Ph. D.). Universidad de Málaga, Spain

Montesinos, E., & López, M.M. (1996). Métodos de control de las bacteriosis, In: *Patología Vegetal*, Llácer, G., López, M.M., Trapero, A., & Bello, A., pp. 653-678, Sociedad Española de Fitopatología- Phytoma España, S.L.- Grupo Mundi-Prensa, S.A., ISBN 84-921910-0-7, Valencia, Spain

Morris, C.E., Monier, J., & Jacques, M. (1997). Methods for observing microbial biofilms directly on leaf surfaces and recovering them for isolation of culturable microorganisms. *Applied and Environmental Microbiology*, Vol. 63, No. 4, (April 1997), pp. 1570-1576, ISSN 0099-2240

Morris, C.E., & Kinkel, L.L. (2002). Fifty years of phylosphere microbiology: significant contributions to research in related fields, In: *Phyllosphere Microbiology*, Lindow, S.E., Hecht-Poinar, E.I. & Elliott, V.J., pp. 365-375, APS Press, ISBN 978-0-89054-286-6, USA

Muñoz, J., Benítez, Y., Trapero, A., Caballero, J.L., & Dorado, G. (2005). Identificación de genes expresados diferencialmente en la interacción entre olivo *Olea europaea* L. y el hongo parásito causante del repilo *Spilocaea oleagina*, In: *Variedades de Olivo en España*, Rallo, L., Barranco, D., Caballero, J.M., Del Río, C., Martín, A., Tous, J., & Trujillo, I., pp. 459-470, Junta de Andalucía, MAPA and Ediciones Mundi- Prensa, ISBN 848474146X, Madrid, Spain

Ordax, M., Marco-Noales, E., López, M.M., & Biosca, E.G. (2005). Survival strategy of *Erwinia amylovora* against copper: induction of the viable-but-nonculturable state. *Applied and Environmental Microbiology*, Vol. 72, No. 5, (May 2006), pp. 3482-3488, ISSN 0099-2240

Ouzari, H., Khsairi, A., Raddadi, N., Jaoua, L., Hassen, A., Zarrouk, M., Daffonchio, D., & Boudabous, A. (2008). Diversity of auxin-producing bacteria associated to

Pseudomonas savastanoi -induced olive knots. *Journal of Basic Microbiology*, Vol. 48, No. 5, (October 2008), pp. 370-377, ISSN 1521-4028

Padilla, V. (1997). Virosis en olivo. Problemática del material vegetal. *Fruticultura Profesional*, No. 88, pp. 56-58, ISSN 1131-5660

Panagopoulos, C.G. (1993). Olive knot disease in Greece. *Bulletin OEPP / EPPO Bulletin*, Vol. 23, No. 3, (September 1993), pp. 417-422, ISSN 1365-2338

Paoletti, V. (1933). Osservasioni ed esperimenti orientativi di lotta contro la rogna dell'olivo. *Rivista di patologia vegetale*, 23, pp. 47-50, ISSN 0035-6441

Penyalver, R., García, A., Ferrer, A., & López, M.M. (1998). La tuberculosis del olivo: diagnóstico, epidemiología y control. *Phytoma-España*, No. 102, (October 1998), pp. 177-179, ISSN 1131-8988

Penyalver, R., García, A., Del Río, C., Caballero, J.M., Pinochet, J., Piquer, J., & López, M.M. (2003). Sensibilidad varietal del olivo a la tuberculosis causada por *Pseudomonas savastanoi* pv. *savastanoi*. *Agrícola vergel: Fruticultura, horticultura, floricultura*, No. 253, pp. 13-17, ISSN 0211-2728

Penyalver, R., García, A., Pérez-Panadés, J., Del Río, C., Caballero, J.M., Pinochet, J., Piquer, J., Carbonell, E., & López, M.M. (2005). Resistencia y susceptibilidad a la tuberculosis, In: *Variedades de Olivo en España*, Rallo, L., Barranco, D., Caballero, J.M., Del Río, C., Martín, A., Tous, J., & Trujillo, I., pp. 339-346, Junta de Andalucía, MAPA and Ediciones Mundi- Prensa, ISBN 848474146X, Madrid, Spain

Penyalver, R., García, A., Ferrer, A., Bertolini, E., Quesada, J.M., Salcedo, C.I., Piquer, J., Pérez-Panadés, J., Carbonell, E.A., del Río, C., Caballero, J.M., & López, M.M. (2006). Factors affecting *Pseudomonas savastanoi* pv. *savastanoi* plant inoculations and their use for evaluation of olive cultivar susceptibility. *Phytopathology*, Vol. 96, No. 3, (March 2006), pp. 313-319, ISSN 0031-949X

Pérez-Martínez, I., Rodríguez-Moreno, L., Matas, I.M., & Ramos, C. (2007). Strain selection and improvement of gene transfer for genetic manipulation of *Pseudomonas savastanoi* isolated from olive knots. *Research in Microbiology*, Vol. 158, No. 1, (January 2007), pp. 60-69, ISSN 0923-2508

Pérez-Martínez, I., Zhao, Y., Murillo, J., Sundin, G.W., & Ramos C. (2008). Global genomic analisis of *Pseudomonas savastanoi* pv. *savastanoi* plasmids. *Journal of Bacteriology*, Vol. 190, No. 2, (January 2008), pp. 625-635, ISSN 0021-9193

Pérez-Martínez, I., Rodríguez-Moreno, L., Lambertsen, L., Matas, I.M., Murillo, J., Tegli, S., Jiménez, A.J., & Ramos, C. (2010). Fate of a *Pseudomonas savastanoi* pv. *savastanoi* type III secretion system mutant in olive plants (*Olea europaea* L.). *Applied and Environmental Microbiology*, Vol. 76, No. 11, (June 2010), pp. 3611-3619, ISSN 0099-2240

Protta, U. (1995). Le malattie dell' olivo. *Informatore Fitopatologico*, No. 12, pp. 16-26, ISSN 0020-0735

Quesada, J.M., García, A., Bertolini, E., López, M.M., & Penyalver, R. (2007). Recovery of *Pseudomonas savastanoi* pv. *savastanoi* from symptomless shoots of naturally infected olive trees. *International Microbiology*, Vol. 10, No. 2, (June 2007), pp. 77-84, ISSN 1139-6709

Quesada, J.M., Penyalver, R., Pérez-Panadés, J., Salcedo, C.I., Carbonell, E.A., & López, M.M. (2010a). Dissemination of *Pseudomonas savastanoi* pv. *savastanoi* populations and subsequent appearance of olive knot disease. *Plant Pathology*, Vol. 59, No. 2, (April 2010), pp. 262–269, ISSN 1365-3059

Quesada, J.M., Penyalver, R., Pérez-Panadés, J., Salcedo, C.I., Carbonell, E.A., & López, M.M. (2010b). Comparison of chemical treatments for reducing epiphytic *Pseudomonas savastanoi* pv. *savastanoi* populations and for improving subsequent control of olive knot disease. *Crop Protection*, Vol. 29, No. 12, (December 2010), pp. 1413-1420, ISSN 0261-2194

Rodríguez-Moreno, L., Barceló-Muñoz, A., & Ramos, C. (2008). In vitro analysis of the interaction of *Pseudomonas savastanoi* pvs. *savastanoi* and *nerii* with micropropagated olive plants. *Phytopathology*, Vol. 98, No. 7, (July 2008), pp. 815-822, ISSN 0031-949X

Rodríguez-Moreno, L., Jiménez, A.J., & Ramos, C. (2009). Endopathogenic lifestyle of *Pseudomonas savastanoi* pv. *savastanoi* in olive knots. *Microbial Biotechnology*, Vol. 2, No. 4, (July 2009), pp. 476-488, ISSN 1751-7907

Rodríguez-Palenzuela, P., Matas, I.M., Murillo, J., López-Solanilla, E., Bardaji, L., Pérez-Martínez, I., Rodríguez-Moskera, M.E., Penyalver, R., López, M.M., Quesada, J.M., Biehl, B.S., Perna, N.T., Glasner, J.D., Cabot, E.L., Neeno-Eckwall, E., & Ramos, C. (2010). Annotation and overview of the *Pseudomonas savastanoi* pv. *savastanoi* NCPPB 3335 draft genome reveals the virulence gene complement of a tumour-inducing pathogen of woody hosts. *Environmental Microbiology*, Vol. 12, No. 6, (June 2010), pp. 1604-1620, ISSN 1462 2920

Rojas, A.M. (1999). Análisis fenotípicos, genéticos y filogenéticos de flora endofítica asociada a *Pseudomonas savastanoi*. Thesis (Ph. D.). Universidad San Pablo CEU, Madrid, Spain

Rojas, A.M., García de los Ríos, J.E., Fischer-Le Saux, M., Jiménez, P., Reche, P., Bonneau, S., Sutra, L., Mathieu-Daudé, F., & McCelland, M. (2004). *Erwinia toletana* sp. nov., associated with *Pseudomonas savastanoi*-induced tree knots. *International Journal of Systematic and Evolutionary Microbiology*, Vol. 54, No. 6, (November 2004), pp. 2217-2222, ISSN 1466 5026

Rokni-Zadeh, H., Khavazi, K., Asgharzadeh, A., Hosseini-Mazinani, M., & De Mot, R. (2008). Biocontrol of *Pseudomonas savastanoi*, causative agent of olive knot disease: antagonistic potential of non-pathogenic rhizosphere isolates of fluorescent *Pseudomonas*. *Communications in Agricultural and Applied Biological Sciences*, Vol. 73, No. 1, pp. 199-203, ISSN 1379 1176

Roussos, P.A., Pontikis, C.A., & Tsantili, E. (2002). Root promoting compounds detected in olive knot extract in high quantities as a response to infection by the bacterium *Pseudomonas savastanoi* pv. *savastanoi*. *Plant Science*, Vol. 163, No. 3, (September 2002), pp. 533-541, ISSN 0168-9452

Scheck, H.J., Pscheidt, J.W., & Moore, L.W. (1996). Copper and streptomycin resistance in strains of *Pseudomonas syringae* from Pacific Northwest nurseries. *Plant Disease*, Vol. 83, No. 9, (September 1996), pp. 1034-1039, ISSN 0191-2917

Schiff-Giorgini, R. (1906). Untersuchungen über die tuberkelkrankheit des oelbaumes. Centralb. Bakteriol., Parasitenk. Abt. 2. 15, pp. 200-211

Schroth, M.N., & Hildebrand, D.C. (1968). A chemotherapeutic treatment for selectively eradicating crown gall and olive knot neoplasms. *Phytopathology*, Vol. 58, No. 6, (June 1968), pp. 848-854, ISSN 0031-949X

Schroth M.N., Hildbrand, D.C., & Reilly, H.J. (1968). Off-flavor of olives from trees with olive knot tumors. *Phytopathology*, Vol. 58, pp. 524-525, ISSN 0031-949X

Schroth, M.N., Osgood, J.W., & Miller, T.D. (1973). Quantitative assessment of the effect of the olive knot disease on olive yield and quality. *Phytopathology*, Vol. 63, No. 8, (August 1973), pp. 1064–1065, ISSN 0031-949X

Scortichini, M. (2002). Bacterial canker and decline of European hazelnut. *Plant Disease*, Vol. 86, No. 7, (July 2002), pp. 704-709, ISSN 0191-2917

Scrivani, P., & Bugiani, A. (1955). Messa a punto de un metodo per la riproduzione artificiale della rogna dell'Olivo e risultati dei primi saggi terapeutici a mezo di sostanze antibiotiche. *L'Italia Agrícola*, Vol. 92, pp. 361-369, ISSN 0021-275X

Sisto, A., Cipriani, M.G., & Morea, M. (2004). Knot formation caused by *Pseudomonas syringae* subsp. *savastanoi* on olive plants is hrp-dependent. *Phytopathology*, Vol. 94, No. 5, (May 2004), pp. 484-489, ISSN 0031-949X

Sisto, A., & Iacobellis, N.S. (1999). La "Rogna dell' olivo": aspetti patogenetici, epidemiologici e strategie di lotta. *Olivo & olio*, Vol. 2, No. 12, pp. 32-38, ISSN 1127-0713

Smidt, M., & Kosuge, T. (1978). The role of indole-3-acetic acid accumulation by alphamethyl tryptophan-resistant mutants of *Pseudomonas savastanoi* in gall formation on oleanders. *Physiological Plant Pathology*, Vol. 13, No. 2, (September 1978), pp. 203-214, ISSN 0885-5765

Smith, E.F. (1908). Recent studies on the olive-tubercle organism. U.S. Dept. Agr. Bur. Plant Indust. Bull. No. 131 Part, IV.

Smith, E.F. (1920). Pathogenicity of the olive knot organism on hosts related to the olive. *Phytopathology*, Vol. 12, pp. 271-278, ISSN 0031-949X

Smith, E.F., & Rorer, J.B. (1904). The olive tubercule. *Science N.Y.*, 19, pp. 416-417

Smith, I.M., Dunez, J., Lelliot, R.A., Phillips, D.H., & Archer, S.A. (1991). *Manual de enfermedades de las plantas*, Editorial Mundi-Prensa, ISBN 84-7114-358-5, Madrid, Spain

Sundin, G.W., Jones, A.L., & Fulbright, D.W. (1989). Copper resistance in *Pseudomonas syringae* pv. *syringae* from cherry orchards and its associated transfer in vitro and in planta with a plasmid. *Phytopathology*, Vol. 79, No. 8, (August 1989), pp. 861-865, ISSN 0031-949X

Sundin, G.W., Demezas, D.H., & Bender, C.L. (1994). Genetic and plasmid diversity within natural populations of *Pseudomonas syringae* with various exposures to copper and streptomycin bactericides. *Applied and Environmental Microbiology*, Vol. 60, No. 12, (December 1994), pp. 4421-4431, ISSN 0099-2240

Surico, G. (1977). Histological observations on olive knots. *Phytopathologia Mediterranea*, Vol. 16, pp. 109-125, ISSN 0031-9465

Surico, G., Iacobellis, N.S., & Sisto, S. (1985). Studies on the role of indole-3-acetic acid and citokinins in the formation of knots on olive and oleander plants by *Pseudomonas syringae* pv. *savastanoi. Physiological Plant Pathology*, Vol. 26, No. 3, (May 1985), pp. 309-320, ISSN 0885-5765

Surico, G. (1993). Scanning electron microscopy of olive and oleander leaves colonized by *Pseudomonas syringae* subsp. *savastanoi. Journal of Phytopathology*, Vo. 138, No. 1, (May 1993), pp. 31-40, ISSN 1439-0434

Temsah, M., Hanna, L., & Saad, A.T. (2007a) Anatomical observations of *Pseudomonas savastanoi* on *Rhamnus alaternus. Forest Pathology*, Vol. 37, No. 1, (February 2007), pp. 64–72, ISSN 1439-0329

Temsah, M., Hanna, L., & Saad, A.T. (2007b) Histological pathogenesis of *Pseudomonas savastanoi* on *Myrtus communis. Journal of Plant Pathology*, Vol. 89, No. 2, (July 2007), pp. 241-249, ISSN: 1125-4653

Teviotdale, B.L., & Krueger, W.H. (2004). Effects of timing of copper sprays, defoliation, rainfall, and inoculum concentration on incidence of olive knot disease. *Plant Disease*, Vol. 88, No. 2, (February 2004), pp. 131-135, ISSN 0191-2917

Tjamos, E.C., Graniti, A., Smith, I.M., & Lamberti, F. (1993). Conference on olive diseases. *Bulletin OEPP / EPPO Bulletin*, Vol. 23, No. 3, (September 1993), pp. 365-550, ISSN 1365-2338

Tous, J., Romero, A., & Hermoso, J.F. (2007). The hedgerow system for olive growing. *Olea FAO Olive Network*, No. 26, pp. 20-26, ISSN 0214-6614

Trapero, A., & Blanco, M.A. (1998). Enfermedades, In: *El cultivo del olivo*, Barranco, D., Fernández-Escobar, D., & Rallo, L., pp. 461-507, Junta de Andalucía- Mundi-Prensa, ISBN 978-84-8474-234-0, Madrid, Spain

Vallad, G.E., & Goodman, R.M. (2004). Systemic acquired resistance and induced systemic resistance in conventional agriculture. *Crop Science*, Vol. 44, No. 6, (November 2004), pp. 1920-1934, ISSN 0011-183X

Varvaro, L., & Ferrulli, M. (1983). Sopravvivenza di *Pseudomonas syringae* pv. *savastanoi* (Smith) Young et al. sulle foglie di due varietà di olivo (*Olea europea* L.). *Phytopathologia Mediterranea*, Vol. 22, No. 1/2, (April 1983), pp. 1-4, ISSN 0031-9465

Varvaro, L., & Martella, L. (1993). Virulent and avirulent isolates of *Pseudomonas syringae* subsp. *savastanoi* as colonizers of olive leaves: evaluation of possible biological control of the olive knot pathogen. *Bulletin OEPP / EPPO Bulletin*, Vol. 23, No. 3, (September 1993), pp. 423-427, ISSN 1365-2338

Varvaro, L., & Surico, G. (1978). Comportamento di diverse cultivars di olivo *Olea europaea* L. alla inoculazione artificiale con *Pseudomonas savastanoi* E. F. Smith Stevens. *Phytopathologia Mediterranea*, Vol. 17, No. 3, (December 1978), pp. 174-178, ISSN 0031-9465

Vivian, A., & Mansfield, J. (1993). A proposal for a uniform genetic nomenclature for avirulence genes in phytopathogenic pseudomonads. *Molecular Plant Microbe Interactions*, Vol. 6, No. 1, (January 1993), pp. 9-10, ISSN 0894-0282

Wilson, E.E. (1935). The olive knot disease: its inception, development and control. *Hilgardia*, 9, pp. 233-264

Wilson, E.E. (1965). Pathological histogenesis in oleander tumors induced by *Pseudomonas savastanoi*. *Phytopathology*, Vol. 55, pp. 1244-1249, ISSN 0031-949X

Wilson, E.E., & Magie, A.R. (1964). Systemic invasion of the host plant by the tumor-inducing bacterium, *Pseudomonas savastanoi*. *Phytopathology*, Vol. 54, pp. 576-579, ISSN 0031-949X

Wilson, E.E., & Ogawa, J.M. (1979). *Fungal, bacterial, and certain nonparasitic diseases of fruit and nut crops in California*, Division of Agricultural Sciences, University of California, ISBN 093187629X, Berkeley, USA

Wilson, M., & Lindow, S.E. (1992). Relationship of total viable and culturable cells in epiphytic populations of *Pseudomonas syringae*. *Applied and Environmental Microbiology*, Vol. 58, No. 12, (December 1992), pp. 3908-3913, ISSN 0099-2240

Permissions

The contributors of this book come from diverse backgrounds, making this book a truly international effort. This book will bring forth new frontiers with its revolutionizing research information and detailed analysis of the nascent developments around the world.

We would like to thank Dr. Christian Joseph R. Cumagun, for lending his expertise to make the book truly unique. He has played a crucial role in the development of this book. Without his invaluable contribution this book wouldn't have been possible. He has made vital efforts to compile up to date information on the varied aspects of this subject to make this book a valuable addition to the collection of many professionals and students.

This book was conceptualized with the vision of imparting up-to-date information and advanced data in this field. To ensure the same, a matchless editorial board was set up. Every individual on the board went through rigorous rounds of assessment to prove their worth. After which they invested a large part of their time researching and compiling the most relevant data for our readers. Conferences and sessions were held from time to time between the editorial board and the contributing authors to present the data in the most comprehensible form. The editorial team has worked tirelessly to provide valuable and valid information to help people across the globe.

Every chapter published in this book has been scrutinized by our experts. Their significance has been extensively debated. The topics covered herein carry significant findings which will fuel the growth of the discipline. They may even be implemented as practical applications or may be referred to as a beginning point for another development. Chapters in this book were first published by InTech; hereby published with permission under the Creative Commons Attribution License or equivalent.

The editorial board has been involved in producing this book since its inception. They have spent rigorous hours researching and exploring the diverse topics which have resulted in the successful publishing of this book. They have passed on their knowledge of decades through this book. To expedite this challenging task, the publisher supported the team at every step. A small team of assistant editors was also appointed to further simplify the editing procedure and attain best results for the readers.

Our editorial team has been hand-picked from every corner of the world. Their multi-ethnicity adds dynamic inputs to the discussions which result in innovative outcomes. These outcomes are then further discussed with the researchers and contributors who give their valuable feedback and opinion regarding the same. The feedback is then collaborated with the researches and they are edited in a comprehensive manner to aid the understanding of the subject.

Apart from the editorial board, the designing team has also invested a significant amount of their time in understanding the subject and creating the most relevant covers. They scrutinized every image to scout for the most suitable representation of the subject and create an appropriate cover for the book.

The publishing team has been involved in this book since its early stages. They were actively engaged in every process, be it collecting the data, connecting with the contributors or procuring relevant information. The team has been an ardent support to the editorial, designing and production team. Their endless efforts to recruit the best for this project, has resulted in the accomplishment of this book. They are a veteran in the field of academics and their pool of knowledge is as vast as their experience in printing. Their expertise and guidance has proved useful at every step. Their uncompromising quality standards have made this book an exceptional effort. Their encouragement from time to time has been an inspiration for everyone.

The publisher and the editorial board hope that this book will prove to be a valuable piece of knowledge for researchers, students, practitioners and scholars across the globe.

List of Contributors

Genhua Yang and Chengyun Li
Key Laboratory of Agro-Biodiversity and Pest Management of Education, Ministry of China, Yunnan Agricultural University, Kunming, Yunnan, China

M. Mahamuda Begum, Teresita U. Dalisay and Christian Joseph R. Cumagun
Crop Protection Cluster, College of Agriculture, University of the Philippines Los Baños, Philippines

Anastasios I. Darras
Department of Greenhouse Cultivation and Floriculture, Technological Educational Institute of Kalamata, Greece

Nickolas J. Panopoulos
Department of Biology, University of Crete, Heraklion, Greece
University of California, Berkeley, CA, USA

Emmanouil A. Trantas
Department of Plant Sciences, School of Agricultural Technology, Technological Educational Institute of Crete, Heraklion, Greece

Nicholas Skandalis
Benaki Phytopathological Institute, Kifisia, Athens, Greece

Anastasia P. Tampakaki
Department of Agricultural Biotechnology, Agricultural University of Athens, Athens, Greece

Maria Kapanidou
Department of Biology, University of Crete, Heraklion, Greece

Panagiotis F. Sarris
Department of Biology, University of Crete, Heraklion, Greece
Department of Plant Sciences, School of Agricultural Technology, Technological Educational Institute of Crete, Heraklion, Greece

Michael Kokkinidis
Department of Biology, University of Crete, Heraklion, Greece
Institute of Molecular Biology and Biotechnology, Foundation for Research and Technology-Hellas, Heraklion, Greece

Jing Yang and Chengyun Li
Key Laboratory of Agro-Biodiversity and Pest Management of Education, Ministry of China, Yunnan Agricultural University, Kunming, Yunnan, China

Nieves Capote, Ana María Pastrana and Ana Aguado
IFAPA Las Torres-Tomejil, Junta de Andalucía, Alcalá del Río, Sevilla, Spain

Paloma Sánchez-Torres
IVIA, Generalitat Valenciana, Moncada, Valencia, Spain

Kenichi Ikeda, Kanako Inoue, Hiroko Kitagawa, Hiroko Meguro, Saki Shimoi and Pyoyun Park
Kobe University, Japan

Kurt Brunner, Andreas Farnleitner and Robert L. Mach
Vienna University of Technology, Institute of Chemical Engineering, Vienna, Austria

Dartanhã José Soares
Empresa Brasileira de Pesquisa Agropecuária, Embrapa Algodão, Campina Grande, Brazil

R. Thangavelu and M.M. Mustaffa
National Research Centre for Banana, Trichirapalli, India

Peter McMahon
Department of Botany, La Trobe University, Bundoora Vic, Australia

P.J. Keane
Department of Botany, La Trobe University, Victoria, Australia

Ramón Penyalver and María M. López
Centro de Protección Vegetal y Biotecnología, Instituto Valenciano de Investigaciones Agrarias (IVIA), Valencia, Spain

José M. Quesada
Departamento de Protección Ambiental, Estación Experimental del Zaidín, Consejo Superior de Investigaciones Científicas (CSIC), Granada, Spain